OPTICAL CONSTANTS

Complex index of refraction

$$\mathfrak{N} = \eta + i\kappa \tag{2.81}$$

Absorption coefficient

$$\alpha = 2\alpha_E = \frac{4\pi\kappa}{\lambda} \tag{2.84}$$

Frequency-dependent relationships for optical constants

$$\eta^2 - \kappa^2 = 1 + \frac{Ne^2}{m\epsilon_0}\left(\frac{\omega_0^2 - \omega^2}{(\omega_0^2 - \omega^2)^2 + \gamma^2\omega^2}\right) \tag{2.87}$$

$$2\eta\kappa = \frac{Ne^2}{m\epsilon_0}\left(\frac{\gamma\omega}{(\omega_0^2 - \omega^2)^2 + \gamma^2\omega^2}\right) \tag{2.88}$$

Sellmeier's formula

$$\mathfrak{N}^2 \equiv \eta^2 = 1 + \frac{Ne^2}{m\epsilon_0}\sum_j\left(\frac{f_j}{\omega_j^2 - \omega^2}\right) \tag{2.90}$$

COHERENCE

Temporal coherence (longitudinal coherence)

$$l_c = \lambda\left(\frac{\lambda}{\Delta\lambda}\right) = \frac{\lambda^2}{\Delta\lambda} \tag{2.98}$$

Spatial coherence (transverse coherence)

$$l_t = \frac{r\lambda}{s} = \frac{\lambda}{\theta_s} \tag{2.99}$$

RELATION BETWEEN OSCILLATOR STRENGTH AND TRANSITION PROBABILITY

$$A_{ul} = \frac{e^2\omega_{ul}^2}{2\pi\epsilon_0 m_e c^3}\left(\frac{g_l}{g_u}\right)f_{lu} = \frac{2\pi e^2 \nu_{ul}^2}{\epsilon_0 m_e c^3}\left(\frac{g_l}{g_u}\right)f_{lu}$$

$$= \frac{2\pi e^2}{\epsilon_0 m_e c\lambda_{ul}^2}\left(\frac{g_l}{g_u}\right)f_{lu} \tag{4.78}$$

Relation between absorption and emission oscillator strengths

$$f_{ul} = -\frac{g_l}{g_u}f_{lu} \tag{4.79}$$

Empirical expression for relationship between A_{ul} and f_{lu}

$$A_{ul} = \frac{10^{-4}(f_{lu})}{1.5(g_u/g_l)\lambda_{ul}^2}\ \mathrm{s}^{-1}\ [\lambda\ \text{in m}] \tag{4.7}$$

HOMOGENEOUS BROADENING

Homogeneous linewidth

$$\Delta\nu_{ul}^H = \frac{\gamma_{ul}^T}{2\pi} = \frac{1}{2\pi}\left[\left(\sum_i A_{ui} + \sum_j A_{lj}\right)\right.$$
$$\left. + \frac{1}{T_1^u} + \frac{1}{T_1^l} + \frac{2}{T_2}\right] \tag{4.44}$$

Homogeneous lineshape

$$I(\nu) = I_0\frac{\gamma_{ul}^T/4\pi^2}{(\nu - \nu_0)^2 + (\gamma_{ul}^T/4\pi)^2} \tag{4.37}$$

DOPPLER (INHOMOGENEOUS) BROADENING

Average velocity

$$\bar{v} = \sqrt{\frac{8kT}{M\pi}} \tag{4.49}$$

Doppler width

$$\Delta\nu^D = 2\nu_0\sqrt{\frac{2(\ln 2)kT}{Mc^2}}$$
$$= (7.16\times 10^{-7})\nu_0\sqrt{\frac{T}{M_N}} \tag{4.59}$$
$$[T\ \text{in K},\ M_N\ \text{is mass number}]$$

Doppler lineshape

$$I(\nu) = \frac{2(\ln 2)^{1/2}}{\pi^{1/2}\Delta\nu^D}I_0\exp\left\{-\left[\frac{4(\ln 2)(\nu - \nu_0)^2}{(\Delta\nu^D)^2}\right]\right\} \tag{4.60}$$

SELECTION RULES FOR ALLOWED ELECTRIC DIPOLE TRANSITIONS

For atoms

$\Delta l = \pm 1$ for the changing electron (change in parity)

$$\Delta S = 0, \qquad \Delta L = 0, \pm 1 \tag{4.104}$$

$\Delta J = 0, \pm 1$ but $J = 0 \not\to J = 0$

$\Delta M_J = 0, \pm 1$ but $M_J = 0 \not\to M_J = 0$ if $\Delta J = 0$

Also parity must change

For molecules

Rotational transitions

$$\Delta J = 0, \pm 1, \qquad \Delta K = \pm 1 \tag{5.9}$$

Rotational–vibrational transitions

$$\Delta v = \pm 1 \tag{5.11}$$
$$\Delta J = 0, \pm 1 \tag{5.12}$$

Branch definitions

$$\begin{aligned}\Delta J &= +1 \quad \text{P branch}\\ \Delta J &= 0 \qquad \text{Q branch}\\ \Delta J &= -1 \quad \text{R branch}\end{aligned} \tag{5.13}$$

Electronic transitions

$$\Delta\Lambda = 0, \pm 1 \tag{5.15}$$
$$\Delta S = 0 \tag{5.16}$$

BLACKBODY RADIATION (intensity per unit λ)

$$I_{BB}(\nu) = \frac{2\pi h \nu^3}{c^2} \frac{1}{e^{h\nu/kT}-1} \tag{6.38}$$

$$I_{BB}(\lambda, T) = \frac{2\pi c^2 h}{\lambda^5(e^{ch/\lambda kT}-1)} \tag{6.41}$$

$$I_{BB}(\lambda, T) = \frac{3.75 \times 10^{-22}}{\lambda^5(e^{0.0144/\lambda T}+1)} \quad \text{W/m}^2\text{-}\mu\text{m}$$
$$[\lambda \text{ in m, } T \text{ in K}] \tag{6.43}$$

$$I_{BB}(\lambda, T) = \frac{3.75 \times 10^{-25}}{\lambda^5(e^{0.0144/\lambda T}-1)} \quad \text{W/m}^2\text{-nm}$$
$$[\lambda \text{ in m, } T \text{ in K}] \tag{6.44}$$

EINSTEIN A AND B COEFFICIENTS

$$\frac{g_l B_{lu}}{g_u B_{ul}} = 1 \quad \text{or} \quad g_l B_{lu} = g_u B_{ul} \tag{6.49}$$

$$B_{ul} = \frac{c^3}{8\pi h \nu^3} A_{ul} \tag{6.51}$$

Ratio between stimulated emission rate and spontaneous emission rate

$$\frac{B_{ul} u(\nu)}{A_{ul}} = \frac{1}{e^{h\nu_{ul}/kT}-1} \tag{6.56}$$

GAIN COEFFICIENTS AND STIMULATED EMISSION CROSS SECTION

Homogeneous broadening

$$g^H(\nu) = \left[N_u - \frac{g_u}{g_l} N_l\right] \frac{c^2}{8\pi\nu^2}$$
$$\times \left[\frac{\gamma_W^{ul}/4\pi^2}{(\nu-\nu_0)^2 + (\gamma_W^{ul}/4\pi)^2}\right] A_{ul} \tag{7.11}$$

$$\Delta N_{ul} = \left[N_u - \frac{g_u}{g_l} N_l\right] \tag{7.12}$$

$$\sigma_{ul}^H(\nu) = \frac{c^2}{8\pi\nu^2} A_{ul}(\nu)$$
$$= \frac{c^2}{8\pi\nu^2}\left[\frac{\gamma_W^{ul}/4\pi^2}{(\nu-\nu_0)^2 + (\gamma_W^{ul}/4\pi)^2}\right] A_{ul} \tag{7.13}$$

$$g^H(\nu = \nu_0) \equiv g^H(\nu_0)$$
$$\equiv g_0^H = \frac{c^2}{2\pi\nu_0^2 \gamma_{ul}} A_{ul}\left[N_u - \frac{g_u}{g_l} N_l\right] \tag{7.15}$$

$$\sigma_{ul}^H(\nu_0) = \frac{c^2 A_{ul}}{2\pi\nu_0^2 \gamma_{ul}} = \frac{\lambda_{ul}^2 A_{ul}}{4\pi^2 \Delta\nu_{ul}^H} \tag{7.16}$$

Exponential growth

$$I = I_0 e^{g^H(\nu)z} = I_0 e^{\sigma_{ul}^H(\nu)[N_u-(g_u/g_l)N_l]z}$$
$$= I_0 e^{\sigma_{ul}^H(\nu)\Delta N_{ul}z} \tag{7.18}$$

Doppler broadening

$$g^D(\nu) = \sqrt{\frac{\ln 2}{16\pi^3}} \frac{c^2 A_{ul}}{\nu_0^2 \Delta\nu_D}\left[N_u - \frac{g_u}{g_l} N_l\right]$$
$$\times \exp\left\{-\left[\frac{4\ln 2(\nu-\nu_0)^2}{\Delta\nu_D^2}\right]\right\} \tag{7.25}$$

$$g^D(\nu = \nu_0) \equiv g^D(\nu_0)$$
$$\equiv g_0^D = \sqrt{\frac{\ln 2}{16\pi^3}} \frac{\lambda_{ul}^2 A_{ul}}{\Delta\nu_D}\left[N_u - \frac{g_u}{g_l} N_l\right] \tag{7.26}$$

$$g^D(\nu) = \sigma_{ul}^D(\nu)\left[N_u - \frac{g_u}{g_l} N_l\right] = \sigma_{ul}^D(\nu)\Delta N_{ul} \tag{7.27}$$

$$\sigma_{ul}^D(\nu_0) = \sqrt{\frac{\ln 2}{16\pi^3}} \frac{\lambda_{ul}^2 A_{ul}}{\Delta\nu_D} \tag{7.28}$$

$$\sigma_{ul}^D(\nu_0) = (1.74 \times 10^{-4})\lambda_{ul}^3 A_{ul}\sqrt{M_N/T} \tag{7.29}$$
$$[\lambda \text{ in m, } A_{ul} \text{ in s}^{-1}, T \text{ in K, } M_N \text{ is mass number}]$$

Exponential growth

$$I = I_0 e^{g^D(\nu)z} = I_0 e^{\sigma_{ul}^D(\nu)[N_u-(g_u/g_l)N_l]z}$$
$$= I_0 e^{\sigma_{ul}^D(\nu)\Delta N_{ul}z} \tag{7.30}$$

SATURATION INTENSITY

$$I_{\text{sat}} = \frac{h\nu_{ul}}{\sigma_{ul}^H(\nu)\tau_u} \tag{7.42}$$

$$E_{\text{sat}} = I_{\text{sat}}\Delta\tau_p = \frac{h\nu_{ul}}{\sigma_{ul}^H(\nu)}\left(\frac{\Delta\tau_p}{\tau_u}\right) \tag{7.43}$$

GAIN SATURATION

$$g(\nu) = \frac{g^0(\nu)}{1+(I/I_{\text{sat}})} = \frac{\sigma_{ul}(\nu)\Delta N_{ul}^0}{1+(I/I_{\text{sat}})} \tag{7.71}$$

THRESHOLD CONDITIONS FOR LASERS

No mirrors

$$\sigma_{ul}\Delta N_{ul}L \cong 12 \pm 5 \tag{7.56}$$

One mirror

$$\sigma_{ul}\Delta N_{ul}(2L) \cong 12 \pm 5 \tag{7.57}$$

Two mirrors

$$g(\nu_0) = \frac{1}{2L}\ln\frac{1}{R^2} \tag{7.59}$$

$$g(\nu_0) = \frac{1}{2L}\left[\frac{1}{R_1 R_2(1-a_1)(1-a_2)}\right] + \alpha \tag{7.61}$$

$$t_s = m[\eta_C(d-L)+\eta_L L]/c \tag{7.65}$$

LASER FUNDAMENTALS

LASER
FUNDAMENTALS

WILLIAM T. SILFVAST

Center for Research and Education in Optics and Lasers
and
Department of Physics and Electrical/Computing Engineering
University of Central Florida

Published by the Press Syndicate of the University of Cambridge
The Pitt Building, Trumpington Street, Cambridge CB2 1RP
40 West 20th Street, New York, NY 10011-4211, USA
10 Stamford Road, Oakleigh, Melbourne 3166, Australia

First published 1996

Printed in the United States of America

Silfvast, William Thomas, 1937–
Laser fundamentals / William T. Silfvast.
p.   cm.
Includes bibliographical references.
ISBN 0-521-55424-1 (hc). – ISBN 0-521-55617-1 (pb)
1. Lasers. I. Title.
TA1675.S52   1996
621.36′6 – dc20   95-44207
                                                   CIP

A catalog record for this book is available from the British Library.

ISBN 0-521-55424-1 Hardback
      0-521-55617-1 Paperback

To my wife, Susan, and my three children, Scott, Robert and Stacey, all of whom are such an important part of my life.

CONTENTS

CONTENTS

V SPECIFIC LASER SYSTEMS

PREFACE

I wrote *Laser Fundamentals* with the idea of simplifying the explanation of how lasers operate. It is designed to be used as a senior-level or first-year graduate student textbook and/or as a reference book. The first draft was written the first time I taught the course "Laser Principles" at the University of Central Florida. Before that, I authored several general laser articles and taught short courses on the subject, giving careful consideration to the sequence in which various topics should be presented. During that period I adjusted the sequence, and I am now convinced that it is the optimal one.

Understanding lasers involves concepts associated with light, viewed either as waves or as photons, and its interaction with matter. I have used the first part of the book to introduce these concepts. Chapters 2 through 6 include fundamental wave properties, such as the solution of the wave equation, polarization, and the interaction of light with dielectric materials, as well as the fundamental quantum properties, including discrete energy levels, emission of radiation, emission broadening (in gases, liquids, and solids), and stimulated emission. The concept of amplification is introduced in Chapter 7, and further properties of laser amplifiers dealing with inversions and pumping are covered in Chapters 8 and 9. Chapter 10 discusses cavity properties associated with both longitudinal and transverse modes, and Chapters 11 and 12 follow up with Gaussian beams and special laser cavities. Chapters 13 and 14 provide descriptions of the most common lasers. The book concludes in Chapter 15 with a brief overview of some of the nonlinear optical techniques for laser frequency conversion.

Some of the unique aspects of the book are the treatment of emission linewidth and broadening in Chapter 4, the development of a simple model of a laser amplifier in Chapter 7, the discussion of special laser cavities in Chapter 12, and the laser summaries in Chapters 13 and 14. Throughout the book, whenever a particular concept is introduced, I have tried to relate that concept to all the various types of laser amplifiers including gas lasers, liquid (dye) lasers, and solid-state lasers. My intention is to give the reader a good understanding, not just of one specific type of laser but rather of all types of lasers, as each concept is introduced.

The book can be used in either a one- or two-semester course. In one semester the topics of Chapters 2 through 12 would be emphasized. In two semesters, extended coverage of the specific lasers of Chapters 13 and 14, as well as the frequency multiplication in Chapter 15, could be included.

In a one-semester course I have been able to cover a portion of the material in Chapters 13 and 14 by having each student write a report about one specific laser and then give a ten- or fifteen-minute classroom presentation about that laser. The simple quantum mechanical descriptions in Chapters 3 and 4 were introduced to describe how radiative transitions occur in matter. If the instructor chooses to avoid quantum mechanics in the course, it would be sufficient to stress the important results that are highlighted at the ends of each of those sections.

Writing this book has been a rewarding experience for me. I have been associated with lasers since shortly after their discovery in 1960 when, as an undergraduate student at the University of Utah, I helped build a ruby laser for a research project under Professor Frank Harris. He was the first person to instill in me an enthusiasm for optics and light. I was then very fortunate to be able to do my thesis work with Professor Grant Fowles, who encouraged me to reduce ideas to simple concepts. We discovered many new metal vapor lasers during that period. I also thank Dr. John Sanders for giving me the opportunity to do postdoctoral work at the Clarendon Laboratory at Oxford University in England, and Dr. Kumar Patel for bringing me to Bell Laboratories in Holmdel, New Jersey. Being a part of a stimulating group of researchers at Bell Laboratories during the growth of the field of lasers was an unparalleled opportunity. During that period I was also able to spend an extremely rewarding sabbatical year at Stanford University with Professor Steve Harris. Finally, to round out my career I put on my academic hat at the University of Central Florida as a member of the Center for Research and Education in Optics and Lasers (CREOL) and the Department of Physics and of Electrical and Computer Engineering. Working in the field of lasers at several different institutions has provided me with a broad perspective that I hope has successfully contributed to the manner in which many of the concepts are presented in this book.

ACKNOWLEDGMENTS

I first acknowledge the support of my wife, Susan. Without her encouragement and patience, I would never have completed this book.

Second, I am deeply indebted to Mike Langlais, an undergraduate student at the University of Central Florida and a former graphics illustrator, who did most of the figures for the book. I provided Mike with rough sketches, and a few days later he appeared with professional quality figures. These figures add immensely to the completeness of the book.

Colleagues who have helped me resolve particular issues associated with this book include Michael Bass, Peter Delfyett, Luis Elias, David Hagan, James Harvey, Martin Richardson, and Eric Van Stryland of CREOL; Tao Chang, Larry Coldren, Dick Fork, Eric Ippen, Jack Jewell, Wayne Knox, Herwig Kogelnik, Tingye Li, David Miller, Peter Smith, Ben Tell, and Obert Wood of Bell Laboratories; Bob Byer, Steve Harris, and Tony Siegman of Stanford University; Boris Stoicheff of the University of Toronto; Gary Eden of the University of Illinois; Ron Waynant of the FDA; Arto Nurmiko of Brown University; Dennis Matthews of Lawrence Livermore National Laboratories; Syzmon Suckewer of Princeton University; Colin Webb of Oxford University; John Macklin of Stanford University and Bell Labs; Jorgé Rocca of Colorado State University; Frank Tittle of Rice University; Frank Duarte of Kodak; Alan Petersen of Spectra Physics; Norman Goldblatt of Coherent, Inc.; and my editor friend, Irwin Cohen. I also thank the many laser companies who contributed figures, primarily in Chapters 13 and 14. I'm sure that I have left a few people out; for that, I apologize to them. In spite of all the assistance, I accept full responsibility for the final text.

I thank my editor, Philip Meyler, at Cambridge University Press for convincing me that CUP was the best publishing company and for assisting me in determining the general layout of my book. I also thank editor Matt Darnell for doing such a skillful job in taking my manuscript and making it into a "real" book.

I am indebted to several graduate students at CREOL. Howard Bender, Jason Eichenholz, and Art Hanzo helped with several of the figures. In addition, Jason Eichenholz assisted me in taking the cover photo, Howard Bender and Art Hanzo helped with the laser photo on the back cover, and Marc Klosner did a careful proofreading of one of the later versions of the text. I am also indebted to Al Ducharme for suggesting the title for the book.

Finally, I thank the students who took the "Laser Principles" course the first year I taught it (Fall 1991). At that point I was writing and passing out drafts of my chapters to the students at a frantic pace. Because those students had to suffer through that first draft, I promised all of them a free copy of the book. I stand by that promise and hope those students will get in touch with me to collect.

INTRODUCTION

SUMMARY A laser is a device that amplifies light and produces a highly directional, high-intensity beam that most often has a very pure frequency or wavelength. It comes in sizes ranging from approximately one tenth the diameter of a human hair to the size of a very large building, in powers ranging from 10^{-9} to 10^{20} W, and in wavelengths ranging from the microwave to the soft–X-ray spectral regions with corresponding frequencies from 10^{11} to 10^{17} Hz. Lasers have pulse energies as high as 10^4 J and pulse durations as short as 6×10^{-15} s. They can easily drill holes in the most durable of materials and can weld detached retinas within the human eye. They are a key component of some of our most modern communication systems and are the "phonograph needle" of our compact disc players. They perform heat treatment of high-strength materials, such as the pistons of our automobile engines, and provide a special surgical knife for many types of medical procedures. They act as target designators for military weapons and provide for the rapid check-out we have come to expect at the supermarket. What a remarkable range of characteristics for a device that is in only its fourth decade of existence!

INTRODUCTION

There is nothing magical about a laser. It can be thought of as just another type of light source. It certainly has many unique properties that make it a special light source, but these properties can be understood without knowledge of sophisticated mathematical techniques or complex ideas. It is the objective of this text to explain the operation of the laser in a simple, logical approach that builds from one concept to the next as the chapters evolve. The concepts, as they are developed, will be applied to all classes of laser materials, so that the reader will develop a sense of the broad field of lasers, while still acquiring the capability to study, design, or simply understand a specific type of laser system in detail.

DEFINITION OF THE LASER

The word *laser* is an acronym for Light Amplification by Stimulated Emission of Radiation. The laser makes use of processes that increase or amplify light signals after those signals have been generated by other means. These processes include (1) stimulated emission, a natural effect that was deduced by considerations relating to thermodynamic equilibrium, and (2) optical feedback (present in most lasers) that is usually provided by mirrors. Thus, in its simplest form, a laser consists of a gain or amplifying medium (where

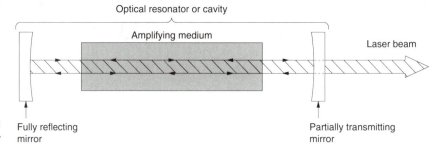

Optical resonator or cavity

Amplifying medium

Laser beam

Figure 1-1. Simplified schematic of typical laser

Fully reflecting mirror

Partially transmitting mirror

stimulated emission occurs), and a set of mirrors to feed the light back into the amplifier for continued growth of the developing beam, as seen in Figure 1-1.

SIMPLICITY OF A LASER

The simplicity of a laser can be understood by considering the light from a candle. Normally, a burning candle radiates light in all directions, and therefore illuminates various objects equally if they are equidistant from the candle. A laser takes light that would normally be emitted in all directions, such as from a candle, and concentrates that light into a single direction. Thus, if the light radiating in all directions from a candle were concentrated into a single beam of the diameter of the pupil of your eye (approximately 3 mm), and you were standing a distance of 1 m from the candle, then the light intensity would be 1,000,000 times as bright as the light that you normally see radiating from the candle! That is essentially the underlying concept of the operation of a laser. However, a candle is not the kind of medium that produces amplification, and thus there are no candle lasers. It takes relatively special conditions within the laser medium for amplification to occur, but it is that capability of taking light that would normally radiate from a source in all directions – and concentrating that light into a beam traveling in a single direction – that is involved in making a laser. These special conditions, and the media within which they are produced, will be described in some detail in this book.

UNIQUE PROPERTIES OF A LASER

The beam of light generated by a typical laser can have many properties that are unique. When comparing laser properties to those of other light sources, it can be readily recognized that the values of various parameters for laser light either greatly exceed or are much more restrictive than the values for many common light sources. We never use lasers for street illumination, or for illumination within our houses. We don't use them for searchlights or flashlights or as headlights in our cars. Lasers generally have a narrower frequency distribution, or much higher intensity, or a much greater degree of collimation, or much shorter pulse duration, than that available from more common types of light sources. Therefore, we do use them in compact disc players, in supermarket check-out scanners, in surveying instruments, and in medical applications as a surgical knife or for welding detached retinas. We also use them in communications systems and in radar and military targeting applications, as well as many other areas. *A laser is a specialized light source that should be used only when its unique properties are required.*

THE LASER SPECTRUM AND WAVELENGTHS

A portion of the electromagnetic radiation spectrum is shown in Figure 1-2 for the region covered by currently existing lasers. Such lasers span the wavelength range from the far infrared part of the spectrum ($\lambda = 1,000\ \mu m$) to the soft–X-ray region ($\lambda = 3$ nm), thereby covering a range of wavelengths of almost six orders of magnitude. There are several types of units that are used to define laser wavelengths. These range from micrometers or microns (μm) in the infrared to nanometers (nm) and angstroms (Å) in the visible, ultraviolet (UV), vacuum ultraviolet (VUV), extreme ultraviolet (EUV or XUV), and soft–X-ray (SXR) spectral regions.

WAVELENGTH UNITS

$1\ \mu m = 10^{-6}$ m;
$1\ \text{Å} = 10^{-10}$ m;
$1\ \text{nm} = 10^{-9}$ m.

Consequently, 1 micron (μm) = 10,000 angstroms (Å) = 1,000 nanometers (nm). For example, green light has a wavelength of 5×10^{-7} m = 0.5 μm = 5,000 Å = 500 nm.

WAVELENGTH REGIONS

Far infrared: 10 to 1,000 μm;
middle infrared: 1 to 10 μm;
near infrared: 0.7 to 1 μm;

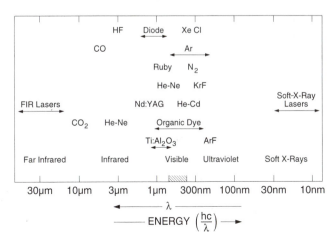

Figure 1-2. Wavelength range of various lasers

visible: 0.4 to 0.7 μm, or 400 to 700 nm;
ultraviolet: 0.2 to 0.4 μm, or 200 to 400 nm;
vacuum ultraviolet: 0.1 to 0.2 μm, or 100 to 200 nm;
extreme ultraviolet: 10 to 100 nm;
soft X-rays: 1 nm to approximately 20–30 nm (some overlap with EUV).

A BRIEF HISTORY OF THE LASER

Charles Townes took advantage of the stimulated emission process to construct a microwave amplifier, referred to as a *maser*. This device produced a coherent beam of microwaves to be used for communications. The first maser was produced in ammonia vapor with the inversion between two energy levels that produced gain at a wavelength of 1.25 cm. The wavelengths produced in the maser were comparable to the dimensions of the device, so extrapolation to the optical regime – where wavelengths were five orders of magnitude smaller – was not an obvious extension of that work.

In 1958, Townes and Schawlow published a paper concerning their ideas about extending the maser concept to optical frequencies. They developed the concept of an optical amplifier surrounded by an optical mirror resonant cavity to allow for growth of the beam. Townes and Schawlow both received Nobel Prizes for their work in this field.

In 1960, Theodore Maiman of Hughes Research Laboratories produced the first laser using a ruby crystal as the amplifier and a flashlamp as the energy source. The helical flashlamp surrounded a rod-shaped ruby crystal, and the optical cavity was formed by coating the flattened ends of the ruby rod with a highly reflecting material. An intense red beam was observed to emerge from the end of the rod when the flashlamp was fired!

The first gas laser was developed in 1961 by A. Javan, W. Bennett, and D. Harriott of Bell Laboratories, using a mixture of helium and neon gases. At the same laboratories, L. F. Johnson and K. Nassau demonstrated the first neodymium laser, which has since become one of the most reliable lasers available. This was followed in 1962 by the first semiconductor laser, demonstrated by R. Hall at the General Electric Research Laboratories. In 1963, C. K. N. Patel of Bell Laboratories discovered the infrared carbon dioxide laser, which is one of the most efficient and powerful lasers available today. Later that same year, E. Bell of Spectra Physics discovered the first ion laser, in mercury vapor. In 1964 W. Bridges of Hughes Research Laboratories discovered the argon ion laser, and in 1966 W. Silfvast, G. R. Fowles, and B. D. Hopkins produced the first blue helium–cadmium metal vapor laser. During that same year, P. P. Sorokin and J. R. Lankard of the IBM Research Laboratories developed the first liquid laser using an organic dye dissolved in a solvent, thereby leading to the category of broadly tunable lasers. Also at that time, W. Walter and co-workers at TRG reported the first copper vapor laser.

The first vacuum ultraviolet laser was reported to occur in molecular hydrogen by R. Hodgson of IBM and independently by R. Waynant et al. of the Naval Research Laboratories in 1970. The first of the well-known rare-gas–halide excimer lasers was observed in xenon fluoride by J. J. Ewing and C. Brau of the Avco–Everett Research Laboratory in 1975. In that same year, the first quantum-well laser was made in a gallium arsenide semiconductor by J. van der Ziel and co-workers at Bell Laboratories. In 1976, J. M. J. Madey and co-workers at Stanford University demonstrated the first free-electron laser amplifier operating in the infrared at the CO_2 laser wavelength. In 1979, Walling and co-workers at Allied Chemical Corporation obtained broadly tunable laser output from a solid-state laser material called alexandrite, and in 1985 the first soft–X-ray laser was successfully demonstrated in a highly ionized selenium plasma by D. Matthews and a large number of co-workers at the Lawrence Livermore Laboratories.

In 1961, Fox and Li described the existence of resonant transverse modes in a laser cavity. That same year, Boyd and Gordon obtained solutions of the wave equation for confocal resonator modes. Unstable resonators were demonstrated in 1969 by Krupke and Sooy, and described theoretically by Siegman. Q-switching was first obtained by McClung and Hellwarth in 1962, and described later by Wagner and Lengyel. The first mode-locking was obtained by Hargrove, Fork, and Pollack in 1964. Since then, many special cavity arrangements, feedback schemes, and other devices were also developed to improve the control, operation, and reliability of lasers.

OVERVIEW OF THE BOOK

Isaac Newton described light as small bodies emitted from shining substances. This view was no doubt influenced by the fact that light appears to propagate in a straight line. Christian Huygens, on the other hand, described light as a wave motion in which a small source spreads out in all directions; most observed effects - including diffraction, reflection, and refraction - can be attributed to the consideration of the expansion of primary waves and of secondary wavelets. This dual nature of light is still a useful concept, whereby the choice of particle or wave explanation depends upon the effect to be considered.

Section I of this book deals with the fundamental *wave* properties of light, including Maxwell's equations, the interaction of electromagnetic radiation with matter, absorption and dispersion, and coherence. Section II deals with the fundamental *quantum* properties of light. Chapter 3 describes the concept of discrete energy levels in atomic laser species and also how the periodic table of the elements evolved. Chapter 4 deals with radiative transitions and emission linewidths and the probability of making transitions between energy levels. Chapter 5 considers energy levels of lasers in molecules, liquids, and solids - both dielectric solids and semiconductors. Chapter 6 then considers radiation in equilibrium and the concepts of absorption and stimulated emission of radiation. At this point the student has the basic tools to begin building a laser.

Section III considers laser amplifiers. Chapter 7 describes the theoretical basis for producing population inversions and gain. Chapter 8 describes how population inversions are produced, and Chapter 9 considers how sufficient amplification is achieved to make an intense laser beam. Section IV deals with laser resonators. Chapter 10 considers both longitudinal and transverse modes within a laser cavity, and Chapter 11 investigates the properties of stable resonators and Gaussian beams. Chapter 12 considers a variety of special laser cavities and effects, including unstable resonators, Q-switching, mode-locking, pulse narrowing, ring lasers, and spectral narrowing.

Section V covers specific laser systems. Chapter 13 describes eleven of the most well-known gas and plasma laser systems. Chapter 14 considers eleven well-known dye lasers and solid-state lasers, including both dielectric solid-state lasers and semiconductor lasers. The book concludes with Section VI (Chapter 15), which provides a brief overview of frequency multiplication with lasers and other nonlinear effects.

FUNDAMENTAL WAVE PROPERTIES OF LIGHT

WAVE NATURE OF LIGHT
THE INTERACTION OF LIGHT WITH MATERIALS

2

SUMMARY This chapter will consider some of the wave properties of light that are relevant to the understanding of lasers. We will first briefly review the derivation of Maxwell's wave equations based upon the experimentally obtained laws of electricity and magnetism. We will consider Maxwell's equations in a vacuum, and will demonstrate the equivalence of light and electromagnetic radiation. We will describe both the phase velocity and group velocity of light, and show how polarized light occurs in these transverse electromagnetic waves.

We will then use the wave equations to consider the interaction of electromagnetic waves with transparent and semitransparent materials such as those used for the gain media of solid-state lasers. We will derive expressions for the optical constants η and κ (the real and imaginary components of the index of refraction) that suggest the presence of strong absorptive and dispersive regions at particular resonant wavelengths or frequencies of the dielectric materials. These are regions where η and κ vary quite significantly with frequency. These resonances will later (in Chapters 4, 5, and 6) be related to optical transitions (both emission and absorption) between energy levels in those materials. We will conclude the chapter with a brief description of coherence.

2.1 MAXWELL'S EQUATIONS

Maxwell's wave equations, predicting the propagation – even in a vacuum – of transversely oscillating electromagnetic waves, were the first indication of the true nature of light. His wave equations predicted a velocity c for such a wave in a vacuum ($c = (\mu_0 \epsilon_0)^{-1/2}$) that agreed with independent measurements of the velocity of light. This velocity was based solely upon the value of two previously known constants: μ_0, the permeability of the vacuum; and ϵ_0, the permittivity of the vacuum. These constants had arisen from totally separate investigations of electricity and magnetism that had nothing to do with studies of light! By definition, the exact value of μ_0 is $4\pi \times 10^{-7}$ H/m; the measured value of ϵ_0 is 8.854×10^{-12} F/m.

The fundamental electric and magnetic field vectors **E** and **B** (respectively) can produce forces on physical entities. These vectors are measurable quantities that led to the experimentally derived laws of Gauss, Biot–Savart, Ampere, and Faraday, all of which are briefly outlined in what follows. These laws in turn formed the foundation upon which Maxwell built to develop his equations predicting the existence and properties of electromagnetic radiation.

In addition to **E** and **B**, we must consider properties associated with the electromagnetic state of matter. Such matter can be described at a given location in space by four quantities:

(1) the charge density ρ (charge per unit volume);
(2) the polarization **P** (electric dipole moment per unit volume), in particular for a dielectric material;
(3) the magnetization **M** (magnetic dipole moment per unit volume); and
(4) the current density **J** (current per unit area).

For our purposes, all of these quantities are assumed to be average values. Other relationships that will be useful include:

(5) the Lorentz force law,

$$\mathbf{F} = q(\mathbf{v} \times \mathbf{B}), \tag{2.1}$$

where **F** is the force resulting from a charge q and **v** is the velocity of the moving charge; and
(6) a form of Ohm's law given by

$$\mathbf{J} = \sigma \mathbf{E}, \tag{2.2}$$

which describes the response of electrons in a conducting medium of conductivity σ to an electric field vector **E**.

Maxwell defined the electric displacement vector **D** as follows:

$$\mathbf{D} = \epsilon_0 \mathbf{E} + \mathbf{P} \quad \text{(for general use);} \tag{2.3}$$

D is used so as to avoid explicit inclusion of the charge associated with the polarization **P** in Gauss's flux law.

For the case of free space,

$$\mathbf{D} = \epsilon_0 \mathbf{E}, \tag{2.4}$$

and for an isotropic linear dielectric,

$$\mathbf{D} = \epsilon \mathbf{E}, \tag{2.5}$$

which describes the aggregate response of the bound charges to the electric field.

An alternate way of doing this is to write

$$\mathbf{P} = (\epsilon - \epsilon_0)\mathbf{E} = \chi \epsilon_0 \mathbf{E}, \tag{2.6}$$

which gives the relationship between the polarization **P** and the electric field **E** that produces **P**. The factor χ, known as the *electric susceptibility*, is given by

$$\chi = (\epsilon/\epsilon_0) - 1; \tag{2.7}$$

χ is a useful parameter when considering effects in the optical frequency range. For spherically symmetric materials such as glass, χ is a simple scalar quantity, but for anisotropic materials, χ is expressed as a tensor to account for the variation in polarization response for different directions of the applied field. If (2.6) is generalized by expressing the polarization **P** as a power series in the field strength **E**, higher-order values of χ are obtained that describe the nonlinear optical properties of materials. Such properties are used for frequency conversion and other effects associated with lasers, and will be described in more detail in Chapter 15.

The general expression for **B** can be written as

$$\mathbf{B} = \mu_0(\mathbf{H} + \mathbf{M}), \tag{2.8}$$

where **H** is a convenient vector quantity that provides for a different expression of the magnetic flux **B**, that is, in a form that allows the parallel treatment of the electric and magnetic flux laws. For a vacuum,

$$\mathbf{B} = \mu_0 \mathbf{H}, \tag{2.9}$$

and for isotropic linear magnetic media,

$$\mathbf{B} = \mu \mathbf{H}. \tag{2.10}$$

Using these quantities, Maxwell's equations can be derived from the following experimentally determined relationships.

GAUSS'S LAW   This law states that the total electric flux Φ through any closed surface, or (equivalently) the surface integral over the normal component of the electric field vector **E** over that closed surface, equals the net charge $\sum_n q_n$ inside the surface:

$$\Phi = \oint \mathbf{E} \cdot d\mathbf{S} = \frac{1}{\epsilon_0} \sum_n q_n = \frac{1}{\epsilon_0} \int_V \rho \, dV, \tag{2.11}$$

where **dS** is the surface element vector at any point p on the surface surrounding the volume V. If the charge is distributed within the volume, ρ is the localized charge density within a volume element dV. This law can be expressed in differential form as

$$\nabla \cdot \mathbf{D} = \rho. \tag{2.12}$$

This is considered the *first* of Maxwell's equations.

BIOT–SAVART LAW   This law is expressed here in a form similar to that of Coulomb's law relating the force of attraction of two charges:

$$\mathbf{B} = \frac{\mu_0}{4\pi} \int_V \mathbf{J}(V) \times \frac{\mathbf{r}}{|\mathbf{r}|^3} \, dV, \tag{2.13}$$

where $\mathbf{J}(V)$ is the current density within volume element dV and **r** is the position vector from volume element dV to the point of measurement of **B**. It can be expressed in differential form as

$$\mathbf{B} = \frac{\mu_0}{4\pi} \nabla \times \int \frac{\mathbf{J}(V)}{|\mathbf{r}|} \, dV, \tag{2.14}$$

which leads directly to

$$\mathbf{V} \cdot \mathbf{B} = 0, \tag{2.15}$$

since **B** is the curl of another vector. This is considered the *second* of Maxwell's equations.

AMPERE'S LAW   This law, in its simple form, states that the line integral of the magnetic induction vector **B** around any closed path is equal to the product of the permeability μ_0 and the net current I flowing across the area bounded by the path:

$$\oint \mathbf{B} \cdot \mathbf{dl} = \mu_0 I. \tag{2.16}$$

In differential form, this law can be written as

$$\mathbf{V} \times \mathbf{B} = \mu_0 \left(\mathbf{J} + \frac{\partial \mathbf{D}}{\partial t} \right), \tag{2.17}$$

and represents the *third* of Maxwell's equations.

FARADAY'S LAW   This law is analogous to Ampere's law in stating that the line integral of the electric field vector **E** around any closed path l is equal to the time rate of change of magnetic flux Φ_M passing through the area defined by that path:

$$\oint \mathbf{E} \cdot \mathbf{dl} = -\frac{d\Phi_M}{dt}. \tag{2.18}$$

In differential form, this law is written as

$$\mathbf{V} \times \mathbf{E} = -\frac{\partial \mathbf{B}}{\partial t}, \tag{2.19}$$

and historically represents the *fourth* of Maxwell's equations.

The third equation (eqn. 2.17) was modified by Maxwell to satisfy the law of continuity of charge. Maxwell's four equations relate charge density, current density, and field quantities at a single point in space through their time and space derivatives.

2.2   MAXWELL'S WAVE EQUATIONS

MAXWELL'S WAVE EQUATIONS FOR A VACUUM

Maxwell's equations will first be considered for a vacuum, in order to obtain the simplest form of the electromagnetic wave equation, and also to demonstrate that the predicted waves require no medium to support their existence. In that case the equations reduce to

$$\mathbf{V} \cdot \mathbf{E} = 0, \tag{2.20}$$

$$\mathbf{V} \cdot \mathbf{H} = 0, \tag{2.21}$$

$$\mathbf{\nabla} \times \mathbf{H} = \epsilon_0 \frac{\partial \mathbf{E}}{\partial t}, \tag{2.22}$$

$$\mathbf{\nabla} \times \mathbf{E} = -\mu_0 \frac{\partial \mathbf{H}}{\partial t}. \tag{2.23}$$

These four equations constitute Maxwell's equations for a vacuum (the absence of matter). In rewriting these equations, $\mathbf{B}$ has been replaced by its equivalent $\mu_0 \mathbf{H}$ (eqn. 2.8) since $\mathbf{M} = 0$. Equations (2.20) and (2.21) indicate the absence of charge at the point of consideration.

The solutions for $\mathbf{E}$ and $\mathbf{H}$ can be separated by taking the curl of one and the time derivative of the other. Then, using the fact that the order of differentiation can be reversed, one can obtain the following parallel equations:

$$\mathbf{\nabla} \times (\mathbf{\nabla} \times \mathbf{E}) = -\mu_0 \epsilon_0 \frac{\partial^2 \mathbf{E}}{\partial t^2}, \tag{2.24}$$

$$\mathbf{\nabla} \times (\mathbf{\nabla} \times \mathbf{H}) = -\mu_0 \epsilon_0 \frac{\partial^2 \mathbf{H}}{\partial t^2}. \tag{2.25}$$

Because $\mathbf{\nabla} \times (\mathbf{\nabla} \times \mathbf{A}) = \mathbf{\nabla}(\mathbf{\nabla} \cdot \mathbf{A}) - \nabla^2 \mathbf{A}$ for any vector $\mathbf{A}$, (2.24) and (2.25) lead to the following two Maxwell wave equations:

$$\nabla^2 \mathbf{E} = \mu_0 \epsilon_0 \frac{\partial^2 \mathbf{E}}{\partial t^2} \tag{2.26}$$

and

$$\nabla^2 \mathbf{H} = \mu_0 \epsilon_0 \frac{\partial^2 \mathbf{H}}{\partial t^2}. \tag{2.27}$$

SOLUTION OF THE GENERAL WAVE EQUATION – EQUIVALENCE OF LIGHT AND ELECTROMAGNETIC RADIATION

Equations (2.26) and (2.27) are wave equations of the form

$$\nabla^2 \mathbf{A} = \frac{1}{v^2} \frac{\partial^2 \mathbf{A}}{\partial t^2}, \tag{2.28}$$

where the vector $\mathbf{A}$ is a function of x, y, z, and t that may be expressed as $\mathbf{A}(x, y, z, t)$. In our situation, $\mathbf{A}$ could represent either the electric field vector $\mathbf{E}(x, y, z, t)$ or the magnetic field vector $\mathbf{H}(x, y, z, t)$. We will show that $\mathbf{A}$ is an oscillatory wave function with an amplitude that is transverse to the direction of propagation, and that v is a constant associated with the velocity of that wave.

The left side of (2.28) involves derivatives of $\mathbf{A}$ with respect to the spatial variables x, y, and z. For simplicity we will consider only a single spatial direction, the z direction, and consequently only the z component $A(z, t)$ of $\mathbf{A}$. We can then rewrite (2.28) as

$$\frac{d^2 A(z, t)}{dz^2} = \frac{1}{v^2} \frac{d^2 A(z, t)}{dt^2}. \tag{2.29}$$

We can express the wave function $A(z, t)$ as a product of functions $A_z(z)$ and $A_t(t)$ as follows:

$$A(z, t) = A_z(z) A_t(t) \quad \text{or} \quad A(z, t) = A_z A_t. \tag{2.30}$$

Substitution into (2.29) then leads to

$$A_t \frac{d^2 A_z}{dz^2} - \frac{A_z}{v^2} \frac{d^2 A_t}{dt^2} = 0 \tag{2.31}$$

or

$$\frac{v^2}{A_z} \frac{d^2 A_z}{dz^2} = \frac{1}{A_t} \frac{d^2 A_t}{dt^2}. \tag{2.32}$$

The left side of (2.32) is dependent only upon z and the right side only upon t, and so – in order to satisfy the equation – both sides must be equal to the same constant, which we will arbitrarily denote as $-\omega^2$. This leads us to the following two equations:

$$\frac{d^2 A_z}{dz^2} + \frac{\omega^2}{v^2} A_z = 0, \tag{2.33}$$

$$\frac{d^2 A_t}{dt^2} + \omega^2 A_t = 0. \tag{2.34}$$

These equations are of a familiar form and have the following solutions:

$$A_z = C_1 e^{i(\omega/v)z} + C_2 e^{-i(\omega/v)z}, \tag{2.35}$$

$$A_t = D_1 e^{i\omega t} + D_2 e^{-i\omega t}, \tag{2.36}$$

where the constants C_1, C_2, D_1, and D_2 are determined by the boundary conditions. We can now express the general solution $A(z, t)$ as

$$A(z, t) = A_z(z) A_t(t) \propto e^{\pm i(\omega/v)z} e^{\pm i\omega t} = e^{\pm i[(\omega/v)z + \omega t]}. \tag{2.37}$$

This general solution involves a complex wave function. For our purposes, consider a wave, traveling from left to right, that is a function of the form

$$A(z, t) = C e^{-i(k_z z - \omega t)}, \tag{2.38}$$

where we have defined k_z as

$$k_z = \frac{\omega}{v}. \tag{2.39}$$

The quantity k_z is called the *propagation constant* or the *wave number* (the number of waves per unit length), and has dimensions of 1/length. The wavelength λ is the distance over which the maximum amplitude of the

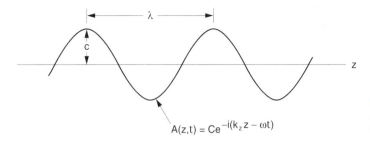

$$A(z,t) = Ce^{-i(k_z z - \omega t)}$$

Figure 2-1. Wavelength of a sinusoidally oscillating wave

wave travels during the time it makes one complete cycle of oscillation, or the distance between successive peaks of a wave frozen in time as shown in Figure 2-1.

$$\lambda = \frac{v}{\nu} = \frac{v}{(\omega/2\pi)} = \frac{2\pi v}{\omega}, \qquad (2.40)$$

where ν is the frequency or the number of complete cycles of oscillation per unit time and λ is related to k_z by

$$k_z = \frac{2\pi}{\lambda}. \qquad (2.41)$$

• PLANE WAVES   The wave described in (2.38) represents a wave that, for any value of z, has the same amplitude value for all values of x and y. It is thus referred to as a *plane wave* since it represents planes of constant value that are of infinite lateral extent, as indicated in Figure 2-2.

EXAMPLE

Write out the expression for a hypothetical plane wave, traveling in the z direction, that has a maximum amplitude of 1 and a wavelength in the visible portion of the spectrum at 514.5 nm.

We will use (2.38) to write the following function for the plane wave:

$$E(z, t) = Ce^{-i(k_z z - \omega t)} = e^{-i(k_z z - \omega t)}.$$

We therefore need values for k_z and ω. From (2.40) we know that $\omega = 2\pi v/\lambda$. For light traveling at a velocity of $v = c = 3 \times 10^8$ m/s we have

$$\omega = \frac{2\pi(3 \times 10^8 \text{ m/s})}{5.145 \times 10^{-6} \text{ m}} = 3.66 \times 10^{14} \text{ rad/s}.$$

Also, we know from (2.39) that $k_z = \omega/v$ and thus

$$k_z = \frac{3.66 \times 10^{14} \text{ rad/s}}{3 \times 10^8 \text{ m/s}} = 1.22 \times 10^6 \text{ m}^{-1}.$$

EXAMPLE (cont.)

Thus there are approximately 1,220,000 waves per meter at this particular wavelength. The plane wave can now be expressed as

$$E = e^{-i[(1.22\times10^6)z-(3.66\times10^{14})t]},$$

where z is expressed in meters and t is in seconds.

WAVE VELOCITY – PHASE AND GROUP VELOCITIES

PHASE VELOCITY   Equation (2.38) describes an oscillatory function that has a sinusoidal variation with either z or t. If we set t equal to a constant (we choose $t = 0$) then we have a sinusoidally varying function of the form $e^{-ik_z z}$, as shown in Figure 2-3(a), that is frozen in time. If we set z equal to a constant (we choose $z = 0$) then we have a similar function of the form $e^{i\omega t}$, which is shown in Figure 2-3(b). In either case, it is understood that we take the real part of the function to obtain the value of the amplitude.

The argument of (2.38) is a phase factor ϕ such that

$$\phi \equiv k_z z - \omega t. \tag{2.42}$$

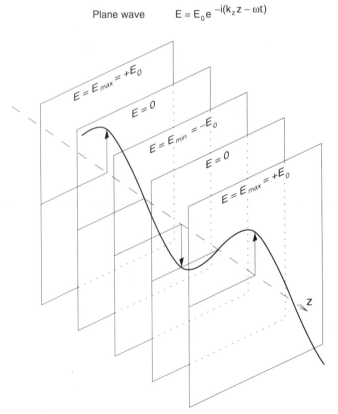

Plane wave $E = E_0 e^{-i(k_z z - \omega t)}$

Figure 2-2. Diagram of a plane wave

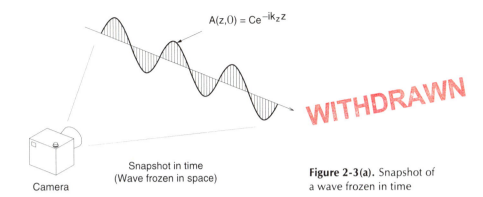

Snapshot in time
(Wave frozen in space)

Camera

Figure 2-3(a). Snapshot of a wave frozen in time

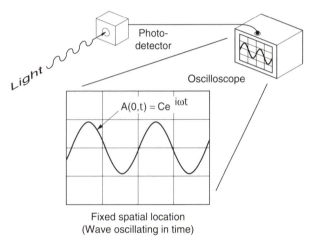

Photo-detector

Oscilloscope

$A(0,t) = Ce^{i\omega t}$

Fixed spatial location
(Wave oscillating in time)

Figure 2-3(b). Temporal display of a wave traveling past a fixed location in space

By considering the value of the wave for $\phi = $ constant we can track how the wave moves with time, since the amplitude of the wave function remains constant when the argument of the wave function is constant. The velocity at which the wave form propagates is called the *phase velocity*. We obtain this velocity by taking the derivative of ϕ and equating it to zero, which denotes the maximum value of the wave form. For $d\phi = 0$ we are led to

$$k_z\,dz = \omega\,dt, \tag{2.43}$$

from which we obtain the phase velocity of the wave as

$$V_{\text{phase}} = \frac{dz}{dt} = \frac{\omega}{k_z} = v. \tag{2.44}$$

This tells us that the constant v in (2.28) is the velocity at which the wave moves in time.

Comparing (2.26) and (2.27) with (2.28) suggests that $v = 1/(\mu_0\epsilon_0)^{1/2}$. A precise electrical measurement of the value of $1/(\mu_0\epsilon_0)^{1/2}$ provided a value of $v = 299{,}784 \pm 10$ km/s with a precision of 1 part in 30,000. This compares favorably with the most accurate direct measurement of the speed of light of $c = 299{,}792{,}457.4 \pm 1.1$ m/s. We can thus conclude that $v \equiv c$.

In this section, for simplicity, we have considered a special case of waves propagating in the z direction with the associated propagation constant or wave number k_z. Later we will consider a wave propagating in an arbitrary direction, with a propagation vector **k** of magnitude k that is a description of a more general wave.

GROUP VELOCITY An infinite wave train of a single frequency has no starting and stopping point; it can't be modulated. Any attempt at modulation, to start and stop a wave train, requires a wave involving more than one frequency (see Chapter 4). For a wave consisting of more than one frequency, the envelope of the combined waves moves with a velocity referred to as the *group velocity*. The group velocity is expressed as

$$V_{\text{group}} = \frac{d\omega}{dk}. \tag{2.45}$$

A finite group velocity implies a change in velocity with a change in wavelength. A medium with such characteristics is referred to as a *dispersive* medium. In a nondispersive medium we can see from (2.44) that the group velocity $d\omega/dk = v$, so the phase velocity and the group velocity are identical. In a dispersive medium, the group velocity is different from the phase velocity. The group velocity is the velocity at which information can be transmitted because sending information involves modulation.

GENERALIZED SOLUTION OF THE WAVE EQUATION

The solution to the wave equation given in (2.38) can be expressed in a more general form, involving all of the spatial coordinates, as

$$\mathbf{A}(x, y, z, t) = \mathbf{A}_0 e^{-i(k_x x + k_y y + k_z z + \omega t)} \tag{2.46}$$

or

$$\mathbf{A}(x, y, z, t) = \mathbf{A}_0 e^{-i(\mathbf{k} \cdot \mathbf{r} - \omega t)}, \tag{2.47}$$

where **k** is referred to as the *propagation wave vector* that defines the direction of propagation of the wave ($\mathbf{k} = k_x \hat{i} + k_y \hat{j} + k_z \hat{k}$), **r** is the *position vector* identifying a particular location in space such that $\mathbf{r} = x\hat{i} + y\hat{j} + z\hat{k}$, and where $\hat{i}$, $\hat{j}$, and $\hat{k}$ are unit vectors. Thus the wave vector **k** defines the normal to the planes of constant amplitude and phase as they propagate with velocity $v = c$.

Comparing (2.28) with (2.26) and (2.27), we conclude that the solution for the electric vector **E** in (2.26) can be expressed as

$$\mathbf{E} = \mathbf{E}_0 e^{-i(\mathbf{k} \cdot \mathbf{r} - \omega t)} \tag{2.48}$$

and that the velocity of the wave is

$$v = c = \frac{1}{(\mu_0 \epsilon_0)^{1/2}} = 2.99792 \times 10^8 \text{ m/s}. \tag{2.49}$$

TRANSVERSE ELECTROMAGNETIC WAVES AND POLARIZED LIGHT

Equation (2.48) represents the electric component of an electromagnetic wave whose amplitude is a vector quantity. A similar relationship can be expressed for the magnetic component $\mathbf{H}$. Since $\nabla \cdot \mathbf{E} = 0$ according to (2.20) and since $\nabla \cdot \mathbf{E} = i\mathbf{k} \cdot \mathbf{E}$, we can conclude that $i\mathbf{k} \cdot \mathbf{E} = 0$ and thus $\mathbf{E}$ is perpendicular to $\mathbf{k}$, the direction of propagation. A similar argument can be made for the magnetic portion of the wave, since $\nabla \cdot \mathbf{H} = 0$ from (2.21). Therefore, both the amplitudes $\mathbf{E}_0$ and $\mathbf{H}_0$ are perpendicular to the direction of the propagation vector $\mathbf{k}$ and thus to the propagation of the wave. Hence these electromagnetic waves are referred to as *transverse* waves.

Because the amplitude factor in an electromagnetic wave is a vector quantity that lies in a plane perpendicular to the direction of propagation, that amplitude can be resolved into independent orthogonal components. Thus, for example, a wave traveling in the z direction, such as that expressed as

$$\mathbf{E} = \mathbf{E}_0 e^{-i(k_z z - \omega t)}, \tag{2.50}$$

can be resolved into its two orthogonal components or polarizations:

$$\mathbf{E} = E_{0x} \hat{i} e^{-i(k_z z - \omega t)} + E_{0y} \hat{j} e^{-i(k_z z - \omega t + \phi)}, \tag{2.51}$$

where $\hat{i}$ and $\hat{j}$ are unit vectors in the x and y directions, respectively. The relative phase factor ϕ suggests that the two independent polarizations need not be in exact phase. If $\phi = 0$, or integral multiples of 2π, then the combined waves represent a single wave with the transverse amplitude vector always oriented in the same direction. Such a wave is referred to as *plane* or *linearly polarized* light, as shown in Figure 2-4. Other values of ϕ cause the amplitude vector of the wave to spiral in either an elliptical or a circular manner, as indicated in Figure 2-4. Such waves are referred to as either *circularly* polarized or *elliptically* polarized light.

FLOW OF ELECTROMAGNETIC ENERGY

The time rate of flow of electromagnetic energy $\mathbf{S}$ is described by Poynting's theorem, which states that

$$\mathbf{S} = \mathbf{E} \times \mathbf{H}. \tag{2.52}$$

The term $\mathbf{S}$ is known as the *Poynting vector,* which has MKS units of W/m^2. This vector indicates both the magnitude and direction of the radiation.

For a plane wave whose electric vector is expressed as in (2.48) and whose magnetic vector is expressed similarly, it can be easily shown that the *average value* of the Poynting vector, written as $\langle \mathbf{S} \rangle$, can be expressed as

$$\langle \mathbf{S} \rangle = \tfrac{1}{2} \mathbf{E}_0 \times \mathbf{H}_0. \tag{2.53}$$

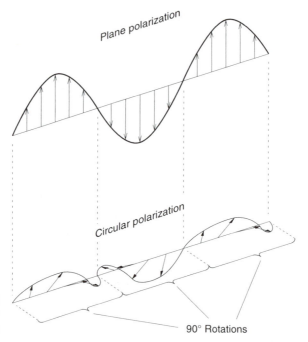

Figure 2-4. Examples of both plane and circularly polarized waves

90° Rotations

RADIATION FROM A POINT SOURCE (Electric Dipole Radiation)

Radiation from a point source, such as that from an isolated atom, is a more complicated situation. It is analyzed in terms of a vector potential **A** and a scalar potential ϕ such that, for example, the electric field vector **E** is obtained from

$$\mathbf{E} = -\nabla\phi - \frac{1}{c}\frac{\partial\mathbf{A}}{\partial t}. \tag{2.54}$$

Since we are dealing with oscillating fields from a point charge, it is not surprising that the vector and scalar potentials are oscillating functions that can be obtained from Maxwell's equations (2.12), (2.15), (2.17), and (2.19). The solution for the scalar potential is of the form

$$\phi = e^{i\omega t}\int\rho\frac{e^{-ikr}}{r}\,dV, \tag{2.55}$$

where ρ is the charge density of the radiating charge and dV is the volume element of that charge under consideration. The spatial variation of the oscillatory potential is of the form e^{ikr}/r, where r is the distance between a volume element of the radiating charge and the point of observation. The solution for the vector potential is similar. The solution for ϕ involves a function that decreases with $1/r$ as the distance from the source is increased, and consequently the electric field vector decreases with a function that involves $1/r^2$. For distances far from the source, where $r \gg \lambda \gg$ [source size] or $kr \gg 1$, it is conventional to expand ϕ and **a** in powers of kr in the form e^{ikr}/kr and to obtain values for **E** from (2.54) and **H** from the equivalent equation for **A**. These fields **E** and **H** are then used to obtain the energy flow or the Poynting vector ⟨**S**⟩ from a point source using (2.53). The results of such an expansion lead to an expression for the Poynting vector (at a significant distance from the source) involving a dominant term that is referred to as the electric dipole term. The next term in the expansion involves both a magnetic dipole term and an electric quadrapole term, both of which have values that are typically many orders of magnitude lower than the electric dipole term. As we develop a quantum mechanical description of radiation in Chapter 4, we will generally consider only electric dipole radiation, since generally it is by far the most dominant component of the radiation.

2.3 INTERACTION OF ELECTROMAGNETIC RADIATION (LIGHT) WITH MATTER

When considering either the design or use of lasers, one encounters the interaction of laser beams (electromagnetic waves) with matter. This interaction occurs within the gain medium and also at windows, mirrors, optical modulators installed within the laser cavity, and other associated optics. It is therefore appropriate to review some of the simple relationships entailed by that interaction. We will consider the propagation of light through transparent or partially transparent materials.

SPEED OF LIGHT IN A MEDIUM

Maxwell's equations for isotropic nonconducting media are identical to those for a vacuum, with the exception that μ_0 and ϵ_0 are replaced by their media values μ and ϵ; these provide a velocity of $v = (\mu\epsilon)^{-1/2}$. The terms μ and ϵ can be related to μ_0 and ϵ_0 by

$$K = \epsilon/\epsilon_0 \quad \text{(relative permittivity)} \tag{2.56}$$

and

$$K_m = \mu/\mu_0 \quad \text{(relative permeability).} \tag{2.57}$$

Thus

$$v = (\mu\epsilon)^{-1/2} = (K_m\mu_0 K\epsilon_0)^{-1/2} = c(K_m K)^{-1/2}, \tag{2.58}$$

and the index of refraction η is defined as

$$\eta = c/v = (K_m K)^{1/2}. \tag{2.59}$$

Most transparent media are nonmagnetic, yielding $K_m = 1$. The index of refraction is therefore given by

$$\eta = K^{1/2}. \tag{2.60}$$

This value is quite good for most nonpolar materials, including most gases and nonpolar solids such as polystyrene. For polar molecules (such as water) the agreement is poor, owing to their high static polarizability. The index of refraction for a given medium varies significantly with wavelength over the near-infrared, visible, and shorter wavelengths, and is responsible for velocity dispersion effects for light that has broad spectral content. This topic will be described in Section 2.4.

MAXWELL'S EQUATIONS IN A MEDIUM

In applying Maxwell's equations to dielectric materials, we can write them in a more compact form using the fact that $\mathbf{M}$ and ρ are both zero:

$$\mathbf{\nabla}\cdot\mathbf{E} = -\frac{1}{\epsilon_0}\mathbf{\nabla}\cdot\mathbf{P}, \tag{2.61}$$

$$\mathbf{\nabla}\cdot\mathbf{H} = 0, \tag{2.62}$$

$$\mathbf{\nabla}\times\mathbf{H} = \epsilon_0\frac{\partial\mathbf{E}}{\partial t} + \frac{\partial\mathbf{P}}{\partial t} + \mathbf{J}, \tag{2.63}$$

$$\mathbf{\nabla}\times\mathbf{E} = -\mu_0\frac{\partial\mathbf{H}}{\partial t}. \tag{2.64}$$

Taking the curl of (2.64) and the time derivative of (2.63) and eliminating $\mathbf{H}$, one obtains the general wave equation for the electric field $\mathbf{E}$:

$$\mathbf{\nabla}\times(\mathbf{\nabla}\times\mathbf{E}) + \frac{1}{c^2}\frac{\partial^2\mathbf{E}}{\partial t^2} = -\mu_0\frac{\partial^2\mathbf{P}}{\partial t^2} - \mu_0\frac{\partial\mathbf{J}}{\partial t}. \tag{2.65}$$

The left-hand side of this equation is the familiar wave equation for a vacuum, as described in (2.24). The two additional terms on the right-hand side are called *source terms*. The first source term involves polarization charges, relating to localized charge effects in dielectric (nonconducting) media; the second term involves conduction charges that are applicable to metallic materials. The first term, for dielectric media, can be used to explain many optical effects, including dispersion, absorption, double refraction, optical activity, and so forth. The second term, for metals, can explain the high absorption and large reflectivity of such materials. The evaluation of semiconductor materials requires the use of both equations, and hence increases the complexity of obtaining solutions to the wave equation. This discussion will consider the effects of only the first source term, involving the polarization $\mathbf{P}$.

APPLICATION OF MAXWELL'S EQUATIONS TO DIELECTRIC MATERIALS – LASER GAIN MEDIA

Laser gain media are typically made from partially transparent materials, referred to here as *dielectric* materials. Such materials could be crystalline or glass solids, organic dyes dissolved in liquid solvents, or various types of gases or gas mixtures. All of these materials behave as dielectric materials in that they have essentially no significant conductivity under normal conditions. All of these materials will conduct when a sufficiently high electric field is applied to them, but this is not the case under normal circumstances. In gas discharges, even when an electrical current is generated in the gas, the atoms that are not ionized (i.e., most of the atoms) still behave like dielectric media, so the following analysis is still applicable.

It is important to understand the response of these materials to electromagnetic waves in terms of optical constants and wavelength-dependent absorptive and refractive features. We will therefore carry out a simple wave analysis that leads to an explanation of some of the optical properties of laser media. We will consider only the simplest case, that of an isotropic dielectric medium (which could be a glass material or a gaseous material) in which the electrons are localized to each atom in a symmetric fashion and thus there is no preferred orientation of the material. We assume that the medium is capable of having a macroscopic polarization $\mathbf{P}$ that is determined by

$$\mathbf{P} = -Ne\mathbf{x}, \tag{2.66}$$

where $\mathbf{x}$ is a vector representing the distance the charge $-e$ is displaced from its equilibrium position (as shown in Figure 2-5) and N is the number of charges displaced per unit volume. If this displacement vector $\mathbf{x}$ is due to a static electric field $\mathbf{E}$, and if k is the restoring-force constant acting on the electron to return it to its central equilibrium position, then the equation balancing the force due to the electric field and the restoring force can be written as

$$-e\mathbf{E} = k\mathbf{x}. \tag{2.67}$$

This leads to the following formula for the static polarization:

$$\mathbf{P} = \frac{Ne^2}{k}\mathbf{E}. \tag{2.68}$$

If the applied field is not static – that is, if it varies with time – then (2.68) is no longer applicable. The differential equation of motion of this localized charge then becomes

$$m\frac{d^2\mathbf{x}}{dt^2} + m\gamma\frac{d\mathbf{x}}{dt} + k\mathbf{x} = -e\mathbf{E}. \tag{2.69}$$

The second term, $m\gamma(d\mathbf{x}/dt)$, is the frictional damping term, which is assumed to be proportional to the instantaneous velocity of the charge with a constant of proportionality of $m\gamma$. Here γ is the specific damping coefficient for a given material, and has dimensions of $1/\text{time}$; it represents the rate at which the polarization will decay after the applied field is removed. The reciprocal of that rate is referred to as the *polarization decay time* $\tau = 1/\gamma$. We will assume that the applied electric field varies harmonically in a form $\mathbf{E} = \mathbf{E}_0 e^{-i\omega t}$ where ω is the angular frequency, and which is related to

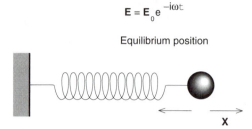

Equilibrium position

Stretched position

Compressed position

Figure 2-5. Harmonic oscillator model of an electron oscillating within a crystalline structure

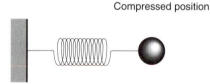

the oscillation frequency by $\omega = 2\pi\nu$. We will also assume that the time dependence of the electron motion will be similar so that $\mathbf{x} = \mathbf{x}_0 e^{-i\omega t}$. Substituting these functions into (2.69) leads to an equation of the form

$$(-m\omega^2 - i\omega m\gamma + k)\mathbf{x} = -e\mathbf{E}. \qquad (2.70)$$

Using this equation to solve for $\mathbf{x}_0$ and substituting into (2.66) by replacing $\mathbf{x}$ with $\mathbf{x}_0$ provides a value for $\mathbf{P}$ of

$$\mathbf{P} = \left(\frac{Ne^2}{-m\omega^2 - i\omega m\gamma + k}\right)\mathbf{E}. \qquad (2.71)$$

This solution for $\mathbf{P}$ reduces to the static value of (2.68) for $\omega = 0$, as would be expected. It also suggests that the polarization has a significant frequency dependence in response to an electric field. The presence of an imaginary component leads to a phase shift in the response under certain conditions.

If we divide the numerator and denominator of the right-hand side of (2.71) by m and define a resonant angular frequency ω_0 and oscillating frequency ν_0 as

$$\omega_0 = \left(\frac{k}{m}\right)^{1/2} = 2\pi \nu_0, \tag{2.72}$$

then we can obtain a more convenient form of this expression as follows:

$$\mathbf{P} = \frac{Ne^2/m}{\omega_0^2 - \omega^2 - i\omega\gamma}\mathbf{E}. \tag{2.73}$$

The resonant frequency ω_0 corresponds to that of a simple harmonic oscillator, and is strictly associated with the properties of the material and not of the electric field interacting with the material. If there is such a resonant effect in a given material, this analysis implies that the electrons are most easily perturbed or excited at that resonant frequency.

We can expect to observe the effects of this resonance frequency because they are associated with the bound electrons of the material. The actual value of the resonant frequency will be related to the strength of the effective binding force and its associated force constant k for each type of material. We will see that this resonance frequency determines the variation of both the absorptive and the refractive characteristics of the material as the applied oscillating electric field (the optical wavelength) incident upon the material is varied.

We can now include this formula for the polarization $\mathbf{P}$ in the wave equation, as given in (2.65) for $\mathbf{J} = 0$ (nonconducting medium), to obtain

$$\mathbf{\nabla} \times (\mathbf{\nabla} \times \mathbf{E}) + \frac{1}{c^2}\frac{\partial^2 \mathbf{E}}{\partial t^2} = -\frac{\mu_0 Ne^2}{m}\left(\frac{1}{\omega_0^2 - \omega^2 - i\gamma\omega}\right)\frac{\partial^2 \mathbf{E}}{\partial t^2}. \tag{2.74}$$

For a medium in which there is no localized charge density, Gauss's law (eqn. 2.12) shows that $\mathbf{\nabla} \cdot \mathbf{D} = 0$. Using (2.3) we can thus show that $\epsilon_0 \mathbf{\nabla} \cdot \mathbf{E} + \mathbf{\nabla} \cdot \mathbf{P} = 0$. For a spatially uniform medium, $\mathbf{\nabla} \cdot \mathbf{P} = 0$ and thus $\mathbf{\nabla} \cdot \mathbf{E} = 0$. Therefore $\mathbf{\nabla} \times (\mathbf{\nabla} \times \mathbf{E}) = -\mathbf{\nabla}^2 \mathbf{E}$. Hence (2.74) reduces to the simplified wave equation (using $c = (\mu_0\epsilon_0)^{-1/2}$)

$$\boxed{\mathbf{\nabla}^2 \mathbf{E} = \frac{1}{c^2}\left(1 + \frac{Ne^2}{m\epsilon_0}\left[\frac{1}{\omega_0^2 - \omega^2 - i\gamma\omega}\right]\right)\frac{\partial^2 \mathbf{E}}{\partial t^2}. } \tag{2.75}$$

COMPLEX INDEX OF REFRACTION –
OPTICAL CONSTANTS

We will use a trial solution for (2.75) of the form

$$\mathbf{E} = \mathbf{E}_0 e^{i(\mathcal{K}_z z - \omega t)}, \tag{2.76}$$

where we have allowed for a complex propagation constant $\mathcal{K}_z$. Equation (2.76) represents a plane wave traveling in the z direction, and is a solution of (2.75) if

$$\mathcal{K}_z^2 = \frac{\omega^2}{c^2}\left(1 + \frac{Ne^2}{m\epsilon_0}\left[\frac{1}{\omega_0^2 - \omega^2 - i\gamma\omega}\right]\right). \tag{2.77}$$

The constant $\mathcal{K}_z$ is a complex number that can be separated into real and imaginary parts as follows:

$$\mathcal{K}_z = k_z + i\alpha_E. \tag{2.78}$$

In this relationship, the subscript **E** refers to the electric field. We can now rewrite the solution for **E**, given previously in (2.50), as

$$\mathbf{E} = \mathbf{E}_0 e^{-\alpha_E z} e^{i(k_z z - \omega t)}. \tag{2.79}$$

The first exponential factor of (2.79) involves α_E, the *extinction index*. This factor represents an exponential decay of the electric field **E** with distance z into the medium. The energy absorption is proportional to $|\mathbf{E}|^2$, which leads to an exponential decay of the energy that is proportional to $e^{-2\alpha_E z}$. We will define

$$\alpha \equiv 2\alpha_E, \tag{2.80}$$

where α is the *absorption coefficient*. This term has dimensions of $1/\text{length}$ and determines the amount of absorption of energy that occurs within each interval of distance the electromagnetic beam penetrates into a specific material. The reciprocal of the absorption coefficient α is the absorption depth $l_d = 1/\alpha$, the average depth over which the absorption of the beam occurs within the material.

The use of the complex coefficient $\mathcal{K}_z$ is equivalent to using a complex index of refraction

$$\mathfrak{N} = \eta + i\kappa \tag{2.81}$$

such that

$$\mathcal{K}_z = \frac{\omega}{c} \mathfrak{N}. \tag{2.82}$$

The term α_E can be related to the imaginary coefficient κ of the complex index of refraction $\mathfrak{N}$ as

$$\alpha_E = \frac{\omega}{c}\kappa, \tag{2.83}$$

and so

$$\boxed{\alpha = 2\alpha_E = \frac{4\pi\kappa}{\lambda}.} \tag{2.84}$$

Using (2.81) and (2.82) in (2.76), the wave velocity $v = \omega/k_z$ can be shown to be $v = c/\eta = \lambda\nu$. Also, the effective energy absorption equation is

$$|\mathbf{E}|^2 = |\mathbf{E}_0|^2 e^{-2\alpha_E z} = |\mathbf{E}_0|^2 e^{-\alpha z} = |\mathbf{E}_0|^2 e^{-2(\omega/c)\kappa z}. \tag{2.85}$$

Using (2.77) and (2.82), we can show that

$$\mathfrak{N}^2 = (\eta + i\kappa)^2 = 1 + \frac{Ne^2}{m\epsilon_0}\left(\frac{1}{\omega_0^2 - \omega^2 - i\gamma\omega}\right). \tag{2.86}$$

ABSORPTION AND DISPERSION

If we now equate the real and imaginary parts of (2.86), we obtain

$$\eta^2 - \kappa^2 = 1 + \frac{Ne^2}{m\epsilon_0}\left(\frac{\omega_0^2 - \omega^2}{(\omega_0^2 - \omega^2)^2 + \gamma^2\omega^2}\right), \tag{2.87}$$

$$2\eta\kappa = \frac{Ne^2}{m\epsilon_0}\left(\frac{\gamma\omega}{(\omega_0^2 - \omega^2)^2 + \gamma^2\omega^2}\right). \tag{2.88}$$

EXAMPLE

Determine a relationship for η at the resonance frequency.

At the resonance frequency, $\omega = \omega_0$. From (2.87) and (2.88) we therefore have

$$\eta^2 - \kappa^2 = 1 \quad \text{and} \quad 2\eta\kappa = \frac{Ne^2}{m\epsilon_0\gamma\omega_0}.$$

Thus, solving for κ in the second expression, we have

$$\kappa = \frac{Ne^2}{2\eta m\epsilon_0\gamma\omega_0}.$$

Inserting this into the first expression yields

$$\eta^2 - \left(\frac{Ne^2}{2\eta m\epsilon_0\gamma\omega_0}\right)^2 = 1.$$

This can be rewritten as

$$\eta^4 - \eta^2 - \left(\frac{Ne^2}{2m\epsilon_0\gamma\omega_0}\right)^2 = 0,$$

which has the solution for η^2 of

$$\eta^2 = \frac{1}{2} \pm \frac{1}{2}\sqrt{1 + \left(\frac{Ne^2}{m\epsilon_0\gamma\omega_0}\right)^2}.$$

For a real situation, only the plus of the plus-or-minus results in a valid solution, since a minus would render the value of η^2 less than zero (which has no physical meaning). At very high frequencies, the expression for η^2 and thus also for η can be seen to approach unity.

We will see in Chapter 4 that the γ in both (2.87) and (2.88) can be associated with the decay rate of an atomic energy level, and that the variation of this function with angular frequency ω has a similar shape to that of the radiative emission from such a level.

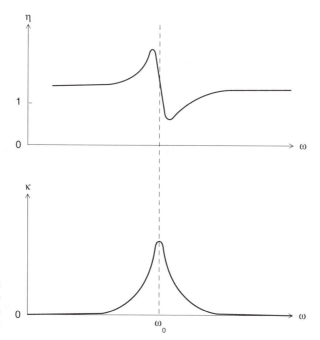

Figure 2-6. Typical values of the real part η and the imaginary part κ of the refractive index

Equations (2.87) and (2.88) are used to solve for the optical coefficients associated with a specific optical material. These coefficients are more commonly known as the optical constants η and κ. Plots of typical values of η and κ versus frequency are shown in Figure 2-6. The value of κ goes through a maximum at the resonance frequency ω_0. Since the wavelength is approximately constant over the region where κ has a significant value, it follows from (2.84) that the absorption coefficient α also goes through a maximum at ω_0. This frequency-sensitive absorption factor α suggests that the material absorption is largely dependent upon the electronic characteristics of the material and how the bound electrons resonate within the material. The resonance frequency turns out to be the frequency at which the incident electric field induces an electron in the material to make a transition from one bound energy state to a higher bound state. Electronic transition processes of this type will be treated in some detail, from a quantum standpoint, in Chapters 3 through 5.

The plot of η versus ω in Figure 2-6 is referred to as the *dispersion curve*. It shows η gradually increasing for increasing ω, passing through a maximum, then decreasing rapidly to a minimum as it passes through ω_0. It then increases again but returns to a lower value than that which it had before it approached the resonance, and again begins a gradual increase toward the next resonance. In any region where η increases with ω, or $d\eta/d\omega > 0$, the dispersion is referred to as *normal* dispersion. This is because $d\eta/d\omega > 0$ for most of the infrared, visible, and ultraviolet spectral regions of most transparent materials. However, in some localized regions where there is a resonance, $d\eta/d\omega < 0$ and we thus have a region of *anomalous* dispersion. This happens only over very narrow frequency regions and so is considered an unusual occurrence. Although applications that involve negative dispersion are potentially very interesting, they generally cannot be implemented: at the value of ω where the negative dispersion exists, the absorption is very high and the electromagnetic radiation is rapidly extinguished.

Real materials generally have electron interactions with surrounding media in which one or more electrons is involved with more than one binding site. This situation is analogous to a mass attached to two or more separate springs, each with different spring constants. Thus it is possible to have more than one resonant frequency at which both the index of refraction and the extinction coefficient exhibit rapid changes. If these bound electrons within a material have more than one resonant oscillation frequency, the relative potential strength of each oscillation is known as the *oscillator strength* f_j, where the total $\sum_j f_j = 1$ for each participating electron. This strength is essentially a relative measure of how easily each particular oscillating mode or frequency can be activated compared to another mode. In considering the division of oscillator strengths among several frequencies, the formula for $\mathfrak{N}^2$, as originally derived for one oscillating frequency in (2.86), can be expressed as

$$\mathfrak{N}^2 = 1 + \frac{Ne^2}{m\epsilon_0} \sum_j \left(\frac{f_j}{\omega_j^2 - \omega^2 - i\gamma_j\omega} \right). \qquad (2.89)$$

The summation extends over all participating electrons and all of their individual interactions that lead to oscillations. The damping coefficient associated with each resonant frequency is given by γ_j.

The implications of (2.89), with respect to how η and κ vary with frequency, are shown in Figure 2-7. It can be seen that over most of the frequency spectrum the value of η increases, even though at each resonance

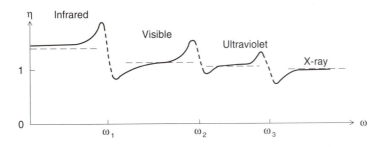

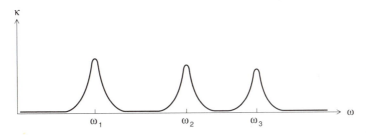

Figure 2-7. A material with absorptive features at several different frequencies across the spectrum

there is a large enough drop to give the total increase with ω a net downward movement. When the frequency reaches the soft–X-ray spectral region, η decreases to a value that is slightly less than unity for all materials.

The absorption α in regions not near a resonance (remember, $\alpha = 4\pi\kappa/\lambda$) also has a significant frequency or wavelength dependence, which is partially due to the $1/\lambda$ factor but more importantly to the strong decrease in κ as the wavelength approaches the X-ray spectral region. For a specific value of the oscillator strength, the maximum absorption decreases for resonances at higher frequencies, as shown in Figure 2-7.

If, within a given frequency range, γ is small enough that it can be neglected, then the index of refraction according to (2.89) is essentially real and can be described as

$$\mathfrak{N}^2 \cong \eta^2 = 1 + \frac{Ne^2}{m\epsilon_0} \sum_j \left(\frac{f_j}{\omega_j^2 - \omega^2} \right). \tag{2.90}$$

This equation, when given as a function of wavelength instead of frequency, is known as *Sellmeier's formula*.

2.4 COHERENCE

Another important consideration in the study of lasers is the interaction of two electromagnetic waves that have only slightly different frequencies, or that originate from points only slightly separated spatially – for example, two closely located but separate laser beams or a single beam illuminating two closely positioned apertures. In such instances the two distinct waves or beams will interfere with each other to produce, in some cases, very dramatic effects. Such effects include the development of longitudinal modes (see Chapter 10), mode locking (Chapter 12), and frequency multiplication and phase matching (Chapter 15).

We will consider two separate waves, with k vector of k_1 and k_2 and with angular frequencies ω_1 and ω_2, both traveling in the z direction. The electric field amplitude of such waves can be expressed as

$$E_1 = E_1^0 e^{-i[k_1 z - \omega_1 t]}, \tag{2.91}$$

$$E_2 = E_2^0 e^{-i[k_2 z - \omega_2 t]}, \tag{2.92}$$

where E_1^0 and E_2^0 represent the maximum value of the field amplitudes during their oscillating cycle. We will consider the two waves to be closely located in space, so that the combined wave E is given as

$$E = E_1^0 e^{-i[k_1 z - \omega_1 t]} + E_2^0 e^{-i[k_2 z + \omega_2 t]}. \tag{2.93}$$

We now wish to consider the intensity resulting from these two distinctly separate waves as

$$I = |E|^2 = (E_1^0)^2 + (E_2^0)^2 + E_1^0 E_2^0 (e^{-i([k_1 - k_2]z + [\omega_1 - \omega_2]t)} + e^{i([k_1 - k_2]z + [\omega_1 - \omega_2]t)})$$

$$= (E_1^0)^2 + (E_2^0)^2 + 2E_1^0 E_2^0 \cos\theta = I_1^2 + I_2^2 + 2E_1^0 E_2^0 \cos\theta, \tag{2.94}$$

where

$$\theta = ([k_1 - k_2]z + [\omega_1 - \omega_2]t). \tag{2.95}$$

Thus, if we examine (2.94) and assume (for simplicity) that $E_1^0 = E_2^0$, we see that for values of θ near or equal to zero (and also near or equal to $2n\pi$), the intensity can be *twice* that of the sum of the intensities of the two individual waves. Also, if $\theta = \pi$ (and 3π, 5π, ...) then the total intensity will be zero. Thus the two waves are interfering with each other to produce effects not associated with each of the waves separately. Looking at θ more closely, we see that if the two waves have nearly identical properties – if k_1 and k_2 are similar and ω_1 and ω_2 are similar – then these interference effects are likely to occur and the two waves are said to be *coherent*. If the frequencies or the k vectors are drastically different, or if there are many waves of different frequencies or originating from very different locations, then the waves do not interfere and are said to be *incoherent*. Thus, for completely coherent waves the combined intensity would be

$$I = I_1 + I_2 \pm 2E_1^0 E_2^0 \quad \text{(coherent)} \tag{2.96}$$

and for completely incoherent waves the total intensity would be

$$I = I_1 + I_2 \quad \text{(incoherent)}. \tag{2.97}$$

When complete coherence is not achieved but some interference still exists, the condition is referred to as *partial coherence*. It is useful to have some guidelines as to how different the beams can be and still achieve some degree of coherence or interference. These conditions can be grouped into two categories: *temporal* coherence, associated with the difference in frequencies of the two waves; and *spatial* coherence, associated with the location of the two waves.

TEMPORAL COHERENCE

The case of temporal coherence refers to the relative phase or the coherence of the two waves at two separate locations along the propagation direction of the two beams. It is sometimes referred to as *longitudinal* coherence. If we assume that the two waves are exactly in phase at the first location, then they will still be at least partially in phase at the second location up to a distance l_c, where l_c is defined as the coherence length as shown in Figure 2-8(a). The coherence length can be determined to be

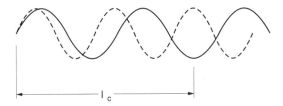

Figure 2-8(a). Temporal coherence length l_c

$$l_c = \lambda\left(\frac{\lambda}{\Delta\lambda}\right) = \frac{\lambda^2}{\Delta\lambda}, \qquad (2.98)$$

where $\Delta\lambda$ represents the difference in wavelength between the two waves and λ is their average wavelength. It can be seen from (2.98) that l_c has a significant value only when $\Delta\lambda \ll \lambda$, that is, when the wavelengths of both waves are nearly identical.

EXAMPLE

What is the temporal coherence length l_c of (1) a mercury vapor lamp emitting in the green portion of the spectrum at a wavelength of 546.1 nm with an emission bandwidth of $\Delta\nu = 6 \times 10^8$ Hz, and (2) a helium–neon laser operating at a wavelength of 632.8 nm with an emission width of $\Delta\nu = 10^6$ Hz?

We will use the expression in (2.98) for l_c, and will therefore need to determine the value of $\Delta\lambda$ for both of these sources. From (2.40) we know that the relationship between the frequency and the wavelength of an electromagnetic wave in terms of the velocity of light c is $\lambda\nu = c$. Thus, differentiating the expression $\nu = c/\lambda$, we obtain $d\nu = (-c/\lambda^2)\,d\lambda$. We wish to obtain an expression for the absolute value of the differential $d\lambda$, which we will call $\Delta\lambda$:

$$|\Delta\lambda| = \frac{\lambda^2\Delta\nu}{c}.$$

Using this expression, we can determine $\Delta\lambda$ for the mercury lamp to be

$$\Delta\lambda = \frac{(546.1 \times 10^{-9}\text{ m})^2(6 \times 10^8\text{ s}^{-1})}{3 \times 10^8\text{ m/s}}$$

$$= 5.96 \times 10^{-13}\text{ m} = 5.96 \times 10^{-4}\text{ nm},$$

and thus the temporal (longitudinal) coherence length can be determined to be

$$l_c = \frac{\lambda^2}{\Delta\lambda} = \frac{(546.1 \times 10^{-9}\text{ m})^2}{5.96 \times 10^{-13}\text{ m}} = 0.50\text{ m}.$$

For the helium–neon laser we find that $\Delta\lambda$ is given by

$$\Delta\lambda = \frac{\lambda^2\Delta\nu}{c} = \frac{(632.8 \times 10^{-9}\text{ m})^2(1 \times 10^6\text{ s}^{-1})}{3 \times 10^8\text{ m/s}}$$

$$= 1.33 \times 10^{-15}\text{ m} = 1.33 \times 10^{-6}\text{ nm}.$$

The temporal coherence length is thus

$$l_c = \frac{\lambda^2}{\Delta\lambda} = \frac{(632.8 \times 10^{-9}\,\text{m})^2}{1.33 \times 10^{-15}\,\text{m}} = 301 \text{ m}.$$

Hence the laser beam in this case has a significantly longer coherence length than that of the mercury lamp.

SPATIAL COHERENCE

Spatial coherence, also referred to as *transverse* coherence, describes how far apart two sources, or two portions of the same source, can be located in a direction transverse to the direction of observation and still exhibit coherent properties over a range of observation points. This is sometimes also referred to as the *lateral* coherence. More specifically, we will ask by what distance l_t can two points be separated in the transverse direction at the region of observation and still have interference effects from the source region over a specific lateral direction of the source. Assume that the two sources are separated by a distance s in the transverse direction to the direction of observation, and are a distance r from the point of observation. If the two sources exhibit interference effects at point a, then the transverse coherence length l_t is the transverse distance from point a to point b, where the two sources cease to show interference effects; see Figure 2-8(b). This distance is given by

$$l_t = \frac{r\lambda}{s} = \frac{\lambda}{\theta_s}, \tag{2.99}$$

where $\theta \cong s/r$.

Source

s

r

b

l_t

a

Figure 2-8(b). Transverse coherence length l_t

EXAMPLE

A laser-produced plasma consisting of a 100-μm–diameter ball radiates very strongly at a wavelength of 10 nm. At a distance of 0.5 m from the source, what is the spatial coherence resulting from light emitted from opposite sides of the plasma?

We will use the expression for the spatial coherence (transverse coherence length) l_t taken from (2.99) to be $l_t = r\lambda/s$. For this example, $r = 0.5$ m, $\lambda = 10$ nm, and $s = 100\ \mu$m $= 10^{-4}$ m. We can therefore compute the spatial coherence to be

EXAMPLE (cont.)

$$l_t = \frac{r\lambda}{s} = \frac{(0.5 \text{ m})(1 \times 10^{-8} \text{ m})}{1 \times 10^{-4} \text{ m}} = 5 \times 10^{-5} \text{ m}.$$

Hence, at any location a distance of 50 μm in any transverse direction from a specific point P that is 0.5 m from the source, the flux will be incoherent with respect to the radiation arriving at the point P.

REFERENCES

J. D. Jackson (1975), *Classical Electrodynamics,* 2nd ed. New York: Wiley, Section 6.
P. W. Milonni and J. H. Eberly (1988), *Lasers.* New York: Wiley, Chapter 3.

PROBLEMS

1. Show how to go from the integral form to the differential form of Maxwell's four equations.

2. Calculate the electrical force of attraction between a positive and negative charge separated by a distance of 5.3×10^{-11} m (the average distance between an electron in its ground state and the nucleus). How does this force compare to the gravitational force an electron would experience at sea level?

3. Show that the function $\mathbf{E} = \mathbf{E}_0 e^{-i(k_z z - \omega t)}$ is a solution to Maxwell's wave equation (eqn. 2.26), assuming that $c = 1/(\mu_0 \epsilon_0)^{1/2}$.

4. Consider two waves, having slightly different angular frequencies ω_1 and ω_2 and wave numbers k_1 and k_2, of the form

$$E_1 = E_1^0 e^{-i(k_1 z - \omega_1 t)} \quad \text{and} \quad E_2 = E_2^0 e^{-i(k_2 z - \omega_2 t)},$$

where $E_1^0 = E_2^0$. Assume their "difference" wave numbers and frequencies can be written as

$$\Delta k = k_1 - k_2 \quad \text{and} \quad \Delta \omega = \omega_1 - \omega_2.$$

Show that the sum of these two waves leads to a slowly varying amplitude function associated with the difference frequencies and wave numbers multiplied by a term associated with the actual frequencies and wave numbers. (*Hint:* Reduce the sum to the product of two cosine functions.)

5. Show that the group velocity and the phase velocity are equal when there is no dispersion, and different when dispersion is present. (*Hint:* Begin with the definitions of group and phase velocity.)

6. Obtain (2.86) from (2.77) and (2.82).

7. Express (2.87) and (2.88) in terms of frequency (instead of angular frequency) and τ (instead of γ).

8. At what value of ω is α a maximum? Solve for η and κ at that value of ω, and give the expression for α.

9. Assume that a dielectric material has a single resonance frequency at $\nu_0 = 3 \times 10^{14}$ Hz, the polarization decay time is $\tau = 2 \times 10^{-7}$ s, and the density of polarizable charges is 5×10^{26} m^{-3}. Determine the full width at half maximum (FWHM) of the absorption resonance in the material.

10. A species of atomic weight 60 is doped into an Al$_2$O$_3$ crystal at a concentration of 5% by weight (cf. Section 5.3). The combined material is found to have an absorbing feature that peaks at 750 nm and a damping constant of 10^{13} s^{-1}. Assume that each atom of the species contributes to the macroscopic polarization associated with that absorbing feature. Make a plot of η and κ versus wavelength in the wavelength region of the absorbing feature.

11. Determine what emission frequency width would be required to have a temporal coherence length of 10 m at a source wavelength of 488 nm.

12. If a photographic film has a minimum resolution of 10 μm, what minimum feature size could be observed at a distance of 2 m without observing coherent effects? (Assume the minimum feature size is equal to the minimum resolution.)

FUNDAMENTAL QUANTUM PROPERTIES OF LIGHT

PARTICLE NATURE OF LIGHT
DISCRETE ENERGY LEVELS

3

SUMMARY In this chapter we will derive equations that indicate the existence of discrete energy levels in atoms. For this derivation we will use a simple model of an atom, consisting of an electron orbiting around a much heavier nucleus. The energy levels will be obtained by using Newton's laws of motion with a modification to include quantization of the angular momentum of the electron. This model describes the energy levels of hydrogen reasonably well, but does not describe those of other atoms; nor does it provide informa-tion concerning the relative intensities of emission at various wavelengths when the electron makes transitions from one energy level to another. We will therefore develop a simplified introduction to the quantum theory of matter that will provide a more accurate description of the energy levels of atoms. We will also devise a labeling method for the energy levels of all atoms that will allow us, in subsequent chapters, to determine the likelihood that radiative transitions might occur between such energy levels in various laser materials.

3.1 BOHR THEORY OF THE HYDROGEN ATOM

HISTORICAL DEVELOPMENT OF THE CONCEPT OF
DISCRETE ENERGY LEVELS

In Chapter 2 we saw how Maxwell's wave equations predicted the existence of transversely oscillating electromagnetic waves with a velocity essentially identical, within experimental error, to that measured for the velocity of light. Researchers demonstrated that the behavior of such waves was con-sistent with such wavelike properties of light as diffraction, reflection, and refraction. However, this wavelike description of light could not explain the results of several other experiments. One such result is the photoelectric effect, which involves the emission of photoelectrons when light is focused onto a solid surface. It was found that the number of electrons emitted was proportional to the intensity of the light, but that the kinetic energies of the electrons were independent of the intensity. Such results, when con-sidered on a microscopic level, implied that the radiant energy impinging upon a surface area comprising approximately 10^8 atoms resulted in the transfer of nearly all of that energy to the kinetic energy of a single photo-electron emitted from one atom within that area. What a remarkable con-centration process! Another difficulty arising from that experiment was in explaining the existence of a sharp threshold wavelength of the incident light, below which the electrons would not be emitted from the surface regardless of the light's intensity.

A third problem became apparent with the observation, using high-resolution spectroscopic instruments, of spectral emission from gaseous discharge tubes. These observations indicated discrete wavelengths or colors of light, as indicated by patterns of closely spaced dark lines on spectroscopic film (the lines were produced by imaging the long, narrow slit of the spectrometer onto the film). Also, when a spectrum of the sun was taken, similar closely spaced spectral sequences were observed on spectroscopic film plates as dark bands, called *Fraunhofer lines,* superimposed upon the bright continuous solar emission spectrum.

ENERGY LEVELS OF THE HYDROGEN ATOM

One of the most far-reaching ideas of modern science was conceived by Niels Bohr in 1913 when he set out to explain anomalies in the emission spectrum of a gas discharge containing hydrogen. At this period in time, electrons had been discovered (1897); their mass, size, and charge had been measured; and, as a result of the Zeeman effect (see Section 3.3) they were shown to participate in the emission of spectral radiation (discrete emission lines) from a gaseous discharge. This spectral radiation was associated with electric dipole radiation resulting from oscillating charged particles within individual atoms of that gas. But the discrete, closely spaced, and orderly sequences of emission lines, as observed with a spectrograph, were unexplained.

In its final form, Bohr's theory used the following concepts, known as *Bohr's hypotheses,* to explain the spectrum of hydrogen.

(1) The hydrogen atom included a positively charged nucleus (proton) and a negatively charged electron orbiting in a circular motion around the nucleus.

(2) The electron could temporarily remain in a particular state (an orbit having a specific radius), provided that the angular momentum of the electron associated with that radius had a value that was an integral multiple of $\hbar$ ($\hbar = h/2\pi$, where h is Planck's constant).

(3) Radiation is emitted from the atom when the electron "jumps" from a higher energy E_2 (larger orbit) to a lower energy E_1 (smaller orbit).

(4) When such radiation is emitted, its frequency is determined by the Einstein frequency condition

$$h\nu_{21} = E_2 - E_1. \tag{3.1}$$

Planck's constant h (initially termed Planck's quantum) was introduced in two ways. First, it was related to the angular momentum of the electron in orbit around the nucleus. Second, it was associated – through Einstein's frequency equation (eqn. 3.1) – with the frequency of the radiation emitted.

Bohr obtained values of the energies of various states or levels of the hydrogen atom by assuming that an electron with velocity v rotates in a circular orbit of radius r around the nucleus with an angular momentum of

$$m_e v r = nh/2\pi = n\hbar, \quad n = 1, 2, 3, \dots . \tag{3.2}$$

This relationship expresses Bohr's second postulate in mathematical form. The electron mass is m_e, and the integer n is called the *principal quantum number.*

Newton's second law of motion, equating the sum of all forces acting on a body to the product of its mass and its acceleration, was then used to equate the electrical force of attraction between the negatively charged orbiting electron and the positively charged nucleus to the radial acceleration associated with the angular rotation of the electron around the much heavier nucleus (a proton):

$$\frac{e \cdot e}{4\pi\epsilon_0 r^2} = \frac{m_e v^2}{r}, \tag{3.3}$$

where e is the electron charge and ϵ_0 is the permittivity of a vacuum.

Using (3.2) and (3.3) to eliminate v leads to the formula for quantized orbits of the electron (Bohr orbits) described by the following radii:

$$r = \frac{\epsilon_0 h^2}{\pi m_e e^2} n^2 = a_H n^2, \tag{3.4}$$

where a_H is the radius of the first Bohr orbit ($n = 1$) given by

$$a_H = \frac{\epsilon_0 h^2}{\pi m_e e^2} = 0.53 \, \text{Å}. \tag{3.5}$$

This value effectively describes the radius of the hydrogen atom in its ground state, or lowest-lying energy state ($n = 1$ in eqn. 3.4).

The total energy for a specific orbit can be obtained by summing the kinetic energy and the potential energy of the electron with respect to the nucleus:

$$E = \frac{1}{2} m_e v^2 - \frac{e^2}{4\pi\epsilon_0 r}, \tag{3.6}$$

where the first term is the kinetic energy of angular motion and the second term is the potential electrical energy of attraction, using the value of r given in (3.4) as the separation distance of the positive and negative charges. Eliminating v by means of (3.3) gives the following value for the energy:

$$E = -\frac{e^2}{8\pi\epsilon_0 r}. \tag{3.7}$$

Substituting the expression (3.4) for r into (3.7) we find that the electron can have only certain discrete negative values of energy E_n associated with the various values of n, where n is a positive integer such that

$$E_n = -\frac{m_e e^4}{8\epsilon_0^2 h^2} \cdot \frac{1}{n^2} = -\frac{E_0}{n^2} \qquad (3.8)$$

and where E_0 is given as

$$E_0 = \frac{m_e e^4}{8\epsilon_0^2 h^2} = 13.595 \text{ eV}. \qquad (3.9)$$

For a slightly more accurate value of the energy we must use the reduced mass μ of the combined electron and proton system,

$$\mu = \frac{m_e}{(1 + m_e/M_P)} = \frac{m_e M_P}{(M_p + m_e)},$$

instead of the mass of the electron as in (3.9). We will use the reduced mass when we solve this problem later in this chapter using quantum mechanics. The reduced mass takes into account the finite size of the nucleus and the fact that both the electron and the proton rotate about the center of mass of the electron–proton system, rather than rotating about the center of mass of the proton as assumed in the Bohr theory.

Thus, from (3.8) we find that the electron can have any one of a series of negative energies, which are referred to as *energy states* or *levels* (these terms are used interchangeably). The lowest value of energy (the most negative) corresponds to setting $n = 1$ in (3.8). Since this is the lowest energy, it is called the *ground state;* every atom has such a lowest energy state or ground state. The ground state is the energy state that normally occurs in

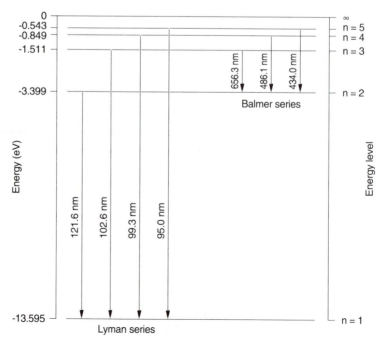

Figure 3-1. Energy-level diagram of atomic hydrogen showing the first five energy levels and some of the radiative transitions between those levels

nature unless additional energy in the form of heat or light is applied to the atom to raise it to a higher state.

The values of the negative energies for the hydrogen atom are shown in Figure 3-1, where the energy levels are indicated as horizontal lines. The vertical lines indicate transitions between such levels that will be discussed later. The actual value of the discrete negative energy indicates the amount of energy required to remove that electron completely away from its orbit around the nucleus, and thus that energy is sometimes referred to as a *binding* energy. It is also referred to as the *ionization* energy, since the result is to create a positively charged ion and a free electron. Once an electron is removed from the attractive force of the ion, it can have any value of velocity and thus can have any value of positive energy associated with its kinetic energy. In summary, all atoms can have only discrete negative values of energy associated with the binding of the electrons to the nucleus, but any value of positive energy of the electrons when they are removed sufficiently far from the nucleus (ionized).

EXAMPLE

Calculate the binding energy of the electron for the first three energy levels of the hydrogen atom ($n = 1, 2,$ and 3).

Because

$$E_n = -\frac{E_0}{n^2} = -\frac{13.595 \text{ eV}}{n^2},$$

we can obtain the first three values of energy by substituting the appropriate value of n as follows:

$$E_1 = -\frac{13.595 \text{ eV}}{1^2} = -13.595 \text{ eV},$$

$$E_2 = -\frac{13.595 \text{ eV}}{2^2} = -3.399 \text{ eV},$$

$$E_3 = -\frac{13.595 \text{ eV}}{3^2} = -1.511 \text{ eV}.$$

Thus the lowest value of the negative binding energy is approximately -13.595 eV (electron volts) or 2.178×10^{-18} J (joules). Using Einstein's relationship of energy and frequency, $E = h\nu$, and the relationship of frequency and wavelength $\lambda\nu = c$ (resulting from Maxwell's wave equation), one could relate this binding energy to an equivalent wavelength of $\lambda = c/\nu = hc/E_1 = 91.4$ nm. Stated another way, this suggests that light of a wavelength 91.4 nm or shorter would have sufficient energy, if absorbed by a hydrogen atom, to remove the electron to an infinitely large distance, or *ionize* the atom, thereby leaving only a positively charged nucleus. This process is known as *photoionization*.

FREQUENCY AND WAVELENGTH OF EMISSION LINES

Using Bohr's fourth hypothesis, one can estimate the frequency ν_{ji} of radiation occurring when an electron makes a transition from a higher energy level E_j to a lower level E_i as

$$\nu_{ji} = \frac{E_j - E_i}{h} = \frac{\Delta E_{ji}}{h} = \frac{E_0}{h}\left(\frac{1}{n_i^2} - \frac{1}{n_j^2}\right)$$
$$= (3.28 \times 10^{15})\left(\frac{1}{n_i^2} - \frac{1}{n_j^2}\right) \, \mathrm{s}^{-1}. \qquad (3.10)$$

This formula relates the frequency of radiation to the energy differences between the various "quantized" negative energy states i and j. It is customary to label higher energy states with larger numbers or with letters higher in alphabetical sequence. Therefore, in this situation we indicated j as having a higher energy value than i.

Using (3.10) and the relationship $\nu = c/\lambda$, we can obtain

$$\frac{1}{\lambda} = R_H\left(\frac{1}{n_i^2} - \frac{1}{n_j^2}\right) = (1.0967758 \times 10^7)\left(\frac{1}{n_i^2} - \frac{1}{n_j^2}\right) \, \mathrm{m}^{-1}, \qquad (3.11)$$

where $R_H = 1.0967758 \times 10^7 \, \mathrm{m}^{-1}$ is referred to as the *Rydberg constant* for hydrogen.

EXAMPLE

Compute the frequency and wavelength of the radiative transition from $n = 3$ to $n = 2$ in the hydrogen atom.

Using (3.10) for the frequency, we have

$$\nu_{32} = (3.28 \times 10^{15})\left(\frac{1}{n_2^2} - \frac{1}{n_3^2}\right) \, \mathrm{s}^{-1} = (3.28 \times 10^{15})\left(\frac{1}{2^2} - \frac{1}{3^2}\right) \, \mathrm{s}^{-1}$$
$$= (3.28 \times 10^{15})\left(\frac{1}{4} - \frac{1}{9}\right) \, \mathrm{s}^{-1} = 4.56 \times 10^{14} \, \mathrm{s}^{-1} = 4.56 \times 10^{14} \, \mathrm{Hz}.$$

We can compute the wavelength, using (3.11), as

$$\frac{1}{\lambda_{ji}} = R_H\left(\frac{1}{n_i^2} - \frac{1}{n_j^2}\right) = (1.0967758 \times 10^7)\left(\frac{1}{n_i^2} - \frac{1}{n_j^2}\right) \, \mathrm{m}^{-1};$$
$$\frac{1}{\lambda_{32}} = (1.0967758 \times 10^7)\left(\frac{1}{2^2} - \frac{1}{3^2}\right) \, \mathrm{m}^{-1}$$
$$= (1.0967758 \times 10^7)\left(\frac{1}{4} - \frac{1}{9}\right) \, \mathrm{m}^{-1} = 1.523299 \times 10^6 \, \mathrm{m}^{-1}.$$

Therefore,

$$\lambda_{32} = \frac{1}{1.523299 \times 10^6} \, \mathrm{m} = 6.5647 \times 10^{-7} \, \mathrm{m} = 656.47 \, \mathrm{nm}.$$

This wavelength value is the wavelength that would be measured in a vacuum. To determine the wavelength measured in an environment of air, the relationship would be $\lambda_{air} = \lambda_{vac}/\eta$, where η is the index of refraction of air at that wavelength. In the visible spectral region this produces a slightly shorter wavelength; spectroscopists measure the wavelength of this transition in air to be 656.27 nm, a difference of 0.2 nm.

The radiation occurring when an electron makes a transition from a higher energy state to a lower energy state will be considered in some detail in Chapter 4. For now, our attention will be addressed to the specific energies described by (3.8). Vertical lines in Figure 3-1 show the jumps (transitions) that electrons can make from one level to another, which – according to Bohr's fourth postulate – produce emission of light having a frequency (or wavelength) corresponding to that change in energy. The location of these transitions, or *emission lines,* can be shown on a wavelength graph (Figure 3-2) since λ is inversely proportional to the energy difference ΔE between the levels, as indicated in (3.11). This wavelength pattern for hydrogen corresponds exactly to that recorded by spectroscopists in the nineteenth century on spectrographic plates taken of the emission from a hydrogen discharge.

As mentioned previously, Figure 3-1 represents the energy-level diagram for neutral atomic hydrogen. The word "neutral" suggests that, for any particular atom under consideration, there are an equal number of positively charged protons and negatively charged electrons (one each in the case of hydrogen) associated with that atom, giving it a net zero or "neutral" charge.

IONIZATION ENERGIES AND ENERGY LEVELS OF IONS

When the electron is completely removed from the nucleus (this corresponds to the electron having a positive energy in Figure 3-1), the hydrogen atom

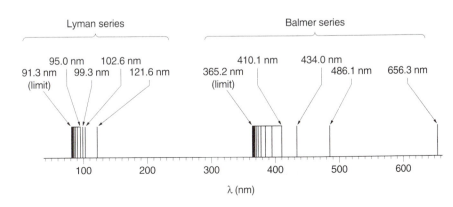

Figure 3-2. Wavelength spectrum of the transitions shown in Figure 3-1

is ionized and becomes a hydrogen ion, denoted H^+. Thus, 13.595 eV is the ionization energy of neutral hydrogen when the electron is in its ground state ($n = 1$).

Other neutral atoms have ionization energies that range from 3.89 to 24.58 eV (see the Appendix). Every atom has negative (bound) energy levels, the location of which are determined in most cases by only the outermost electron moving in an orbit located at one of many possible discrete locations with respect to the nucleus. For atoms with more than one electron, the electrons orbiting closer to the nucleus are not involved in the determination of the energy levels of interest for making lasers, since those electrons are most tightly bound and therefore would generally require more than the ionization energy to be moved to an excited state.

All atoms other than hydrogen have more than one electron, and can therefore also have negative energy states of their ions (one or more electrons removed) as well as those of the neutral atom, because those atoms still have other electrons remaining in orbit around the nucleus after the first electron is removed. The energy states of those atoms and ions, and the radiation spectrum associated with their negative energy states, cannot be as easily described as that of hydrogen, due to the complexities associated with the multiple electrons orbiting around the nucleus. The only exceptions are hydrogen-like ions that have energy levels similar to those of hydrogen; these will be described in the next paragraph. In spite of the difficulties associated with multiple electrons, complex computer codes have recently been developed that consider all of the associated forces on these many electrons circulating around the nucleus, and so reasonably accurate bound energy levels can now be predicted for most atomic and ionic species of all of the elements.

For atoms with many electrons, as each successive electron is removed (ionized), the ionization energy or ionization potential of the next ion stage is found to be much larger than that of the lower ionization stage, owing to the tighter binding energy of what has become the new outer electron. The simplest example of this is to consider the ionization energy of the He^+ ion, which consists of a helium nucleus (two positively charged protons and two neutrons) with one electron rotating about the nucleus. This ion is hydrogen-like, having only one electron in orbit around the nucleus. Hence (3.3) can be used to determine the electron orbits, with the exception that there would be two positive charges in the nucleus instead of one and thus the electron orbit radius would be half that of hydrogen as given in (3.4). A more general equation for the radius of any hydrogen-like (one-electron) ion, such as He^+, Li^{++}, Be^{+++}, B^{4+}, et cetera, can therefore be written as

$$r = \frac{\epsilon_0 h^2}{\pi m_e Z e^2} n^2,$$ (3.12)

where Z represents the number of protons in the nucleus. In a similar manner, the H-like energy levels can be described by

$$E_n = -\frac{m_e Z^2 e^4}{8\epsilon_0^2 h^2} \cdot \frac{1}{n^2} = -\frac{Z^2 E_0}{n^2}. \tag{3.13}$$

If $Z = 1$ then the value of E_n reduces to that of hydrogen, as would be expected. For the helium ion ($Z = 2$), the levels would be in the same proportion to each other as for hydrogen, but the magnitude of the spacings, as well as the ionization limit (the energy for $n = 1$ in eqn. 3.13), would be four times greater since the nuclear charge is twice as great.

The frequencies of emission also have the same relationship as (3.10), once the Z^2 term is included:

$$\nu_{ji} = \frac{E_j - E_i}{h} = \frac{Z^2 E_0}{h}\left(\frac{1}{n_i^2} - \frac{1}{n_j^2}\right). \tag{3.14}$$

EXAMPLE

Compute the frequency for the transition from $n = 2$ to $n = 1$ in hydrogen-like lithium.

For lithium, there are three protons in the nucleus so $Z = 3$. When two of the electrons are stripped from a lithium atom to make a hydrogen-like lithium ion, that ion is referred to a *doubly ionized* lithium or Li^{++}. Using (3.14), we can now compute the frequency of the transition from $n = 2$ to $n = 1$ as

$$\nu_{ji} = \frac{Z^2 E_0}{h}\left(\frac{1}{n_i^2} - \frac{1}{n_j^2}\right),$$

$$\nu_{21}(Li^{++}) = 3^2(3.28 \times 10^{15})\left(\frac{1}{1^2} - \frac{1}{2^2}\right) = 2.214 \times 10^{16} \text{ s}^{-1}.$$

We can compute the wavelength by using the relationship $\lambda_{21} = c/\nu_{21}$ to obtain

$$\lambda_{21} = \frac{3 \times 10^8 \text{ m-s}^{-1}}{2.214 \times 10^{16} \text{ s}^{-1}} = 13.55 \text{ nm};$$

this is a wavelength in the soft–X-ray spectral region.

PHOTONS

When an electron jumps from a higher energy state to a lower energy state and radiates an exact amount of energy ΔE_{ji}, as calculated for the hydrogen atom in (3.10), that exact amount of energy is thought of as a single bundle of energy because of its discrete nature. We referred to light as consisting of electromagnetic waves in Chapter 2; however, when individual atoms radiate discrete values of energy, those small bundles of energy are considered to be particles rather than waves, and are referred to as *photons*. Thus we refer to the process of a photon being emitted from an atom (when the atom changes from a higher energy state to a lower energy state) with the understanding that the photon has an energy ΔE_{ji}, where the frequency and wavelength of the photon correspond to the values given in (3.10) and (3.11).

3.2 QUANTUM THEORY OF ATOMIC ENERGY LEVELS

The Bohr theory of the atom provided a major advance in the understanding of the nature of the atom, and in explaining the general characteristics of the spectral emission from the hydrogen atom. It did not, however, describe the structure of more complex atoms, in which multiple electrons surround the nucleus. It also did not explain the complicated discrete spectral lines radiating from those other atoms when excited in a gas discharge. But perhaps more significant was the unsatisfying nature of the Bohr theory, which used classical mechanics along with an arbitrary quantization of the angular momentum to obtain the correct answer for the energy levels of hydrogen. Scientists began to look for a more universal theory to describe matter, a theory that would also reduce to the well-known classical theory that had proved so successful in describing everyday situations.

WAVE NATURE OF PARTICLES

Light does not behave in only a wavelike fashion; it also has discrete quantized values (photons) when an electron makes a transition between energy levels. In 1924, de Broglie suggested that, conversely, particles such as electrons have wavelike properties also. He theorized that an electron might have a wavelength such that

$$\lambda_e = \frac{h}{p_e}, \tag{3.15}$$

where p_e is the momentum of the electron (in which $p_e = m_e v_e$ for the classical situation). Using this relationship, de Broglie obtained Bohr's condition for quantization of the angular momentum of the electron orbiting around the nucleus by assuming an exact integral number of electron wavelengths fitting into the circular orbit of the electron, of radius r_e, such that

$$2\pi r_e = n\lambda_e. \tag{3.16}$$

Substituting (3.15) for λ_e yields

$$p_e r_e = m_e v r_e = nh/2\pi = n\hbar, \tag{3.17}$$

which is just Bohr's second postulate.

The hypothesis that an electron could have wavelike characteristics was later verified by a double-slit experiment, in which electrons incident on the slits were shown to illuminate a screen positioned beyond the slits with an intensity distribution that resembled a wavelike diffraction pattern. This result had a profound effect on the way people thought of particles. If electrons that passed through a double slit produced a distribution corresponding to that of an optical diffraction pattern, then there was obviously some uncertainty as to where those electrons would end up. They could therefore not be described by normal classical trajectories, as Bohr had postulated, when circulating around a positively charged nucleus.

HEISENBERG UNCERTAINTY PRINCIPLE

At about the same time that the wavelike properties of electrons were verified, Heisenberg put forth his uncertainty principle. This principle suggested that canonical variables, such as distance x and momentum p, or energy E and time t, have an uncertainty in their product, below which the value of the product could never be determined. This minimum uncertainty $\hbar$, the value of Planck's constant divided by 2π, is expressed as

$$\Delta x \Delta p \approx \hbar \quad \text{or} \quad \Delta E \Delta t \approx \hbar, \tag{3.18}$$

where Δ represents the uncertainty of that variable. This relationship was arrived at by considering a thought experiment in which an electron was illuminated with a photon that was subsequently observed on a screen. The very act of illumination would cause the electron to be displaced, so it would no longer arrive at the same location as it would had the illumination not occurred. In addition, there would be some uncertainty in where the photon arrived on the measuring screen, owing to its wavelike properties.

WAVE THEORY

The two considerations just summarized – of particles having wavelike properties, and of uncertainties in determining exact values of various parameters – suggested the need for a theory leading to *probability distributions* of such parameters as position, momentum, energy, and time in place of *precise values* such as exact x, y, z coordinates. Thus evolved the theory of quantum mechanics. It was developed independently at about the same time (1926–1927) by Schrödinger and by Heisenberg, using what were then thought to be two completely different techniques: wave mechanics and matrix mechanics. These techniques were later shown to be equivalent.

The development of quantum mechanics is beyond the scope of this book. We will only outline briefly some of the results that are important for understanding the atomic theory of energy levels and the radiation that occurs when electrons jump from a higher to a lower energy level, using the wave mechanics method of Schrödinger. This will be useful in understanding the interaction of light with various types of laser gain media, including gaseous, liquid, and solid materials.

WAVE FUNCTIONS

Wave functions Ψ are used in quantum mechanics to describe various atomic parameters, including the properties of an electron associated with an atom. They are mathematical functions that depend upon the variables necessary to describe the particular features of that electron. For example, the location of the electron would be described with the three spatial variables x, y, z and the time t, or $\Psi(x, y, z, t)$. The term Ψ is a complex quantity that has no physical meaning in itself, but the square of its absolute value, $|\Psi|^2$ or $\Psi^*\Psi$, associated with any specific set of variables, is defined as the probability distribution function associated with those variables (where Ψ^* is the complex conjugate of Ψ). For example, in describing the location of the electron in space, $|\Psi|^2 \, dV$ is the probability that the electron is located within a specific volume element dV. Remember, it is a probability and not an exact value! It is equivalent to the one-dimensional probability $f(x) \, dx$ in statistics, where $f(x)$ is the probability distribution function and dx is the incremental region over which the probability is being evaluated.

If $\Psi^*\Psi$ denotes the probability distribution function of an electron bound to an atom, circulating around the nucleus in a given orbit, then the probability that the electron could be found at a specific location between x and $x+dx$, y and $y+dy$, and z and $z+dz$ at time t could be given by

$$\Psi^*(x, y, z, t)\Psi(x, y, z, t) \, dx \, dy \, dz. \tag{3.19}$$

This probability is completely consistent with Heisenberg's uncertainty principle, and also with the concept that particles can be wavelike.

As in all probability distribution functions, the probability that the electron is located *some*where must be unity, which leads to the normalizing definition for Ψ of

$$\int_x \int_y \int_z \Psi^*\Psi \, dx \, dy \, dz = 1. \tag{3.20}$$

QUANTUM STATES

Two kinds of quantum mechanical states of the electron will be described here. The first is the characteristic state or eigenstate, which has a well-defined energy and is therefore referred to as a *stationary* state. It corresponds to the electron residing in one of the discrete energy levels described by (3.8) in the Bohr theory. The second is a state of an electron in transition, referred to as a *coherent* state, which describes the movement of the electron from one stationary state to another.

STATIONARY STATES   An electron circulating around a nucleus can exist in one of many different stationary energy states, as deduced by Bohr. Each of the n stationary states is described by a wave function Ψ_n that can be used to determine the probability of finding the electron at a specific location with respect to the nucleus. Let E_n denote the energy of state n. Quantum theory suggests that the time dependence of the wave function in describing that state n is given by the exponential factor $e^{-iE_n t/\hbar}$, where $\hbar = h/2\pi$. Although no formal justification will be made here for including this time-dependence factor, we note that the resulting description is not an unreasonable one since $E = h\nu = \hbar\omega$ and thus $E/\hbar = \omega$. Therefore $e^{-iE_n t/\hbar} = e^{-i\hbar\omega t/\hbar} = e^{-i\omega t}$ and consequently the time dependence becomes a factor $e^{-i\omega t}$, an oscillatory behavior that will have relevance later in defining coherent states.

The description for Ψ_n, the nth state of the electron, now becomes

$$\Psi_n(x, y, z, t) = \psi_n(x, y, z)e^{-i\omega t}, \tag{3.21}$$

where Ψ_n represents only the spatial distribution of the electron. The probability of locating the electron at a point x, y, z can now be obtained as

$$\Psi_n^*(x, y, z, t)\Psi_n(x, y, z, t)\,dx\,dy\,dz$$
$$= \psi_n^*(x, y, z)e^{i\omega t}\psi_n(x, y, z)e^{-i\omega t}\,dx\,dy\,dz = \psi_n^*\psi_n\,dx\,dy\,dz. \tag{3.22}$$

We can see that the time dependence is effectively eliminated, which is the basis for referring to this state as a stationary state. Thus the probability distribution function described by $\Psi^*\Psi$ for any location in space is independent of time. When the electron is in a stationary state, it does not radiate and it gives up no energy to its surrounding medium. The electron in this state can therefore be described as a "charge cloud" having a constant distribution of electron density surrounding the atom that does not change with time.

COHERENT STATES   When the electron moves into a mode of changing from one specific state to a lower state, the wave function must incorporate both the state where it initially resides and the state where it finally ends up. The intermediate state during this transition is therefore a linear combination of the initial and final states:

$$\Psi = C_1\psi_1 e^{-iE_1 t/\hbar} + C_2\psi_2 e^{-iE_2 t/\hbar}. \tag{3.23}$$

The constant factors are slowly varying functions of time when compared to the time-dependent exponential factors. In the coherent state, the energy is not well defined. The probability distribution function of this state can be represented as

$$\Psi^*\Psi = C_1^*C_1\psi_1^*\psi_1 + C_2^*C_2\psi_2^*\psi_2$$
$$+ C_1^*C_2\psi_1^*\psi_2 e^{-i\omega_{21} t} + C_2^*C_1\psi_2^*\psi_1 e^{i\omega_{21} t}, \tag{3.24}$$

where

$$\omega_{21} = \frac{E_2 - E_1}{\hbar} \quad \text{or} \quad \nu_{21} = \frac{E_2 - E_1}{h}. \tag{3.25}$$

From (3.24) and (3.25) it can be seen that an oscillation between the two states occurs at a frequency corresponding to the energy separation between the states, which agrees exactly with the fourth Bohr condition. Thus the quantum mechanical description of an atom changing states, with no external interference, consists of an electron oscillating between the initial and final states at a frequency corresponding to the second Bohr condition $E_i - E_f = h\nu_{if}$, where i refers to the initial state and f refers to the final state. The oscillating electron produces the corresponding electric dipole radiation from its cyclical motion at the frequency ν_{if}.

THE SCHRÖDINGER WAVE EQUATION

It was suggested in previous sections that wave functions would form the basis for predicting the probability functions for locating electrons as they exist in various possible orbits around the nucleus. However, no mention was made of how to derive the electron energies associated with those orbits. The derivation of such energies and their related wave functions is one of the basic applications of quantum mechanics, involving the solution of a differential equation referred to as the *Schrödinger equation*. Seeking stable or stationary energy states of the electron implies the use of a time-independent form of the wave function as described earlier.

In considering only the time-independent form of the wave function, it is not unreasonable to assume that ψ must also have a wavelike or sinusoidal variation of the form $e^{i\mathbf{k}\cdot\mathbf{r}}$, just as we assumed a sinusoidal variation of the time dependence, since electrons were observed to have spatial wavelike properties including the generation of a spatial diffraction pattern when passed through a double slit. For such a spatial dependence, $\mathbf{k}$ is the propagation vector ($|\mathbf{k}| = 2\pi/\lambda$) in the direction of the wave and $\mathbf{r}$ is the position vector associated with a particular coordinate location in space, as discussed in Section 2.2. Therefore, ψ must satisfy a standard time-independent wave equation of the form

$$\nabla^2\psi + k^2\psi = 0 \quad \text{or} \quad \nabla^2\psi + (2\pi/\lambda)^2\psi = 0, \tag{3.26}$$

where we recall that ∇^2 is just a symbol for the second partial derivative with respect to the spatial coordinates.

Now consider the case of an electron orbiting a nucleus. If we invoke de Broglie's hypothesis that the electron has a wavelength $\lambda = h/p$ (eqn. 3.15), then (3.26) for that electron can be rewritten as

$$\nabla^2\psi + (2\pi p/h)^2\psi = 0. \tag{3.27}$$

Consider now the energy of that electron. Its energy E would include the kinetic energy plus the potential energy V, or $E = \frac{1}{2}m_e v^2 + V$, where v is the velocity of the electron. Since the momentum $p = m_e v$, the energy can be rewritten as $E = p^2/2m_e + V$, or p^2 can be expressed as

$$p^2 = 2m_e(E - V). \tag{3.28}$$

Using (3.28) in (3.27) leads to the Schrödinger wave equation for an electron,

$$\nabla^2\psi + \frac{8\pi^2 m_e}{h^2}(E - V)\psi = 0, \tag{3.29}$$

which can be written in terms of $\hbar$ as

$$\nabla^2 \psi + \frac{2m_e}{\hbar^2}(E-V)\psi = 0. \tag{3.30}$$

This is the famous wave equation developed by Schrödinger in 1926 and referred to by his name. It is a linear second-order partial differential equation, the solutions for which are obtained by inserting the appropriate potential function V into the equation. The wave functions must be quadratically integrable and thereby finite in extent, as suggested by (3.20).

The Schrödinger equation is often written in the following form:

$$\hat{H}\psi = E\psi, \tag{3.31}$$

where

$$\hat{H} = -\frac{\hbar^2}{2m_e}\nabla^2 + V. \tag{3.32}$$

The term $\hat{H}$ is known as the *Hamiltonian operator* of the system, or the *energy operator*. When it operates upon ψ, by taking the product of $\hat{H}$ and ψ, it yields the solutions for E or the *energy eigenvalues*.

The goal is to obtain the energy E and the wave functions ψ for the various states of the electron as described by solutions to the Schrödinger equation, which is known mathematically as a *characteristic value* equation. Such solutions are of finite extent only if the characteristic values, in this case the values of the energy E, have finite discrete eigenvalues E_n. Then a solution for the wave function ψ_n associated with each value of the energy E_n can be obtained. The solutions for ψ_n are known as *eigenfunctions*. Thus, using the Schrödinger equation, it is possible to determine first the electron energies (eigenvalues) and then the wave functions (eigenfunctions). These wave functions can then be used to determine the probability distribution function $\psi_n^*\psi_n$ of the electron, for various discrete energies, as it circulates around the nucleus.

TIME-DEPENDENT SCHRÖDINGER EQUATION   The time-dependent Schrödinger equation, which is the basic expression for describing nonstationary functions, can be written as

$$i\hbar\frac{\partial\Psi}{\partial t} = -\frac{\hbar^2}{2m_e}\nabla^2\Psi + V\Psi. \tag{3.33}$$

Expressing this in operator form leads to

$$i\hbar\frac{\partial\Psi}{\partial t} = \hat{H}\Psi. \tag{3.34}$$

We will not need to consider solutions to this more complicated time-dependent equation, since they are not necessary for understanding the fundamentals of radiative emission from excited states of matter. At this point we will return to the time-independent Schrödinger equation and obtain solutions of the energy values and wave functions for the various electronic states of the hydrogen atom.

SCHRÖDINGER'S EQUATION FOR THE HYDROGEN ATOM Solving Schrödinger's equation for the hydrogen atom is most easily carried out by using polar coordinates, which take advantage of the spherical symmetry of the combined system of an electron circulating around a proton. In such a situation, the Laplacian operator becomes (in polar coordinates)

$$\nabla^2 = \frac{1}{r^2}\left[\frac{\partial}{\partial r}\left(r^2\frac{\partial}{\partial r}\right) + \frac{1}{\sin\theta}\frac{\partial}{\partial\theta}\left(\sin\theta\frac{\partial}{\partial\theta}\right) + \frac{1}{\sin^2\theta}\frac{\partial^2}{\partial\phi^2}\right]; \qquad (3.35)$$

using this in (3.29) yields

$$\frac{1}{r^2}\left[\frac{\partial}{\partial r}\left(r^2\frac{\partial}{\partial r}\right) + \frac{1}{\sin\theta}\frac{\partial}{\partial\theta}\left(\sin\theta\frac{\partial}{\partial\theta}\right)\right.$$

$$\left. + \frac{1}{\sin^2\theta}\frac{\partial^2}{\partial\phi^2}\right]\psi + \frac{8\pi^2\mu}{h^2}(E-V)\psi = 0 \qquad (3.36)$$

for the Schrödinger equation for the hydrogen atom. Here μ represents the reduced mass of the combined electron mass m_e and proton mass M; $\mu = m_e M/(m_e + M)$. For this situation, the potential V is that of a positive and negative point charge separated by a distance r,

$$V = -\frac{e^2}{4\pi\epsilon_0 r}, \qquad (3.37)$$

as was used for the Bohr theory in (3.6).

ENERGY AND WAVE FUNCTION FOR THE GROUND STATE
OF THE HYDROGEN ATOM

It is appropriate to solve for the energy E_1 and the wave function ψ_1 of the ground state ($n = 1$) of the hydrogen atom first, since this is the simplest electronic energy state of any atom and perhaps even of all matter. The solution for ψ_1 turns out to be completely spherically symmetric and relatively simple. Instead of investigating a general solution, we will use a trial solution for ψ_1 which, if it works, will be at least one real solution to the equation. It turns out that it is the only solution, but we will proceed without that assumption. Thus we substitute a solution for ψ_1 that is radially symmetric and decays exponentially:

$$\psi_1 = e^{-\alpha r}, \qquad (3.38)$$

where α is a constant to be determined. Using this solution for ψ_1, Schrödinger's equation becomes

$$\left(\alpha^2 + \frac{8\pi^2\mu E_1}{h^2}\right)e^{-\alpha r} + \left(\frac{2\pi\mu e^2}{\epsilon_0 h^2} - 2\alpha\right)\frac{e^{-\alpha r}}{r} = 0. \qquad (3.39)$$

In order for this equation to be valid for all values of r, both factors in parentheses must separately be equal to zero. Equating the right-hand terms to zero leads to a solution for α of

$$\alpha = \frac{\pi \mu e^2}{\epsilon_0 h^2} = \frac{1}{a_H}. \tag{3.40}$$

As seen from (3.40), α represents the reciprocal of the value of the first Bohr radius, as previously defined in (3.5), with the replacement of m_e by the reduced mass μ. Using this value of α within the left-hand parentheses of (3.39), one can solve for the energy E_1 to obtain

$$E_1 = -\frac{\mu e^4}{8\epsilon_0^2 h^2}. \tag{3.41}$$

This value is exactly the same as that obtained in (3.8) using Bohr's theory, again with the exception that the reduced electron–proton mass μ is used instead of the electron mass m_e.

Equation (3.38) gives only the radial dependence of the wave function ψ_1. To be complete it would have to include a normalizing constant C: $\psi = C_1 e^{-\alpha r} = C_1 e^{-r/a_H}$. To satisfy the normalization condition of (3.20) for a spherically symmetric volume element $4\pi r^2 \, dr$ (which would be a shell of radius r and thickness dr), we have

$$\int_0^\infty C_1^2 e^{-2r/a_H} 4\pi r^2 \, dr = 1, \tag{3.42}$$

which leads to $C_1 = 1/(\pi^{1/2} a_H^{3/2})$. Thus the wave function for the ground state of hydrogen can be expressed as $\psi_1 = (1/\pi^{1/2} a_H^{3/2}) e^{-r/a_H}$. This wave function is spherically symmetric and thus has only an r dependence, which is shown in Figure 3-3(a). It has a maximum at $r = 0$ and is reduced by a

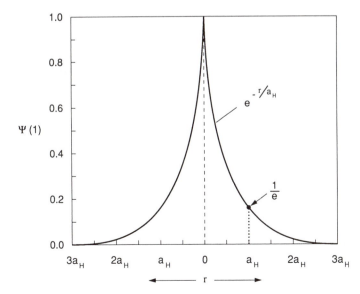

Figure 3-3(a). Wave function ψ_1 of the ground state of the hydrogen atom

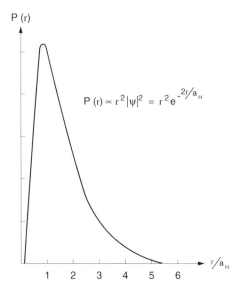

P (r)

$$P(r) \propto r^2 |\psi|^2 = r^2 e^{-2r/a_H}$$

r/a_H

1   2   3   4   5   6

Figure 3-3(b).
Probability of finding an
electron at any specific
distance from the center
of the nucleus in units of
r/a_H

factor of $1/e$ at $r = a_H$ (the first Bohr radius). As mentioned earlier, this wave function does not tell us where we would actually find the electron.

The probability of finding an electron at any radial location within a spherically symmetric shell of thickness dr and volume $4\pi r^2 \, dr$ is obtained by considering

$$\psi_1^* \psi_1 \, dV = \psi_1^* \psi_1 4\pi r^2 \, dr = 4\pi r^2 C_1^2 e^{-2r/a_H} \, dr. \tag{3.43}$$

The probability of finding the electron within a shell of thickness dr is thus given by

$$4\pi r^2 C_1^2 e^{-2\alpha r} = 4\alpha^3 r^2 e^{-2\alpha r}, \tag{3.44}$$

which is plotted in Figure 3-3(b). It can be seen that the maximum value of this probability occurs ar $r = a_H$, the first Bohr radius.

Thus the Schrödinger equation yields a solution for the lowest energy state $n = 1$ (ground state) of the hydrogen atom that is consistent with the Bohr theory, yet yields a picture of the location of the electron that is distinctly different from that of the Bohr theory. It suggests a "smeared out" distribution in which the electron spends a very high percentage of its time near the Bohr radius, as shown in the radial plot of $r^2 \psi_1^*(r) \psi_1(r)$ in Figure 3-3(b), but is almost never in that exact location – contrary to what was suggested by the Bohr theory.

EXCITED STATES OF HYDROGEN

The ground state of hydrogen was assumed to be spherically symmetric. It was therefore possible to use a trial solution for ψ_1 (the correct one) that had no angular dependence; in effect, ψ_1 was independent of θ and ϕ. The excited-state wave functions are much more complicated than the wave function for the ground state, and therefore the complete Schrödinger equation (eqn. 3.36) must be solved.

It turns out that the Schrödinger equation for $\psi(r, \theta, \phi)$ for the hydrogen atom is separable. This means that a trial solution for ψ can be written as the product of three separate functions:

$$\psi(r, \theta, \phi) = CR(r)\Theta(\theta)\Phi(\phi), \tag{3.45}$$

where C is a constant. Inserting this solution into (3.36) yields

$$\frac{1}{\Phi} \frac{d^2\Phi}{d\phi^2} + \frac{\sin\theta}{\Theta} \frac{d}{d\theta}\left(\sin\theta \frac{d\Theta}{d\theta}\right)$$

$$+ \frac{\sin^2\theta}{R} \frac{d}{dr}\left(r^2 \frac{dR}{dr}\right) + \sin^2\theta \frac{8\pi^2\mu r^2}{h^2}(E - V) = 0, \tag{3.46}$$

which is a simpler equation to deal with than (3.36). Since only the left term in (3.46) contains ϕ, the only way that a solution can exist for all values of r, θ, and ϕ is if the left-hand term equals a constant. Using foresight, we will let that constant be $-m^2$. Substituting that value for the left-hand term in (3.46) leaves an equation that can be separated into two components: a radial component with variable r, and an angular component with variable θ. Using the same arguments as before, a solution for this equation for all values of r, θ, and ϕ can be obtained by equating both components to the same constant, which we choose (again with foresight) to be $l(l+1)$. This leads to the following set of differential equations:

$$\frac{1}{\Phi} \frac{d^2\Phi}{d\phi^2} = -m^2, \tag{3.47}$$

$$\frac{m^2}{\sin^2\theta} - \frac{1}{\Theta \sin\theta} \frac{d}{d\theta}\left(\sin\theta \frac{d\Theta}{d\theta}\right) = l(l+1), \tag{3.48}$$

$$\frac{1}{R} \frac{d}{dr}\left(r^2 \frac{dR}{dr}\right) + \frac{8\pi^2\mu r^2}{h^2}(E - V) = l(l+1). \tag{3.49}$$

As in the solution for the ground state, the potential function V that we have used here is the coulomb potential of (3.37).

In examining these three equations, (3.47) appears to be relatively simple whereas the other two are rather complex. The first equation, rewritten as

$$\frac{d^2\Phi}{d\phi^2} + m^2\Phi = 0, \tag{3.50}$$

has a solution of the form

$$\Phi = e^{im\phi}. \tag{3.51}$$

This function must have the same value at $\phi = 0, 2\pi, 4\pi, 6\pi, \ldots$, since that would be one complete revolution around the atom. This would occur only if the constant m is either 0 or an integer. We thus have the following restriction on m:

$$m = 0, \pm 1, \pm 2, \pm 3, \ldots. \tag{3.52}$$

Solving (3.48) and (3.49) is somewhat more complex, so in this treatment we simply provide the solutions. More detailed discussions can be found in any text on quantum mechanics. The solutions of both, however, are equations of the type that had been solved prior to the development of the Schrödinger equation, and therefore did not pose extensive difficulties for Schrödinger in obtaining his results.

Equation (3.48) has the form of Legendre's differential equation, which has solutions only if l is a positive integer with a value greater than or equal to $|m|$. This integer l is called the *azimuthal* quantum number. The solutions to Legendre's equation involve terms known as *Legendre polynomials*. These polynomials are of the form $P_l^m(\cos\theta)$, and can be generated by the function

$$\Theta(\theta) = P_l^{|m|}(x) = \frac{(1-x^2)^{1/2|m|}}{2^l l!} \frac{d^{|m|+l}}{dx^{|m|+l}}(x^2-1)^l, \tag{3.53}$$

where in this case x represents $\cos\theta$. Several of these functions for values of both l and m of 0, 1, and 2 are listed as follows:

$$P_0^0(x) = 1, \qquad\qquad P_1^1(x) = (1-x^2)^{1/2},$$
$$P_1^0(x) = x, \qquad\qquad P_2^1(x) = 3x(1-x^2)^{1/2}, \tag{3.54}$$
$$P_2^0(x) = \tfrac{1}{2}(3x^2-1), \qquad P_2^2(x) = 3(1-x^2).$$

Now we must address the solution of (3.49), the radial portion of the equation. This equation is of a form known as *Laguerre's differential equation*. Solutions are given by the formula

$$R(\rho) = \rho^l e^{-\rho/2} L_{n+l}^{2l+1}(\rho), \tag{3.55}$$

where ρ is a simple linear function of r,

$$\rho = \frac{2r}{na_H}. \tag{3.56}$$

From the standpoint of solutions to the Schrödinger equation, n is referred to as the *principal* quantum number and can have the values $n \geq l+1$. The functions $L_{n+l}^{2l+1}(\rho)$ are known as *associated Laguerre polynomials* and can be generated by the following equation:

$$L_{n+l}^{2l+1}(\rho) = \left(\frac{d}{d\rho}\right)^{2l+1}\left[e^\rho\left(\frac{d}{d\rho}\right)^{n+l}(\rho^{n+l}e^{-\rho})\right]. \tag{3.57}$$

As with the Legendre functions, we list a few examples of the low-order polynomials:

$$L_1^1(\rho) = -1, \qquad L_2^2(\rho) = 2,$$
$$L_2^1(\rho) = 2\rho-4, \qquad L_3^3(\rho) = -6. \tag{3.58}$$

The radial equation (3.49) contains the energy E, from which the energy eigenvalues can be obtained as

$$E_n = -\frac{\mu e^4}{8\epsilon_0^2 h^2}\left(\frac{1}{n^2}\right). \tag{3.59}$$

This can be viewed as identical to (3.8), the Bohr equation for the energy of various electronic levels or states of the hydrogen atom, except for its use of the more accurate reduced mass μ in place of m_e.

ALLOWED QUANTUM NUMBERS FOR THE HYDROGEN ATOM WAVE FUNCTIONS

We can now summarize the constraints placed upon the possible wave functions in terms of the restrictions on various quantum numbers that are dictated by solutions to the preceding equations:

$$n \geq 1, \tag{3.60}$$

$$n \geq l+1, \tag{3.61}$$

$$l \geq 0, \tag{3.62}$$

$$l \geq |m|, \tag{3.63}$$

$$m = 0, \pm 1, \pm 2, ..., \pm l. \tag{3.64}$$

We are familiar with the quantum number n from the Bohr theory, but l and m are new at this point. The l quantum number, which is the *angular momentum* quantum number, can have n values ranging from 0 to $n-1$. Historically these values were given names of s, p, d, f, ... respectively for increasing values of l; for example, $l = 0$ refers to an s eigenstate. This labeling will be dealt with in more detail in the next section.

The quantum number m is referred to as the *magnetic* quantum number. In examining (3.64) one can deduce that there are $2l+1$ values of m for every value of l. Thus, using the rules just summarized, we can write down all the allowed eigenfunctions describing the various possible quantum states (eigenstates) for the hydrogen atom by using the notation $\psi(n, l, m)$ for each state. In doing this it can be seen that there is one possible state for $n = 1$ ($\psi(1, 0, 0)$), four states for $n = 2$ ($\psi(2, 0, 0)$, $\psi(2, 1, 1)$, $\psi(2, 1, 0)$, and $\psi(2, 1, -1)$), nine for $n = 3$, and so forth, suggesting that there are n^2 states for every possible value of n (taking into account all allowed values of l and m). We thus have – according to (3.59) – several states all with the same energy for any value of $n > 1$. These energy states are therefore said to be *degenerate*.

We can now provide the entire wave function for various states of the hydrogen atom as follows:

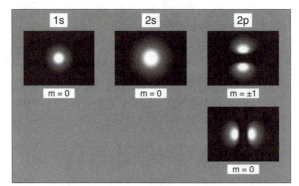

Figure 3-4. Probability distribution function of the ground state ($n = 1$) and the possible sublevels of the first excited state ($n = 2$)

$$\psi_{n,l,m} = C\rho^l e^{-\rho/2} L_{n+l}^{2l+1}(\rho) P_l^{|m|}(\cos\theta)e^{im\phi}, \qquad (3.65)$$

where $\rho = 2r/na_H$ and C is a normalizing constant. Substituting this wave function into Equation (3.20) leads to the following value for C:

$$C = -\left\{\left(\frac{2Z}{na_H}\right)^3 \frac{(n-l-1)!}{2n[(n+l)!]^3}\right\}^{1/2}\left[\frac{2l+1}{4\pi}\frac{(l-|m|)!}{(l+|m|)!}\right]^{1/2}. \qquad (3.66)$$

The term Z represents the charge on the nucleus, which for hydrogen would be $Z = 1$. The value of the probability distribution function $|\psi|^2$ is shown in Figure 3-4 for $n = 1$ (1s) and also for $n = 2$ (2s and 2p). Note that this is not $|\psi|^2 dV$, the probability of finding the electron at a specific location. It can be seen that both 1s and 2s are spherically symmetric, whereas 2p has a dumbbell shape for $m = 0$ and a doughnut shape for $m = \pm 1$.

3.3 ANGULAR MOMENTUM OF ATOMS

We will now investigate how different values of an atom's angular momentum, associated with different angular momentum quantum numbers, lead to different energies for various quantum states within the atom. The angular momentum of an electron moving around a positively charged nucleus can be derived with quantum mechanics by solving the Schrödinger equation, using the appropriate eigenvalue equation (eqn. 3.31) but with the angular momentum operator replacing the energy operator. The solution to this equation provides an angular momentum that is quantized, as in the Bohr theory, but is a more complete and accurate result than that obtained from the Bohr theory. The total angular momentum is divided into two components, orbital angular momentum and spin angular momentum.

ORBITAL ANGULAR MOMENTUM

The orbital angular momentum **L** is a vector quantity that is found to have a magnitude given by

$$|\mathbf{L}| = [l(l+1)]^{1/2}\frac{h}{2\pi} = [l(l+1)]^{1/2}\hbar, \tag{3.67}$$

where l is the azimuthal quantum number. For $l = 0$ (s states), there is no orbital angular momentum, which implies that there is no net rotation of the electron charge cloud. This does not imply that there is no electron motion but only no *net* motion for this particular s state.

The solution for the angular momentum (using the Schrödinger equation) also suggests that the z component of the angular momentum L_z is quantized and of the form

$$L_z = m\frac{h}{2\pi} = m\hbar. \tag{3.68}$$

The magnetic quantum number m is associated with rotation of the electron about the z axis through the angular rotation ϕ, with integral values up to $m = \pm l$. The various possible values of angular momentum $m\hbar$ represent the z components of the angular momentum, for which there will be $2l+1$ different values. Thus, various states will have their angular momentum vectors precisely oriented in certain discrete orientations, as shown in Figure 3-5.

If a magnetic field is applied to an ensemble of atoms, the levels will split into the associated angular momentum states (except for s states, in which m is always zero). This will provide $2l+1$ states of slightly different energy for each l value associated with the nth quantum number. This is due to the alignment of electrons with the magnetic field and the splitting of energy into components associated with the field. The multiple values for the energies of these states lead to many more possible transitions between states. Thus, when electrons jump from one state to another, more wavelengths are emitted when a magnetic field is present than when it is not present; this is known as the *Zeeman effect*. We will not discuss it further because it does not concern our study of lasers.

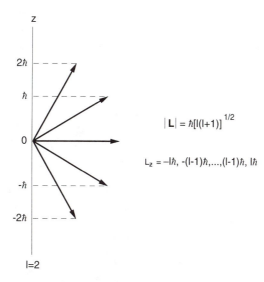

Figure 3-5. Possible values of the orbital angular momentum vector for the case $l = 2$, along with the components projected onto the z axis

SPIN ANGULAR MOMENTUM

When the spectral lines of hydrogen (produced when radiative transitions occur between energy levels) were first observed with an extra–high-resolution spectrograph, it was seen that there was a splitting of the emission lines similar to that observed with the Zeeman effect, except that there was no magnetic field present. This implied that the energy levels were split into several components, and that the theoretical interpretation of this splitting would require another component in the energy Hamiltonian in Schrödinger's equation. Pauli suggested that the electron had an intrinsic spin angular momentum around its own axis with a value of either $+\hbar/2$ or $-\hbar/2$.

The value of the electron spin angular momentum vector $\mathbf{S}$ was obtained from the appropriate Schrödinger equation to be

$$|\mathbf{S}| = [s(s+1)]^{1/2}\hbar, \tag{3.69}$$

where s is the spin quantum number with a value of $\frac{1}{2}$. The z components have values of

$$S_z = \pm\hbar/2 = \pm s\hbar. \tag{3.70}$$

This suggestion of only two quantum numbers (eigenvalues) for the electron spin is unusual in quantum theory, where most eigenfunctions have many possible eigenvalues. The justification for having only two half-integral values is the successful prediction of experimental evidence.

TOTAL ANGULAR MOMENTUM

In classical terms, the total angular momentum vector $\mathbf{J}$ consists of the vector sum of the orbital angular momentum vector $\mathbf{L}$ and the spin angular momentum vector $\mathbf{S}$:

$$\mathbf{J} = \mathbf{L} + \mathbf{S}. \tag{3.71}$$

Thus, the total angular momentum could have any possible value between $\mathbf{L}+\mathbf{S}$ and $\mathbf{L}-\mathbf{S}$. However, the result is slightly more complicated from a quantum mechanical point of view. Since both $\mathbf{L}$ and $\mathbf{S}$ are quantized, their directions are restricted. The quantum mechanical rules for combining these angular momenta are difficult to derive but not difficult to understand and use. The magnitude of $\mathbf{J}$ is given by

$$|\mathbf{J}| = [j(j+1)]^{1/2}\hbar, \tag{3.72}$$

where j has values for a one-electron system of either

$$j = l + s = l + \tfrac{1}{2} \tag{3.73}$$

or

$$j = l - s = l - \tfrac{1}{2} \quad (l \neq 0). \tag{3.74}$$

Spin and orbit angular momenta - same direction

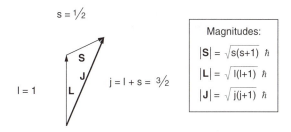

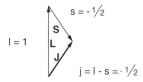

Spin and orbit angular momenta - opposite direction

Figure 3-6. The two possible values of the total angular momentum **J** for a one-electron atom

If $l = 0$, the total quantum number j is just equal to s. The vector model of the total angular momentum associated with a one-electron atom is shown in Figure 3-6 for $l = 1$ for the two possible cases of $j = l + \frac{1}{2}$ and $j = l - \frac{1}{2}$. Recall that the lengths of the vectors are proportional to $[l(l+1)]^{1/2}$, $[s(s+1)]^{1/2}$, and $[j(j+1)]^{1/2}$, as indicated in (3.67), (3.69), and (3.72). Even though **L** and **S** are not exactly aligned, as shown in the figure, they are referred to as being *parallel* when $j = l + s$ and *antiparallel* when $j = l - s$. A few examples of total angular momentum quantum numbers are: for $l = 1$, $j = \frac{1}{2}$ or $\frac{3}{2}$; for $l = 2$, $j = \frac{3}{2}$ or $\frac{5}{2}$; and so on. For $l = 0$, a special case exists in which there is only one value of j ($j = \frac{1}{2}$).

3.4 ENERGY LEVELS ASSOCIATED WITH ONE-ELECTRON ATOMS

FINE STRUCTURE OF SPECTRAL LINES

With the Bohr theory, energy levels of the hydrogen atom are arranged according to the value of the n quantum number (eqn. 3.8). We have also seen that hydrogen-like ions have a similar relationship associated with the n quantum number, but with a scale factor associated with the total number of protons or positive charge in the nucleus; this factor is designated Z, as indicated in (3.13). From the foregoing derivation using quantum mechanics, we realize that the electron also has quantum numbers associated with the orbital angular momentum l and the magnetic quantum number m, as indicated in (3.65) for the wave functions of the hydrogen atom. This derivation did not consider the electron spin angular momentum, which we later added as a relevant quantum number of either $+\frac{1}{2}$ or $-\frac{1}{2}$. We will see in the following discussions that the most useful four quantum numbers for designating energy levels are n, l, s, and j rather than n, l, m, and *s*, since *j* indicates the relative orientation of the electron orbital and spin angular momenta. Thus *j* determines the energy of the state (as will be shown next), whereas *m* is not relevant except when an external magnetic field is present.

When we referred to the electron spin angular momentum in the previous paragraph, we referred to the splitting of emission lines that was observed by spectroscopists, suggesting that in many cases there were at least two energy levels corresponding to each value of *n*. A simple explanation for this splitting of energy levels due to the electron spin can be envisioned by assuming that, as the electron orbits around the nucleus, it produces a magnetic field that in turn interacts with the magnetic moment of the electron produced by its own spin. This field then shifts the electron's energy in a manner that depends upon whether the electron spin is aligned with or against the orbital magnetic field. This is the equivalent of the alignment of the electron spin angular momentum **S** with the electron orbital angular momentum **L**. The energy-level splitting due to this interaction is shown in Figure 3-7.

Thus we realize that there may be many possible states associated with each value of *n*, and that these states can have different energy values (referred to as *sublevels*) depending upon the values of the quantum numbers *j*, *l*, and *s*. For example, the energy is higher for *l* = 1 when the alignment of the orbital and spin angular momenta is parallel than when it is antiparallel, as shown in Figure 3-7. Table 3-1 lists the energy levels associated with a one-electron atom (in the absence of a magnetic field) for *n* = 1, 2, and 3. It can be seen that for *n* = 1 (the ground state) there is only one energy level. For *n* = 2 there are three levels, for *n* = 3 there are five levels, and so forth. Also note that there is but one value of *s* and therefore one level for *l* = 0 for all values of *n*. This is because there is no orbital angular momentum with which to reference both a $+\frac{1}{2}$ and a $-\frac{1}{2}$ spin state.

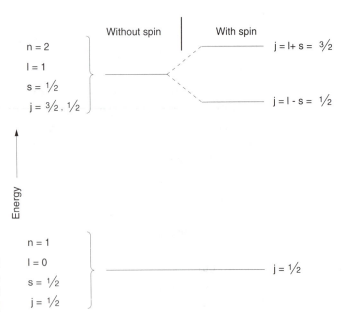

Figure 3-7. Energy-level splitting due to the electron spin for *n* = 2 and *l* = 1

TABLE 3-1

ENERGY LEVELS ASSOCIATED WITH A ONE-ELECTRON ATOM

n	l	Spin	j
1	0	$+\frac{1}{2}$	$+\frac{1}{2}$
2	0	$+\frac{1}{2}$	$+\frac{1}{2}$
2	1	$+\frac{1}{2}$	$+\frac{3}{2}$
2	1	$-\frac{1}{2}$	$+\frac{1}{2}$
3	0	$+\frac{1}{2}$	$+\frac{1}{2}$
3	1	$+\frac{1}{2}$	$+\frac{3}{2}$
3	1	$-\frac{1}{2}$	$+\frac{1}{2}$
3	2	$+\frac{1}{2}$	$+\frac{5}{2}$
3	2	$-\frac{1}{2}$	$+\frac{3}{2}$

Therefore, the spin and total angular momentum are equal and have only one value. We will later establish a notation for the levels that will help distinguish between the different angular momentum states.

As mentioned previously, each of the energy levels having a specific value of l actually consists of $2l+1$ energy sublevels, each associated with a specific value of m. Normally these sublevels all have the same energy, but when a magnetic field is applied to the atom the sublevels split apart into distinguishably separate levels. We will not be concerned further with this splitting of energies according to m, since this effect is not applicable to most lasers, but it is important to know that such an effect can occur. Note that each of these sublevels count as part of the *degeneracy* g_i (the total number of such sublevels) that is associated with each energy level i.

PAULI EXCLUSION PRINCIPLE

Pauli also suggested that no two electrons could occupy the same state. This means that no two electrons could have all of the same quantum numbers, including the n and l and m quantum numbers as well as the spin quantum number of either $+\frac{1}{2}$ or $-\frac{1}{2}$. This is referred to as the Pauli exclusion principle; we will use it when we develop the periodic table of the elements, as well as when we determine the designation for states or levels involving more than one electron (multi-electron states).

3.5 PERIODIC TABLE OF THE ELEMENTS

QUANTUM CONDITIONS ASSOCIATED WITH MULTIPLE ELECTRONS ATTACHED TO NUCLEI

The periodic table is built up of elements (atoms) that have various numbers of protons, neutrons, and electrons. In order to consider the excited

TABLE 3-2a

QUANTUM NUMBERS FOR VARIOUS SHELLS AND SUBSHELLS

Shell	K	L		M			N				O					P		
n	1	2		3			4				5					6		
Subshell	1s	2s	2p	3s	3p	3d	4s	4p	4d	4f	5s	5p	5d	5f	5g	6s	6p	6d
l	0	0	1	0	1	2	0	1	2	3	0	1	2	3	4	0	1	2
Number of electrons	2	2	6	2	6	10	2	6	10	14	2	6	10	14	18	2	6	10

states of those various kinds of atoms from which laser action can take place, we must first know their ground-state electronic configurations. The periodic table of the elements can be viewed as sequentially adding protons, neutrons, and electrons such that charge neutrality is satisfied. Each element is uniquely determined by the number of protons in the nucleus. Different isotopes of each element result from the different numbers of neutrons that are also within the nucleus. Only certain specific numbers of neutrons can lead to stable atom configurations for a given element. The number of electrons must balance the number of protons in order to have a neutral atom. Those electrons can be excited to higher neutral energy levels or removed (ionized) from the atom to produce ions, but they will eventually decay back to the neutral atom ground state. In developing the periodic table of the elements, as more protons and electrons are added to produce new elements, the electrons are added sequentially into shells according to their n, l, m, and s quantum numbers, beginning with the lowest allowed values and working toward higher values, with the maximum number of electrons determined by the number of protons in the nucleus.

From the rules of equations (3.60)–(3.64), we know of the restrictions on quantum numbers n, l, and m. We also have the restrictions on s of $\pm\frac{1}{2}$, and also the Pauli exclusion principle indicating that no two electrons can occupy the same state (i.e., all four quantum numbers cannot be the same). Keeping these restrictions in mind, Table 3-2(a) shows how the various elements evolve according to the way electrons fill the possible quantum states. For people working with X-rays, the value of n for a given energy level is specified alphabetically in a sequence that begins with the letter K. Thus $n = 1$ is referred to as the K shell, $n = 2$ is the L shell, $n = 3$ is the M shell, et cetera. Each shell is made up of various subshells according to the possible values of l. We remind ourselves that in this discussion we are referring to the electronic configuration of the ground state of the specific element or type of atom; we will refer to excited states later. Since from (3.61) we know that $n \geq l + 1$, the possible values of l are given as the subshell designation in Table 3-2(a). Table 3-2(a) also lists the number of electrons that can fill each subshell according to the number of m values for each l, and the two allowed values of s for each value of m. The notation discussed previously for a single electron is used to describe each of the electrons of a specific element – namely, s, p, d, f, ... for $l = 0, 1, 2, 3, ...,$ respectively. These electrons fill various shells in the sequence shown in Table 3-2(a), going from left to right.

A hydrogen atom contains one 1s electron. Helium atoms consist of two 1s electrons ($1s^2$) such that the K shell is completely filled. The spins of the two electrons must be in opposite directions for the ground state, since the n and l quantum numbers of both electrons are the same. Lithium has three electrons: two 1s electrons and one 2s electron ($1s^2 2s$). Beryllium has four electrons ($1s^2 2s^2$), while boron has five electrons: two 1s electrons, two 2s electrons, and one 2p electron ($1s^2 2s^2 2p$). Carbon has six electrons, two in the K shell and four in the L shell; its ground state would therefore be $1s^2 2s^2 2p^2$ for a total of six electrons. The completely filled p subshell ($l = 1$) for $n = 2$ would contain six electrons, since we have $m = \pm 1$ and 0 for $l = 1$ and two values of spin for each value of m. Neon is the atom with a completely filled p subshell; its electronic configuration is $1s^2 2s^2 2p^6$. It also has a completely filled L shell since, according to (3.61), there are no d electrons allowed in the $n = 2$ shell.

Moving further along in the periodic table, we find for example that silver has 47 electrons; its ground state would consist of 2 electrons in the K shell, 8 in the L shell, 18 in the M shell, 18 in the N shell, and 1 in the O shell. Its ground state would therefore be $1s^2 2s^2 2p^6 3s^2 3p^6 3d^{10} 4s^2 4p^6 4d^{10} 5s$, which consists of 47 electrons. Cadmium, with one more electron than silver, would have two electrons in the O shell ($n = 5$) instead of one. Hence the shell designation for cadmium would be the same as that for silver except that the final 5s electron for silver would be replaced by two 5s electrons ($5s^2$) for cadmium.

When a shell or a subshell is filled, that configuration is referred to as *closed*. Thus $1s^2$ (helium) and $1s^2 2s^2 2p^6$ (neon) are closed shells, and $2s^2$ is a closed subshell. Elements with closed shells in their ground state are extremely stable and unreactive. These include the "noble" gases of helium, neon, argon, krypton, and xenon, as shown in Table 3-2(b). If an electron is energized or moved to an excited state of one of these noble gases, the element is no longer in a closed shell and can thus be reactive. We will see later how this relates to the development of excimer lasers.

Figure 3-8 shows the periodic table of the elements that is obtained from the rules just described. When we reach the O shell, the f electrons do not fill that shell in exactly the normal sequence. The first f electron subshell (4f) does not begin to fill its slot until *after* the two 6s electrons fill their slots, as can be seen in the table. This pattern results from slight deviations from the sequential rules outlined here that are due to the complex electron interactions in an atom having a large number of electrons. We need not be concerned with this minor effect when considering the energy levels associated with various kinds of lasers.

TABLE 3-2b

CLOSED SUBSHELLS CORRESPONDING TO THE NOBLE GASES

Subshells	Number of electrons in outer shell	Total number of electrons	Element
$(1s)^2$	2	2	He
$(2s)^2 + (2p)^6$	8	10	Ne
$(3s)^2 + (3p)^6$	8	18	Ar
$(4s)^2 + (3d)^{10} + (4p)^6$	18	36	Kr
$(5s)^2 + (4d)^{10} + (5p)^6$	18	54	Xe
$(6s)^2 + (4f)^{14} + (5d)^{10} + (6p)^6$	32	86	Em

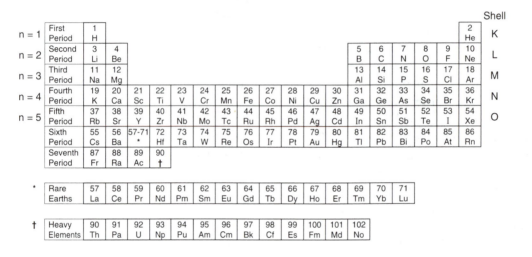

Figure 3-8. Periodic table of the elements

SHORTHAND NOTATION FOR ELECTRONIC CONFIGURATIONS OF ATOMS HAVING MORE THAN ONE ELECTRON

In general, only the electrons of an unfilled subshell are involved in determining radiative transitions between the electronic levels of an atom or ion. Inner-shell electrons can be excited to higher levels, but such levels are higher than the ionization potential of that particular atom and therefore population in these states will rapidly ionize (lose an electron); the state that it converts to is a state belonging to the next higher ionization stage. Inner-shell states ionize very rapidly (typically in a time of 10^{-14} s) and are consequently not suitable for producing lasers. Therefore, from the perspective of understanding how lasers operate, the electronic designation

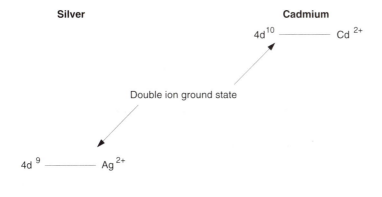

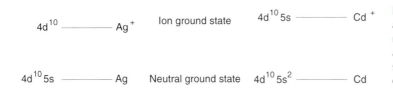

Figure 3-9. Electron arrangements for the neutral-atom, single-ion, and double-ion ground states of silver and cadmium

associated with labeling energy levels can be significantly shortened to the last one or two unfilled subshells of the atom or ion. Thus the silver neutral ground state mentioned previously would be designated $4d^{10}5s$, and the singly ionized silver (Ag^+) ground state would be designated as $4d^{10}$. Neutral cadmium would be $4d^{10}5s^2$ and singly ionized cadmium (Cd^+ in its ground state) would be $4d^{10}5s$. These electron arrangements are shown in Figure 3-9.

An even simpler level designation is also used. When electrons are promoted to higher energy levels above the ground state in atoms and ions, the excitation process most often affects only the outer or final electron of the sequence, even if there might be more than one electron in the unfilled subshell. Hence the state of this final electron is quite often the only one referred to in the level designation. For helium, whose ground state is $1s^2$, an excited state designation might be 1s2s or 1s3p. However, in most situations reference is made only to the outer electron (2s or 3p in this case). For neutral carbon, levels such as $1s^2 2s^2 2p3s$ or $1s^2 2s^2 2p5d$ would simply be designated as 3s or 5d. For doubly ionized carbon (C^{2+}) the ground state would be $1s^2 2s^2$ or just $2s^2$ for the shortened notation, since both of the p electrons were removed in the ionizing process. An excited state of C^{2+} might be $1s^2 2s3p$ or simply 3p. Fifth-ionized carbon, C^{5+}, would have only a single electron. Its ground state would be designated 1s, the same as hydrogen; this ion is therefore referred to as hydrogen-like or H-like.

3.6 ENERGY LEVELS OF MULTI-ELECTRON ATOMS

ENERGY-LEVEL DESIGNATION FOR
MULTI-ELECTRON STATES

The energies of the electronic levels or states, as mentioned in the previous paragraph, are determined primarily by the particular electrons that exist in the unfilled subshells. For hydrogen (and for hydrogen-like ions) this creates no problem, because we already know how to solve the Schrödinger equation for one electron circulating around a nucleus. For two or more electrons in unfilled subshells the solutions are much more complicated, so we will not concern ourselves here with how to obtain such solutions. However, it is important to understand how the energy levels evolve and how they are designated (labeled), so that potential transitions between energy levels can be identified and understood within a simple framework.

For the hydrogen atom, we used both the kinetic and potential energy of the electron in the Hamiltonian when determining the values of bound energy states or levels of the atom. This led to a dependence of the energy on the value of the quantum number n only. We later suggested that the electron spin and orbital angular momenta interact to produce two different possible energy states for a given value of l. There are a number of possible interactions that must be included in the Hamiltonian in order to obtain a more exact solution for the possible energy levels of atoms, especially for multi-electron atoms. These interactions are listed as follows.

(1) Kinetic energy of the electrons – the energy associated with the electron velocities.
(2) Electrostatic interaction energy of the electrons with the nucleus – for hydrogen, this was the coulomb potential between the proton and the electron.
(3) Mutual electrostatic energy of the electrons, or the electrons being repulsed by each other.
(4) Spin–orbit interaction – the alignment of the electron spin angular momentum with respect to the orbital angular momentum.
(5) Spin–spin interactions of the electrons – the perturbations caused by the spin of one electron interacting with the spin of another electron.
(6) Interaction between the orbital magnetic moments of different electrons.
(7) Interactions between the electron spins and the nuclear spin.
(8) Nuclear spin interaction with the orbital angular momentum of the electrons.
(9) Relativistic effects.
(10) An "exchange correlation" that tends to align electronic spins parallel to each other.

Items 1 and 2 were used for the solution of the hydrogen atom earlier in this chapter. For most atoms, in addition to the kinetic and potential energies (items 1 and 2), the residual electrostatic energy (item 3) and the spin–spin correlations (item 10) are the largest. The technique that has been the most successful in determining energy levels when such interactions are considered is referred to as *Russell–Saunders* or *LS coupling*. For some atoms, mainly heavier atoms, the spin–orbit energy (item 4) dominates; that condition is referred to as *j–j* coupling.

We must be aware of these different couplings and their designations in order to understand the notation for various energy levels and to appreciate why there are so many different energy levels. In Chapter 4 we will learn how to predict when electronic transitions can occur between certain levels by learning selection rules that relate to the specific energy-level designations.

RUSSELL–SAUNDERS OR *LS* COUPLING – NOTATION FOR ENERGY LEVELS

Because the *LS* coupling model assumes that the electrostatic interaction (item 3) dominates for atoms with two electrons in unfilled subshells, the orbital angular momenta of the two electrons couple with each other such that $\mathbf{L} = \mathbf{L}_1 + \mathbf{L}_2$. Separately, the spins also couple such that $\mathbf{S} = \mathbf{S}_1 + \mathbf{S}_2$. Thus the total angular momentum, represented by $\mathbf{J}$, is the vector sum $\mathbf{J} = \mathbf{L} + \mathbf{S}$. The radiating properties of a level, and to some extent the relative energy of a level, can be deduced from the spectroscopic level notation associated with the quantum numbers of $\mathbf{L}$, $\mathbf{S}$, and $\mathbf{J}$. This notation is

$$^{2S+1}L_J, \tag{3.75}$$

where S is the total spin quantum number, L is the total orbital angular momentum quantum number, and J is the total angular momentum quantum number of that particular level or state. This notation is applicable both to one-electron and multi-electron atoms. *When we use J, L, and S without the* (**bold**) *vector notation, we mean the quantum number instead of the actual magnitude of the angular momentum.* (The magnitude of, e.g., the angular momentum is designated $|\mathbf{J}|$.) In so doing we use designations as they have historically evolved for the classification of atomic energy levels. The total orbital angular momentum quantum number L carries the same historical classification that the individual ls do in the sense that S is used for $L = 0$, P for $L = 1$, D for $L = 2$, and so on:

$$\begin{array}{llllllll} L: & 0 & 1 & 2 & 3 & 4 & 5 & 6 \ldots, \\ \text{term designation:} & \text{S} & \text{P} & \text{D} & \text{F} & \text{G} & \text{H} & \text{I} \ldots. \end{array} \tag{3.76}$$

We must wait until the next chapter to appreciate the usefulness of this designation for energy levels.

When energy levels involving more than one electron are split into more than one sublevel (owing to various interactions of the electrons with each other and with the nucleus), the number of component energy levels associated with the splittings is called the *multiplicity* of the levels. The actual multiplicity is given by either $2S+1$ or $2L+1$, whichever is smaller. In most cases $2S+1$ is smaller than $2L+1$, so the superscript $2S+1$ in the term designation of (3.75) generally indicates the multiplicity of that level. Thus, for example, a $^1D_2$ level would indicate that the total orbital angular momentum has a value of 2, the total spin angular momentum has a value of 0, and the total angular momentum has a value of 2 (which is obtained by adding $2+0=2$). In this case the multiplicity is 1, indicating only one sublevel. Such a level could result, for instance, from a d electron combining with an s electron in such a way that the spins of the two electrons are opposite (antiparallel); in this case, the orbital angular momentum is determined by that of the d electron only. We will see how to determine the energy-level designations for various multi-electron configurations in the next section.

ENERGY LEVELS ASSOCIATED WITH TWO ELECTRONS IN UNFILLED SHELLS

We have seen that when two electrons in unfilled shells interact with each other it leads to a number of energy sublevels. Radiative transitions can occur from these various sublevels for a given n value, to various sublevels having a lower value of the n quantum number. Since we examine such transitions in the next chapter, it is important to understand how the splitting of the energy levels into these various sublevels occurs. We will first consider the simplest example of a two-electron system, that of helium. In its ground state we have two 1s electrons, or $1s^2$. For this system $l=0$ for both electrons, and the only way the spins can align is antiparallel (according to the Pauli exclusion principle). Thus $S=0$ for the combined electrons and $L=0+0=0$ for the orbital angular momentum. This also makes $J=0$, since $J=L+S$. Therefore the term designation for the ground state of helium, according to (3.75), is $^1S_0$.

Designations for the excited states of helium become a little more complicated, since there are many possible ways of combining the electrons. Let us assume that one of the electrons is excited to the 2s level, yielding a configuration of 1s2s ($n=1$ for the first electron and $n=2$ for the second electron). This level also has $l=0$ for both electrons so that $L=0+0=0$, and the orbital angular momentum term designation would be S for $L=0$, according to (3.76). In this case, however, the spins can be either parallel or antiparallel; the Pauli exclusion principle does not apply, because each electron has a different value of n. We therefore have two possible states for the spin: $S=\frac{1}{2}+\frac{1}{2}=1$ and $S=\frac{1}{2}-\frac{1}{2}=0$. In the first case ($S=1$), the multiplicity (number of levels) is 1 even though $2S+1=3$, since the multiplicity is determined by whichever is the smaller of $2S+1$ and $2L+1$. The second case ($S = 0$) leads to a multiplicity of 1. We thus have two possible states, $^3S_1$ and $^1S_0$, for the electron configuration 1s2s. When the superscript has a value of 3 we refer to it as a triplet, whereas for a value of 1 it is called a singlet. Similarly, for other numbers and combinations of electrons we might have doublets, quartets, et cetera.

For the excited state 1s2p of helium, $l = 0$ for one electron and $l = 1$ for the other electron. This leads to only one possible value for L of $L = 0 + 1 = 1$ and so to an angular momentum designation of P, according to (3.76). The spins combine with two possibilities, as just discussed for the configuration 1s2s, leading to total spin quantum numbers of 1 and 0. Thus, for $S = 1$ and $L = 1$, we can have values of $J = 1 + 1 = 2$ (spin and orbital aligned), $J = 1 + 0 = 1$ (spin perpendicular to orbital), and $J = 1 - 1 = 0$ (spin antiparallel to orbital). This is summarized in Figure 3-10(a).

For reasons beyond the scope of our present discussion, the possible values of J vary by integral values. Thus we always start with the maximum value of J ($J = L + S$), and decrease it by 1 for each state until we reach $J = L - S$. Thus for $L = 1$ and $S = 1$ we have the states $^3P_2$, $^3P_1$, and $^3P_0$. For $L = 1$ and $S = 0$ we have $J = L + S = 1 + 0 = 1$, leading to a single state of $^1P_1$. All of these levels of helium are diagrammed in Figure 3-10(b), where we show the singlets and triplets in separate columns. We do so because there tends to be very little interaction between singlets and triplets, as we

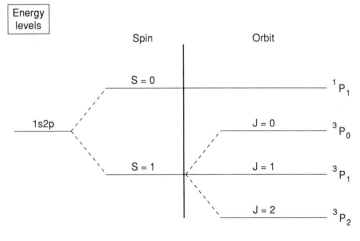

Helium

1s ($l = 0$, $s = \frac{1}{2}$)   2p ($l = 1$, $s = \frac{1}{2}$)

$L = l + l = 0 + 1 = 1$ P states

$S = s_1 + s_2 = 1$

$S = s_1 - s_2 = 0$

$S = 1$ $J = \bar{L} + \bar{S} = 1 + 1 = 2 \Rightarrow {}^3P_2$

$= 1 + 0 = 1 \Rightarrow {}^3P_1$

$= 1 - 1 = 0 \Rightarrow {}^3P_0$

$S = 0$ $J = \bar{L} + \bar{S} = 1 + 0 = 1 \Rightarrow {}^1P_1$

Energy levels

Spin   Orbit

$S = 0$ $^1P_1$

1s2p

$J = 0$ $^3P_0$

$S = 1$ $J = 1$ $^3P_1$

$J = 2$ $^3P_2$

Figure 3-10(a). Singlet and triplet energy levels of the 1s2p state of the helium atom

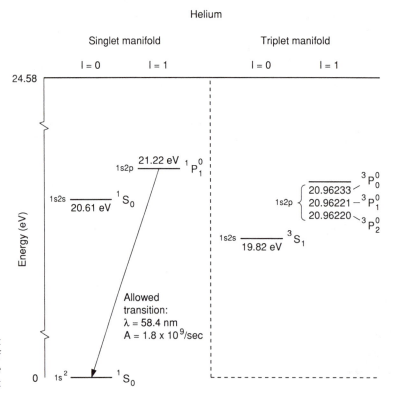

Figure 3-10(b). The first several energy levels of helium, along with the ionization limit

shall find out in Section 4.4 when we consider radiative transition probabilities between various energy levels. We show the approximate relative values of the energies for these various levels without actually deriving them, since the derivation is too complicated to be included in this text. The value of the energy for the triplet levels of equivalent electrons is almost always lower than that for the singlet, as can be seen in Figure 3-10(a). We will see later that these combinations of singlet and triplet levels will occur in the energy levels associated with organic dye solutions, and will play an important role in the operation of a dye laser. The first several energy levels of helium are shown in Figure 3-10(b) (not to scale). The only allowed transition from any of the energy levels associated with $n = 2$ to the single level with $n = 1$ is the transition $^1P_1 \rightarrow {}^1S_0$, as shown on the diagram. Determining the probability of such transitions occurring between any pair of levels will be examined in the next chapter.

We now consider the case of neutral carbon, in which two 2p electrons are in an unfilled shell. Remember that the ground state for carbon is $1s^2 2s^2 2p^2$ or simply $2p^2$. Thus, in order to determine the structure of this state, we must consider how these two 2p electrons combine. We will first consider a more general case, where one of the electrons is excited to the 3p state so that we have an arrangement of 2p3p. Figure 3-11 shows the splitting of the resulting energy levels according to LS coupling. The two different total values of spin, $S = 0$ and $S = 1$, have the largest effect on the

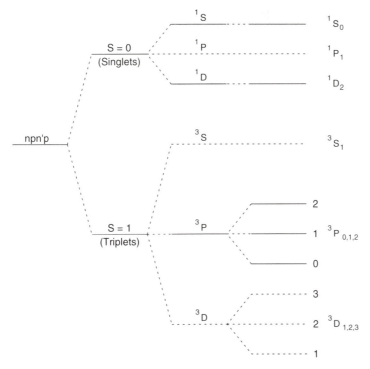

Figure 3-11. Energy levels associated with two p electrons, where the dashed lines indicate the levels that are absent when the two p electrons are equivalent (have the same value of n)

splitting of the energy levels that yields singlet and triplet states. Next, the splitting of the L values occurs to produce S, P, and D states, and then the spin and orbital numbers are combined to give all of the allowed J values on the extreme right. Now for the slightly more complicated case of two equivalent p electrons, or 2p2p, we must *exclude* $^3D_{1,2,3}$, $^1P_1$, and $^3S_1$ states owing to the Pauli exclusion principle. This is a general rule that always applies for two identical p electrons. The states that are excluded are shown as dashed lines in Figure 3-11.

EXAMPLE

Assume that one of the 2p electrons of carbon is excited to the 3d state. The unfilled outer shell would then consist of one 2p and one 3d electron, a 2p3d state. Obtain the splitting and designation for all of the energy levels from this combination of 2p3d electrons surrounding the closed shell and subshell of $1s^2 2s^2$.

Let us first consider the electron spins. For this situation, the two electrons can have their spins aligned either parallel or antiparallel. If they are parallel then $S = \frac{1}{2} + \frac{1}{2} = 1$; if they are antiparallel then $S = \frac{1}{2} - \frac{1}{2} = 0$. Thus the state splits into two spin values of $S = 1$ and $S = 0$. This is shown as the first splitting in Figure 3-12. The quantum numbers for the orbital angular momenta of the two electrons are 1 and 2 for p and d, respectively. Thus the total orbital angular momentum

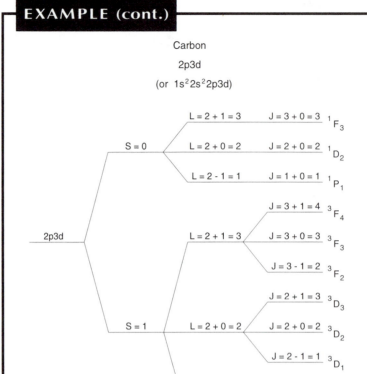

Carbon

2p3d

(or $1s^2 2s^2 2p3d$)

Figure 3-12. Energy levels of the carbon atom associated with the electron configuration of 2p3d

can take on values ranging from $2+1$ to $2-1$ (changing in integral values) – that is, 3, 2, and 1 – which are decoded as F, D, and P states from (3.76). This splitting of orbital angular momentum is shown as the next splitting in Figure 3-12.

Finally, the total angular momentum is equal to $L+S$ for each of the spin states. For $S = 0$, there are only three states as shown, each with the same value for J and L since $S = 0$. For $S = 1$, there are three J values for each L value, where J ranges from $L+S$ to $L-S$. Thus, for $L = 1$, $J = 0$, 1, and 2; for $L = 2$, $J = 1$, 2, and 3; and for $L = 3$, $J = 2$, 3, and 4. This total splitting is shown as the final splitting in Figure 3-12. Thus, what might originally have been considered one energy level – that associated with the quantum numbers $n = 2$ and $n = 3$ for the two electrons – was actually split into 12 levels owing to the various interactions between the two electrons.

j–j COUPLING

Using the *j–j* coupling scheme for two electrons involves, first, coupling the spin and orbit values so that $\mathbf{J}_1 = \mathbf{L}_1 + \mathbf{S}_1$ and $\mathbf{J}_2 = \mathbf{L}_2 + \mathbf{S}_2$ for each of the two electrons. Then the total angular momentum is obtained by combining $\mathbf{J}_1$ and $\mathbf{J}_2$ such that $\mathbf{J} = \mathbf{J}_1 + \mathbf{J}_2$. The quantum numbers would therefore take on values ranging from $|j_1 + j_2|$ to $|j_1 - j_2|$. The reader is referred to references at the end of this chapter and other books on spectroscopy for further details about *j–j* coupling. We will use *LS* notation for energy levels throughout the rest of this text since *LS* coupling is more common for laser transitions.

ISOELECTRONIC SCALING

The occurrence of similar energy-level relationships among atoms having identical electron configurations but different quantities of nuclear charge is referred to as *isoelectronic scaling* of energy levels. A graph of energy-level relationships for H atoms, He$^+$ ions, and Li^{++} and other ions, indicating the transition from $n = 3$ to $n = 2$ in each case, is shown in Figure 3-13; this is referred to as an *isoelectronic sequence* of atoms. It can be seen that, for higher values of Z, the wavelengths of the transition from $n = 3$ to $n = 2$ become increasingly shorter, corresponding to the greater separation of energy levels mandated by (3.13). Such isoelectronic scaling

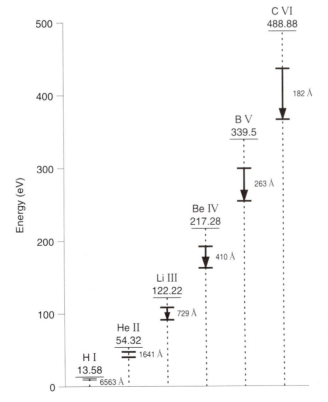

Figure 3-13. Isoelectronic sequence of the hydrogen atom along with hydrogen-like helium, lithium, beryllium, boron, and carbon

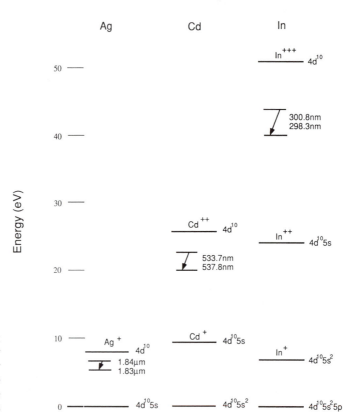

Figure 3-14. Isoelectronic sequence of laser transitions in neutral silver atoms, cadmium single ions, and indium double ions

will be used extensively in discussing X-ray lasers (see Section 13.10). As mentioned previously, the greatest separation between energy levels for a neutral atom is 24.6 eV for helium, which would correspond to a wavelength of 50.4 nm; this is not an X-ray wavelength. Therefore, in order to make X-ray wavelengths available from transitions between bound states, one must remove (strip) electrons from an atom until one obtains an ionic species of that atom that contains energy-level separations – and consequently wavelengths – that are available for making X-ray lasers. Hydrogen and helium do not have energy levels that could lead to X-ray wavelengths for transitions, and are therefore not candidates for X-ray lasers.

Another example of an isoelectronic sequence is shown in Figure 3-14, which includes some of the energy levels of silver, cadmium, and indium. In this situation the energy levels of neutral silver, singly ionized cadmium, and doubly ionized indium have exactly the same total electronic configuration and thus have similar energy-level arrangements and designations. However, the separation between levels, and consequently the wavelength for transitions between the identical electronic configurations, is shorter with increasing ionization stage, as shown in Figure 3-14.

REFERENCES

C. Cohen-Tannoudji, B. Diu, and F. Laloe (1977), *Quantum Mechanics,* vol. 1. New York: Wiley.

R. B. Leighton (1959), *Principles of Modern Physics.* New York: McGraw-Hill.
L. I. Schiff (1955), *Quantum Mechanics,* 2nd ed. New York: McGraw-Hill.

PROBLEMS

1. Using the Bohr theory, carry out the derivation of the electron orbit radius r (eqn. 3.4) and the orbital energy E_n (eqn. 3.8) for the hydrogen atom.

2. Compute the wavelengths for the following transitions in the hydrogen atom, using the Bohr theory:

(a) $n = 2$ to $n = 1$;
(b) $n = 3$ to $n = 1$;
(c) $n = 4$ to $n = 2$;
(d) $n = 5$ to $n = 3$;
(e) $n = 5$ to $n = 4$.

3. Compute the ionization potential for five-times ionized carbon (C^{5+}) and also compute the wavelength of the transition from $n = 3$ to $n = 2$ for that species.

4. In the isoelectronic scaling of hydrogen-like species, what is the lowest-atomic number element that would have a possible X-ray laser transition at a wavelength shorter than 4.4 nm? (Here, consider the transition from $n = 3$ to $n = 2$ as the laser transition.)

5. Solve the Schrödinger equation for the energy eigenvalue of the ground state of the hydrogen atom, assuming a solution of the form $\psi_1(r) = Ce^{-\alpha r}$ (eqn. 3.38).

6. Write out the wave function for the hydrogen atom for $n = 3$, $l = 2$, and $m = +1$.

7. Obtain the possible term designations for a hydrogen atom in which the electron is in the 2d level.

8. Give the electron configuration for a lithium atom in its ground state, and write out the term designation for that state.

9. List the electron configurations for the following elements in their ground states: aluminum, fluorine, calcium, germanium, and krypton.

10. Determine the number of states and write out the term designations (eqn. 3.75) for those states that result from an outer electron configuration of 2s4p.

4

RADIATIVE TRANSITIONS AND EMISSION LINEWIDTH

SUMMARY In Chapter 3 we provided a theoretical basis for the existence of discrete, bound, negative energy levels of atoms in which the lowest energy level was referred to as the ground state. When hydrogen atoms were excited in a low-pressure gaseous discharge, experimental observation of radiative emission at various discrete wavelengths was found to agree with the predicted wavelengths determined by computing differences in energy between the theoretically derived negative energy levels. This provided strong justification for concluding that atoms radiate electromagnetic energy when an excited electron jumps from an upper energy state (u) to a lower energy state (l), thereby emitting a photon of frequency ν_{ul} and wavelength λ_{ul} corresponding to the exact energy difference ΔE_{ul} between the two levels such that $\Delta E_{ul} = h\nu_{ul} = hc/\lambda_{ul}$.

Some of the characteristics of radiating atoms can be most easily visualized if classical theory is used rather than the quantum theory. We will first consider radiation from a specific level u from the standpoint of how it affects the decay in population residing in that level. We will also consider the increased decay rate from that level due to interaction with nearby atoms. We will then calculate classically the decay in population due to the radiating electron, assuming that the electron is oscillating sinusoidally at a specific frequency and radiating electromagnetic radiation at that frequency. We will show that the radiation intensity coming from the electron will decrease exponentially, with a decay time of the order of τ_u, due strictly to the loss of energy of the electron by the emission of that radiation.

Broadening of the emission wavelength will be shown to be an inevitable consequence of the decay of population from an excited level. The emission linewidth of the radiation resulting from a transition from level u to level l will be shown quantum mechanically to be associated with the radiative decay from both levels u and l. This linewidth is referred to as the natural linewidth $\Delta\nu_{ul}^N$. The shape of the emission linewidth will be shown to be either Lorentzian (for homogeneous broadening) or Gaussian (for most forms of inhomogeneous broadening). We will discuss how collisions while the species is radiating can reduce the decay time of the emission and cause increased broadening of the emission linewidth. Another form of interaction with the radiating species merely interrupts its phase, leading to broadening without increasing the decay rate. These effects can be associated with interactions (collisions) of other atoms or electrons with a species and its oscillating electron, as in a gas (e.g. a gas laser), or phonons in a crystal matrix (such as a solid-state laser material), or in a liquid solvent (such as an organic dye laser medium).

Thermal motion will be shown to produce Doppler broadening, particularly in gaseous laser media. Also, if the gain medium is composed of several isotopic species of a material, there may be frequency broadening due to the isotopic shifts in the energy levels for the different isotopic species.

We will conclude this chapter with a quantum mechanical analysis of how radiative transitions occur between energy levels, and how we can estimate the probability for such transitions. We will then obtain the selection rules for predicting when transitions will occur between pairs of levels. This will allow us to predict which energy levels might be favorable for making lasers. We will show the relationship between the transition probability and the oscillator strength of a transition, since the concept of oscillator strength is useful in comparing emission strengths of various transitions of different wavelengths.

4.1 DECAY OF EXCITED STATES

If we consider a group of atoms at room temperature, all of the electrons essentially reside in the ground state of those atoms. If energy is applied to the atoms – in the form of particle collisions, electric current, or absorbed light – then the outer electrons, those with the highest quantum numbers, are "pumped" to excited energy states or levels (we generally use the words "level" and "state" interchangeably throughout the text). These excited electrons eventually decay back down to the ground state when the energy source or pumping source is removed. This section will discuss the many processes that determine the decay of those electrons from higher energy states to lower energy states.

RADIATIVE DECAY OF EXCITED STATES OF ISOLATED ATOMS – SPONTANEOUS EMISSION

Let us assume that a group of identical atoms are in their ground levels, and that some form of excitation pulse is applied to transfer many of them rapidly to an excited energy level u. The population density in level u is expressed as N_u (number per unit volume). If we were to measure N_u as a function of time after the excitation pulse, we would observe that it typically decays with an exponential decay rate, as shown in Figure 4-1(a). This population is transferred to lower energy levels, which could be the ground state or other low-lying energy levels. In the simplest case, let the population decay to a single lower energy level l, as shown in Figure 4-1(b). If N_u

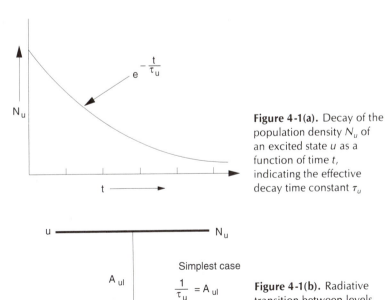

Figure 4-1(a). Decay of the population density N_u of an excited state u as a function of time t, indicating the effective decay time constant τ_u

Figure 4-1(b). Radiative transition between levels u and l, with its associated transition probability A_{ul}

is low enough (the atoms are sufficiently isolated from each other) such that none of the excited atoms collide while the population in level u is decaying, then the only way the energy loss $\Delta E_{ul} = E_u - E_l$ can be accounted for – and still conserve energy and momentum – is if radiation is emitted during the decay process. This radiative decay process is referred to as *spontaneous emission,* since it is a natural process that occurs without any external stimulus.

Each "photon" radiated when one of the atoms decays would have a frequency of $\nu_{ul} = (E_u - E_l)/h$ and a wavelength of $\lambda_{ul} = c/\nu_{ul}$, where c is the velocity of light in a vacuum. If the photons are emitted in a medium such as a gas, the measured wavelength would be shifted to a value of $\lambda_{ul} = c/\eta\nu_{ul}$, where η is the index of refraction of the medium, because the velocity v of the photons (light waves) would be reduced to a value of $v = c/\eta$. For a gas, at most laser wavelengths the index of refraction is near unity and thus the wavelength shift is very slight.

We can write a simple equation to express the change of the population density N_u as the population is transferred to level l:

$$\frac{dN_u}{dt} = -A_{ul}N_u, \tag{4.1}$$

where A_{ul} is the rate at which the population is being transferred. A_{ul} has units of 1/time, and is referred to as the *radiative transition rate* or radiative transition probability. The solution to (4.1) is

$$N_u = N_u^0 e^{-A_{ul}t}, \tag{4.2}$$

where N_u^0 is the initial population density in level u at the time that the excitation pulse occurs. Equation (4.2) indicates that the population density in level u decays exponentially with a rate A_{ul}, as we suggested previously.

We can define a time τ_u according to the following relationship:

$$N_u = N_u^0 e^{-t/\tau_u}. \tag{4.3}$$

Such an exponential decay was shown in Figure 4-1. From (4.3) we can see that when $t = \tau_u$ the population N_u has decayed to $1/e$ of its original value. This time τ_u is referred to as the *lifetime* of level u. Comparing (4.2) with (4.3) suggests that $1/\tau_u = A_{ul}$.

For the more general case, in which the population in level u decays radiatively to several lower-lying levels, the expression for the decay is written as

$$\frac{dN_u}{dt} = -(A_{ui} + A_{uj} + A_{uk} + \cdots)N_u = -\left(\sum_i A_{ui}\right)N_u, \tag{4.4}$$

which has as its solution

$$N_u = N_u^0 \exp\left\{-\left(\sum_i A_{ui}\right)t\right\}. \tag{4.5}$$

Comparing (4.3) and (4.5) yields

$$\tau_u = \frac{1}{\sum_i A_{ui}}. \tag{4.6}$$

Equation (4.6) indicates that the lifetime τ_u of energy level u is completely determined by the reciprocal of the sum of all the possible radiative decay rates to lower-lying levels.

EXAMPLE

The upper and lower levels of two common atomic laser transitions are shown in Figure 4-2 along with other lower levels to which the upper level can radiatively decay. These are the 325.0-nm and 353.6-nm laser transitions in cadmium vapor (associated with the helium–cadmium laser) and the 632.8-nm red laser transition in neon (associated with the helium–neon laser). Compute the lifetime of the upper energy levels involved in these laser transitions. The wavelengths and transition probabilities are shown next to each transition originating from the upper laser level for both cadmium and neon.

(a) The upper laser level of the two ultraviolet He–Cd lasers decays radiatively via the two laser transitions that occur at 325.0 nm (with a transition probability of $A = 7.8 \times 10^5$ s^{-1}) and at 353.6 nm (with $A = 1.6 \times 10^5$ s^{-1}), as shown in Figure 4-2(a). Therefore, using (4.6) to compute the lifetimes, we have

$$\tau_u = \frac{1}{\sum_i A_{ui}} = \frac{1}{7.8 \times 10^5 \text{ s}^{-1} + 1.6 \times 10^5 \text{ s}^{-1}}$$

$$= \frac{1}{9.40 \times 10^5 \text{ s}^{-1}} = 1.06 \times 10^{-6} \text{ s}.$$

(b) The upper laser level of the red He–Ne laser decays radiatively via ten transitions as shown in Figure 4-2(b), including the well-known

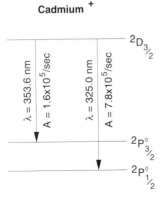

Figure 4-2(a). Two transitions of the He–Cd laser, along with their associated radiative transition probabilities

EXAMPLE (cont.)

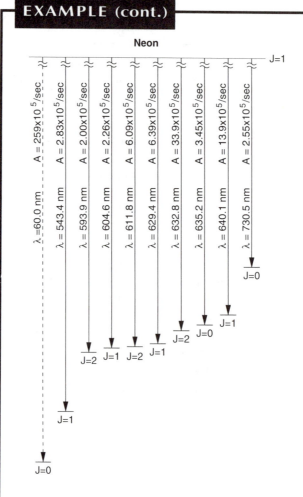

Figure 4-2(b). Radiative transitions from the upper laser level of the 632.8-nm helium–neon laser transition, along with the relevant radiative transition probabilities

632.8-nm transition. Using (4.6) and the calculated transition probabilities shown in the figure, we have

$$\tau_u = \frac{1}{\sum_i A_{ui}}$$

$$= \frac{1}{(259 + 2.55 + 13.9 + 3.45 + 33.9 + 6.39 + 6.09 + 2.26 + 2.00 + 2.83) \times 10^5 \text{ s}^{-1}}$$

$$= \frac{1}{332.37 \times 10^5 \text{ s}^{-1}} = 3.01 \times 10^{-8} \text{ s} = 30.1 \text{ ns}.$$

In this example, the dominant decay occurs on the short wavelength transition to the neon ground state at 60 nm. We could therefore have made a reasonable approximation of the lifetime of 38.6 ns simply by including the transition probability from that strong transition. Including all of the other nine transitions decreased the lifetime by another 8 ns, or 22%. *Note:* There is also an eleventh transition (in fact a laser transition) from this level, occurring at 3.39 μm, that we are ignoring since it has a very low transition probability.

We can see that the laser levels of these two examples have lifetimes that are in the submicrosecond region. The neon upper laser level lifetime is much shorter than the Cd upper laser level lifetime because the radiative decay from the cadmium level involves a change in quantum numbers of two of the electrons ($4d^9 5s^2 \rightarrow 4d^{10} 5p$), leading to a much smaller transition probability than that for neon, whose decay involves a change in only one electron ($2p^5 5s \rightarrow 2p^6$).

The spontaneous emission rate A_{ul} from level u to level l is referred to as either the radiative transition probability or the spontaneous transition probability (or simply the transition probability). Every atom transferred to an excited energy level will eventually decay by spontaneous emission if left isolated for a sufficient time. The rate at which radiative decay occurs is unique to each transition. The exact value of the transition probability A_{ij} for a transition from level i to j can be calculated by using the quantum mechanical techniques summarized later in this chapter.

SPONTANEOUS EMISSION DECAY RATE – RADIATIVE TRANSITION PROBABILITY

We will later obtain an explicit expression for the radiative transition probability A_{ul} in terms of various atomic parameters. For now, we will use a simplified expression that includes only the absorption oscillator strength f_{lu}, the wavelength λ_{ul}, and the statistical weights g_u and g_l of (resp.) the upper and lower energy levels of the transition:

$$A_{ul} = \frac{10^{-4}(f_{lu})}{1.5(g_u/g_l)\lambda_{ul}^2}. \tag{4.7}$$

This expression is applicable only if the various parameters are used with the correct units. (A_{ul} is in units of 1/seconds, λ_{ul} is expressed in meters, and f_{lu}, g_u, and g_l are dimensionless numbers.) The value of the oscillator strength f_{lu} is unique for each specific transition, and in general $f_{lu} \leq 1$. For most radiative transitions $f_{lu} \ll 1$, and for most laser transitions $0.1 > f_{lu} > 0.01$. Some solid-state laser transitions, such as that for the ruby laser, have oscillator strengths f_{lu} ranging approximately from 10^{-4} to 10^{-5}. Later in this chapter, we will show how quantum mechanics is used explicitly to calculate values of probabilities and oscillator strengths for various transitions. It is useful at this point to obtain a general idea of the value of A_{ul}

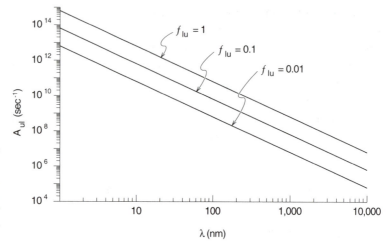

Figure 4-3. Plot of the radiative transition probability as a function of wavelength for several values of oscillator strength f (assuming $g_1 = g_2 = 1$)

for various wavelengths and oscillator strengths. It can be seen from (4.7) that, for a given oscillator strength, A_{ul} is strongly dependent upon the wavelength λ_{ul}. Figure 4-3 shows a plot of the radiative transition probability versus wavelength for values of oscillator strength of 1.0, 0.1, and 0.01 (eqn. 4.7), where we have assumed that $g_u/g_l \approx 1$. The ratio g_u/g_l will vary from 0.5 to 2 for most transitions, so the graphs of Figure 4-3 provide good estimates for f_{lu}.

LIFETIME OF A RADIATING ELECTRON – THE ELECTRON AS A CLASSICAL RADIATING HARMONIC OSCILLATOR

We know from electromagnetic theory that a charged particle oscillating back and forth radiates electromagnetic waves. We will therefore use this model to obtain an expression for the decay time of an electron residing in an excited energy level as it makes the transition to a lower level. Use of such a model is justified because we know that electromagnetic radiation is emitted from atoms, we can measure the frequency ω_0 of that radiation, and we know from observations of emission spectra that an electron jumping from one orbit to another can produce radiation at that frequency. We need not know the exact nature of the radiation process within the atom to obtain a classical expression for the decay time of the radiating electron. We will use an electron model that is similar to that used in Chapter 2 for simulating the interaction of electromagnetic radiation with a dielectric solid material. In the present situation, we consider the electron to be radiating at its oscillating frequency.

We assume that such an electron of mass m_e is oscillating at an angular frequency ω_0 with the constraints of a classical harmonic oscillator. Suppose the electron is displaced from an equilibrium position by an amount x_0, and has a restoring force due to the coulomb attractive force of the positively charged nucleus that is proportional (with proportionality constant k) to the displacement of the electron from the nucleus. It thus travels

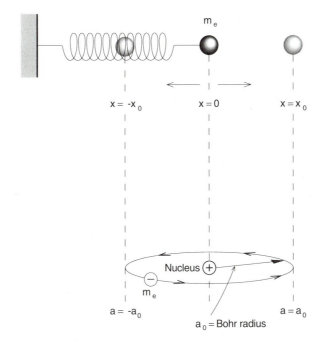

Figure 4-4. A classical radiating–harmonic oscillator model of an electron circulating around a nucleus

initially through an amplitude excursion from $+x_0$ to $-x_0$, as shown in Figure 4-4. When the electron is displaced from the nucleus, it forms a dipole. We can obtain the equation of motion for this dipole by equating the restoring force $-kx$ to the product of the mass m_e and the acceleration d^2x/dt^2. This leads to the following equation of motion:

$$m_e \frac{d^2x}{dt^2} + kx = 0. \tag{4.8}$$

The solution for the oscillatory motion of the electron is $x = x_0 e^{-i\omega_0 t}$, with an oscillatory frequency of

$$\omega_0 = \left(\frac{k}{m_e}\right)^{1/2} = 2\pi\nu_0. \tag{4.9}$$

Equation (4.9), wherein ν_0 is the frequency at which the oscillation makes one complete cycle, leads to a value for the "spring constant" $k = 4\pi^2\nu_0^2 m_e$.

We know also that the solution for the total energy E_T of a harmonic oscillator can be expressed as the sum of the kinetic and potential energies: $E_T = \frac{1}{2}m_e v^2 + \frac{1}{2}kx^2$. When the harmonic oscillator is at the extreme turning points during its oscillating cycle ($x = \pm x_0$), the velocity v at those points is zero and thus the total energy may be written as

$$E_T = \frac{1}{2}kx_0^2 = 2\pi^2\nu_0^2 m_e x_0^2. \tag{4.10}$$

In this simple model we have neglected the loss in energy due to the power radiated by the oscillating electron. The power P_R radiated from a classical dipole (representing the oscillating electron) can be found in any book on electricity and magnetism; it is obtained from the Poynting vector discussed in Section 2.2. This power can also be thought of as the loss of energy per unit time, and is given by

$$P_R = -\frac{dE_T}{dt} = \frac{16\pi^3 v_0^4 (ex_0)^2}{3\epsilon_0 c^3} = \frac{\omega_0^4 (ex_0)^2}{3\pi\epsilon_0 c^3}. \tag{4.11}$$

In this equation, e is the charge of the electron and ϵ_0 is the permittivity of a vacuum. Comparing (4.10) and (4.11), we can express (4.11) in the form

$$\frac{dE_T}{dt} = -\gamma_0 E_T, \tag{4.12}$$

which is a reasonable expression in that it suggests that the radiated energy is proportional to the instantaneous energy at a given time. Here γ_0 is the proportionality factor, representing the rate at which energy is lost. The solution for E_T from (4.12) is

$$E_T = E_T^0 e^{-\gamma_0 t} = E_t^0 e^{-t/\tau_0}, \tag{4.13}$$

which is an exponentially decreasing function of time that decays with a time constant τ_0 such that

$$\tau_0 = 1/\gamma_0. \tag{4.14}$$

It is possible to solve for an explicit expression of γ_0 from equations (4.9), (4.11), and (4.12), but we will not do so at this time. Later in this chapter we will solve this expression quantum mechanically to obtain the correct expression for γ_0, which we will refer to as the *radiative decay rate A_{ul}* as defined in (4.7).

We will now perform a somewhat more careful analysis to show that the envelope of the electromagnetic wave decays with an overall rate γ_0 (with a corresponding decay time of τ_0), while oscillating at a much higher frequency ω_0 within that envelope. We begin by rewriting the equation of motion for the oscillating electron to include a damping factor γ_0, representing the radiated energy loss that is proportional to the velocity. This more closely represents the realistic situation of an oscillating electron. The revised equation for electron motion then becomes

$$m\frac{d^2x}{dt^2} + \gamma_0 \frac{dx}{dt} + 4\pi^2 v_0^2 m_e x = 0, \tag{4.15}$$

where we have replaced k with the value of $4\pi^2 v_0^2 m_e = \omega_0^2 m_e$ as the coefficient for x, and where γ_0 is the decay rate derived previously. In (4.15), the solution for x (based upon initial conditions in which the oscillator is displaced a distance x_0 and then released) and for $\gamma_0 \ll \omega_0$ has the form of a damped oscillation similar to that obtained for the decaying electric field in Chapter 2 (eqn. 2.79):

$$x \cong x_0 e^{-\gamma_0 t/2} e^{-i\omega_0 t}. \tag{4.16}$$

It follows from electromagnetic theory that the electric field $E(t)$ radiated by that electron has the same time dependence and can thus be expressed as indicated in (4.17) and also as illustrated in Figure 4-5:

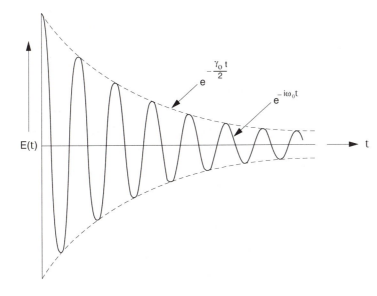

Figure 4-5. Time-dependent decay of the electric field produced by an oscillating electron when the loss of energy due to radiation is taken into account

$$E(t) = \begin{cases} E_0 e^{-\gamma_0 t/2} e^{-i\omega_0 t} & \text{if } t \geq 0, \\ 0 & \text{if } t < 0. \end{cases} \qquad (4.17)$$

This formula assumes that the electron begins to radiate at a time $t = 0$ and that the subsequent decay of the electric field occurs with an exponential decay rate of $\gamma_0/2$. The decay of the intensity of the radiation $I(t)$, which is equal to $|E(t)|^2$, would be

$$I(t) = \begin{cases} |E(t)|^2 = E_0^2 (e^{-(\gamma_0/2)t})^2 = I_0 e^{-\gamma_0 t} & \text{if } t \geq 0, \\ 0 & \text{if } t < 0; \end{cases} \qquad (4.18)$$

not surprisingly, this formula has the same time dependence as that of the energy decay in (4.13). This radiation process can be described as spontaneous emission, since it is initiated with no external stimulus.

NONRADIATIVE DECAY OF THE EXCITED STATES – COLLISIONAL DECAY

Atoms are seldom isolated, and in many cases the population in excited states will interact (collide) with the surrounding atoms. These collisions can cause the excited atoms to make a transition to a lower level without radiating, thereby decreasing the decay time of the excited energy level. The energy lost by the excited atom due to this collisional decay is given to the colliding atom in order to conserve energy and momentum. In some instances a collision will not produce decay of the excited level but instead will interrupt the phase of the radiating atom. This effect will produce a broadening of the emission frequency, as described in Section 4.2.

A general term for the decay rate of an excited state (level) u is the symbol γ_u, which has units of 1/seconds. The general expression for the decay of the excited state u would thus be

$$N_u = N_u^0 e^{-\gamma_u t}, \tag{4.19}$$

where

$$\gamma_u = \gamma_u^{rad} + \gamma_u^{coll}. \tag{4.20}$$

In this expression, $\gamma_u^{rad} = \sum_i A_{ui}$ represents the radiative decay rate and $\gamma_u^{coll} = 1/T_1^u$ represents the collisional decay rate; T_1^u, the collisional decay time from level u, will be explained in more detail later in this chapter. It can be seen from (4.20) that if the collisional decay rate is negligible then the decay rate is simply the radiative decay rate. A more general expression than (4.6) for the decay time may thus be given as

$$\tau_u = \frac{1}{\gamma_u} = \frac{1}{\gamma_u^{rad} + \gamma_u^{coll}}. \tag{4.21}$$

Collisional decay occurs in all types of media. In atoms and molecules it generally occurs in the form of binary collisions, or collisions between two atoms or molecules. In liquids and solids, it occurs owing to the interactions of the closely located surrounding material with the excited atoms. In solids these are generally referred to as *phonon* collisions.

DECAY TIME ASSOCIATED WITH COLLISIONAL DEPOPULATION IN ATOMIC OR MOLECULAR GASES Collisional depopulation of excited atoms in a gaseous state can occur when electrons, atoms, or molecules colliding with the excited species move the population of that species to another level (generally a lower level). Electron collisional de-excitation is more common at very high electron densities, and is very sensitive also to the energy difference between the excited level and the level to which the excited population decays. For example, the electron collisional depopulation rate is proportional to the inverse cube of the energy difference between the two levels. This process will more appropriately be treated in Chapter 8. Collisional deactivation by atomic or molecular collisions is a process that is sensitive to the gas pressure because at higher pressures the colliding species will travel shorter distances before making collisions, which will therefore occur more frequently. For example, the collisional deactivation of the upper laser level of the 632.8-nm helium–neon laser (via collisions of ground-state helium atoms with neon atoms in the upper laser level) increases the decay rate of the neon upper laser level by a factor of 7.5×10^6/s-torr (He). In other words, the decay rate of $\sum_i A_{ui} = 3.32 \times 10^7$ s^{-1} (calculated for that level in the foregoing example) would increase by a factor of 0.75×10^7 s^{-1} for every torr of helium that is added to the laser gas. This in effect decreases the lifetime of the laser level and thereby detrimentally reduces the population in that level. Such collisions with helium also favorably reduce the lower laser level population in the CO_2 laser (this will be discussed in more detail in Section 13.5). Such collisional deactivation can also occur at a much higher rate if the deactivating atom has an excited energy level that is in near coincidence with the energy difference between the level being de-excited and the level to which it decays.

DECAY TIME ASSOCIATED WITH LATTICE RELAXATIONS IN HIGH-DENSITY MATERIALS Collisional interactions in liquids and solids are due to rapid short-range movement of the closely spaced atoms of the dense medium, vibrating at velocities associated with the temperature of the medium. Thermal velocities of atoms at or near room temperature are similar to those mentioned for gases (10^2–10^3 m/s). Therefore, for movements of the order of the atomic dimensions of liquid and solid materials (10^{-10} m), the interaction time is approximately 10^{-10} m/(10^2–10^3 m/s) $= 10^{-12}$–10^{-13} s. Consequently, if these movements cause collisional decay of an excited energy level, the decay rate is significantly enhanced over the normal radiative decay rate of the atoms. Collisional interactions in solids are referred to as *phonon* or *lattice relaxations* and are due to the crystalline lattice vibrations. Whether or not collisional decay occurs is determined by how well the radiating species is shielded from the surrounding material. For example, in many cases a d or an f electron will orbit well within the radius of an s or a p electron of the same atom and thus the s or p electron forms a shield around the inner d or f electron, preventing the lattice vibrations from collisionally deactivating the inner shell electron when it is excited. This shielding occurs in many solid-state laser materials, with the result that the relevant excited state of the laser species has a decay time of the order of 10^{-3}–10^{-4} s rather than the 10^{-12}–10^{-13} s associated with collisional decay; this means that the collisional decay time is reduced by at least a factor of 10^9 for those species.

We will find in Chapter 8 that it is desirable for the upper level u of a laser transition to have a long lifetime and the lower laser level l to have a short lifetime. Since most excited states of liquids and solids have the rapid decay described in the preceding paragraph, satisfying the requirements for the lower laser level is generally not difficult. The problem in liquids and solids is to obtain materials that have long-lived excited states, with upper laser levels that decay radiatively rather than collisionally, and that also have other desirable properties such as stability, good thermal conduction, and so forth. Figure 4-6 shows the decay time of the upper laser level of a titanium sapphire laser material composed of titanium atoms located within a crystalline structure of sapphire (Al_2O_3). It can be seen that, at very low temperatures, the only decay process is the radiative one, with a decay time of 3.9 μs. As the temperature of the medium is increased, the lattice interactions become more pronounced and the decay time is decreased by approximately 25% to a value of 3.0 μs at room temperature. It would thus be desirable not to heat this material much above room temperature during laser operation.

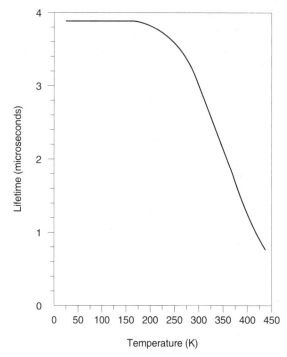

Figure 4-6. Decay time of the upper laser level of a titanium sapphire laser material as a function of temperature (K)

4.2 EMISSION BROADENING AND LINEWIDTH DUE TO RADIATIVE DECAY

Radiating atoms possess a characteristic frequency $\omega_0 = \Delta E/\hbar$ or $\nu_0 = \Delta E/h$, corresponding to the energy separation ΔE between the levels involved in the radiating transition. Such atoms also have a finite emission width or emission linewidth $\Delta\omega$ or $\Delta\nu$. This width is associated with the characteristic decay time of the radiating levels, as well as with other emission-line-broadening mechanisms that are described in this section.

CLASSICAL EMISSION LINEWIDTH OF A RADIATING ELECTRON

The decaying electric field component of the electromagnetic wave described by (4.17) is not an infinitely long wave and therefore cannot be represented by a single pure frequency ω_0. Rather, it has a finite starting point at $t = 0$ and then decays exponentially with a time constant $\tau_0 = 2/\gamma_0$. However, we can obtain the frequency components of the wave represented by (4.17) by taking its Fourier transform as follows:

$$E(\omega) = \frac{1}{\sqrt{2\pi}} \int_{-\infty}^{\infty} E(t)e^{i\omega t}\, dt. \tag{4.22}$$

Using the value of $E(t)$ of (4.17) in (4.22), we have

$$E(\omega) = \frac{E_0}{\sqrt{2\pi}} \int_0^\infty e^{i[(\omega - \omega_0) + i\gamma_0/2]t} \, dt$$

$$= -\frac{E_0}{\sqrt{2\pi}} \frac{1}{i[(\omega - \omega_0) + i\gamma_0/2]}. \tag{4.23}$$

The intensity distribution per unit frequency $I(\omega)$ for this wave is proportional to $|E(\omega)|^2$, and is therefore given by

$$I(\omega) = |E(\omega)|^2 = I_0 \frac{\gamma_0/2\pi}{[(\omega - \omega_0)^2 + \gamma_0^2/4]}. \tag{4.24}$$

Equation (4.24) has been normalized such that

$$I_0 = \int_0^\infty I(\omega) \, d\omega, \tag{4.25}$$

where I_0 represents the total intensity of the emission integrated over the entire frequency width of the emission line. We must make the approximation $\gamma_0/\omega \ll 1$ in order to carry out the normalization integral. This is a reasonable approximation for the wavelength regions we are dealing with in this text.

A graph of $I(\omega)$ versus $\omega - \omega_0$ is shown in Figure 4-7. We will refer to this spectral distribution as the *classical* theoretical emission shape (or broadening) of the radiation emitted by the oscillating electron. The form of the lineshape (eqn. 4.24) is known as a *Lorentzian* distribution and is symmetrical with respect to ω_0. The full width of this emission at half maximum intensity $\Delta\omega^{\text{FWHM}}$ is obtained by setting (4.24) equal to half the value of $I(\omega)$ at $\omega = \omega_0$,

$$I(\omega) = \left(\frac{1}{2}\right)\left(\frac{2}{\pi\gamma_0}\right)I_0,$$

and solving for $\omega - \omega_0$. We can then obtain the FWHM emission linewidth as

$$\Delta\omega^{\text{FWHM}} = 2(\omega - \omega_0) = \gamma_0 = 1/\tau_0 = 2\pi\Delta\nu_C, \tag{4.26}$$

where $\Delta\nu_C$ denotes the classical linewidth.

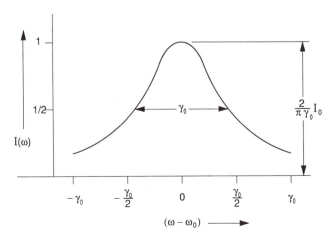

Figure 4-7. Graph of the intensity $I(\omega)$ versus ω from a classical analysis of a decaying and radiating electron as it makes a transition from one energy level to a lower-lying level

Although the lineshape function of (4.24) was derived for a single electron, it can be shown that an identical lineshape function will result for an assembly of N atoms in the same upper level, all radiating on the same transition with random phases, with the intensity of the emission increased by a factor of N. This type of emission broadening occurs when every atom of the same species making the same transition produces an identical emission lineshape and width. Such a situation leads to the Lorentzian lineshape function of (4.24), and is referred to as *homogeneous* broadening.

NATURAL EMISSION LINEWIDTH AS DEDUCED BY QUANTUM MECHANICS (Minimum Linewidth)

Equation (4.26) showed the relationship between the frequency width $\Delta\omega^{\text{FWHM}}$, the decay rate γ_0, the decay time τ_0, and the linewidth $\Delta\nu_C$ of a classically radiating electron. As we have seen previously, classical analysis is not always complete when considering atomic interactions. In determining the true emission linewidth when an electron radiates while decaying from an upper energy level u to a lower energy level l, the classical width $\Delta\nu_C$ is not the only factor that is associated with an accurate description of the observed emission width. We must turn to the uncertainty principle to obtain what we refer to as the *natural* emission linewidth, or $\Delta\nu_N$, of the transition.

The uncertainty principle (eqn. 3.18) states that the uncertainty in determining the energy width ΔE of an energy level that has a minimum uncertainty in its lifetime of Δt is obtained from the relationship

$$\Delta E \cdot \Delta t \approx \hbar = h/2\pi. \tag{4.27}$$

Thus an energy level u will have an uncertainty in energy ΔE_u associated with the uncertainty in the radiative lifetime τ_u, as described by (4.21), such that $\Delta t \approx \tau_u$. We can then express ΔE_u as

$$\Delta E_u = \frac{\hbar}{\tau_u} = \hbar\gamma_u = \hbar \sum_i A_{ui}. \tag{4.28}$$

In considering a radiative transition from energy level u to energy level l, as seen in Figure 4-8, the lower level would also have a finite width, given by

$$\Delta E_l = \frac{\hbar}{\tau_l} = \hbar\gamma_l = \hbar \sum_j A_{lj}. \tag{4.29}$$

As a result, total effective energy width for the two levels would thus be

$$\Delta E_T = \Delta E_u + \Delta E_l = \hbar(\gamma_u + \gamma_l) = \hbar\left(\sum_i A_{ui} + \sum_j A_{lj}\right). \tag{4.30}$$

We can also relate this energy width ΔE_T to the frequency width of radiative emission that occurs from level u to level l (as suggested in Figure 4-8) such that

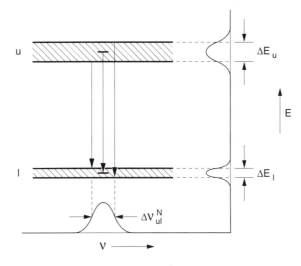

Figure 4-8. Quantum mechanical description of the natural linewidth of emission resulting from a radiating transition between two levels, showing the contribution from both the lower level l and the upper level u

$$\Delta E_T = \hbar \Delta \omega_{ul} = \frac{h}{2\pi} \cdot 2\pi \Delta \nu_{ul} = h \Delta \nu_{ul}. \tag{4.31}$$

We can now equate the right-hand expressions of (4.30) and (4.31) to obtain

$$\hbar \left(\sum_i A_{ui} + \sum_j A_{lj} \right) = h \Delta \nu_{ul}. \tag{4.32}$$

From (4.32) we can thus define the natural emission linewidth $\Delta \nu_{ul}^N$ such that

$$\Delta \nu_{ul}^N = \frac{\sum_i A_{ui} + \sum_j A_{lj}}{2\pi}. \tag{4.33}$$

We can also define an expression γ_{ul}^T for the combined *total* decay rate of the upper and lower levels such that

$$\gamma_{ul}^T = \gamma_u + \gamma_l. \tag{4.34}$$

Thus, from (4.30) and (4.33) we can obtain the following relationship between γ_{ul}^T and $\Delta \nu_{ul}^N$:

$$\gamma_{ul}^T = 2\pi \Delta \nu_{ul}^N = \sum_i A_{ui} + \sum_j A_{lj}. \tag{4.35}$$

Taking into account this total decay rate γ_{ul}^T describing the minimum emission linewidth, the lineshape function of (4.24) becomes

$$I(\omega) = I_0 \frac{\gamma_{ul}^T / 2\pi}{(\omega - \omega_0)^2 + (\gamma_{ul}^T)^2/4}. \tag{4.36}$$

If the lower level is a ground state or an excited state that has a very long lifetime (i.e., a "metastable" state), then the total emission linewidth is just that determined by the energy width of the upper level since $\gamma_l \approx 0$.

We can express (4.36) in terms of decay γ_{ul}^T and frequency ν (instead of angular frequency ω) as

$$I(\nu) = I_0 \frac{\gamma_{ul}^T/4\pi^2}{(\nu - \nu_0)^2 + (\gamma_{ul}^T/4\pi)^2}. \qquad (4.37)$$

4.3 ADDITIONAL EMISSION-BROADENING PROCESSES

When atoms are relatively isolated and an energy level u above the ground state is populated by an excitation process, radiation occurs from level u in the manner just described, at a rate associated with the radiative decay rate A_{ul} due to spontaneous emission and with an emission linewidth $\Delta\nu_{ul}^N$ associated with natural broadening. When a large number of atoms are concentrated in a small volume to produce a high density of atoms, their interactions with each other can produce significant broadening of the emission linewidth in addition to the effect they have upon the decay time of the level (collisional decay). This collisional broadening can be divided into two categories.

The first category is that which decreases the decay time τ_u of the atoms residing in the excited level u. This decreased decay time occurs when other atoms, molecules, or free electrons (or phonons in solids) collide with the excited atom and "knock" its electron down from the excited level u before that electron has the opportunity to radiate spontaneously. This decay process is very common, and is responsible for the broad emission linewidths associated with some solid-state laser transitions. It is usually a homogeneous process, thereby producing a Lorentzian-shaped emission spectrum.

The second category involves processes that do not affect the lifetime but do affect the linewidth. These processes include collisional broadening due to dephasing collisions, amorphous crystal broadening, Doppler broadening, and isotope broadening. The first of these is a homogeneous type of broadening, whereas the last three are inhomogeneous broadening processes. A comparison of the relative lineshapes for both homogeneous (Lorentzian) and inhomogeneous (Gaussian) broadening processes having the same effective linewidth and total emission intensity will be made later in this chapter (see Figure 4-15).

BROADENING DUE TO NONRADIATIVE
(Collisional) DECAY

An increase in the decay rate of population in level u (and a corresponding decrease in the decay time) due to collisional interactions with surrounding atoms, as described earlier in this chapter, produces increased broadening that is referred to as T_1 broadening. If we define an additional decay rate due to collisions in terms of the decay time T_1 associated with those collisions such that $\gamma_1 = 1/T_1$, then using (4.20) we can express the increased decay rate γ_u of a specific level u due to collisions as

$$\gamma_u = \frac{1}{\tau_u} = \sum_i A_{ui} + \frac{1}{T_1^u}, \tag{4.38}$$

where T_1^u refers specifically to the increased collisional decay time of level u. A similar equation would apply to the decay of the lower level l with a collisional decay time T_1^l:

$$\gamma_l = \frac{1}{\tau_l} = \sum_j A_{lj} + \frac{1}{T_1^l}. \tag{4.39}$$

Thus we can obtain an expression for the total emission linewidth when taking into account the T_1 decay by using the relationships of (4.35) to obtain the following expression relating γ_{ul} and $\Delta\nu_{ul}$:

$$\hbar(\gamma_u + \gamma_l) = \hbar\gamma_{ul} = \frac{h}{2\pi}\gamma_{ul} = h\Delta\nu_{ul}. \tag{4.40}$$

The increased emission linewidth $\Delta\nu_{ul}$ can then be written as

$$\Delta\nu_{ul} = \frac{1}{2\pi}\left[\left(\sum_i A_{ui} + \sum_j A_{lj}\right) + \frac{1}{T_1^u} + \frac{1}{T_1^l}\right]$$

$$= \Delta\nu_{ul}^N + \frac{1}{2\pi}\left(\frac{1}{T_1^u} + \frac{1}{T_1^l}\right). \tag{4.41}$$

Thus as T_1 becomes shorter than the radiative decay time for a given level, the collisional decay rate $1/T_1$ begins to dominate over the radiative decay rate of that level, thereby increasing the emission linewidth as described by (4.41).

The more rapid decay of the upper laser level due to T_1 has a negative effect upon the efficiency of the laser amplifier: it reduces the population in the upper laser level and also increases the emission linewidth, both of which have a deleterious effect upon the laser amplifier gain (as will be seen in Chapter 7). Examples of this type of decay include electron collisions in gases and phonon collisions in solids. In summary, this T_1 broadening and decay is a process that one generally attempts to minimize by changing the conditions or constituents of the medium.

BROADENING DUE TO DEPHASING COLLISIONS

Broadening due to dephasing collisions is generally referred to as T_2 broadening. It involves a process that interrupts the phase of the radiating atoms at a rate γ_2, without increasing their population decay rate. The term T_2 represents the average time that occurs between phase-interrupting collisions. When this process dominates over other broadening mechanisms, it

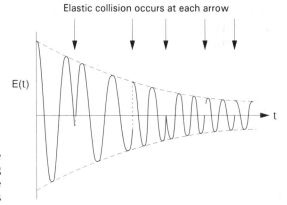

Elastic collision occurs at each arrow

E(t)

Figure 4-9. Phase interruptions of a decaying electromagnetic wave due to dephasing collisions

usually produces a homogeneous emission broadening with a Lorentzian distribution.

Figure 4-9 illustrates T_2 broadening due to collisional dephasing in the time domain. Instead of a radiation decay pattern as shown in Figure 4-5, the decaying wave exhibits an interrupted sinusoidal pattern in which the duration of each interrupted sinusoidal segment varies, but has some average time T_2. Each of these segments represents a sine wave of the same frequency ω_0 that is cut off or interrupted, and then continues radiating at the same frequency until it is again interrupted. The average phase interruption time T_2 that occurs during the decay of level u, as shown in Figure 4-9, would contribute to an increased decay rate γ_2 for level u of

$$\gamma_2^u = 1/T_2^u \tag{4.42}$$

and similarly, for the lower level l, of

$$\gamma_2^l = 1/T_2^l. \tag{4.43}$$

Typically, $T_2^u = T_2^l \equiv T_2$ and hence $\gamma_2^u + \gamma_2^l = 2/T_2$. As with the inclusion of T_1 broadening in (4.40) and (4.41), T_2 broadening can be incorporated into the following expression for the total homogeneous emission linewidth:

$$\Delta\nu_{ul}^H = \frac{\gamma_{ul}^T}{2\pi} = \frac{1}{2\pi}\left[\left(\sum_i A_{ui} + \sum_j A_{lj}\right) + \frac{1}{T_1^u} + \frac{1}{T_1^l} + \frac{2}{T_2}\right]. \tag{4.44}$$

Of course, T_2 broadening would be observable only if T_2 is short enough that the factor $2/T_2$ in (4.44) becomes significant when compared to the other terms. This (4.44) process is the major broadening mechanism for many solid-state laser transitions, including those of broadband tunable solid-state laser materials such as $Ti:Al_2O_3$ (see Chapter 5).

We must remember that γ_2 is a rate that is strictly associated with phase interruption, leading to broadening of the emission associated with a radiative transition between levels u and l as described in (4.44). It is not a factor associated with the decay rate of a level, in contrast with γ_1.

AMORPHOUS CRYSTAL BROADENING

Glass materials are composed of a molasses-type mixture of molecules frozen into a solid, leaving various small regions of the material oriented in slightly different directions. Thus, each of the glass molecules can have slightly different energy levels depending upon the strains that are placed upon them in the frozen matrix. The different energy values of excited states for various portions of this "molasses" lead to different radiating frequencies for different regions. Since the emission line is composed of the sum of all of the individual lines, this leads to a much broader emission spectrum than would occur if they were composed of one crystalline structure. Thus the emission from a glass material such as Nd:glass is inhomogeneously broadened with a Gaussian shape and is usually much broader than the emission line of a pure crystalline material such as Nd:YAG, which is homogeneously broadened with a Lorentzian shape. Thus Nd doped in glass (cf. Section 5.3) will have a broadening of $\Delta\nu = 7.5 \times 10^{12}$ Hz compared to the much narrower linewidth of $\Delta\nu = 1.2 \times 10^{11}$ Hz for Nd doped in a crystalline matrix. Consequently, for the same doping concentration, the maximum emission at the center frequency of the Nd:YAG material is much greater than that of Nd:glass, since there are the same number of radiating species and the radiative rates from the upper laser level are similar (but not identical, since they have different upper-level lifetimes). A comparison of Nd:glass and Nd:YAG emission linewidths and shapes is shown in Figure 4-10.

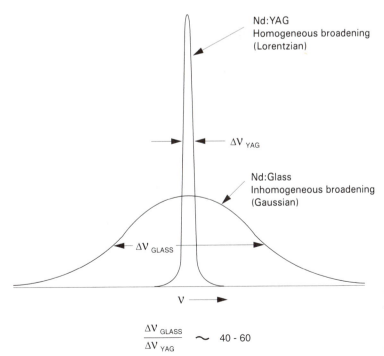

Figure 4-10. Relative emission linewidths of a radiating Nd ion doped into either a YAG crystal or a glass material

DOPPLER BROADENING IN GASES

We are all familiar with the Doppler effect associated with sound waves, such as when a train is approaching and sounding its horn. As the train approaches, the sound is first heard at a higher frequency and then changes to a lower frequency as the train passes.

This same frequency shift is characteristic of the emission of light waves from moving atoms. Consider light of frequency ν_0 and velocity c being emitted over a time interval Δt from an atom moving toward you with a velocity v. At the end of the period Δt, those waves will cover a distance, in your direction, of $(c-v)\Delta\tau$, whereas if the source had not been moving with respect to you (the observer), they would have been emitted into a distance of $c\Delta t$. In effect, the separation between wave peaks, which is the wavelength λ, is compressed for waves traveling in your direction, which thereby raises their frequency in order to satisfy the requirement that $\lambda\nu = c$.

The value of this frequency shift, for a source moving toward the observer, can be obtained by letting the observed frequency ν shift with respect to the original frequency ν_0 as just described; that is,

$$\nu = \left(\frac{c\Delta t}{(c-v)\Delta t}\right)\nu_0 = \left(\frac{1}{1-v/c}\right)\nu_0 \cong \left[1 + \frac{v}{c} + \left(\frac{v}{c}\right)^2 + \cdots\right]\nu_0. \qquad (4.45)$$

The same argument would work for the source moving away from the observer, yielding a frequency shift of

$$\nu = \left(\frac{c\Delta t}{(c+v)\Delta t}\right)\nu_0 = \left(\frac{1}{1+v/c}\right)\nu_0 \cong \left[1 - \frac{v}{c} + \left(\frac{v}{c}\right)^2 - \cdots\right]\nu_0. \qquad (4.46)$$

For nonrelativistic velocities, where the ratio v/c is very small, terms involving powers of v/c (i.e. $(v/c)^2$ or higher) can be neglected. In this case the preceding equations may be recast as follows:

$$\nu = \left(1 + \frac{v}{c}\right)\nu_0 \quad \text{(moving toward the observer);} \qquad (4.47)$$

$$\nu = \left(1 - \frac{v}{c}\right)\nu_0 \quad \text{(moving away from the observer).} \qquad (4.48)$$

This Doppler shift can be significant when observing radiation from atoms in the gaseous state. In such a case the atoms are in thermal equilibrium, and are therefore moving randomly in all directions with a range of velocities that are related to the gas temperature. In this situation the average velocity is given by

$$\bar{v} = \sqrt{\frac{8kT}{M\pi}}, \qquad (4.49)$$

where T is the average atom temperature, k is Boltzmann's constant, and M is the mass of the atom. For atoms at or slightly above room temperature, the velocities typically range from 10^2 to 10^3 m/s, with lighter atoms (such as helium) having the higher velocities.

For such a thermal distribution of velocities, some of the atoms are moving directly toward you or away from you at any given instant, while some are moving only partially toward you or away from you. These groups of atoms will appear to have shifted frequencies if you observe the radiation with an instrument having sufficient spectral resolution. Another group of atoms will be moving directly to one side or the other with respect to the observer, and will therefore have no component of their velocity in a direction toward or away from you. These atoms will be observed to have frequencies radiating at the center frequency ν_0 according to (4.47) and (4.48), since ν represents the *component* of velocity moving either toward or away from the observer and for this situation $\nu = 0$.

We have shown that all radiating transitions have a characteristic natural emission linewidth, so we would observe the Doppler broadening only if it were significantly larger than the natural broadening for that transition. We will therefore calculate the Doppler width of an emission line and compare it to the natural linewidth for several different atoms in various situations.

If we define the direction of atoms moving in a straight line toward the observer as the x direction, then from (4.47) the frequency shift seen by the observer would be

$$\nu = \nu_0 \left(1 + \frac{v_x}{c} \right), \tag{4.50}$$

where v_x is the component of the velocity in the x direction.

If the radiating atoms are in thermal equilibrium, then they have a velocity distribution that is spherically symmetric, and the probability $P(v_x)$ that an atom moving in the x direction has a velocity between v_x and $v_x + dv_x$ is given by the Maxwellian probability distribution function

$$P(v_x) = \left(\frac{M}{2\pi kT} \right)^{1/2} \exp\left\{ -\frac{M}{2kT} v_x^2 \right\} dv_x, \tag{4.51}$$

which has a Gaussian shape of the form e^{-ax^2} where a is a positive constant. (Recall that $\exp\{x\}$ denotes e^x.) In more general terms, the probability for finding the atom within a velocity range of $(v_x + dv_x, v_y + dv_y, v_z + dv_z)$ is given by

$$P(v_x, v_y, v_z)$$
$$= \left(\frac{M}{2\pi kT} \right)^{3/2} \exp\left\{ -\frac{M}{2kT} (v_x^2 + v_y^2 + v_z^2) \right\} dv_x \, dv_y \, dv_z. \tag{4.52}$$

We have accounted for the fact that each atom must be located somewhere by using the normalizing factor

$$\int_{-\infty}^{\infty} \int \int P(v_x, v_y, v_z) \, dv_x \, dv_y \, dv_z = 1. \tag{4.53}$$

From (4.50), we can deduce that the probability $G(\nu)\,d\nu$ that the transition frequency is between ν and $\nu+d\nu$ is equal to the probability that v_x will be found between $v_x = (\nu-\nu_0)(c/\nu_0)$ and $v_x+dv_x = (\nu+d\nu-\nu_0)(c/\nu_0)$, regardless of the values of v_y and v_z, since the Doppler shift is based upon only that component of velocity moving either toward or away from the observer. Thus $G(\nu)\,d\nu$ can be obtained by inserting $v_x = (\nu-\nu_0)(c/\nu_0)$ and $dv_x = (c/\nu_0)\,d\nu$ into (4.52) for $P(v_x, v_y, v_z)$ and integrating over all values of v_y and v_z. This leads to an integral of the form

$$G(\nu)\,d\nu = \left(\frac{M}{2\pi kT}\right)^{3/2} \int_{-\infty}^{\infty} \int_{-\infty}^{\infty} \exp\left\{-\left(\frac{M}{2kT}\right)(v_y^2 + v_z^2)\right\} dv_y\,dv_z$$

$$\exp\left\{-\left(\frac{M}{2kT}\right)\left(\frac{c^2}{\nu_0^2}\right)(\nu-\nu_0)^2\right\}\left(\frac{c}{\nu_0}\right) d\nu, \quad (4.54)$$

where $G(\nu)\,d\nu$ denotes the probability of the frequency occurring between ν and $\nu+d\nu$. The solution for the definite integral

$$\int_{-\infty}^{\infty} \exp\left\{-\left(\frac{M}{2kT}\right)^{1/2} v_z^2\right\} dv_z = \left(\frac{2\pi kT}{M}\right)^{1/2} \quad (4.55)$$

leads to the following solution of (4.54):

$$G(\nu) = \frac{c}{\nu_0}\left(\frac{M}{2\pi kT}\right)^{1/2} \exp\left\{-\left(\frac{M}{2kT}\right)\left(\frac{c^2}{\nu_0^2}\right)(\nu-\nu_0)^2\right\} \quad (4.56)$$

for the Doppler lineshape of an assembly of atoms radiating at the same frequency ν_0. This Gaussian lineshape function, normalized in a similar way to that of (4.53), represents the distribution of atoms having various velocity components in the x direction (both positive x and negative x).

The intensity of the emission line as a function of frequency can thus be expressed as

$$I(\nu) = \left(\frac{M}{2\pi kT}\right)^{1/2}\left(\frac{c}{\nu_0}\right)I_0 \exp\left\{-\left(\frac{M}{2kT}\right)\left(\frac{c^2}{\nu_0^2}\right)(\nu-\nu_0)^2\right\}, \quad (4.57)$$

which is normalized as

$$\int_0^{\infty} I(\nu)\,d\nu = I_0, \quad (4.58)$$

so that the total emission intensity from the transition is I_0. It should be noted that (4.57) is the inhomogeneous (Gaussian) shape function, which dominates the homogeneous (Lorentzian) lineshape function of (4.37) if the Doppler broadening is larger than the homogeneous broadening.

The effective width or broadening of the emission line can be derived by determining the frequency shift $(\nu-\nu_0)$ with respect to the center frequency ν_0 at which the distribution falls to half its maximum value. By taking twice this value (for both positive and negative shifts from the center frequency) and adding them to obtain the FWHM, one arrives at a value in Hertz for the *Doppler width* of

$$\Delta \nu^D = 2\nu_0 \sqrt{\frac{2(\ln 2)kT}{Mc^2}} = (7.16 \times 10^{-7})\nu_0 \sqrt{\frac{T}{M_N}}, \qquad (4.59)$$

where in the second equality T is temperature in degrees Kelvin, M_N denotes mass number (see the Appendix), and ν_0 has units of 1/seconds. Equation (4.57) can then be rewritten in terms of $\Delta \nu^D$ as

$$I(\nu) = \frac{2(\ln 2)^{1/2}}{\pi^{1/2}\Delta \nu^D} I_0 \exp\left\{-\left[\frac{4(\ln 2)(\nu-\nu_0)^2}{(\Delta \nu^D)^2}\right]\right\}. \qquad (4.60)$$

The shape of this emission line is shown in Figure 4-11. It consists of a sum of Lorentzian function of atoms traveling in different directions that are thus observed to radiate at different frequencies. Those frequencies are all superimposed to produce the total Gaussian emission spectrum of Figure 4-11.

A few examples of Doppler versus natural broadening are given in Table 4-1 for a number of well-known laser transitions in various gases. It can be seen that in most instances the Doppler broadening dominates the natural broadening by a significant amount.

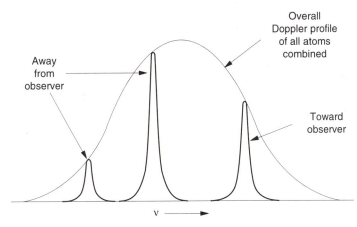

Overall Doppler profile of all atoms combined

Away from observer

Toward observer

ν ⟶

Natural emission linewidth (Lorentzian profile) of individual atoms traveling in different directions

Figure 4-11. Shape of a Doppler-broadened emission line, indicating the natural emission linewidths of individual atoms radiating while traveling in various directions

TABLE 4-1

DOPPLER VERSUS NATURAL BROADENING

Laser Species	λ (nm)	f	A (s^{-1})	$\Delta \nu_N$ (Hz)	$\Delta \nu_D$ (Hz)
Neon (He–Ne)	632.8	0.012	3.4×10^6	5.4×10^5	1.5×10^9
Argon ion	488.0	0.418	7.8×10^7	1.2×10^7	2.7×10^9
Cadmium (He–Cd)	441.6	0.006	1.4×10^6	2.2×10^5	1.1×10^9
Copper	510.5	0.005	2.0×10^6	2.2×10^7	2.3×10^9

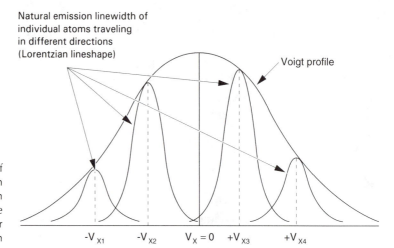

Figure 4-12. Shape of an emission line in which the natural emission linewidths of the atoms are comparable to the Doppler width

VOIGT LINESHAPE PROFILE

There are many transitions in gases for which the emission linewidth due to natural (homogeneous) broadening and that due to Doppler (inhomogeneous) broadening are comparable or nearly comparable. In such cases the line profile would be a convolution of both these effects. When these two effects are comparable for a given transition, each subgroup of atoms with a specific velocity component in the viewer's direction radiates with its natural linewidth, and all the different velocity components are superimposed to produce the total lineshape shown in Figure 4-12. The resulting line profile, a combination of Doppler and Lorentzian shapes, is known as a Voigt profile. We can express this spectral profile as the sum of individual groups of Lorentzian shapes over a Doppler distribution of velocities. If we define the Doppler parameter as $\Delta = (4\pi\nu_0/c)(2kT/M)^{1/2}$ then, using (4.37) and (4.57), the intensity distribution can be written in terms of the ratio γ_{ul}^T/Δ as

$$I\left(\nu_0 - \nu, \frac{\gamma_{ul}^T}{\Delta}\right) = I_0 \int_0^\infty \frac{1/\pi^{3/2}}{(\nu - \nu_0')^2 + (\gamma_{ul}^T/4\pi)^2}\left(\frac{\gamma_{ul}^T}{\Delta}\right)$$

$$\exp\left\{-\left(\frac{16\pi^2}{\Delta^2}\right)(\nu_0' - \nu_0)^2\right\}d\nu_0'. \quad (4.61)$$

The Voigt profile cannot be expressed in analytical form. It can, however, be evaluated numerically; see Corney (1977) for tabulations and references.

The natural linewidth is generally much smaller than the Doppler linewidth, so Doppler broadening is the dominant broadening for most gas laser transitions. The natural linewidth $\Delta\nu^N$ is proportional to the transition probability (eqn. 4.33), which in turn is proportional to the reciprocal of the square of the wavelength of the transition for a given oscillator strength (eqn. 4.7). The Doppler width is proportional to the frequency or the reciprocal of the wavelength, as well as to the square root of the ratio of the gas temperature to the atomic weight (see eqn. 4.59). The ratio of the natural linewidth to the Doppler linewidth thus increases inversely with wavelength. It also increases for heavier elements and for lower gas temperatures. Thus, since the Voigt profile must be considered when the ratio of γ/Δ becomes significant, this would more likely occur for transitions in heavier elements at lower temperatures and at shorter wavelengths.

BROADENING IN GASES DUE TO ISOTOPE SHIFTS

Another form of inhomogeneous broadening is that determined by the frequency shifts due to the presence of more than one isotopic form of the species in the laser gas mixture. Different isotopic forms of the same element exist in various relative abundances when the element is obtained in its naturally occurring form. These different isotopes consist of atoms having the same number of protons and electrons (which identifies them as belonging to the same element) but with different numbers of neutrons. Naturally occurring quantities of various elements contain certain specific fractions of various isotopes. For example, helium exists naturally in a relative abundance of 99.0008% He^4 (2 neutrons) and 0.00013% He^3 (1 neutron), where the superscripts 3 and 4 denote the mass number. Neon exists naturally as 90.8% Ne^{20} (10 neutrons), 0.26% Ne^{21} (11 neutrons), and 8.9% Ne^{22} (12 neutrons).

Atoms with slightly different numbers of neutrons within their nuclei exhibit small differences in energy-level values. Such energy shifts cause a slight difference in center frequency of the emission lines for the different isotopes. In many cases these different values of energy are too small to be noticed from the standpoint of the emission linewidth in a laser medium. However, in some cases the energy differences, which produce an *isotope shift* in the emission spectrum between different isotopes, is significant. We are concerned only with isotope shifts that are of an amount comparable to the Doppler width of the radiative transition. Shifts that are significantly smaller than the Doppler width are inconsequential, as far as lasers are concerned.

In general, there are two causes for energy-level shifts that result in frequency shifts of the emission. The first is a mass effect that occurs mostly for the lighter atoms (elements with lower atomic numbers). This effect produces a shift in the radiative transition to a higher frequency for the heavier isotope (with the most neutrons), which has a greater mass. The second is a volume shift produced by a difference in volume of the nucleus for atoms with different numbers of neutrons. This effect is applicable primarily to elements with high atomic numbers. The magnitude of this shift depends upon the specific electronic configuration of the energy levels involved in the transition. Appreciable shifts are observed only when the number of s electrons is different between those two levels. As in the mass shift, this shift also leads to a higher frequency of emission for the heavier isotopes. The even isotopes produce approximately equal shifts whereas the odd isotopes split into several levels, making their emission spectrum slightly more complicated. This odd-isotope effect will be seen in the example for cadmium discussed later in this section.

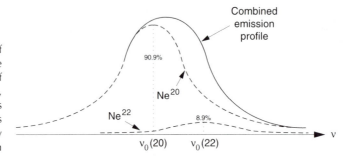

Figure 4-13. Shape of emission line for the 632.8-nm transition of the helium–neon laser, showing the contributions from two different isotopes associated with a naturally occurring mixture of neon

The slightly different energy-level values for different isotopes provide slightly different frequencies for the transitions. When observed in a high-resolution spectrogram the result is a broadening of the specific radiative transition (if the frequencies are shifted but still blend together) or, if the shifts are large enough, the result is distinctly separate but very closely spaced emission lines. For most transitions in most atoms, the isotope shifts are small. The effect depends upon both the specific atom involved and the type of transition. For the 632.8-nm laser transition in Ne, the line-width due to the Doppler broadening is of the order of 1.5 GHz whereas the isotope shift between Ne^{20} and Ne^{22} is approximately 1 GHz. Thus the line shape for the 632.8-nm transition in a naturally occurring mixture of Ne would appear as shown in Figure 4-13. Since the abundance of isotope 22 is less than 9%, it has only a small effect on the spontaneous emission profile; this is indicated by the combined lineshape in Figure 4-13.

One of the most interesting isotope distributions is that occurring for the He–Cd laser operating at 441.6 nm; here the isotope shifts are as large as the Doppler broadening. The naturally occurring isotope distribution for the 441.6-nm emission line in Cd is shown in Figure 4-14(a), with transitions associated with individual isotopes ranging from Cd^{106} to Cd^{116}. The relative abundance of each isotope is shown along with the isotope shifts between adjacent isotopes. The total emission spectrum at 441.6 nm for a gas discharge involving cadmium would thus comprise the sum of all of the individual emission spectra of each isotope, as shown in the figure. The two isotopes with the most abundance are Cd^{112} and Cd^{114}. Because they have the greatest abundance, they contribute to most of the laser output, thereby producing a laser spectrum as shown in Figure 4-14(b). If a special uniform mixture of isotopes is used, then the emission spectrum will be distributed equally over all isotopes and the consequent laser frequency spectrum will look like that of Figure 4-14(c). A single isotopic species of cadmium will produce the narrowest emission spectrum, which is just the Doppler width or approximately 2 GHz, thereby providing the highest laser gain (as we shall learn in Chapter 7). Cadmium is therefore a very obvious case in which inhomogeneous broadening, due to the combined effects of Doppler broadening and isotope broadening, plays the dominant role in determining both the width and shape of the emission spectrum.

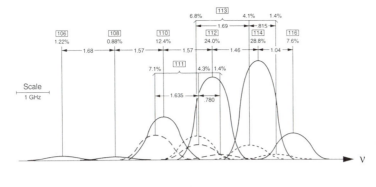

Figure 4-14(a). Emission lineshapes of the various isotopes of the cadmium 441.6-nm laser transition from a naturally occurring isotopic mixture

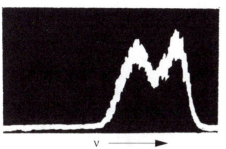

Figure 4-14(b). Laser output at 441.6 nm from a natural isotopic mixture of Cd

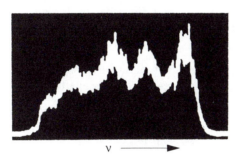

Figure 4-14(c). Laser output at 441.6 nm from a special isotopic mixture of Cd in which the contributions from the various isotopes are uniform

COMPARISON OF VARIOUS TYPES OF EMISSION BROADENING

In this section we will investigate the differences between the Lorentzian- and Gaussian-shaped emission on two equivalent radiative transitions from a hypothetical level u to a level l. We will assume that excitation has occurred within a medium that contains energy levels u and l, such that a population density N_u (number of species per unit volume) has been produced in level u. The total number of transitions downward to level l per unit volume per unit time is then $N_u A_{ul}$, as described in (4.1). This expression does not, however, describe the frequency distribution of this process. Since we now know that emission broadening occurs, we can consider the details of both the Lorentzian- and Gaussian-shaped processes.

HOMOGENEOUS BROADENING It was stated previously that, in the homogeneous emission processes that lead to a Lorentzian distribution of emitting frequencies, all of the atoms in level u have an equal probability of participating in the emission at any frequency ν of that emission linewidth. The factor that varies to give the Lorentzian emission lineshape is the transition probability. We can thus express the radiative transition probability as having a Lorentzian-shaped frequency dependence of

$$A_{ul}(\nu) = \frac{C_1}{(\nu - \nu_0)^2 + (\gamma_{ul}^T/4\pi)^2}, \tag{4.62}$$

where ν_0 is the center frequency of the emission and γ_{ul}^T is related to the frequency width $\Delta\nu_{ul}$ of the emission line at full width half maximum, according to (4.44). The term C_1 is a constant that is determined by the relationship

$$A_{ul} = \int_0^\infty A_{ul}(\nu)\,d\nu. \tag{4.63}$$

Equation (4.63) states that the total transition probability A_{ul} from level u to level l is the sum of all of the probability at all frequencies over the emission linewidth for that transition. Using this relationship, the expression for $A_{ul}(\nu)$ can be written as

$$A_{ul}(\nu) = \frac{\gamma_{ul}^T/4\pi^2}{(\nu - \nu_0)^2 + (\gamma_{ul}^T/4\pi)^2}\, A_{ul}. \tag{4.64}$$

We can thus write the number of transitions per unit volume per unit frequency for a homogeneously broadened (Lorentzian-shaped) transition as

$$N_u A_{ul}(\nu) = \frac{\gamma_{ul}^T/4\pi^2}{(\nu - \nu_0)^2 + (\gamma_{ul}^T/4\pi)^2}\, N_u A_{ul}. \tag{4.65}$$

INHOMOGENEOUS BROADENING For inhomogeneous emission processes that lead to a Gaussian distribution of emitting frequencies, we described how specific portions of the population density N_u contribute to different portions of the emission linewidth. These differences were shown to occur, for example, in Doppler broadening for atoms that are moving in different directions with respect to the location from which the emission is observed. Each of those separate directions was described as having different velocities v_x with respect to the observer. For an incrementally small velocity segment Δv_x ranging from v_x to $v_x + \Delta v_x$, all of the atoms within that segment will radiate with their homogeneously broadened emission linewidth. But since it represents only a small portion of the entire frequency distribution if Doppler broadening is dominant, this homogeneous broadening is not apparent when the entire emission line is observed. What is apparent is the consequence of all of the individual velocity segments adding up over the linewidth, yielding the Gaussian shape that is determined by the population distribution according to velocity as shown in Figure 4-11.

We can thus express the population distribution as a function of frequency ν as

$$N_u(\nu) = C_2 \exp\left\{-\left[\frac{4\ln 2(\nu - \nu_0)^2}{\Delta\nu_D^2}\right]\right\}, \tag{4.66}$$

where the Gaussian profile was obtained from (4.60). The normalizing constant can be derived by assuming that the sum of all of the population densities $N_u(\nu)$ radiating at specific frequencies over the Doppler–broadened emission linewidth must equal the total population density N_u in level u:

$$N_u = \int_0^\infty N_u(\nu)\, d\nu. \tag{4.67}$$

Carrying out this normalization integral using (4.66) for $N_u(\nu)$, we find that

$$C_2 = 2\sqrt{\frac{\ln 2}{\pi}}\,\frac{1}{\Delta\nu_D} N_u \tag{4.68}$$

and consequently

$$N_u(\nu) = 2\sqrt{\frac{\ln 2}{\pi}}\,\frac{1}{\Delta\nu_D} N_u \exp\left\{-\left[\frac{4\ln 2(\nu - \nu_0)^2}{\Delta\nu_D^2}\right]\right\}. \tag{4.69}$$

Thus we have

$$N_u(\nu)A_{ul} = 2\sqrt{\frac{\ln 2}{\pi}}\,\frac{1}{\Delta\nu_D} \exp\left\{-\left[\frac{4\ln 2(\nu - \nu_0)^2}{\Delta\nu_D^2}\right]\right\} N_u A_{ul}. \tag{4.70}$$

The two expressions (eqns. 4.65 and 4.70) for the number of radiative transitions per unit volume per unit frequency per unit time emphasize the differences in the radiative process for cases with homogeneous versus Doppler broadening. They will be used in Chapter 7 to derive the expression for the gain or amplification of a beam of light that can occur in a medium when specific conditions are satisfied.

At this point it is interesting to compare the emission intensity from a homogeneously broadened transition to that from a Doppler-broadened transition. We will consider atoms in level u with a population density N_u making a transition to level l and radiating at a center frequency ν_0. We will compute the power radiated per unit volume for each case at the center of the emission line over a frequency width $\Delta\nu$ significantly narrower than the emission linewidth of those radiating atoms. We will then take the ratio of those powers for the two types of broadening.

The power per unit volume radiated from a homogeneously broadened transition at $\nu = \nu_0$ can be expressed using the results of (4.65) as

$$N_u A_{ul}(\nu = \nu_0)\Delta\nu h\nu_{ul} = \frac{\gamma_{ul}/4\pi^2}{\gamma_{ul}^2/16\pi^2} N_u A_{ul}\Delta\nu h\nu_{ul}$$

$$= \frac{4}{\gamma_{ul}} N_u A_{ul}\Delta\nu h\nu_{ul}, \tag{4.71}$$

where the rate γ_{ul}^T has been abbreviated as γ_{ul}. Similarly, using (4.70), the power radiated from a Doppler-broadened transition at $\nu = \nu_0$ per unit volume can be expressed as

$$N_u(\nu = \nu_0)A_{ul}\Delta\nu h\nu_{ul} = 2\left(\frac{\ln 2}{\pi}\right)^{1/2}\frac{1}{\Delta\nu^D}N_u A_{ul}\Delta\nu h\nu_{ul}. \tag{4.72}$$

Thus, the ratio of powers (per unit volume) P_{ratio} at the center of the emission line for homogeneous broadening compared to Doppler broadening is obtained by taking the ratio of (4.71) to (4.72) as follows:

$$P_{\text{ratio}} = \frac{2(\pi)^{1/2}\Delta\nu^D}{(\ln 2)^{1/2}\gamma_{ul}} = \frac{2(\pi)^{1/2}\Delta\nu^D}{(\ln 2)^{1/2}2\pi\Delta\nu^H}$$

$$= \frac{1}{(\pi\ln 2)^{1/2}}\frac{\Delta\nu^D}{\Delta\nu^H} = 0.85\frac{\Delta\nu^D}{\Delta\nu^H}. \tag{4.73}$$

Thus, the ratio of the power at the *line center* for the two types of broadening is approximately the inverse ratio of their emission linewidths. For the case where $\Delta\nu^D \ll \Delta\nu^H$, we can ignore Doppler broadening. For the case where $\Delta\nu^H \ll \Delta\nu^D$, the Doppler effect essentially spreads the emission over a much broader range of frequencies than would normally occur for homogeneous broadening, especially if that homogeneous broadening were determined by natural broadening.

We can now compare the shapes of the two types of emission lines, Lorentzian and Gaussian, for the equivalent emission linewidths $\Delta\nu^H = \Delta\nu^D$, where they have the same total power or emission intensity integrated over the entire emission line as derived in the foregoing. This is shown in Figure 4-15. It can be seen that the center of the homogeneously broadened line is lower (by the factor 0.85 obtained in eqn. 4.73) and higher in the wings of the emission.

Table 4-2 gives a summary of the various types of broadening and decay processes.

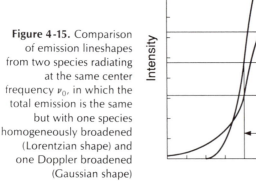
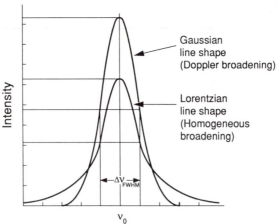

Figure 4-15. Comparison of emission lineshapes from two species radiating at the same center frequency ν_0, in which the total emission is the same but with one species homogeneously broadened (Lorentzian shape) and one Doppler broadened (Gaussian shape)

TABLE 4-2

SUMMARY OF BROADENING AND DECAY PROCESSES

Type of Broadening	Nature of Broadening	Special Features
Natural	$\Delta E_T = \Delta E_u + \Delta E_l,$ $h\Delta\nu_{ul}^N = h(\Delta\nu_u + \Delta\nu_l) = h\left(\dfrac{A_u}{2\pi} + \dfrac{A_l}{2\pi}\right),$ $\Delta\nu_{ul}^N = \dfrac{1}{2\pi}\left[\sum_i A_{ui} + \sum_j A_{lj}\right].$	Determines the minimum emission linewidth of a radiating transition. It involves decay rates of both the upper and lower levels. It is homogeneous with a Lorentzian profile.
T_1	$T_1^{u,l}$ are the decay times of levels u and l due to outside disturbances; $\Delta\nu_{T_1} = \dfrac{1}{2\pi}\left[\dfrac{1}{T_1^u} + \dfrac{1}{T_1^l}\right].$ It will be relevant only if $\Delta\nu_{T_1} > \Delta\nu_N$.	Both T_1 and T_2 determine the increased broadening on a transition from level u to level l due to outside influences, if T_1 or $T_2 > 2\pi\Delta\nu_{ul}$. We cannot determine which form is dominant in a given instance merely by measuring the broadening. T_2 broadening is less detrimental because it does not destroy population. However, it still reduces the gain due to the increased broadening. Both are homogeneous with a Lorentzian profile.
T_2	T_2 is the average phase interruption time of level u or l for outside disturbances. It usually occurs equally for levels u and l. Therefore $\Delta\nu_{T_2} = \dfrac{1}{2\pi}\left[\dfrac{2}{T_2}\right].$ It will be relevant only if $\Delta\nu_{T_2} > \Delta\nu_N$.	
Doppler	$\Delta\nu_D = 2\nu_0\sqrt{\dfrac{2(\ln 2)kT}{Mc^2}}.$ It will be relevant only if $\Delta\nu_D > \Delta\nu_N$.	This is due entirely to the motion of atoms. It has nothing to do with decay rates or phase interruptions. Each atom still radiates with its own natural width. The total Doppler width is the sum of the shifted natural emission widths of all the atoms.
Isotope	If there is more than one isotope present, each isotope behaves independently and has its own broadening (usually Doppler). The linewidth and shape are determined by the sum of the emissions from each isotope. It is relevant only if the isotope shift is significant compared to linewidths of individual isotopes.	This is essentially a sum of separate emitters that have slight shifts in frequency. The overall emission envelope is therefore an unusually shaped sum of many separate Doppler widths.

4.4 QUANTUM MECHANICAL DESCRIPTION OF RADIATING ATOMS

In this chapter we first considered the classical model of an electron oscillating back and forth over a constrained distance d in a sinusoidal or harmonic motion. Such a description is exactly the motion one would observe if an electron were rotating about a positive charge at a radius $r = d/2$ and viewed in the plane of rotation. This classical radiation picture was developed to illustrate how radiation from an oscillating electron can lead to an exponential decay of the population of the radiating state as well as how some of the emission-broadening processes occur. It was not, however, a sufficient model to explain the details of the radiation emitted from atoms, including the relative strengths or intensities of various emission lines and how to determine the probability of emission when electron transitions occur between various energy levels associated with the electron–nucleus system.

We will therefore use the techniques available from quantum mechanics to derive some atomic emission parameters. We will give only a brief outline of the technique, because the main emphasis of this book is not on the details of atomic emission but rather on how various energy levels and transitions between those levels are used to produce population inversions and gain in various media.

ELECTRIC DIPOLE RADIATION

In (3.21) and (3.23) we described the electronic wave functions for energy states of atoms that were identified as either stationary or coherent states, the latter occurring when the electron jumps from an upper state u to a lower state l. We argued that when this transition occurs, the electronic wave function must be made up of a mixture of both the initial and final states. It is useful to know how to determine if a transition will occcur between two specific energy states, and also how likely it is to occur. We can surmise that such a transition must involve a calculation using both the initial and final state wave functions. Whether or not a transition will occur can be determined by various selection rules concerning the angular momentum quantum numbers of the various states, as well as the parity of the states (all of which will be outlined in this chapter). We will first determine how to estimate the likelihood or probability of a particular transition occurring – its *emission strength,* or oscillator strength.

In the Chapter 3 derivation of energy eigenvalues using the Schrödinger equation, the Hamiltonian (representing the energy of the system) incorporated only the energy associated with the atomic particles, including the kinetic energy of the electron rotating about the nucleus and the potential energy of attraction between positive and negative charges. One of the most important phenomena associated with atomic physics is the consideration of the emission (and absorption, to be treated later) of electromagnetic radiation. In order to consider the emission of radiation in the form of an electromagnetic wave from such a system, it is necessary to include in the Hamiltonian an additional term expressing the interaction of the atom with such a radiation field. The solution involves factors associated with both the electromagnetic field and the electron involved in the transition. Valid quantum solutions of the electromagnetic component of that problem are possible only if the radiation field increases (or decreases, for absorption) by a value of one photon when the electron makes a transition to a lower level, which of course results from the emission of one photon from the atom. We will not delve into that analysis in this treatment, but only bring it to the reader's attention and suggest that details, for those who are interested, can be found in books on quantum mechanics.

ELECTRIC DIPOLE MATRIX ELEMENT

We now look more closely at the interaction associated with the bound electron that changes energy states during the transition. That interaction involves computation of the electric dipole matrix element M_{ul}, which is defined as

$$M_{ul} = \langle e\mathbf{r} \rangle = \int \psi_u^* e\mathbf{r} \psi_l \, dV \qquad (4.74)$$

for $\mathbf{r} = x\hat{i} + y\hat{j} + z\hat{k}$, where $\hat{i}, \hat{j}, \hat{k}$ are unit vectors, e is the electron charge, and the integral is taken over the entire volume V associated with the radiating atom. The factor $e\mathbf{r}$ represents the electric dipole moment of the electron–proton system that is equivalent to the term ex_0 in the classical expression of (4.11). Equation (4.74) is essentially a calculation of the amplitude of the oscillating dipole moment of the coherent state formed by the two stationary state wave functions ψ_u and ψ_l. Since the dipole moment operator is "sandwiched" between the two wave functions ψ_u and ψ_l, one can think of the integral of (4.74) as calculating the spatial overlap of the electron wave functions of the two states weighted with the dipole moment. The evaluation of the extent of that overlap provides the probability that the transition will occur.

ELECTRIC DIPOLE TRANSITION PROBABILITY

We can write a quantum mechanical expression for the power radiated *per atom* as

$$P_R = \frac{dE}{dt} = A_{ul} h\nu_{ul}, \qquad (4.75)$$

where A_{ul} is the transition probability per second and $h\nu_{ul}$ is the energy of the radiated photon. Using (4.11), this expression can be equated to the classical power radiated as

$$P_R = \frac{dE}{dt} = A_{ul} h\nu_{ul} = \frac{16\pi^3 \nu_{ul}^4 (ex_0)^2}{3\epsilon_0 c^3}, \qquad (4.76)$$

where we have set $\nu_0 \equiv \nu_{ul}$. If we replace the classical dipole moment (ex_0) by the quantum mechanical dipole matrix element M_{ul} and solve for the transition probability A_{ul}, we obtain

$$A_{ul} = \frac{16\pi^3 \nu_{ul}^3 M_{ul}^2}{3h\epsilon_0 c^3} = \frac{\omega_{ul}^3 M_{ul}^2}{3\pi\hbar\epsilon_0 c^3}, \tag{4.77}$$

where the last term follows by converting from ν to ω and from h to $\hbar$.

OSCILLATOR STRENGTH

We introduced the concept of oscillator strength f in Chapter 2 (eqn. 2.89) and provided an expression relating A_{ul} and f_{lu} in (4.7). We can now define the transition probability in terms of the oscillator strength and either the angular frequency ω_{ul}, the frequency ν_{ul}, or the wavelength λ_{ul} as follows:

$$A_{ul} = \frac{e^2 \omega_{ul}^2}{2\pi\epsilon_0 m_e c^3}\left(\frac{g_l}{g_u}\right)f_{lu} = \frac{2\pi e^2 \nu_{ul}^2}{\epsilon_0 m_e c^3}\left(\frac{g_l}{g_u}\right)f_{lu}$$

$$= \frac{2\pi e^2}{\epsilon_0 m_e c\lambda_{ul}^2}\left(\frac{g_l}{g_u}\right)f_{lu}, \tag{4.78}$$

where f_{lu} is called the *absorption oscillator strength* of the transition. The expression on the right-hand side in terms of λ_{ul} is that used to obtain (4.7). There is also an *emission oscillator strength,* which has a value of

$$f_{ul} = -\frac{g_l}{g_u}f_{lu}. \tag{4.79}$$

For historical reasons, most tables and expressions involving oscillator strengths use the absorption oscillator strength instead of the emission oscillator strength. This is due to the fact that, before the advent of lasers, oscillator strengths were most often used in calculating absorption effects in various materials.

If we assume that a classical oscillator has a total strength of unity, then we can make the analogy that the sum of the oscillator strengths to all other energy levels from a specific level should add up to unity. Therefore, we can define f as being the fraction of either the radiated or absorbed energy that is associated with the electron making a transition from one level to any other specific energy level. The concept of oscillator strengths is a useful tool for estimating how strong a transition is compared to other transitions originating from the same energy level. It can be shown, for example, that if level l is the lowest level in a one-electron atom, then the values of the absorption oscillator strength to all allowed higher energy levels k obey the sum rule

$$\sum_k f_{lk} = 1. \tag{4.80}$$

This summation indicates that a single electron, when undergoing an absorptive transition from the ground state to many possible excited states, divides its absorption in proportion to the values of the individual oscillator strengths to each of those excited states, and that the total oscillator strength is unity. A similar rule applies to an excited state u, except that we must consider both absorption and emission from state u such that

$$\sum_{i<u} f_{ui} + \sum_{k>u} f_{uk} = 1. \tag{4.81}$$

This is known as the Thomas–Kuhn–Reiche sum rule. Since the emission oscillator strengths f_{ui} are negative according to (4.79), the separate summations in (4.81) will partially cancel each other. If more than one electron is involved in making a transition (which is usually not the case), then the total oscillator strength could be greater than unity.

SELECTION RULES FOR ELECTRIC DIPOLE TRANSITIONS INVOLVING ATOMS WITH A SINGLE ELECTRON IN AN UNFILLED SUBSHELL

Obtaining the transition probability from (4.77) involves calculating the square of the electric dipole matrix element M_{ul} as defined in (4.74). We thus have

$$M_{ul}^2 = e^2(X^2 + Y^2 + Z^2), \tag{4.82}$$

where X, Y, and Z each involve three separate integrals as follows:

$$X = \int \psi_u^*(n', l', m') \cdot x \cdot \psi_l(n, l, m) \, dV$$

$$= \int_0^\infty \int_0^\pi \int_0^{2\pi} \psi_{n'l'm'}^* r \sin\theta \cos\phi \psi_{nlm} r^2 \sin\theta \, d\theta \, d\phi \, dr; \tag{4.83}$$

$$Y = \int \psi_u^*(n', l', m') \cdot y \cdot \psi_l(n, l, m) \, dV$$

$$= \int_0^\infty \int_0^\pi \int_0^{2\pi} \psi_{n'l'm'}^* r \sin\theta \sin\phi \psi_{nlm} r^2 \sin\theta \, d\theta \, d\phi \, dr; \tag{4.84}$$

$$Z = \int \psi_u^*(n', l', m') \cdot z \cdot \psi_l(n, l, m) \, dV$$

$$= \int_0^\infty \int_0^\pi \int_0^{2\pi} \psi_{n'l'm'}^* r \cos\theta \psi_{nlm} r^2 \sin\theta \, d\theta \, d\phi \, dr. \tag{4.85}$$

For example, the X-matrix element from (4.83), using the values of ψ from (3.65) and (3.66), is

$$X = C_{n', l', m'}^* C_{n, l, m} \left(\int_0^\infty R_{n'l'} r R_{nl} r^2 \, dr \right)$$

$$\times \left[\int_0^\pi P_{l'}^{m'}(\cos\theta) \sin\theta P_m^l(\cos\theta) \sin\theta \, d\theta \right]$$

$$\times \left(\int_0^{2\pi} e^{-im'\phi} \cos\phi \, e^{im\phi} \, d\phi \right). \tag{4.86}$$

Substituting $\cos\phi = \frac{1}{2}(e^{i\phi} + e^{-i\phi})$ into the integral for ϕ leads to

$$\frac{1}{2}\int_0^{2\pi} [e^{i(m-m'+1)\phi} + e^{i(m-m'-1)\phi}]\,d\phi = \begin{cases} \pi & \text{if } m' = m \pm 1, \\ 0 & \text{otherwise.} \end{cases} \quad (4.87)$$

Using this result in (4.86) while also making use of the recurring relation

$$\sin\theta P_{l'}^{m-1}(\cos\theta) = \frac{P_{l'+1}^m(\cos\theta) - P_{l'-1}^m(\cos\theta)}{2l'+1} \quad (4.88)$$

yields an integral of a product of $P_l^m(\cos\theta)$ functions having the same upper index. This suggests that

$$\int_0^{\pi} P_{l'}^{m\pm1}\sin\theta P_l^m \sin\theta\,d\theta = 0 \quad \text{unless } l' = l \pm 1 \quad (4.89)$$

and

$$\text{there exist no restrictions on the } R \text{ integral} \quad \text{if } l' = l \pm 1. \quad (4.90)$$

If we carried out the evaluation of Z^2 we would find also that the dipole matrix element is allowed only for $m' = m$ or $\Delta m = 0$. This type of analysis leads to the selection rules for electric dipole transitions, which we will not develop any further but only summarize. These selection rules, which were obtained by ignoring electron spin, indicate that transitions can occur between two levels only if $\Delta l = \pm 1$ and $\Delta m = \pm 1$ or 0. If the spin is taken into account, then the selection rules for an atom with a single electron in an unfilled subshell become

$$\begin{array}{l} \Delta l = \pm 1, \\ \Delta s = 0, \\ \Delta j = \pm 1 \text{ or } 0 \quad \text{but } j = 0 \nrightarrow j = 0, \\ \Delta m_j = \pm 1 \text{ or } 0 \quad \text{but } m_j = 0 \nrightarrow m_j = 0 \text{ if } \Delta j = 0. \end{array} \quad (4.91)$$

An arrow with a cancellation slash indicates that such transitions are not allowed for the cases specified. For instance, it is not possible to have a transition from one state with $j = 0$ to another state with $j = 0$.

EXAMPLE

Calculate the transition probability for the transition in hydrogen from one of the levels of the first excited state defined by the quantum numbers $n = 2$, $l = 1$, $m = +1$ ($2p\,^2P_{3/2}^o$) to the ground level $n = 1$, $l = 0$, $m = 0$ ($1s\,^2S_{1/2}$). (The superscript o denotes odd parity.) Then obtain the oscillator strength, determine the radiative lifetime of the upper level, and obtain the natural linewidth of the transition. The 2p level is sometimes referred to as the *resonance* level (or state) because it has the greatest probability of making a transition to the ground state compared to other possible transitions from higher energy levels.

This is the equivalent of driving a harmonic oscillator at its most likely resonance frequency.

We must remember that we are here considering only the transition from the $m = +1$ sublevel of the 2p energy level. The other values of m (i.e. $m = 0$ and $m = -1$) must also be calculated to obtain the total transition probability from $n = 2$, $l = 1$ to $n = 1$, $l = 0$. As before, we will designate the upper p state with a subscript u for upper state, and the ground state s with an l for lower state. We will use (4.74) and (4.82) to obtain the transition probability. From (3.65) and (3.66), the wave functions for the upper and lower levels involved in the transition are

$$2p\,^2P^o_{3/2}\,(2, 1, +1) \Rightarrow \psi_u^{2,1,+1} = \frac{1}{8\pi^{1/2}a_H^{5/2}}re^{-r/2a_H}\sin\theta e^{i\phi} \qquad (4.92)$$

and

$$1s\,^2S_{1/2}\,(1, 0, 0) \Rightarrow \psi_l^{1,0,0} = \frac{1}{\pi^{1/2}a_H^{3/2}}e^{-r/a_H}. \qquad (4.93)$$

Using (4.77), we have

$$A_{ul}(2p_{+1} \rightarrow 1s) = \frac{\omega_{21}^3}{3\pi\epsilon_0 \hbar c^3}\left|\int \psi_u^{2,1,+1*}\mathbf{er}\psi_l^{1,0,0}\,dV\right|^2. \qquad (4.94)$$

Using the wave function values of (4.92) and (4.93) together with

$$\mathbf{r} = x\hat{i} + y\hat{j} + z\hat{k}, \qquad (4.95)$$

where

$$x = r\sin\theta\cos\phi, \qquad y = r\sin\theta\sin\phi, \qquad z = r\cos\theta, \qquad (4.96)$$

and

$$dV = r^2 \sin\theta\,dr\,d\theta\,d\phi, \qquad (4.97)$$

we can proceed to calculate $A_{ul}(2p_{+1} \rightarrow 1s)$ using (4.82) to obtain

$$\left|\int \psi_u^*\mathbf{er}\psi_l\,dV\right|^2 = M_{ul}^2 = e^2\left[\left|\int \psi_u^*x\psi_l\,dV\right|^2 \right.$$
$$\left. + \left|\int \psi_u^*y\psi_l\,dV\right|^2 + \left|\int \psi_u^*z\psi_l\,dV\right|^2\right]. \qquad (4.98)$$

We will compute the integral associated with X, and then merely state the values of the integrals for Y and Z, to obtain the total value for the dipole matrix element as follows:

$$X = \int \psi_u^*x\psi_l\,dV$$

$$= \int \frac{1}{8\pi^{1/2}a_H^{5/2}}re^{-r/2a_H}\sin\theta e^{-i\phi}r$$

$$\sin\theta\cos\phi\frac{1}{\pi^{1/2}a_H^{3/2}}e^{-r/a_H}r^2\sin\theta\,dr\,d\theta\,d\phi$$

EXAMPLE (cont.)

$$= \frac{1}{8\pi a_H^4} \int \int \int r^4 e^{-3r/2a_H}\, dr (\cos^2\theta - 1)\, d(\cos\theta) e^{-i\phi} \left(\frac{e^{i\phi} + e^{-i\phi}}{2} \right) d\phi$$

$$= \frac{1}{8\pi a_H^4} \int_0^\infty r^4 e^{-3r/2a_H}\, dr \int_0^\pi (\cos^2\theta - 1)\, d(\cos\theta) \int_0^{2\pi} \frac{1 + e^{-2i\phi}}{2}\, d\phi$$

$$= 4 \left(\frac{2}{3} \right)^5 a_H.$$

Similarly,

$$\int \psi_u^* y \psi_l\, dV = \frac{4}{i} \left(\frac{2}{3} \right)^5 a_H \quad \text{and} \quad \int \psi_u^* z \psi_l\, dV = 0.$$

Therefore, from (4.98) we obtain

$$M_{ul}^2 = \left| \int \psi_u^* e\mathbf{r}\psi_l\, dV \right|^2 = e^2 \left[\left| 4 \left(\frac{2}{3} \right)^5 a_H \right|^2 + \left| \frac{4}{i} \left(\frac{2}{3} \right)^5 a_H \right|^2 + |0|^2 \right]$$

$$= \left[4^2 \left(\frac{2}{3} \right)^{10} a_H^2 + 4^2 \left(\frac{2}{3} \right)^{10} a_H^2 \right] = 32 \left(\frac{2}{3} \right)^{10} e^2 a_H^2. \tag{4.99}$$

Hence

$$A_{ul}(2\mathrm{p}_{+1} \to 1\mathrm{s}) = \frac{\omega_{ul}^3}{3\pi\epsilon_0 \hbar c^3} 32 \left(\frac{2}{3} \right)^{10} e^2 a_H^2 = 6.2 \times 10^8\ \mathrm{s}^{-1}. \tag{4.100}$$

We have only calculated the transition probability from a specific sublevel of $n = 2$ for hydrogen. Energy level 2 has four levels, all with the same energy since we have $l = 1$ ($m_l = +1$, $m_l = 0$, and $m_l = -1$) and $l = 0$; therefore this state is referred to as *degenerate*. Thus, when a transition occurs from such a state, we do not know which level the electron actually originated from. It is therefore useful in such situations to calculate an average transition probability from $n = 2$ to $n = 1$. We know from the selection rules that the transition probability is zero from the sublevel in which $l = 0$. Therefore, in order to determine the average transition probability of all possible transitions from $n = 2$ to $n = 1$, we use the value obtained in (4.100) for one of the 2p states ($m = +1$), weighted by the statistical weights $g = 6$ for all of the 2p states (since they all have that same transition probability), divided by the combined statistical weights for all of the 2p and 2s states ($g = 6 + 2 = 8$), leading to

$$A(2 \to 1) = (6/8)(6.2 \times 10^8\ \mathrm{s}^{-1}) = 4.7 \times 10^8\ \mathrm{s}^{-1}. \tag{4.101}$$

Using (4.7), we can then obtain a value of the oscillator strength for the $(2 \to 1)$ transition as follows:

$$f_{12} = (1.5 \times 10^4)\left(\frac{g_2}{g_1}\right)\lambda_{21}^2 A_{21}. \tag{4.102}$$

From Chapter 3 we know that $\lambda_{21} = 1.2157 \times 10^{-7}$ m, $g_2 = 8$, and $g_1 = 2$. Hence

$$f_{12} = 0.416. \tag{4.103}$$

SELECTION RULES FOR RADIATIVE TRANSITIONS INVOLVING ATOMS WITH MORE THAN ONE ELECTRON IN AN UNFILLED SUBSHELL

At the end of Chapter 3, the splittings of electron levels were described for atoms with more than one electron in an unfilled subshell. In doing so, the notation for labeling the energy levels was introduced in terms of the angular momentum quantum numbers of the system. The values of the spin, orbital, and total angular momentum quantum numbers for the combined electrons of the unfilled shell were given in the energy-level notation with the capital letters S, L, and J (eqn. 3.74) for each energy level. For such multiple electron levels, the following selection rules apply:

$$\Delta l = \pm 1 \quad \text{for the changing electron (change in parity),}$$
$$\Delta S = 0, \qquad \Delta L = 0, \pm 1,$$
$$\Delta J = 0, \pm 1 \quad \text{but } J = 0 \nrightarrow J = 0, \tag{4.104}$$
$$\Delta M_J = 0, \pm 1 \quad \text{but } M_J = 0 \nrightarrow M_J = 0 \text{ if } \Delta J = 0,$$

where again $J = 0 \nrightarrow J = 0$ indicates that a particular level with quantum number $J = 0$ cannot radiate and decay to a lower level also having a value of $J = 0$.

We generally need to use only the ΔL, ΔS, and ΔJ selection rules, as well as the parity selection rule, when we are using LS coupling notation. We can ignore the ΔM rules in most cases.

PARITY SELECTION RULE

The wave functions for the various electronic states possess either even or odd symmetry, as viewed when reflected through the origin (the nucleus of the atom or center of symmetry). The states with odd orbital angular momentum quantum numbers (L) possess odd symmetry, or *parity*, while those with even quantum numbers possess even symmetry or parity. If an overlap integral of the form $\int \psi^*(A)\psi(B)\,dV$ is made between two states A and B, the integral will have a nonzero value only if A and B have the same parity. Thus, when the dipole matrix element M_{AB} in the form of

$\int \psi^*(A) \cdot e\mathbf{r} \cdot \psi(B) \, dV$ (eqn. 4.74) is determined, that integral will have non-zero values only when the wave functions $\psi(A)$ and $\psi(B)$ have opposite parity, since the vector $\mathbf{r}$ is an odd function and therefore introduces odd symmetry to the integral. This effect is reflected in the foregoing selection rules by requiring that $\Delta l = \pm 1$, which entails that the electron involved in the transition must change parity.

The energy-level parity is determined by the number of electrons having odd parity, those with $l = 1, 3, 5, \ldots$ designated as p, f, h, ... If there are an even number of odd-parity electrons then the energy level has even parity. If there are an odd number of odd-parity electrons then the energy level has odd parity. In determining parity, it is necessary to count the number of odd-parity electrons in unfilled subshells only, since all filled shells and subshells have an even number of odd electrons. For example, an energy level having an unfilled subshell with 2p or $(2p)^3$ or 4f would have an odd number of odd-parity electrons and would therefore be designated as an odd state. Energy levels with an even number of odd-parity electrons (such as $(2p)^2$, $(4f)^4$, 2p4f, etc.) would all have an even number of odd-parity electrons and would thus be recognized as even-parity states. For example, an electron might make a transition from a 2p to a 1s electronic state; the net change is to go from an odd number of odd electrons (one 2p electron) to an even number of odd electrons (zero p electrons) in the unclosed subshell. This transition would be allowed by the parity selection rule.

For convenience, the energy levels with odd parity are often denoted in the energy-level notation with a superscript o in the upper right-hand corner, such as $^3P_1^o$ or $^1D_2^o$. Thus, a transition $2p\,^2P_{3/2}^o \rightarrow 1s\,^2S_{1/2}$ would be allowed from the standpoint of the angular momentum selection rules ($\Delta S = 0$, $\Delta L = -1$, and $\Delta J = -1$) as well as from the standpoint of the parity selection rule ($\Delta l = -1$). On the other hand, transitions of the form $1p3p\,^3D_2^o \rightarrow 1p2p\,^3P_1^o$ or $2p3p\,^3D_2 \rightarrow 2p3f\,^3F_3$, which would be allowed according to the angular momentum selection rules, would not be allowed because parity does not change from one level to the other in the transition.

INEFFICIENT RADIATIVE TRANSITIONS – ELECTRIC QUADRUPOLE AND OTHER HIGHER-ORDER TRANSITIONS

Electric dipole transition probabilities are those involving matrix elements having the dipole moment operator $e\mathbf{r}$ sandwiched between the wave functions of the two states involved in the transition. This involves the linear dependence r of the separation between the positive and negative charges of the atom. Higher-order radiating functions, such as those leading to magnetic dipole and electric quadrupole transitions, are described theoretically by inserting higher-order factors (such as r^2, etc.) associated with the separation between the charges, giving rise to much weaker transition probabilities than those of electric dipole transitions. Electric quadrupole transitions, for example, can be shown to have selection rules of $\Delta L = \pm 2$.

Radiative lifetimes of states that can decay only by quadrupole radiation are as much as four to six orders of magnitude longer than for electric dipole radiation, leading to decay rates four to six orders of magnitude slower than for electric dipole transitions. These "slow" transitions do play a role in some laser systems, so it is therefore important to be aware of their existence.

REFERENCES

A. Corney (1977), *Atomic and Laser Spectroscopy.* Oxford: Clarendon Press.
H. G. Kuhn (1969), *Atomic Spectra,* 2nd ed. New York: Academic Press.
R. B. Leighton (1959), *Principles of Modern Physics.* New York: McGraw-Hill.

PROBLEMS

1. Show how to solve the differential equation (4.1) for N_u to obtain (4.2). Show that your answer is in fact a solution of the equation.

2. Compute the power radiated from an electron oscillating at an angular frequency of $\omega = 10^{15}$ rad/s. Assume that x_0 is equal to the Bohr radius of an orbiting electron.

3. If 10 torr of He gas is added to an electrical discharge composed of H atoms, how much would the decay time of the $n = 2$ state of H decrease if the pressure-broadening factor due to He is 10^7/s-torr? What would it be if the He pressure were increased to 100 torr? (1 torr $= 1/760$ of atmospheric pressure.) *Hint:* Use the average value of the transition probability for $n = 2$ to $n = 1$ of H calculated in the example at the end of this chapter.

4. With reference to (4.24), obtain values of the frequency at which the intensity drops to one half its maximum value. What is the FWHM (full width at half maximum) value for the frequency width?

5. Using the value of $I(\omega)$ in (4.24), show that $\int_0^\infty I(\omega)\, d\omega = I_0$ (eqn. 4.25). *Hint:* You need to assume that $\gamma_{ul} \ll \omega_0$, a reasonable approximation.

6. Compute the Doppler broadening for the 632.8-nm laser transition in the He–Ne laser, assuming a single isotope of Ne^{20} and that the laser operates at a discharge-bore temperature of 100°C. Compare this broadening to the natural broadening obtained in the example at the beginning of the chapter.

7. Assume that you have the following two special isotope mixtures of cadmium:

(a) equal amounts of Cd^{112} and Cd^{114};
(b) equal amounts of Cd^{110} and Cd^{114}.

Assume that the He–Cd laser is operating in a gaseous discharge at a temperature of 300°C. Make a plot of the emission envelope of the 441.6-nm transition for isotope mixtures (a) and (b). Use the following values for the difference in the center frequencies of the emission of the different isotopes on the 441.6-nm transition:

$$\Delta\nu(Cd^{110} - Cd^{112}) = 1.57 \times 10^9 \text{ Hz};$$
$$\Delta\nu(Cd^{112} - Cd^{114}) = 1.46 \times 10^9 \text{ Hz}.$$

Note: Both isotopes radiate at a wavelength of approximately 441.6 nm, with only a very slight difference in their center frequencies as indicated here.

8. The pulsed lead laser operates on several transitions in the visible and ultraviolet spectrum, including the 722.9-nm and 405.7-nm transitions. Compute the natural linewidth, the Doppler width, and $\Delta\nu_D/\nu_0$ for these two laser transitions. Assume that the lead vapor is at a temperature of 1,300 K, providing a lead vapor pressure of approximately 1 torr. The energy-level diagram associated with those transitions is shown in the figure. Why would the transition from $6p7s\ ^3P_1^o \to 6p^2\ ^1D_2$

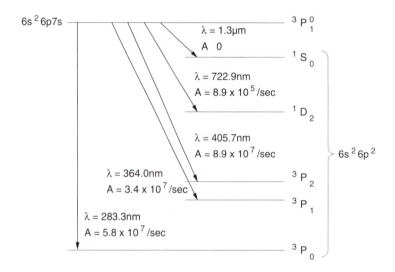

The dominant isotope for lead is Pb^{208}.

not normally be allowed according to electric dipole transition selection rules? (This is a case where the rules break down; there is a finite value of the electric dipole matrix element connecting these two energy levels, because the $^1D_2$ level is actually not a pure singlet state but instead has some triplet wave function mixed in with the singlet wave function. This often occurs for heavier atoms.)

9. The selenium soft–X-ray laser operates at a wavelength of 20.6 nm in Se^{24+} ions. Those ions are produced at a plasma temperature of approximately 10^7 K. How large is the Doppler width of that laser transition compared with the width of the He–Ne 632.8-nm laser determined in Problem 6?

10. If two emission lines in two different species have the same center frequency ν_0 and the same emission linewidth $\Delta\nu_{FWHM}$, but one is homogeneously broadened and the other is Doppler broadened, how much greater is the value of the emission from the homogeneously broadened line at a frequency displaced from ν_0 by a factor of $\nu = 5\Delta\nu_{FWHM}$?

11. Obtain the expression for the oscillator strength of (4.7) from (4.78).

12. The first nine levels of atomic calcium are tabulated as follows.

4s4p	———————	$^1P_1^o$	2.3652×10^6 m^{-1}	
	———————	$^1D_2$	2.1850×10^6 m^{-1}	
	———————	$^3D_3$	2.0370×10^6 m^{-1}	
4s3d	———————	$^3D_2$	2.0349×10^6 m^{-1}	
	———————	$^3D_1$	2.0335×10^6 m^{-1}	
	———————	$^3P_2^o$	1.5316×10^6 m^{-1}	
4s4p	———————	$^3P_1^o$	1.5210×10^6 m^{-1}	
	———————	$^3P_0^o$	1.5158×10^6 m^{-1}	
4s^2	———————	$^1S_0$	0.00	

Determine which transitions are most likely to occur based upon the selection rules given in this chapter. Also determine the wavelengths of those transitions. The energy value for a specific level i above the ground level 1 is given on the right-hand side in wavenumbers σ_{i1}. The wavenumber σ_{ij} of the transition from level i to level j is determined from the equation for the energy difference between the two levels involved in the transition: $\Delta E_{ij} = E_i - E_j = h\nu_{ij} = hc\sigma_{ij} = hc(\sigma_{i1} - \sigma_{j1})$. Therefore, the wavelength of the transition would be $\lambda_{ij} = 1/\sigma_{ij} = 1/(\sigma_{i1} - \sigma_{j1})$.

5 ENERGY LEVELS AND RADIATIVE PROPERTIES OF MOLECULES, LIQUIDS, AND SOLIDS

SUMMARY In Chapter 3, the concept of discrete energy levels was introduced, and both the classical and quantum mechanical derivations of the energy levels and state functions (wave functions) associated with those energy levels were outlined. Chapter 4 then dealt with the radiative properties of transitions between energy levels. Various emission-line–broadening mechanisms were also introduced to provide a basis for understanding the widths of emission and absorption lines and, later on in the book, for predicting the spectral width over which gain occurs in the laser medium. Transition probabilities were then discussed, and selection rules were summarized for predicting which transitions were most likely to occur between energy levels.

The present chapter will now extend these concepts of energy levels and radiative properties to more complex types of laser materials, including molecules, liquids (organic dyes), and solids (both dielectric and semiconductor laser materials). Discrete energy levels of molecules will be shown to include rotational and vibrational levels in addition to those determined by various electronic configurations. Energy levels of organic dye molecules will be described and related to the singlet and triplet levels derived in Chapter 3 for the helium atom. Energy levels of solid-state dielectric materials will be outlined, and the notation for those levels will be associated with the atomic level designations previously obtained in Chapter 3. Finally, the energy levels and radiative properties of semiconductors will be discussed. These levels will be shown to be significantly different from those of the other species discussed in this chapter, owing to the regularity and close spacing of the atoms in the crystal matrix of the semiconductor material.

5.1 MOLECULAR ENERGY LEVELS AND SPECTRA

ENERGY LEVELS OF MOLECULES

When two or more atoms are located in proximity of each other (within distances of the order of atomic dimensions), certain short-range forces act between them; in many instances those forces bind the atoms together to make molecules. This binding can take the form of *covalent* binding if the two atoms are similar, such as H_2 or O_2, or *ionic* binding if the atoms are highly dissimilar, such as NaCl or CaF_2. In either situation, the energy levels that result from the molecular binding are found to be quantized as they were for atoms, and, as in the case of atoms, the molecules also emit photons when electrons jump from higher energy states to lower ones. Unlike atoms, however, molecules have energy states that do not necessarily involve electronic transitions. These states, related to rotational and vibrational motions of the molecules, lead to energy levels much more closely spaced

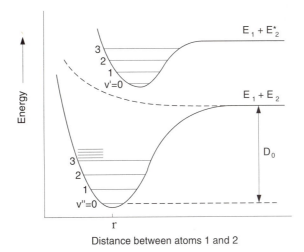

Figure 5-1. Energy levels versus separation distance r between two atoms

than those associated with electronic transitions. Consequently, the radiation resulting from transitions between those levels is at much longer wavelengths, in the middle and far infrared, than that from electronic transitions.

Assume that two atoms 1 and 2 are separated by an infinite distance and have ground-state energies (ionization energies) of E_1 and E_2. We have indicated the combined sum of these energies ($E_1 + E_2$) on the right-hand side of Figure 5-1, a graph of energy versus separation distance r between the two atoms. We have also indicated the ground state of atom 1 with energy E_1 combining with an excited state of atom 2 with energy E_2^*. The effect of bringing the atoms closer together is also shown in Figure 5-1. The resulting electronic energy levels shown at various distances of the combined atoms are in effect the energy levels of the molecule. For many of the electronic energy levels in a large number of atom pairs, there is a significant "dip" in the energy at a specific value of r, which suggests a stable binding of the atoms of that molecule when it resides in a specific energy level that is below the top of the dip and within the allowed range of separation distances r. When the atoms reach that range of values of r, they are held together or *bound* in that region by the potential "well." They could only be pulled apart if enough energy (D_0) is supplied, in the form of either collisional or optical excitation, to "climb out" of the well; D_0 is known as the *dissociation energy.* For some combinations of atoms, the electronic energy levels increase continuously with decreasing r (dashed line of Figure 5-1), thereby representing unbound levels in which the atoms do not have a stable attractive state and are therefore repulsed or pushed apart. Repulsive (unbound) or attractive (bound) states can occur as either the ground electronic state or excited electronic states of a molecule.

Electronic energy levels of molecules typically have the largest energy separations. However, superimposed upon each electronic state of the molecule are the possible vibrational levels, which are equally spaced ladderlike levels contained within the potential wells as shown in Figure 5-1. Associated with each vibrational level are a series of more closely spaced rotational

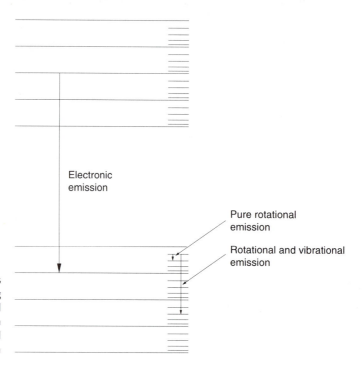

Figure 5-2. Energy levels of a molecule, indicating electronic emission as well as pure rotational emission and rotational–vibrational emission

energy levels that do not have equal spacing but instead are separated in proportion to their rotational quantum numbers.

The dipped curve indicates the extreme values of r that can be attained by atoms of the molecule for any specific electronic state. Thus, when higher vibrational levels are populated within an electronic state, the molecules oscillate through wider excursions of r as shown in Figure 5-1.

Figure 5-2 shows a diagram of two electronic energy levels of a hypothetical molecule, indicating both equally spaced vibrational levels and the more closely spaced rotational levels associated with each electronic level. Several possible types of transitions are shown, including pure rotational transitions (in which only the rotational level changes), rotational–vibrational transitions (in which both the rotational and vibrational levels change), and electronic transitions that could occur between any two sublevels of the two electronic levels. Transitions between electronic states typically produce emission in the visible or ultraviolet spectral regions. Transitions between vibrational states are generally observed in the middle-infrared portion of the spectrum, and pure rotational transitions produce spectra in the far-infrared and microwave regions.

We will now consider the possible values of the various types of energy levels of molecules. We will consider the total energy of the system as the sum of the electronic, vibrational, and rotational energies, which can be expressed as

$$E_{\text{total}} = E_{\text{elect}} + E_{\text{vib}} + E_{\text{rot}}. \tag{5.1}$$

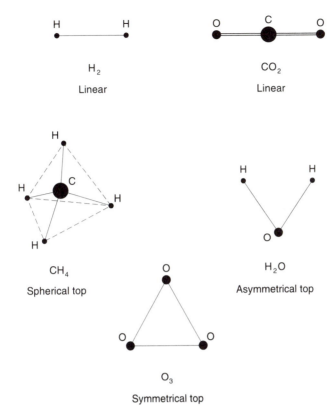

Figure 5-3. Description of five types of simple molecules, along with specific examples of each type

CLASSIFICATION OF SIMPLE MOLECULES

For a molecule containing N atoms, there are $3N$ degrees of freedom. Three of these involve the three translational directions of motion, three involve rotation about the three axes (two for linear molecules), and the remaining number $3N-6$ ($3N-5$ for linear molecules) involve vibrational motions. Simple molecules are categorized into four basic types: linear molecules, spherical-top molecules, symmetrical-top molecules, and asymmetrical-top molecules. Examples of each are shown in Figure 5-3.

In the following sections we will show how to calculate the energy levels for the linear molecule and for the symmetric-top molecule. We will not consider the asymmetric-top molecule or the spherical-top molecule, since they are much more complicated.

ROTATIONAL ENERGY LEVELS OF LINEAR MODELS

The allowed rotational energies of simple molecules are dependent upon which type of molecules are involved. The classical rotational kinetic energy for linear molecules is given by

$$E_{rot} = \frac{1}{2} I\omega^2 = \frac{(I\omega)^2}{2I} = \frac{L^2}{2I}, \tag{5.2}$$

where I is the moment of inertia of rotation, ω is the angular velocity of rotation, and L is the angular momentum $I\omega$. The solution of the Schrödinger equation (see Chapter 3) for the rotation of a rigid body leads to the quantization of the angular momentum:

$$L^2 = J(J+1)\hbar^2, \tag{5.3}$$

where J is the angular momentum quantum number with values of

$$J = 0, 1, 2, 3, \ldots . \tag{5.4}$$

Equation (5.3) has the same form as that obtained for the angular momentum of the hydrogen atom in (3.67), with the angular momentum quantum number l replaced by the symbol J used by molecular spectroscopists. Using (5.2) and (5.3), the rotational energy can be expressed as

$$E_{\text{rot}}^J = \frac{J(J+1)\hbar^2}{2I} = J(J+1)Bhc, \tag{5.5}$$

where B is defined as

$$B = \frac{h}{8\pi^2 cI}. \tag{5.6}$$

Here I is the moment of inertia of the molecule about the axis of rotation and c is the velocity of light. For a diatomic molecule in which the masses are identical, $I = Mb^2$, where $2b$ is the average separation distance between the atoms of the molecule and the mass of each atom is $M/2$.

EXAMPLE

Compute the energy of the first excited rotational energy level ($J = 1$) of the linear molecule H_2. For a hydrogen molecule, the average separation between the two hydrogen atoms in the ground state is 0.107 nm. Also compute the wavelength of a hypothetical radiative transition from the first excited rotational state to the ground rotational state ($J = 0$). (Such a transition cannot actually occur in H_2, since it is a homonuclear molecule and hence has no dipole moment. However we will use H_2 for its simplicity, assuming for the sake of this exercise that it can radiate.)

We will first compute the energy for $J = 1$ from (5.5), with $b = 0.107/2$ nm $= 0.0535$ nm and $M = 2M_H = 2 \times 1.67 \times 10^{-27}$ kg $= 3.34 \times 10^{-27}$ kg:

$$I = Mb^2 = (3.34 \times 10^{-27} \text{ kg})(0.0535 \text{ nm})^2 = 9.56 \times 10^{-48} \text{ kg-m}^2,$$

$$B = \frac{h}{8\pi^2 cI} = \frac{6.62 \times 10^{-34} \text{ J-s}}{8\pi^2 (3 \times 10^8 \text{ m/s})(9.56 \times 10^{-48} \text{ kg-m}^2)}$$

$$= 2.92 \times 10^3 \text{ J-s}^2/\text{kg-m}^3.$$

The values of the rotational energies can therefore be expressed as

$$E_{rot}^J = \frac{J(J+1)\hbar^2}{2I} = J(J+1)Bhc$$

$$= J(J+1)(2.92\times10^3 \text{ J-s}^2/\text{kg-m}^3)(6.62\times10^{-34} \text{ J-s})(3\times10^8 \text{ m/s})$$

$$= J(J+1)(5.80\times10^{-22} \text{ J}^2\text{-s}^2/\text{kg-m}^2) = J(J+1)(5.80\times10^{-22} \text{ J}).$$

Thus, for the first excited energy level $J = 1$,

$$E_{rot}^1 = 1(1+1)(5.58\times10^{-22} \text{ J}) = 1.16\times10^{-21} \text{ J} = 7.25\times10^{-3} \text{ eV}.$$

The transition wavelength between E_{rot}^1 and E_{rot}^0 is

$$\lambda_{10} = \frac{hc}{E_{rot}^1 - E_{rot}^0} = \frac{(6.62\times10^{-34} \text{ J-s})(3\times10^8 \text{ m/s})}{1.16\times10^{-21} \text{ J}}$$

$$= 1.71\times10^{-4} \text{ m} = 171 \ \mu\text{m},$$

which is obviously a transition in the far-infrared portion of the spectrum.

ROTATIONAL ENERGY LEVELS OF SYMMETRIC-TOP MOLECULES

For symmetric-top molecules, two quantum numbers J and K are necessary to specify the rotational energy levels. The values of the rotational energy levels are then given by

$$E_{rot} = J(J+1)Bhc + K^2(C-B)hc, \tag{5.7}$$

where J has the same range of values as for (5.4), K has values of 1, 2, 3, ..., and B and C involve the two moments of inertia of the molecule:

$$B = \frac{h}{8\pi^2 cI_b}, \qquad C = \frac{h}{8\pi^2 cI_c}, \tag{5.8}$$

where I_c is the moment about the symmetry axis and I_b is the moment about the perpendicular axis.

SELECTION RULES FOR ROTATIONAL TRANSITIONS

When considering radiation that involves changes in rotational levels, the general selection rules for rotational states can be summarized as follows:

$$\Delta J = 0, \pm 1, \qquad \Delta K = \pm 1. \tag{5.9}$$

Thus, transitions can occur either between rotational levels with the same J value or between different J values separated by a value of ± 1, whereas K must change by ± 1.

Radiative transitions between rotational levels occur as a result of changes of the rotational or spinning rate of a dipole molecule in which the positive and negative charges are slightly separated. Symmetric molecules with no charge asymmetry have no dipole distribution or dipole moment, and therefore produce no emission from pure rotational transitions.

EXAMPLE

Draw a diagram of the first five rotational energy levels of the ground state of a diatomic molecule and the possible allowed transitions between those levels.

The first five rotational energy levels are spaced according to (5.5) in units of Bhc. Thus the energy levels increase in separation in ratios of

$$\frac{(J+1)(J+2)Bhc}{J(J+1)Bhc} = \frac{J+2}{J}.$$

The energy levels are shown in Figure 5-4 with their appropriate J values. According to the selection rules of (5.9), transitions are allowed for changes in J of $\Delta J = 0, \pm 1$. In this case, the only allowed transitions are those for $\Delta J = -1$ because transitions can move only downward, and downward movements correspond to a decrease in values of J as shown in Figure 5.4. Transitions for $\Delta J = 0$ or $+1$ will

J=4

J=3

J=2

J=1
J=0

Figure 5-4. The first five rotational levels of a diatomic molecule, showing the relative energies between the levels

occur only when they are between rotational levels of different vibrational levels, as will be described in the following sections. For those cases it will be possible to have transitions to lower energy levels that have higher J values.

VIBRATIONAL ENERGY LEVELS

The simplest way to describe the vibrational levels of a molecule is to compare the molecule to a simple harmonic oscillator; this was done in Chapter 3 for the Bohr model of an atom. In the molecular case, the most appropriate model we can use is that of two atoms attached to opposite ends of a vibrating spring, with the system oscillating in various possible normal modes. This harmonic oscillator model for a molecule is more realistic and much more like an actual molecule than the Bohr atomic model was like an atom, especially for the case of a linear diatomic molecule.

The Schrödinger equation can be solved for the eigenvalues of the energy by using the harmonic oscillator term $\sum_i \frac{1}{2} k_i x_i^2$ for the potential energy, where k_i is the effective spring constant for each vibrational mode and x_i is the displacement from the equilibrium position for each "spring." The solution provides values of the vibrational energies of

$$E_{\text{vib}} = (v_1 + \tfrac{1}{2})h\nu_1 + (v_2 + \tfrac{1}{2})h\nu_2 + \cdots; \qquad (5.10)$$

$\nu_1, \nu_2, \ldots$ are the resonant frequencies of the oscillator and $v_1, v_2, \ldots$ are the vibrational quantum numbers, where v_i can have only integral values $0, 1, 2, \ldots$. Equation (5.10) thus describes a molecule having two separate vibrational modes, with vibration quantum numbers v_1 and v_2.

For the linear diatomic molecule consisting of masses M_1 and M_2, there would be only one vibrational quantum number v_1 and one resonant frequency $\nu_1 = (1/2\pi)\sqrt{k_1/\mu}$, where $\mu = M_1 M_2 / (M_1 + M_2)$ is the reduced mass of the molecule system. For the case of a molecule such as H_2, where both atoms are identical, $\mu = M/2$. The vibrational energy levels are thus represented by a series of equidistant levels, with the lowest level having a nonzero energy that is $h\nu_1/2$ above zero energy. This level is referred to as the *zero-point energy,* and is occupied even at absolute zero temperature. The zero-point energy represents the lowest possible value of energy for the system, and cannot be obtained by classical calculations.

SELECTION RULE FOR VIBRATIONAL TRANSITIONS

The selection rule for vibrational transitions is

$$\Delta v = \pm 1, \qquad (5.11)$$

and is strictly applicable only for pure harmonic motion. In many instances the motion is anharmonic (nonsinusoidal); in such cases transitions involving $\Delta v = \pm 2, \pm 3, \ldots$ are also possible, although they are generally much weaker than the fundamental transitions ($\Delta v = \pm 1$), much like electric quadrupole radiation is much weaker than electric dipole radiation for an atom.

Covalent or homonuclear molecules do not have a permanent electric dipole moment, and therefore have no radiative decay mode. With no dipole moment, populated excited levels are therefore most likely to lose their energy by collisions with other particles. On the other hand, ionic or heteronuclear molecules do have a dipole moment and thus produce dipole radiation on pure rotational or on rotational–vibrational transitions.

ROTATIONAL–VIBRATIONAL TRANSITIONS

As a molecule is vibrating, it can also be rotating. Because rotational energy–level spacings are much smaller than vibrational energy–level spacings, each vibrational level is associated with a set of rotational levels. When this occurs, the levels are called rotational–vibrational levels. When radiative transitions occur between such levels, they occur from a rotational sublevel of one vibrational level to a rotational sublevel of another vibrational level. Such a transition is referred to as a rotational–vibrational transition.

An energy diagram for two vibrational levels J_1 and J_2 of a typical molecule along with their rotational sublevels is shown in Figure 5-5, which indicates also the radiative transitions that can occur between the levels. The J selection rule for rotational–vibrational transitions is

$$\Delta J = 0, \pm 1. \tag{5.12}$$

The emission spectrum is normally arranged in three branches, referred to as the P, Q, and R branches. The following selection rules determine which branch a transition is associated with:

$$\begin{aligned}
\Delta J &= +1 \quad \text{P branch,} \\
\Delta J &= 0 \quad\;\; \text{Q branch,} \\
\Delta J &= -1 \quad \text{R branch.}
\end{aligned} \tag{5.13}$$

Thus, a P(4) transition would be from $J = 3$ to $J = 4$. The transition designation is determined by the J value of the level on which the transition terminates.

In most diatomic molecules (and also in some other linear molecules such as CO_2), transitions associated with the Q branch are usually not allowed, as shown by the dashed lines for the Q-branch transitions between J_2 and J_1 in Figure 5-5. This is because the orbital angular momentum is zero in both the upper and lower states for these particular transitions. The molecule nitrous oxide (NO) is the only exception to this rule for diatomic molecules. Also, for triatomic molecules such as CO_2, the selection rule

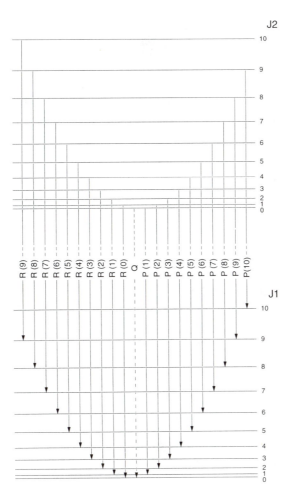

Figure 5-5. Energy-level diagram, showing two vibrational levels of a molecule along with the rotational sublevels, as well as the presence of the P- and R-branch transitions and absence of Q-branch transitions

$\Delta v = \pm 1$ is not relevant since transitions can occur from one vibrational mode to another. For example, we can have a laser transition from $v_1 = 0$, $v_2 = 0$, $v_3 = 1$ $(0, 0, 1)$ to $v_1 = 0$, $v_2 = 2$, $v_3 = 0$ $(0, 2, 0)$.

EXAMPLE

Construct an energy-level diagram of the vibrational energy levels associated with the CO_2 laser operating in the region of 10.6 μm. Also obtain the spacings of the rotational levels associated with each vibrational level.

The CO_2 molecule is a linear triatomic molecule with three normal modes of vibration; it thus has three separate quantum numbers v_1, v_2, v_3, each of which can have values of 0, 1, 2, 3, …. As shown in Figure 5-6(a), these normal modes include the symmetric stretch mode, the bending mode, and the asymmetric stretch mode. (The bending mode actually involves two separate modes each moving in a perpendicular or orthogonal direction with respect to the other, and is thus degenerate since both modes have the same energy.)

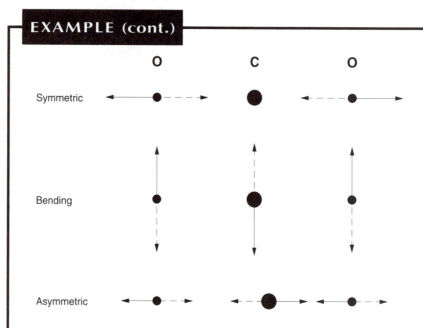

Figure 5-6(a). Vibrational modes of the CO_2 molecule

The total energy associated with a given energy level of this molecule would be described by an expression of the form of (5.10) but with three quantum numbers:

$$E_{total}(v_1, v_2, v_3) = h\nu_1(v_1 + \tfrac{1}{2}) + h\nu_2(v_2 + \tfrac{1}{2}) + h\nu_3(v_3 + \tfrac{1}{2}).$$

For CO_2, the normal mode frequencies are:

$$\nu_1 \approx 4.16 \times 10^{13} \text{ s}^{-1} \quad \text{symmetric stretch mode } (\nu_1),$$
$$\nu_2 \approx 2.00 \times 10^{13} \text{ s}^{-1} \quad \text{bending mode } (\nu_2),$$
$$\nu_3 \approx 7.05 \times 10^{13} \text{ s}^{-1} \quad \text{asymmetric stretch mode } (\nu_3).$$

These would be the frequencies of radiative transitions between energy levels associated with each vibrational mode. The actual energy levels are obtained by the relationship $E_{v_i} = v_i h\nu_i$. For example, the energy associated with $v_1 = 1$ and $v_2, v_3 = 0$ would be

$$E_1 = v_1 h\nu_1 = 1 \cdot (4.14 \times 10^{-15} \text{ eV-s})(4.16 \times 10^{13} \text{ s}^{-1}) = 0.172 \text{ eV}.$$

Similarly,

$$E_2 = v_2 h\nu_2 = 0.083 \text{ eV} \quad \text{for } v_2 = 1, v_1 = 0, v_3 = 0;$$
$$E_3 = v_3 h\nu_3 = 0.292 \text{ eV} \quad \text{for } v_3 = 1, v_1 = 0, v_2 = 0.$$

These three energies represent the energy above the ground state of the molecule. According to (5.10), the ground state has an energy equal to $E(0, 0, 0) = \tfrac{1}{2}(h\nu_1 + h\nu_2 + h\nu_3)$, which is the lowest energy of the molecule. Thus the energy-level diagram for the lowest levels would be that shown in Figure 5-6(b), where $E(0, 0, 0)$ is indicated as zero

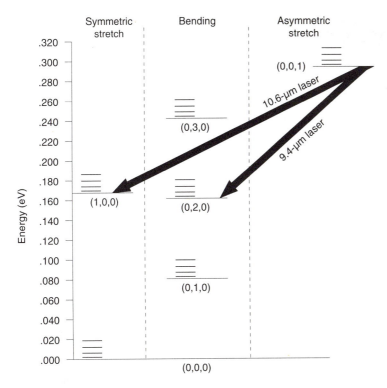

Figure 5-6(b). Laser transitions between vibrational levels of the CO_2 molecule

for reference purposes (instead of having the zero-point energy value just mentioned).

Each of the vibrational levels would have superimposed upon it a set of rotational levels described according to (5.5). The value of B for CO_2 is $B = 39 \text{ m}^{-1}$ and thus

$$E_{rot} = J(J+1)Bhc$$
$$= J(J+1)(39 \text{ m}^{-1})(4.14 \times 10^{-15} \text{ eV-s})(3 \times 10^8 \text{ m/s})$$
$$= J(J+1)(4.72 \times 10^{-5}) \text{ eV}.$$

Hence, the rotational levels would be spaced by energies that are more than three orders of magnitude smaller than the vibrational energies. A typical vibrational level would thus consist of a series of rotational levels spaced according to the preceding formula. Transitions would occur according to the selection rules $\Delta J = 0, \pm 1$ and $\Delta v = \pm 1$, discussed previously. For $\Delta J = +1$ we would have the P branch of the spectrum, and for $\Delta J = -1$ we would have the R branch. As was the case for most diatomic molecules, the Q branch ($\Delta J = 0$) does not occur for CO_2.

The most well-known transitions of the CO_2 laser involve those from $(v_1, v_2, v_3) = (0, 0, 1)$ to $(1, 0, 0)$, as shown in Figures 5-6(b) and 5-6(c). Some of the strongest of those transitions are P(20) and R(20). For example, the designation P(20) completely specifies a particular transition, since the P symbol indicates that $\Delta J = +1$ and thus P(20)

EXAMPLE (cont.)

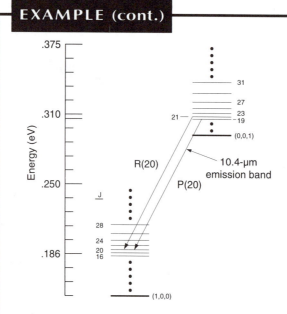

Figure 5-6(c). Two specific P- and R-branch laser transitions of the CO_2 molecule

would be from $J = 19$ of $(0, 0, 1)$ to $J = 20$ of $(1, 0, 0)$. For reasons of symmetry, alternate lines for both the P and R branches are missing and therefore not shown in the figure.

The Doppler linewidths of these transitions can be obtained from (4.55) and the Appendix as

$$\Delta \nu_D = (7.16 \times 10^{-7}) \nu_0 \sqrt{\frac{T}{M_N}}$$

$$= (7.16 \times 10^{-7}) \frac{3 \times 10^8 \text{ m/s}}{10.6 \times 10^{-6} \text{ m}} \sqrt{\frac{500K}{44}}$$

$$= 6.83 \times 10^7 \text{ Hz},$$

which corresponds to an energy width of $\Delta E = h\Delta\nu = 2.83 \times 10^{-7}$ eV. Thus the separation between rotational lines is much greater than the Doppler width, and so the CO_2 laser transitions would all occur as separate discrete laser lines. Pressure broadening occurs in CO_2 and begins to affect the emission linewidth at pressures above 50 torr. At pressures above 8 atmospheres, the pressure broadening becomes so extensive that it causes the emission lines to smear together; in this case the emission would be observed as a continuous spectrum in the region of the laser wavelengths.

The radiative probabilities for the rotational–vibrational laser transitions range from 0.2 to 0.3 s^{-1}. Consequently, the natural linewidths would be extremely narrow, of the order of 1 Hz or less, and thus natural broadening would be of no significance in these molecules.

PROBABILITIES OF ROTATIONAL AND VIBRATIONAL TRANSITIONS

The relative arrangement of the various atoms in the molecule with respect to each other defines the properties of each rotational–vibrational state. Thus, unlike electronic transitions in atoms, rotational and vibrational transitions in molecules involve changes from one specific atomic arrangement (state) to another. Estimating the transition probabilities therefore requires calculating the quantum mechanical matrix element associated with a change in the nuclear moment rather than a change in the electronic dipole moment, as was the case for the atomic transitions described by (4.74). Such calculations indicate that typical decay rates for rotational–vibrational transitions range from 10^{-1} to $10^{3} \, s^{-1}$; pure rotational transition rates are even slower.

ELECTRONIC ENERGY LEVELS OF MOLECULES

The electronic energy levels of molecules, as shown in Figure 5-1, are historically labeled in a way similar to that of atomic levels (Chapter 3), except that capital Greek letters replace English letters for the orbital angular momentum designations. The orbital and spin angular momenta of the electrons interact in much the same way as they do in atoms. The angular momentum quantum number for a diatomic molecule involves the sum of the projections of the orbital angular momentum of each atom onto the axis connecting the two atoms.

The orbital angular momentum quantum number is given the symbol Λ, and the various electronic states are identified by the Greek letters as follows:

$$\begin{array}{llll} \Lambda: & 0 & 1 & 2 & 3, \\ \text{electronic state:} & \Sigma & \Pi & \Delta & \Phi. \end{array} \tag{5.14}$$

For any specific value of Λ, the rotational quantum number J can take on values of $\Lambda, \Lambda+1, \Lambda+2, \dots$.

The spin determines the multiplicity of the electronic state, and is given by $2S+1$ as in the case of atoms; it describes the number of sublevels for a given J value. Thus there are singlet states $(S = 0)$ $^1\Sigma, \, ^1\Pi, \, ^1\Delta, \dots$, doublet states $(S = \frac{1}{2})$ $^2\Sigma, \, ^2\Pi, \, ^2\Delta, \dots$, and so forth. As in the case of atoms, the multiplicity or number of degenerate sublevels is always odd if the total number of electrons is even, and vice versa.

Historically, the lowest electronic "manifold" or ground state is usually labeled X followed by the notation just outlined for the specific angular momentum designation for that configuration. Excited states that have been identified are labeled A, B, C, ..., usually in the order of their identification. For example, a ground-state designation might be $X^2\Sigma$, and an excited state might be labeled $A^2\Pi$. We have omitted the subscript associated with the total angular momentum for the labeling of atomic transitions. For molecular transitions, the subscript notation is either g (gerade), which refers to an even-parity state, or u (ungerade), which denotes an odd-parity state. Thus the two levels listed previously might be expressed as $X^2\Sigma_u$ and $A^2\Pi_g$ states. It must be remembered that the complexity of the interactions – not only of all the electrons of each atom of the molecule with each other, but also with the other electrons and nuclei – lead to extremely complex energy-level structures for all but the simplest of molecules. Thus, the identification of the energy levels (and the radiative transitions associated with them) has not been a readily forthcoming process. There is much work, both experimental and theoretical, still being carried out to identify energy levels of even some of the relatively simple molecules.

ELECTRONIC TRANSITIONS AND ASSOCIATED SELECTION RULES FOR MOLECULES

Electronic energy levels have superimposed upon them the various possible vibrational and rotational sublevels associated with each molecule. This is because, no matter which electronic state the molecule is in, it is still vibrating and rotating. It is therefore in one of the many rotational-vibrational states associated with that particular electronic state, as indicated in Figure 5-1. Thus the spectrum associated with an electronic transition originating from a specific electronic level consists of an array of emission lines from the various rotational–vibrational sublevels of that electronic level to the rotational and vibrational sublevels of a lower electronic level, as allowed by the selection rules associated with the rotational, vibrational, and electronic transitions. The selection rules for electronic transitions are

$$\Delta\Lambda = 0, \pm 1, \tag{5.15}$$

$$\Delta S = 0. \tag{5.16}$$

The transition spectrum would typically appear as orderly sequences of closely spaced emission lines, from which the arrangements of the energy levels can be deduced by spectroscopists. Molecular levels are not as easy to determine from the emission spectra as are atomic levels. Nevertheless, molecular spectroscopists have determined a large number of levels in a variety of molecules. In many cases this has been possible only in recent years owing to the capabilities of tunable lasers, which can provide selective excitation of specific molecular levels. The energy levels associated with the emission spectrum of a molecular nitrogen molecule (N_2) are shown in Figure 5-7. N_2 is one of the more extensively studied molecules, and thus much is known about the arrangement of its energy levels.

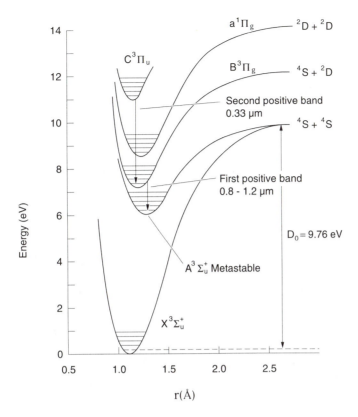

Figure 5-7. Energy levels of molecular nitrogen as a function of separation distance r between the two nitrogen atoms and also the dissociation energy D_0

EMISSION LINEWIDTH OF MOLECULAR TRANSITIONS

The emission linewidth of each molecular transition is determined by the natural width of the transition, by the Doppler width (if that is dominant), or by collisional broadening if the pressure is sufficiently high. The first and last are cases of homogeneous broadening and the Doppler case is of inhomogeneous broadening. If the emission linewidths overlap (as they often do) for adjacent rotational and/or vibrational transitions then the broadening would appear to be much greater, since the emissions from the various levels would, in effect, smear together. For example, the CO_2 laser transition at 10.6 μm is dominated by Doppler broadening for pressures below 50 torr ($\Delta\nu_D = 60$ MHz); above 50 torr, pressure broadening begins to dominate; and above 8 atmospheres (approximately 6,000 torr), the levels begin to overlap and thus appear as one broad continuum of emission.

THE FRANCK–CONDON PRINCIPLE

The Franck–Condon principle is a very important concept for analyzing radiative transitions in molecules, yet it is very simple to understand and use. It states that all electronic transitions, either downward (via radiation or collisions) or upward (via excitation or absorption), must take place in

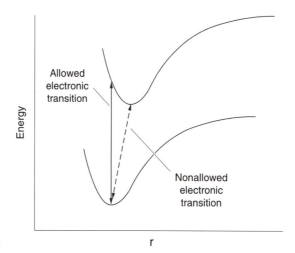

Figure 5-8. The Franck–Condon principle

a vertical direction on a diagram of molecular energy versus atomic separation (see Figure 5-8). The justification for this principle is the assumption that electronic transitions take place so rapidly that internuclear separations do not have time to change during the transition. This rule places certain restrictions on which energy levels and sublevels can be involved in specific electronic transitions. Since vibrating molecules spend most of their time at the endpoints of their vibrational excursions, the Franck–Condon principle suggests that the vibrational levels of excited electronic states produced via excitation from lower electronic levels must have a portion of their allowed excursions in a direct vertical pathway above those endpoints.

EXCIMER ENERGY LEVELS

Diatomic and triatomic molecules that lead to laser output are referred to as *excimer* molecules. The energy-level diagram of such a molecule, the xenon fluoride molecule, is shown in Figure 5-9. These molecules have stable excited energy levels (levels in which there is a potential well), but their ground states are either repulsive, as indicated for the A level in the figure, or very loosely attractive, as indicated by the X level. Thus the molecules do not normally exist in nature, since any time the two ground-state atoms (here, xenon and fluorine) associated with the molecules are in the vicinity of each other, they are repulsed. For these combinations of atoms, the molecules can be formed in their excited states only by various special excitation techniques. After the excited states are formed, the population rapidly decays to the repulsive ground state in a time that typically ranges from 1 to 5 ns, and the atoms subsequently fly apart. Thus the molecules exist only for a duration corresponding to the lifetime of the excited states.

The most common excimer molecules used for lasers are xenon fluoride (XeF), xenon chloride (XeCl), krypton fluoride (KrF), argon fluoride (ArF), mercury bromide (HgBr), and molecular xenon (Xe_2) and argon (Ar_2). Molecular fluorine (F_2) also can produce laser output in the VUV at

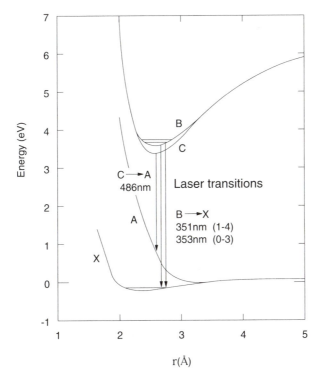

Figure 5-9. Molecular energy levels involved in a xenon–fluoride excimer laser

153 nm when excited in an excimer type of discharge. There are also a few triatomic excimer molecules that have been used to make lasers.

The energy-level diagram for xenon fluoride shown in Figure 5-9 indicates both ultraviolet laser transitions $B \rightarrow X$ and $C \rightarrow X$, as well as visible transitions $C \rightarrow A$. The specific vibrational levels associated with the ultraviolet transitions are shown in parentheses.

5.2 LIQUID ENERGY LEVELS AND THEIR RADIATION PROPERTIES

STRUCTURE OF DYE MOLECULES

Organic dye molecules are very complex molecules that belong to one of several classes. These include: polymethine dyes, which radiate in the red or near infrared spectral regions at 700–1,500 nm; xanthene dyes, which emit in the visible at 500–700 nm; coumarin dyes, which radiate in the blue and green at 400–500 nm; and scintillator dyes, which emit in the ultraviolet (320–400 nm). When they are used as laser media, dye molecules are dissolved in liquid solvents such as water, alcohol, and ethylene glycol. In such solutions the concentration of dye molecules is of the order of one part in ten thousand, so the dye molecules are somewhat isolated from each other and hence are surrounded only by solvent molecules. Laser dyes have strong absorption regions in the ultraviolet and visible spectral regions, as shown for a typical dye molecule (RhB) in Figure 5-10. If they are

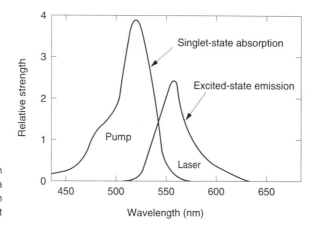

Figure 5-10. Absorption and emission spectrum of a RhB laser dye dissolved in a liquid solvent

irradiated or pumped with light at these wavelengths then the dye molecules radiate very efficiently at somewhat longer wavelengths than the pump wavelengths, as shown in the emission spectrum of Figure 5-10.

Because of this efficient optical absorption of dye molecules, dye lasers are produced by optical pumping techniques using flashlamps or other lasers. The concentration of the dye in the solvent is generally determined by pumping considerations that relate to the geometry of the gain medium and the desired gain of the amplifier. Such considerations will be dealt with in detail in Chapter 9. Suffice it to say for now that, for efficient pumping, the dye concentration is typically adjusted so that all of the pump energy is absorbed within the desired gain region of the laser.

The molecular structure of several dye molecules is shown in Figure 5-11, along with the solvents used and the emission bandwidth. It can be seen that organic dye molecules are complex structures, but the radiating portion of the dye is relatively simple; it is referred to as a *chromophor* located at a specific location on the dye molecule. The energy levels associated with the chromophor are similar to those of some of the simpler molecules discussed in previous sections.

Organic dyes are efficient radiators and thus make good lasers, because the dye molecules are sufficiently shielded from the solvent medium that decay from the first excited state to the ground state of the chromophor occurs predominantly by radiative decay. Otherwise, as in the case of many other liquids and solids, in the absence of shielding the interactions with the surrounding medium would cause nonradiative (collisional) decay to dominate. In that case, the energy lost when the electron jumps to the ground state would – instead of being radiated – be transferred to the surrounding medium, thereby heating up the medium. The ratio of radiating excited dye molecules to total number of excited dye molecules (exhibiting both radiative and nonradiative decay processes) is known as the *quantum yield*. Most dyes that work effectively as dye lasers have quantum yields approaching unity. Such high quantum yields are also desirable for lasers made from solid-state dielectric laser media, as will be described in Section 5.3.

Dye	Structure	Solvent	Wavelength
Acridine red	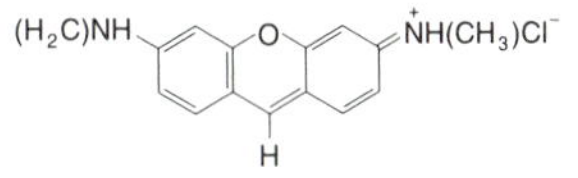 $(H_2C)NH$ ... O ... $\overset{+}{N}H(CH_3)Cl^-$; H	EtOH	Red 600 - 630 nm
Rhodamine 6G	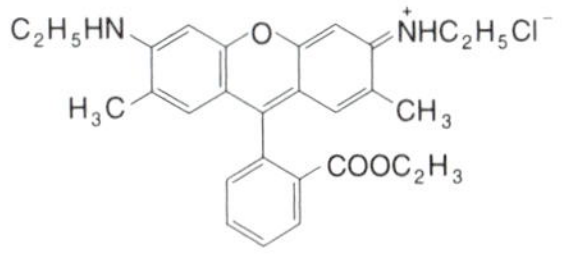 C_2H_5HN ... O ... $\overset{+}{N}HC_2H_5Cl^-$; H_3C, CH_3, $COOC_2H_5$	EtOH MeOH H_2O DMSO Polymethyl-methacrylate	Yellow 570 - 610 nm
Rhodamine B	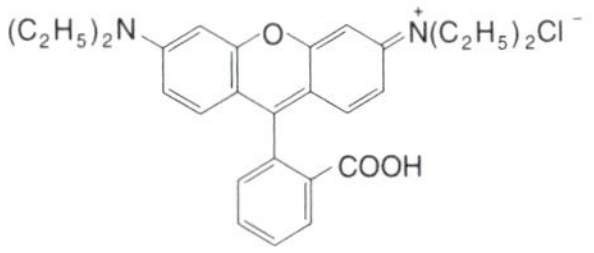 $(C_2H_5)_2N$ ... O ... $\overset{+}{N}(C_2H_5)_2Cl^-$; COOH	EtOH MeOH Polymethyl-methacrylate	Red 605 - 635 nm
Na-fluorescein	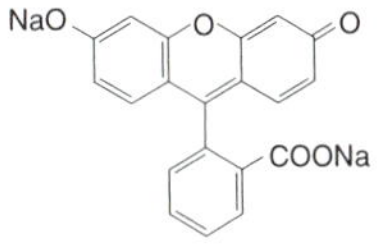 NaO ... O ... O ; COONa	EtOH H_2O	Green 530 - 560 nm
7-Hydroxycoumarin	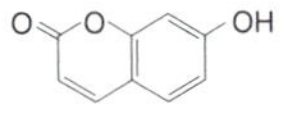 O ... O ... OH	H_2O (pH~9)	Blue 450 - 470 nm

Figure 5-11. Molecular structure of several laser dyes, along with the laser wavelength range for each dye

ENERGY LEVELS OF DYE MOLECULES

The energy levels associated with organic dyes suspended in solvents are shown in Figure 5-12. There are two *manifolds* (sets of energy levels) within which excitation and decay can occur, just as we deduced for atomic helium in Chapter 3; these are the singlet and the triplet manifold, referred to (respectively) as S and T. In the singlet state of the molecule, the spin of the excited electron is antiparallel to the spin of the remaining molecule. In the triplet state, the spins are parallel. As in the case of atoms and simple molecules, transitions between singlet states, or between triplet states, are much more likely to occur than transitions between singlet and triplet states; when these latter do occur, they are referred to as *intersystem* crossings. Thus, efficient absorption and emission of radiation occurs either between singlet levels or between triplet levels. As in the case of helium, the singlet ground state S_0 lies quite a bit lower in energy than the lowest triplet state T_1, so most of the atoms normally reside in the singlet, where laser action generally occurs. In fact, we will later see that the triplet manifold creates a problem for dye lasers when attempts are made to produce gain

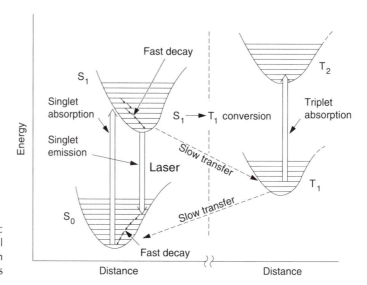

Figure 5-12. Electronic energy levels of a typical laser dye, showing both the singlet and triplet states

in the medium, owing to absorption from the triplet ground state to the triplet excited state at the laser wavelength.

As in the case of simpler molecules, the various electronic energy levels S_0, S_1, T_1, and T_2 of Figure 5-12 are associated with vibrational and rotational energy levels. The vibrational levels are typically spaced by energies that range from 0.1 to 0.15 eV, which would produce emission at wavelengths of 8 to 10 μm; the rotational levels are two orders of magnitude smaller. All of the levels are broadened to the extent that transitions associated with the vibrational and rotational levels for a given electronic transition blur together to form one broad emission spectrum.

Different solvents cause slight shifts in the energy levels of dye molecules, leading to slight variations in the radiative spectrum of those molecules. In order to be useful in a dye laser, the solvent must be transparent both to the pump radiation and to the laser wavelength.

EXCITATION AND EMISSION OF DYE MOLECULES

When excitation occurs in dyes from the ground state S_0 to the excited state S_1, all of the vibrational states of S_1 can be populated if the excitation or pump energy occurs over a broad enough spectrum. However, decay of the rotational and vibrational levels of S_1 occurs very rapidly (in 10^{-12}–10^{-13} s) to the lowest-lying excited level of S_1 via collisional decay. The population temporarily accumulates at this lowest level of S_1 because the decay to S_0, which is predominantly radiative, occurs over a much longer time (1–5×10^{-9} s). If one considers radiative transitions between all of the many possible combinations of allowed level pairs between the lowest level of S_1 and the full manifold of levels of S_0, the resulting emission is a continuous spectrum of radiation, which includes transitions originating from the bottom of S_1 to a spectrum of energies ranging from the top of S_0 to the bottom of S_0, as shown in Figures 5-10 and 5-12. In considering possible transitions one must keep in mind the Franck–Condon principle, which restricts the range of possible states to which the upper level S_1 can decay.

Without the radial shift or displacement of the excited level S_1, the emission spectrum would overlap with the absorption spectrum (S_0 to S_1) of Figure 5-10 and so prevent gain from being obtained in these dye media (for reasons we will investigate later). As it is, with the occurrence of the radial shift of the excited state, the peak of the absorption spectrum is shifted to shorter wavelengths with respect to the emission spectrum, owing to this displaced level S_1 and the Franck–Condon effect (see Figure 5-12).

DETRIMENTAL TRIPLET STATES OF DYE MOLECULES

The triplet states essentially form separate molecules that are only weakly coupled to the singlet states by weak radiative or collisional interactions. Since T_1 lies below S_1 – as can be seen from Figure 5-12 – some of the population of S_1 can decay, mostly collisionally, to T_1 when S_1 is excited. The population will then accumulate in T_1, since the weak decay from T_1 to S_0 occurs within a time frame of 10–100 μs. The wavelength region over which triplet absorption from T_1 to T_2 occurs is similar to that of the emission from S_1 to S_0 at the laser wavelengths shown in Figure 5-10. Thus the triplet state absorption of a dye can reduce or cancel the laser gain produced in the singlets if enough population moves into the triplet state T_1 from S_1. Thus, for continuous wave (cw), it is desirable to "flow the dye" in the excitation region in order to physically remove the triplet state population from that region and thereby enhance laser output from the singlet system (see Section 9.3).

EXAMPLE

A rhodamine-B dye mixture in ethyl alcohol (EtOH) emits over a spectral bandwidth ranging from 525 to 625 nm, as seen in Figure 5-10, with a peak spontaneous emission wavelength of 565 nm. Assume that the emission is homogeneously broadened, and that the emission originates from the lowest energy of the S_1 level. Determine the energy width of the lower level S_0 to which transitions occur from S_1.

If all of the emission originates from the bottom of energy level S_1, then the entire tuning range occurs as a result of transitions from S_1 down to levels ranging from the top to the bottom of S_0. The energy width of S_0 can therefore be obtained as follows:

$$\Delta E = h\Delta\nu = \frac{hc}{\lambda^2}\Delta\lambda$$

since $|\Delta\nu| = c/\lambda^2|\Delta\lambda|$, where λ is the peak emission wavelength and $\Delta\lambda$ is the wavelength spread. For the conditions listed for RhB, $\lambda \cong$

EXAMPLE (cont.)

565 nm and $\Delta\lambda = 625\ \text{nm} - 525\ \text{nm} = 100\ \text{nm}$. Thus ΔE can be calculated to be

$$\Delta E = \frac{hc}{\lambda^2}\Delta\lambda = \frac{(6.63\times10^{-34}\ \text{J-s})(3\times10^8\ \text{m/s})}{(5.65\times10^{-7}\ \text{m})^2}(100\times10^{-9}\ \text{m})$$

$$= 1.55\times10^{-20}\ \text{J} = 0.389\ \text{eV},$$

or approximately 0.4 eV.

5.3 ENERGY LEVELS IN SOLIDS – DIELECTRIC LASER MATERIALS

Laser gain media produced in solid-state dielectric laser materials are in some ways analogous to organic dyes in solution. The laser species generally consist of ionic species, such as chromium ions (Cr^{3+}) grown or *doped* within a host material such as aluminum oxide (Al_2O_3), or neodymium ions (Nd^{3+}) grown within a yttrium aluminum garnet (YAG) host or even a glass host, at a concentration ranging from 0.1 to 10 percent. This is analogous to the organic dye laser in which the dye molecules are suspended in a solvent. In the case of the solid-state host material, the host behaves as a matrix that suspends the laser ions, rendering them isolated and effectively "frozen" at nearly regularly spaced locations, so the ions are approximately equidistant from each other. This isolation of the ions prevents undesired interactions between them, which would produce collisional decay (quenching) of the laser levels instead of the desired radiative decay.

In dielectric solid laser materials, the laser species is selected and manufactured or "grown" to have specific properties: long upper laser level lifetime and specific emission linewidths (either broad or narrow) depending upon the desired application. Thus, solid-state lasers are engineered for particular laser applications (as are dye lasers), whereas atomic and molecular gas lasers are generally used in their naturally occurring form. The specific engineering in the design of these solid-state materials is to produce energy levels with long lifetimes (from 10^{-6} to 10^{-3} s) that can therefore accumulate and store energy to be later extracted as laser output. Almost all laser species other than dielectric solids have energy-level lifetimes (associated with upper laser levels) that are shorter than 10^{-6} s and in many cases shorter than 10^{-8} s. Thus, dielectric solid-state lasers have unique properties as laser gain media because of their long excited-state lifetimes and their compact, high-density, durable structures.

HOST MATERIALS

Host materials for solid-state lasers can be grouped into two categories, crystalline solids and glasses. Crystalline solids provide a homogeneous regular array or matrix within which the laser species is imbedded, providing homogeneous broadening for the laser material. Glasses have very small, irregular structural units that are randomly oriented and generally lead to inhomogeneous broadening of the radiating dopant species.

The host material must possess good optical properties as well as good mechanical and thermal properties. Poor optical properties include variations in index of refraction, which lead to irregular beam quality in the laser output, and impurities, which lead to scattering and undesired absorption of the pump light or laser beam output. Poor mechanical and thermal properties lead to material deformation or fracturing of the host when the laser is operated under such high thermal loading conditions as a high pulse-repetition rate or high steady-state pumping. The host must also be able to accept the dopant material in a way that will not distort the desired properties of the dopant – a long lifetime and an appropriate lineshape. The host must also be capable of being grown with a uniform distribution of the dopant, in sizes that are required for laser gain media.

Crystalline hosts have higher thermal conductivity and narrower emission (fluorescence) linewidths than glasses, and in many cases these materials also have greater hardness and thus greater durability than glasses. The advantages of glasses are their ease of fabrication and growth in large sizes with very good optical quality. Candidates for crystalline hosts are oxides, phosphates, silicates, tungstates, molybdates, vanadates, beryllates, fluorides, and in some cases optical ceramic materials. The oxides include sapphire (Al_2O_3), garnets such as YAG, aluminates, and oxysulfides.

LASER SPECIES – DOPANT IONS

The laser species are specific ions implanted as dopants within the solid-state host, chosen to obtain the desired long lifetime as well as other suitable laser properties. The property of long lifetime is achieved by using ionic species in which the laser levels consist of inner shell states having electron wave functions that are confined to a relatively small spatial extent near the nucleus and are thereby significantly shielded by outer electron shells from interactions with the solid-state host materials. This shielding prevents rapid quenching of the laser levels by collisional effects, which would otherwise destroy the long-lived characteristics of those levels if the shielding were not present.

Dopant ions can be categorized as rare-earth ions, transition metal ions, and actinide ions. The rare-earth ions generally have a number of sharp emission lines due to the atomic-like character of the radiating 4f states. They usually exist in solids in the trivalent form. Rare-earth triple ions include neodymium (Nd^{3+}), erbium (Er^{3+}), holmium (Ho^{3+}), and thulium (Tm^{3+}). Other (less often used) rare-earth ions are praseodymium, gadolinium, europium, ytterbium, and cerium. Divalent rare-earth ions include samarium (Sm^{2+}), dysprosium (Dy^{2+}), and thulium (Tm^{2+}). The most well-known transition metal ions include chromium (Cr^{3+}) and titanium (Ti^{3+}). Others include nickel as Ni^{2+} and cobalt as Co^{2+}. Actinide ions with 5f electrons have been studied, but most of these ions are radioactive and thus have not proved very successful as laser materials.

From the standpoint of laser characteristics, two types of materials have been successfully developed as effective laser species: materials with narrow emission linewidths, and materials with broadband emission linewidths that lead to broadly tunable lasers.

NARROW-LINEWIDTH LASER MATERIALS

When rare-earth materials such as neodymium (Nd^{3+}) (and also some transition metals) are implanted in a suitable host, the inner-shell d or f electrons are sufficiently shielded from the surroundings by the outer electrons so as to allow the species to become efficient radiators with relatively narrow emission linewidths. The unique factor associated with such solid-state lasers is that the transitions generally occur between two ^4F states. Allowed transitions would not normally occur between such states, according to the selection rules for atoms outlined in Chapter 4, because they have the same parity. The radiation that does in fact occur between the upper laser level and the lower laser level results from effects of localized crystal fields that interact with the laser ions. Such interactions effectively mix the 4f electron wave functions with other wave functions, such as d-electron wave functions of opposite parity. This mixing allows a small amount of electric dipole radiation to occur, due to a small but significant overlap in the modified wave functions of the upper and lower laser levels. The narrow emission linewidths provided by these transitions allow the possibility of high laser gain at a relatively specific laser wavelength, which reduces the pumping requirements for this type of laser. This effect will be clarified later when gain requirements are developed.

RUBY LASER The first laser ever produced, the ruby laser, uses a solid-state gain medium that is in the category of narrow-linewidth laser materials. It consists of the transition metal chromium, in the form of Cr^{3+} ions doped in a sapphire (Al_2O_3) host crystal – $Cr:Al_2O_3$ – at a concentration of the order of 1%. In this laser, shown in Figure 5-13, the ground state is a $^4A_2$ state with $g = 4$. Excitation occurs to highly excited $^4T_1$ and $^4T_2$ states (sometimes also designated as $^4F_1$ and $^4F_2$) which rapidly decay by non-radiative transitions to the ^2E state (the upper laser level). (Note: the designations of these states in ruby do not follow the normal notation for angular momentum states as described in Chapter 3; for historical reasons they use group-theory designations.) Laser action is then produced by radiation

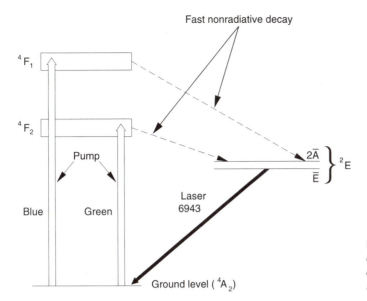

Fast nonradiative decay

$^4F_1$

$^4F_2$

Pump

Blue | Green

Laser
6943

$2\overline{A}$
$\overline{E}$ } ^2E

Ground level ($^4A_2$)

Figure 5-13. Energy levels of ruby crystal involving chromium ions doped into a sapphire crystal

from that state back to the ground state. One of the reasons this laser was selected as a possible first laser candidate is that pumping could occur to the many excited energy levels of $^4F_1$ and $^4F_2$, which are essentially atomic-like levels that have been broadened and smeared out in energy (see Figure 5-13). This broad range of excitation energies allows the pumping light to be provided over a wide spectrum of wavelengths, as in the case of dye lasers, which in effect allows a very high pumping flux from a broadband emitting flashlamp to be used to fill these levels. Rapid collisional decay then occurs from those levels to the narrow upper laser level as described previously. Thus there is a concentrating effect in which the excitation energy is collected over a broad range of energies, such as from a flashlamp, and funneled to a specific narrow energy level from which laser output is produced. The decay to a narrow ground-state energy level leads to a narrow emission linewidth.

NEODYMIUM LASER   Another solid-state laser species in this category of narrow linewidth lasers is the rare-earth Nd^{3+} ion doped in several different host materials. Neodymium is perhaps the most useful laser material developed to date, and is most commonly doped in either YAG or glass hosts. An energy-level diagram for Nd^{3+} doped in a YAG host material is shown in Figure 5-14(a). The relevant laser levels are slightly lowered for doping in glass as compared to YAG, as shown in Figure 5-14(b). The upper laser level is shifted downward by 112 cm^{-1} and the lower laser is shifted downward by 161 cm^{-1}, producing a net decrease in the laser wavelength for glass. The transition in YAG is at 1.064 μm whereas for glass it is typically at 1.054 μm. The excitation bands occur in the blue and the green. The pump energy then transfers nonradiatively (collisionally) to the upper laser level $^4F_{3/2}$. The laser terminates on the $^4I_{11/2}$ lower level, from which

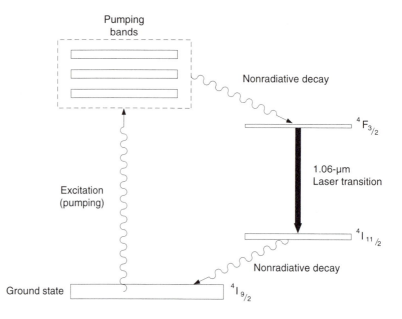

Figure 5-14(a). Energy levels of the Nd:YAG laser crystal

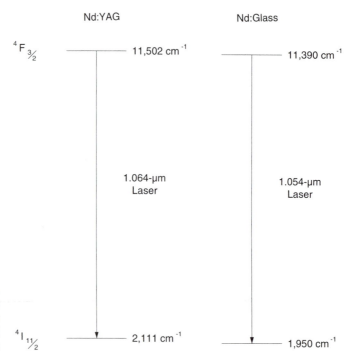

Figure 5-14(b). Comparison of energy levels of the laser transition for Nd doped in YAG and glass

collisional decay takes the population back to the ground state. Two other radiative transitions in the Nd ion are also used as laser transitions. They originate from the same upper laser level as the 1.06-μm transition and decay to the $^4I_{13/2}$ level radiating at 1.35 μm and to the $^4I_{9/2}$ level producing 0.88-μm radiation. These transitions have lower gain than the 1.06-μm transition and therefore are not as easily made to lase.

Nd ions doped in YAG crystals are limited to a maximum concentration of 1.0–1.5%. Higher concentrations lead to increased collisional decay, resulting in a reduced upper laser level lifetime. At room temperature, the 1.06-μm radiative transition is homogeneously broadened with a narrow emission linewidth of 1.2×10^{11} Hz ($\Delta\lambda = 0.45$ nm) and the upper level lifetime is 230 μs. Nd:YAG rods have good heat conduction properties, which make them desirable for high–repetition rate laser operation. They have a disadvantage, however, in that they are limited by crystal growth capabilities to small laser rods, of the order of up to 1 cm in diameter and lengths of the order of 10 cm.

Neodymium can be doped to very high concentrations in glass. Many types of phosphate and silicate glasses have been used for these lasers. The fluorescent lifetime is approximately 300 μs, and the emission linewidth is 4.8–7.5×10^{12} Hz (18–28 nm). This is wider than for Nd:YAG by a factor of 40 to 60, owing to the random arrangement of the structural units of the glass surrounding the dopant ions. This increased emission linewidth (inhomogeneous broadening) in effect reduces the laser gain for reasons that will be explained in Chapter 7. Glass also has a much lower thermal conductivity than does Nd:YAG, and consequently has a limited pulse repetition rate for pulsed lasers and a lower operating power for cw lasers. However, Nd:glass material has good uniformity, and can be manufactured in very large sizes and with diffraction-limited optical quality. Also, the broader emission linewidth compared to Nd:YAG crystals allows for shorter pulse production, which will be discussed in Section 12.3 with respect to mode-locking.

Two other Nd doped materials are Nd:Cr:GSGG and Nd:YLF. The former incorporates the Nd ion doped into a garnet crystal and includes a second dopant, Cr, which is known as a *sensitizer* that provides an additional energy absorption band for pumping. The energy is efficiently collisionally transferred from the excited Cr^{3+} ion to the Nd^{3+} ion to produce laser output, in a similar manner to that of N_2 in the CO_2 laser. In Nd:YLF laser materials, the upper level lifetime is twice as long as that of Nd:YAG, making it useful for diode pumping. Also, the lasing wavelength closely matches that of the wavelengths of many of the Nd:glass materials, thereby making it an attractive oscillator for a Nd:glass amplifier system.

ERBIUM LASER The erbium laser is not as efficient as other solid-state lasers, but it has come into prominence in recent years because it has two particularly useful laser wavelengths: one at 1.54 μm and the other at 2.9 μm. It has been doped in both YAG crystals and in glass materials. Applications for these erbium laser wavelengths include erbium-doped fiber amplifiers for communications (at 1.54 μm) and erbium lasers for medical applications, since the two wavelengths occur within water absorption bands.

EXAMPLE

Nd^{3+} ions are doped into YAG at a 1% atomic concentration, which corresponds to a density of Nd atoms of 1.38×10^{26} per cubic meter in the laser rod. If all of these atoms were instantly pumped to the upper laser level and then began to radiate, what would be the radiated energy per cubic meter and the average power per cubic meter radiated from this material at the emission wavelength of 1.06 μm? If the power radiated from one cubic centimeter could be concentrated into a spot 1 mm in diameter, what would be the intensity (power/m^2)?

Each photon radiated at 1.06 μm would have an energy of

$$E = h\nu = \frac{hc}{\lambda} = \frac{(4.14 \times 10^{-15} \text{ eV-s})(3 \times 10^8 \text{ m/s})}{1.06 \times 10^{-6} \text{ m}} = 1.17 \text{ eV}$$

$$= (1.17 \text{ eV})(1.6 \times 10^{-19} \text{ J/eV}) = 1.87 \times 10^{-19} \text{ J}.$$

Thus, the total energy radiated per cubic meter would be

$$E_R = (1.38 \times 10^{26} \text{ photons/m}^3)(1.87 \times 10^{-19} \text{ J/photon})$$

$$= 2.58 \times 10^7 \text{ J/m}^3.$$

P_R, the total average power radiated per cubic meter, would thus equal the energy radiated per cubic meter multiplied by the transition rate of the level. If the level has a radiative lifetime of 230 μs and we assume that the dominant decay occurs at 1.06 μm (which is only approximate, since there are other low-lying levels to which the upper laser level can decay), then according to (4.6) we have

$$A_{ul} = \frac{1}{\tau_u} = \frac{1}{2.3 \times 10^{-4} \text{ s}} = 4.35 \times 10^3 \text{ s}^{-1};$$

$$P_R = (2.58 \times 10^7 \text{ J/m}^3)(4.35 \times 10^3 \text{ s}^{-1})$$

$$= 1.12 \times 10^{11} \text{ J/m}^3\text{-s} = 1.12 \times 10^{11} \text{ W/m}^3.$$

This is a very high power, but we must remember that it lasts only for an average of 230 μs. The amount of power radiated from one cubic centimeter would thus be

$$(1.12 \times 10^{11} \text{ W/m}^3)(1 \text{ m}^3/10^6 \text{ cm}^3) = 1.12 \times 10^5 \text{ W}.$$

If this power is collected and focused on a 1-mm (diameter) spot, or on an area of $\pi \cdot (5 \times 10^{-4})^2 \text{ m}^2 = 7.85 \times 10^{-7} \text{ m}^2$, then the intensity would be

$$I = (1.12 \times 10^5 \text{ W}/(7.85 \times 10^{-7} \text{ m}^2) = 1.43 \times 10^{11} \text{ W/m}^2,$$

an extremely high intensity. This indicates the kind of intensity that could be obtained in a laser beam from a crystal of this size if all of the radiated energy could be concentrated in a single direction.

BROADBAND TUNABLE LASER MATERIALS

All of the materials in this class of solid-state laser materials produce emission over a very broad linewidth. This provides either significant tunability of the laser output or broadband mode-locking, which leads to very short pulse production. The most noted lasers of this type are the alexandrite lasers ($Cr:BeAl_2O_4$), titanium doped in sapphire ($Ti:Al_2O_3$), and, more recently, chromium doped into either lithium strontium aluminum fluoride (Cr:LiSAF) or lithium strontium calcium fluoride (LiSCaF). In all of these species, the emission bandwidth (homogeneously broadened) is 100–400 nm in width, thereby providing the possibility of laser output over a wide range of frequencies in the near-infrared spectral region from a single laser crystal. The only other laser materials having wide bandwidths are organic dye solutions, but their linewidths are significantly narrower than the broadband solid-state materials. Broadband emission is due to the interaction of the electron with the lattice vibrations or phonons of the crystal. In effect, this is a coupling of the vibrational and the electronic states of the system, which is referred to as a *vibronic* transition.

ALEXANDRITE LASER   Alexandrite is a transition metal, chromium (Cr^{3+}), doped in an oxide host ($BeAl_2O_4$) with a chromium concentration of up to 0.4%. The material is optically and mechanically similar to a ruby laser crystal in that it has strength, chemical stability, very high hardness, and thermal conductivity greater than Nd:YAG. The pumping wavelength band ranges from 380 to 630 nm, with peaks at 410 nm and 590 nm. The emission bandwidth of $\Delta\lambda = 120$ nm ranges from 701 to 818 nm, as can be seen in the energy-level diagram of Figure 5-15. The upper laser level has a radiative lifetime of 260 μs.

TITANIUM SAPPHIRE LASER   The titanium-doped sapphire (or $Ti:Al_2O_3$ crystal) is probably the most well-known material for use as a tunable solid-

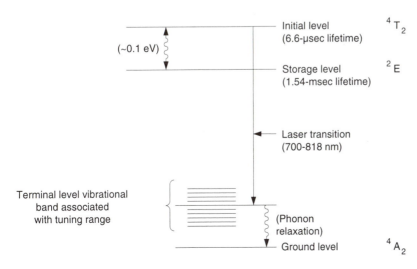

Figure 5-15. Energy levels associated with the alexandrite laser

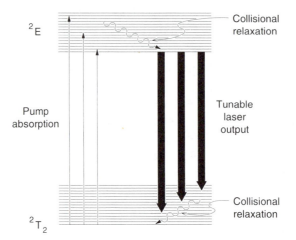

Figure 5-16. Energy levels of the titanium:sapphire laser

state laser; its energy-level diagram is given in Figure 5-16. It has a very broad emission linewidth ($\Delta\lambda = 400$ nm), with the emission ranging from 660 to 1,180 nm and the peak near 800 nm. It also has very minimal excited-state absorption. It has high thermal conductivity, good chemical inertness, and good mechanical stability. The pump band ranges from under 400 nm to just beyond 600 nm. Flashlamp pumping is not as effective in this material as in other solid-state lasers because of the relatively short upper laser level lifetime of 3.2 μs, which does not allow as much pumping flux to accumulate in the upper laser level. Most commercial Ti:Al$_2$O$_3$ lasers are therefore pumped by argon ion lasers and frequency-doubled Nd:YAG lasers.

CHROMIUM LiSAF AND CHROMIUM LiCaF LASERS Chromium-doped fluoride laser crystals such as Cr:LiSrAlF (or LiSAF for short) and Cr:LiCaF (LiCaF) have some advantages over Ti:Al$_2$O$_3$ crystals. For LiSAF these include longer upper-level lifetime (67 μs) and the capability of being grown in large sizes. The long lifetime allows efficient flashlamp pumping. The absorption (pump) band range for LiSAF is 300–720 nm, which also allows the crystal to be diode pumped with short-wavelength semiconductor diodes. However, the emission linewidth is narrower ($\Delta\lambda = 250$ nm) than Ti:Al$_2$O$_3$, with emission occurring from 800 to 1050 nm. As in most fluoride crystals, LiSAF is chemically stable (except for being slightly sensitive to humidity or water), but the mechanical strength and thermal conductivity are not as good as with sapphire-based crystals. LiCaF has a longer upper-level lifetime than LiSAF (190 μs), and operates at shorter wavelengths over a somewhat narrower wavelength range of 700–900 nm.

Other broadband tunable lasers include Tm^{3+}:YAG emitting from 1.8 to 2.2 μm, Co^{2+}:MgF$_2$ emitting from 1.8 to 2.5 μm, and Cr^{4+}:Mg$_2$SiO$_4$ covering the spectral range from 1.2 to 1.35 μm.

TABLE 5-1

SOLID-STATE DIELECTRIC LASER CHARACTERISTICS

Species	Laser Wave-length (nm)	Upper-level Lifetime (μs)	Emission Linewidth (Hz)
$Cr:Al_2O_3$ (Ruby)	694.3	3,000	3.3×10^{11}
Nd:YAG	1,064.1	230	1.2×10^{11}
Nd:Glass	1,054–1,062	≈ 300	7.5×10^{12}
$Cr:BeAl_2O_3$ (Alexandrite)	700–800	260	$\approx 5 \times 10^{13}$
$Ti:Al_2O_3$	660–1,180	3	$\approx 1.5 \times 10^{14}$
Cr:LiSAF	800–1,050	67	$\approx 9 \times 10^{13}$

BROADENING MECHANISM FOR SOLID-STATE LASERS

Table 5-1 shows the energy level lifetimes and emission linewidths for a number of solid-state dielectric laser materials doped in various types of hosts. The long lifetimes of these materials (10^{-3} to 10^{-6} s) implies a relatively small radiative decay rate, which leads to a very narrow natural linewidth of 10KHz to 1MHz according to (4.35). From examination of the experimentally measured broad emission linewidths indicated in Table 5-1, it can be seen that the radiative decay time of the upper level is many orders of magnitude too long to be responsible for the dominant line-broadening mechanism. All of these doped materials (except the glasses) have homogeneously broadened linewidths that are determined by the T_2 broadening associated with crystal interactions (see Section 4.3). Inhomogeneously broadened glass materials are broadened by the randomly arranged and shaped structural units (such as SiO_4) of the glass. The structural units have their own individual emission wavelengths that are homogeneously broadened, with maxima at slightly different wavelengths due to the different shapes and consequent strains of the individual structural units.

5.4 ENERGY LEVELS IN SOLIDS – SEMICONDUCTOR LASER MATERIALS

ENERGY LEVELS IN PERIODIC STRUCTURES

Semiconductors are not located last in this section because of their lack of importance. To the contrary, they are probably one of the most important types of lasers, and will play an even more significant role in the future than they do now. The reason they are located last is that their energy level arrangements have very little in common with the other types of laser materials mentioned in previous sections. Energy levels of molecules, organic dyes, and dielectric solids could all be considered as extensions of the energy levels obtained for isolated atoms in Chapter 3. Levels in the more complex species were seen to retain many atomic-like characteristics, although in most instances the levels were broadened owing to interactions of the specific laser species with the surrounding material. Even the labeling of the levels had parallels to atomic labeling.

Semiconductors have a different type of energy-level structure, one that can be associated with the regularly spaced or periodic arrangement of the materials. As discussed previously, organic dye molecules and solid-state laser ions are approximately uniformly spaced. However, the atoms in semiconductor laser materials are *exactly* spaced in a regular crystalline matrix, with only an occasional impurity or dislocation.

To obtain an expression for the energy of these levels, we will first consider the Schrödinger equation for electrons as derived in Chapter 3:

$$\nabla^2 \psi + \frac{8\pi^2 m_e}{h^2}(E - V)\psi = 0. \tag{3.29}$$

For atomic-like states, the potential energy V we used was negative and led to the development of discrete negative bound energy states or levels, as indicated in (3.8), and their associated eigenfunctions or wave functions such as obtained for hydrogen in (3.65).

In contrast, if V is a positive constant (with respect to a zero-energy reference frame), then the electron moves in a field-free space and is described by a wave function of the type

$$\psi = e^{i\mathbf{k}\cdot\mathbf{r}}, \tag{5.17}$$

which represents a plane wave. Such a wave propagates in a specific direction (the direction of $\mathbf{k}$) with a uniform velocity. The probability distribution function for the electron described by this wave function has an infinite extent and uniform value in any plane perpendicular to the direction of propagation, as discussed in Section 2.2. Using (5.17) as a solution for the Schrödinger equation leads to the following energy value:

$$E = \frac{\hbar^2 k^2}{2m_e} + V. \tag{5.18}$$

Since E is the total energy, $E = E_k + E_p$ where E_k is the kinetic energy and E_p the potential energy of the electron. The kinetic energy can be written as

$$E_k = \frac{1}{2}m_e v^2 = \frac{p^2}{2m_e} = \frac{\hbar^2 k^2}{2m_e}, \tag{5.19}$$

where we have equated this traditional expression for the kinetic energy to the kinetic energy term in (5.18). This leads to a value of the momentum of the electron of

$$p = m_e v = \hbar k, \tag{5.20}$$

where m_e is the mass of the free electron.

When V is a periodic potential (as is the case for crystalline-type solid materials), with regular, repeated alignment or stacking of the atoms the potential must be expressed as

$$V(x) = V(x+a) = V(x+2a) + \cdots + V(x+na), \tag{5.21}$$

where a is the periodic spacing between identical atoms or repeated groups of different atoms and n is an integer. The solution of the Schrödinger equation describing the probability distribution associated with the location of the electron is then of the form

$$\psi = u_k(x+na)e^{ik \cdot r}, \tag{5.22}$$

where $u_k(x+na)$ is a periodic function of x with period a, and where the subscript k is a constant that is a quantum number used to label the specific state of the wave function given in (5.22). This wave function is similar to that of the free electron (eqn. 5.17), but the interpretation of k now becomes more complex. It is not directly proportional to the momentum as in the case of the free electron in (5.20).

However, the relationship between the kinetic energy and k can be approximated in the region of small k by using the energy of a free electron (eqn. 5.18) if the electron mass m_e is replaced with the effective electron mass m_c:

$$E_k = \frac{\hbar^2 k^2}{2m_c}. \tag{5.23}$$

This relationship between E_k and k is a parabola curving upward in energy for increasing values of k. The subscript c for the electron mass in (5.23) refers to this parabolic energy band, which is known as the *conduction* band. The parabolic-shaped conduction band is shown as the upper curved region of Figure 5-17. It describes the range of energies in which electrons can move around freely in the material if induced by an electric field. The downward curve refers to the band of possible energy levels, known as the *valence* band, where electrons reside but cannot easily move around. The negative curve for the valence band can be described by using a negative effective mass m_v in the equation for kinetic energy (eqn. 5.23).

If an electron that normally resides in the valence band at a specific energy is excited up to the conduction band, it leaves a vacancy or a missing electron at the original energy. This vacancy is referred to as a *hole*. When electrons in the conduction band are caused to move by an electric field, they move as a modified plane wave according to (5.22). The effective mass m_c describes the increased resistance of the electron plane wave due to interactions of the wave with the periodic lattice. In a similar fashion, holes can also move as a plane wave in the valence band. The effective masses of the electrons and holes can be associated with the curvature of the energy bands, as suggested in (5.23). If an electric field is applied to the semiconductor, the conduction-band electrons with negative charge move

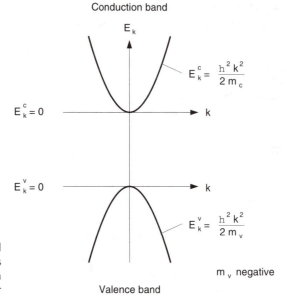

Figure 5-17. Valence and conduction bands associated with a semiconductor laser

in the positive direction of the field. The vacancies or holes in the valence band also move in the same direction, since the holes behave as though they have positive charge but negative mass. Thus, charge motion in the conduction band is entirely due to the conduction electrons, whereas charge motion in the valence band can be completely accounted for by hole motion.

ENERGY LEVELS OF CONDUCTORS, INSULATORS, AND SEMICONDUCTORS

Figure 5-18 shows several different examples of the energy levels associated with one period of the function indicated in (5.23) for a periodically spaced crystalline type of material. For a metallic or conducting material, either the conduction band overlaps with the valence band, as shown in Figure 5-18(a), or the conduction band is partially filled with electrons. Thus, when an electric field is applied to the material, the electrons behave as free electrons that can readily move approximately as a plane wave, as described by the wave function of (5.17).

For an insulating material, the valence band is filled and the conduction band is completely empty; there is a relatively large energy gap, referred to as the *band gap,* between them, as shown in Figure 5-18(b). When an electric field is applied, there are no conduction electrons present and the material behaves as an insulator. The only way that conduction could occur is if the field were high enough to pull electrons into the conduction band. This would require an extremely high electric field, causing a localized conduction and arcing that would most likely damage the material. Because of the high resistance to electrical conduction, these materials are referred to as *insulators.*

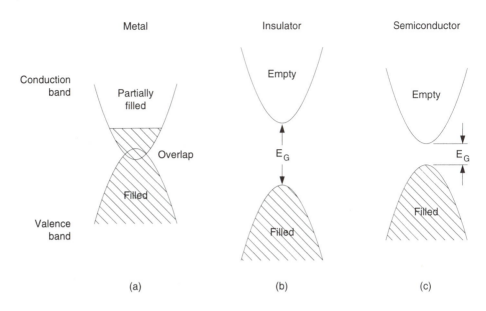

Figure 5-18. Examples of the energy band locations of (a) a metal, (b) an insulator, and (c) a semiconductor

Semiconductor materials have essentially the same structure as insulators – that is, a filled valence band separated from the conduction band by the bandgap energy. The main difference lies in the size of the bandgap. For example, the semiconductor material silicon has a bandgap of the order of 1.1 eV, whereas diamond (an insulator) has a bandgap of approximately 5 eV. These relative bandgap energies are depicted in Figures 5-18(b) and 5-18(c). At low temperatures, the semiconductor acts as an insulator because the valence electrons fill the valence band to the top. If enough energy is added to excite an electron to the conduction band, leaving a hole behind in the valence band, then the material becomes partially conducting. Thus it is the small bandgap of semiconductors that allows the electrons to be easily excited into the conduction band, where they can be used either to conduct current or to produce radiation.

EXCITATION AND DECAY OF EXCITED ENERGY LEVELS – RECOMBINATION RADIATION

Excitation of the semiconductor material, which can be provided either optically or electrically, can transfer electrons from the valence band to the conduction band. As mentioned previously, when the electrons are moved to the conduction band, the vacant sites in the valence band where the electrons were located are referred to as holes. Electrons can decay from the conduction band to the valence band by "recombining" with holes, since the holes behave as positive charges and these positive charges are attracted to the negatively charged electrons in the conduction band. When this occurs, the electrons can either radiate the energy or give it up via interactions or collisions with the semiconductor lattice, which induce a form of

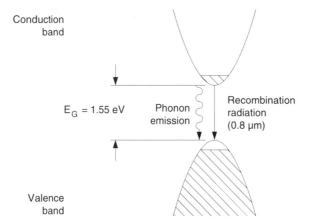

Figure 5-19.
Recombination of
radiation from an excited
GaAs semiconductor
material

lattice vibration or *phonon relaxation* in the material. If recombination radiation occurs then the resulting radiation, shown for a GaAs semiconductor in Figure 5-19, is equivalent to spontaneous emission in atoms. In contrast, the collisional relaxation process reduces the quantum yield of the radiation and produces acoustic waves and unwanted heating of the material.

DIRECT AND INDIRECT BANDGAP SEMICONDUCTORS

There are two basic types of arrangements of the conduction and valence bands in semiconductor materials. In a *direct bandgap* semiconductor, the upper (conduction) band and the lower (valence) band have their minimum and maximum at the same value of k, as suggested in (5.23), and are thus depicted directly above and below each other in a graph of energy versus k; see Figure 5-20. Thus an electron making a transition across the smallest energy gap can do so without a change in k, as indicated in the figure.

On the other hand, *indirect gap* semiconductors have a displacement in k between the minimum of the upper and lower bands that places additional constraints upon the emission of radiation. For such materials a phonon must be emitted (involving a change in momentum) at the same time a photon is emitted, in order to conserve momentum as shown in Figure 5-20. Thus, indirect gap materials are inefficient radiators owing to this restriction. Silicon, probably the most well-known semiconductor material, is an indirect gap material and is therefore not a good radiating species, even though it is used very successfully in making miniature semiconductor switches (transistors) for microelectronic applications that do not require efficient radiative characteristics.

The most studied semiconductor material, from the standpoint of being an efficient radiator, is gallium arsenide, a direct gap material shown in Figure 5-19. The bandgap occurs at an energy of around 1.55 eV, so the recombination radiation occurs at a wavelength of approximately 0.8 μm in the near infrared. Such radiation results when an electron decays from

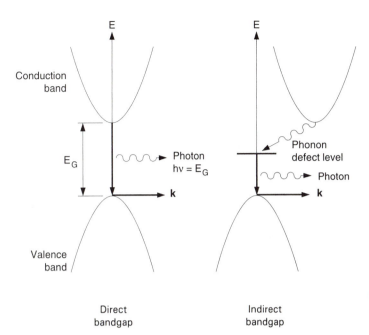

Figure 5-20. Examples of direct gap and indirect gap semiconductor materials

the bottom of the conduction band to the top of the valence band. Energy could decay from a higher portion of the band (as shown by the dashed lines) down to the valence band, if the population existed there long enough to radiate. However, phonon relaxation to the bottom of the conduction band occurs at a much faster rate than recombination, resulting in a build-up of population at that energy.

DENSITY OF STATES IN A DIRECT BANDGAP MATERIAL

Normally the distribution function for electrons versus energy at absolute zero temperature (0 K) in a semiconductor material would be as shown in Figure 5-21(a), where all of the electrons are in the valence band. It is plotted as the number of electrons per unit energy versus energy, or the density of states $\rho(E)$ versus energy. If the electrons are transferred from the valence band to the conduction band via a rapid excitation process (such as laser pumping) while the semiconductor material is at absolute zero temperature, then the density-of-states distribution would look like that of Figure 5-21(b). In this case, all of the conduction electrons fill all of the lowest allowed excited states in the conduction band, so that the distribution has a flat top. If excitation occurs at a temperature greater than zero Kelvin, the density-of-states distribution takes on a different shape that is a function of each specific temperature. The electron thermal energy shifts some of the electrons to higher energy states than that shown in Figure 5-21(b), leaving lower energy states in the conduction band that are not filled to their maximum allowed value, as shown in Figure 5-21(c).

Since referring to electron distribution functions is rather involved, it is sometimes convenient to use the *Fermi energy,* defined as the energy at

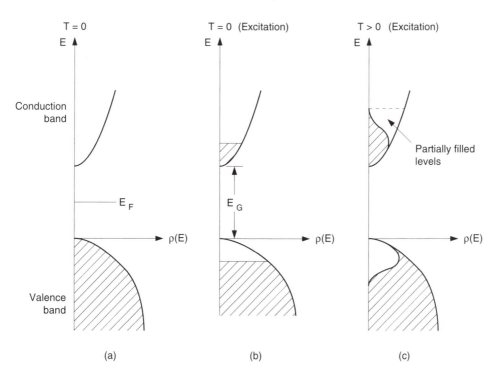

Figure 5-21. Distribution functions of electrons for (a) absolute zero temperature, (b) excitation to the conduction band at absolute zero temperature, and (c) excitation at higher temperatures

which there is a probability of 0.5 that a specific level is occupied. Above that level the population decreases and below that level it rapidly approaches the completely filled value. In Figure 5-21(a) the Fermi energy for a semiconductor at absolute zero therefore occurs halfway between the valence and conduction bands.

INTRINSIC SEMICONDUCTOR MATERIALS

An intrinsic semiconductor material is a crystalline material with no impurities and no lattice defects (no irregularities in the atomic spacing). There are no electrons in the conduction band at 0 K and thus there is no conductivity. At higher temperatures, some electrons are excited into the conduction band, which leads to electron–hole pairs since there is a hole created in the valence band for every electron excited to the conduction band. The electrons decay back to the valence band by recombining with holes as discussed previously. This recombination can occur either in the form of recombination radiation or, more commonly, as an undesirable lattice relaxation that leads to unwanted heating of the crystal.

EXTRINSIC SEMICONDUCTOR MATERIALS – DOPING

Electrical current flow in semiconductor *laser* materials is essential in order to provide excitation energy to the laser levels. It is therefore desirable to control the electrical conductivity associated with the charge-carrier concentration in the material. Pure intrinsic semiconductor materials have relatively low conductivity. Mixing small amounts of "impurity" atoms with the semiconductor material provides additional free electrons and holes that increase the conductivity and thereby allow current to flow more easily. Such a process is referred to as *doping*. The doping is obtained by adding a small percentage of materials that have either one more or one less electron than the semiconductor material species. Such doping produces additional energy levels within the energy band structure, usually within the bandgap, that lie very near the band edge. For an n-type material these levels are near the conduction band, and the process is referred to as n-type doping. The electrons occupying the levels produced by the doping readily move into the conduction band when the material is at room temperature. For p-type materials and p-type doping, the levels are near the valence band. The net effect of the doping is to produce either an excess number of electrons, which begin filling the conduction band, or an excess number of holes, which leave a vacant region of electrons or vacancy in the valence band.

For simplicity we will first consider doping of silicon, even though it is not used as a laser material. A solid crystalline lattice of silicon has four valence electrons. Each electron is shared by two neighboring atoms. Therefore, every atom is surrounded by eight shared electrons, which is sufficient to fill the available energy states in the outer electron shell of the atom. If silicon is doped with an element such as arsenic that has five valence electrons (column 5 of the periodic table), then every arsenic atom will provide one surplus electron to carry charge in the semiconductor material, since there is no room for these electrons in the valence band. Similarly, doping with gallium would provide an excess of holes, since gallium has only three valence electrons. In more general terms, if the semiconductor is silicon then the n-type doping could be phosphorous, arsenic, or antimony, and the p-type doping could be boron, aluminum, gallium, or indium.

Doped and undoped semiconductor materials can also be made out of alloys. The two direct bandgap materials used most often as a basis for semiconductor lasers are gallium arsenide (GaAs) and indium phosphide (InP). For these materials, a doping process similar to that just described for silicon can be carried out. For n-type doping the donor materials would be those with six valence electrons, such as sulfur, selenium, or tellurium. The acceptor materials for p-type doping would use beryllium, zinc, and cadmium. Doping can increase the concentration of conducting electrons, and thereby decrease resistivity, by as much as five orders of magnitude!

The energy-distribution functions or density of states $\rho(E)$ for electrons in both p-type and n-type extrinsic semiconductors are shown in Figure 5-22. Figure 5-22(a) shows the distribution for n-doping and Figure 5-22(b) shows the distribution for p-doping. It can be seen in both cases that a significant number, either of electrons above the bandgap (n-doping) or of holes below the bandgap (p-doping), have been added as compared to the

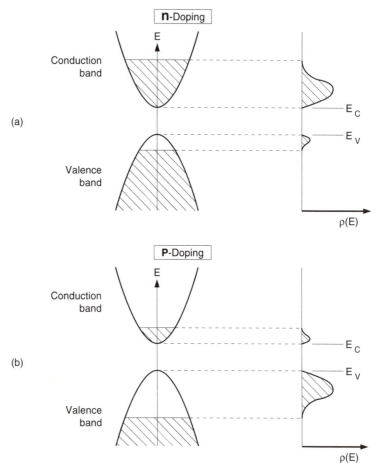

Figure 5-22. Electron distribution functions versus energy for (a) n-doped and (b) p-doped semiconductors at room temperature

distribution function for intrinsic semiconductors shown in Figure 5-21, where the number of conduction electrons and holes are equal. The Fermi level was shown to occur halfway between the conduction band and valence band in Figure 5-21(a). For p-type doping, the Fermi level shifts downward toward the valence band as shown in Figure 5-23(a), since there are excess holes or a lack of electrons which thereby moves the entire distribution downward. For n-type doping, the Fermi level shifts upward toward (or into) the conduction band; this is also shown in Figure 5-23(a).

p–n JUNCTIONS – RECOMBINATION RADIATION DUE TO ELECTRICAL EXCITATION

In order to produce radiation from a semiconductor, it is desirable to cause the excess electrons of the n-type material to come into contact with the excess holes of the p-type material. This is made possible by using separate samples of p-type material and of n-type material, with distinctly different Fermi energies as shown in Figure 5-23(a). A *p–n junction* is formed by bringing the n-type material into contact with the p-type material. When this is done, the excess electrons along the edge of the n-type material flow

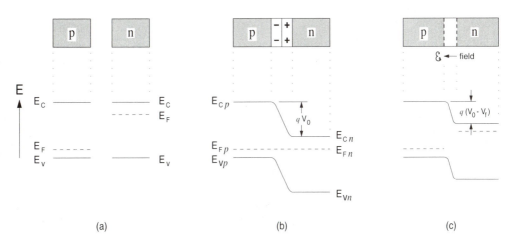

Figure 5-23. Relevant energy levels for valence band, conduction band, and Fermi level of a p–n junction for (a) separated materials, (b) p and n materials in contact, and (c) p and n materials in contact with an applied forward bias

into the p-type material, filling the vacancies or holes in the barrier region. This flow ceases when a space charge builds up owing to the extra doped electrons and holes as indicated in Figure 5-23(b). The consequent voltage V_0 that results from this space charge stops the electron flow. When the initial flow occurs, the electrons and holes produce the minimal amount of recombination, which ceases as soon as the space charge stabilizes the junction region. From the energy-level diagram of this junction region shown in Figure 5-23(b) it can be seen that, where the two materials meet, the Fermi levels of the p- and n-doped materials must be the same in order to produce an equilibrium. Otherwise, the electrons and holes would continue to flow until the Fermi levels were the same. The p–n junction effectively shifts the valence and conduction bands of the p-type material to a higher value relative to the n-type material, as shown in Figure 5-23(b).

We have remarked that the recombination occurring when the materials are brought into contact is a very small amount, since the charges quickly equilibrate and the current flow and recombination cease. However, by applying an electric field (an applied voltage V_f) across the junction, with the positive voltage attached to the p-type material (this is referred to as *forward bias*), the energy barrier between the two materials is reduced significantly, as shown in Figure 5-23(c), and current readily flows between the two materials. Large quantities of electrons then flow through the junction region, and recombination radiation occurs in proportion to the current flow. This radiation leads to semiconductor laser output when the conditions associated with the diode geometry are optimized. Nonradiative decay also occurs in the junction, producing heat that limits the amount of current flow in the material before excessive heating produces permanent damage.

The p–n junction semiconductor diodes are referred to as *homojunction* diodes, since they involve one p-type material and one n-type material. The threshold electrical current density required for these homojunction diodes to produce significant recombination radiation is extremely high. This threshold current density is generally expressed in units of A/cm^2 flowing across the junction. If the junction region has a large surface cross-sectional area then the large volume of the junction region leads to a high ohmic dissipation of heat, which can rapidly lead to damage of the material. It would thus be desirable to minimize the area of the junction region and thereby produce high excitation in a region of small volume; this would preserve the effective excitation within the region and yet minimize the heat produced. The volume can be reduced in the direction of current flow by making heterojunction and quantum-well laser materials, which will be described in the following sections. The volume is reduced in a direction perpendicular to both the current flow and the laser optic axis by using such techniques as gain and index guiding (see Section 14.9).

HETEROJUNCTION SEMICONDUCTOR MATERIALS

In the normal p–n junction, the width of the junction region in the direction of current flow is determined by the distance – which can be many microns – over which the electrons and holes interdiffuse into the neighboring material. Unfortunately, this distance is larger than required to make a laser and thus only provides additional volume in which current flow can lead to additional heat deposition in the material. Also, the current density tapers off gradually away from the junction, so that only near the junction edge is it high enough to produce sufficient excitation to make a laser. The region of current flow that is not right at the junction is too low for laser excitation and produces excess heat, so homojunction lasers can operate only at very low temperatures.

In order to make room-temperature lasers it was necessary to develop a *heterojunction*. Figure 5-24 shows a diagram of a simple heterojunction

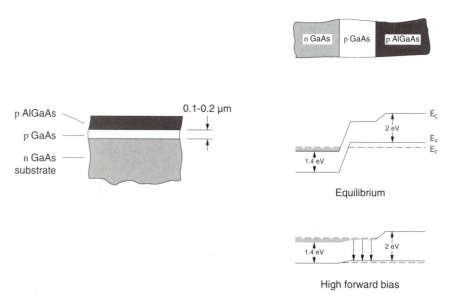

Figure 5-24. A simple heterojunction semiconductor laser material, showing critical dimensions and energy levels both at equilibrium and with an applied forward bias

laser, consisting of an upper p-type layer of AlGaAs followed by a p layer of GaAs and then a substrate of n-type GaAs. In this arrangement the p layer of GaAs has an active region that is only 0.1–0.2 μm thick (or smaller in quantum-well devices). This is the only region where the current can flow, owing to the increased energy of the conduction band for the AlGaAs material that serves as a barrier. Consequently, the thin GaAs layer is also the only region where the excitation and recombination radiation can occur. An energy-level diagram for this arrangement shows the additional 2-eV energy level provided by the p AlGaAs when the material is in equilibrium (with no current flowing), and also when there is a forward-bias voltage applied to the material (causing current to flow). This increased energy prevents current from flowing in the p AlGaAs region, since the energy level remains above the Fermi energy as shown in the figure. The current is thus confined to the p GaAs region, the thickness of which is determined by the fabricated layer instead of by the larger carrier diffusion depth.

Because the charge carriers in heterojunction lasers are confined to a much smaller region than in homojunction lasers, the heat deposition is much lower while the threshold current *density* (and thus the recombination radiation) are still at a sufficiently high level for laser output. In addition, the change in index of refraction at the interface between the p GaAs and the p AlGaAs can provide a guiding effect for the laser beam. This is known as *index guiding*, which will be discussed in Section 14.9.

Double heterojunction structures provide even more control over the size of the active region, and also provide additional index-of-refraction variations that allow for guiding of the optical wave when the semiconductor is operated as a laser. Figure 5-25 shows a simple double heterostructure laser composed of various doping combinations of GaAs and AlGaAs. In addition to the control of layer thickness during fabrication, the side walls of the active region are narrowed via lithographic techniques, confining the laser to a narrow region and thus reducing the threshold current even further.

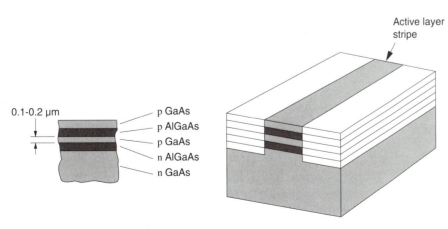

0.1-0.2 μm

p GaAs
p AlGaAs
p GaAs
n AlGaAs
n GaAs

Active layer stripe

Figure 5-25. A double heterostructure gallium arsenide laser showing the active stripe region

QUANTUM WELLS

The thickness of the heterostructure laser active region is from 0.1 to 0.2 μm, which lowers the threshold current required for laser action. If that thickness is reduced even further, to dimensions of 5–10 nm, then the energy levels exhibit quantum-type behavior. For such a narrow active layer, the structure is referred to as a *quantum well*. Since the quantum effects occur only in one dimension, the energy levels can be obtained by considering the electron as a particle in a one-dimensional potential well. The solution of Schrödinger's equation for such a system suggests a series of discrete energy levels instead of the continuous band of levels described by (5.23) for the bulk semiconductor. Using a well of infinite depth as an approximation leads to energy levels described by

$$E_n = \frac{(\pi \hbar)^2 n^2}{2 m_c L^2}, \tag{5.24}$$

where L is the layer thickness and n is a quantum number ($n = 1, 2, 3, \ldots$). Setting the top of the valence band at zero energy, the possible energies of the electron in the conduction band as measured from the top of the valence band are described by

$$E = E_G + E_n, \tag{5.25}$$

where E_G is the bandgap energy of the material. Similarly, for holes the energy would simply be $-E_n$, with m_c replaced by m_v for the valence-band effective mass (effective mass of holes).

For quantum wells, a graph of energy versus density of states does not follow the parabolic description of (5.23). Instead, it has a distribution that consists of steps as shown in Figure 5-26. At the threshold current level, electrons occupy only the location above the first energy level E_1. When transitions occur from that level downward, they occur over narrower wavelength intervals than for a bulk material, since the graph of $\rho(E)$ versus E is very narrow (the electrons are isolated over a much narrower energy range than a normal semiconductor). This leads to higher gain at a much lower threshold current. Threshold currents as low as 0.5 mA/cm^2 have been achieved in quantum-well lasers.

VARIATION OF BANDGAP ENERGY AND RADIATION WAVELENGTH WITH ALLOY COMPOSITION

It is possible to vary the bandgap spacing in semiconductor materials and thereby vary the wavelength for recombination emission. This is done by introducing other selected atoms into the semiconductor to replace the atoms of the original material at randomly located sites. These other atoms must be of nearly identical size to those of the original atoms in order to maintain

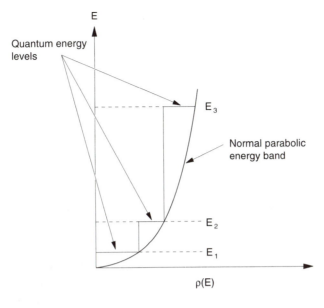

Figure 5-26. Discrete energy levels associated with a quantum-well laser geometry

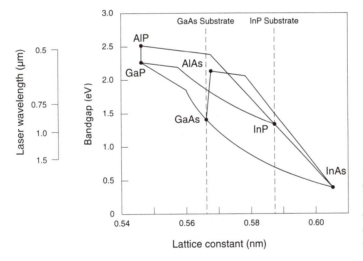

Figure 5-27. Semiconductor laser material mixtures and their associated bandgap energies and potential laser wavelengths

the integrity of the crystal lattice. A diagram showing an example of some laser material mixtures and their associated bandgap energies and potential laser wavelengths is given in Figure 5-27. These alloys are based upon growing the material on either a GaAs substrate or an InP substrate to provide the necessary lattice matching for the various atomic species.

When radiation occurs because of recombination across the bandgap of such materials, the wavelength of that radiation will vary with the concentration of the impurity. The atomic species associated with a gallium arsenide semiconductor material are atomic gallium, which has three 4p electrons, and atomic arsenic, which contains five 4p electrons. Other atoms can be "grown" into the semiconductor lattice to replace either gallium or arsenic, with a slight increase in the spacing between atoms (the lattice constant) of the crystal. For example, aluminum can replace gallium, either partially or completely, which produces an increase in the bandgap energy. For a pure aluminum arsenide (AlAs) system the band energy increases to approximately 2.15 eV, as can be seen in Figure 5-27. For a mixture of aluminum, gallium, and arsenic (AlGaAs), the bandgap energy depends upon the percentage of aluminum in the mixture. This is referred to as a *ternary alloy* and is denoted as $Al_xGa_{1-x}As$, where x refers to the fraction of aluminum replacing gallium in the mixture. Thus the bandgap energy varies between that of pure GaAs (1.55 eV) and pure AlAs (2.15 eV). Similarly, phosphorous can be used to replace arsenic to obtain the ternary compound $GaAs_{1-x}P_x$. However, for $x > 0.45$ this ternary compound becomes an indirect gap material; this makes it a less efficient radiator, but the bandgap is made larger to provide recombination radiation in the visible spectrum. Such materials do not make good lasers, but they are used to make light-emitting diodes (LEDs).

A new class of blue-green semiconductor diode lasers has been demonstrated in quantum-well devices made of layers involving combinations of Zn, Cd, S, and Se. Presently, heterostructure devices of ZnCdSe/ZnSSe/ZnMgSSe have been operated continuously at a wavelength near 509 nm, for short periods at room temperature, with output powers of up to 10 mW. There are significant efforts being employed in the development of such lasers due to their many potential applications as described in Section 14.9.

EXAMPLE

A semiconductor heterojunction laser is grown with a GaAsP active-region layer with a lattice constant of 0.56 nm. Assume that the laser transition occurs from the bottom of the conduction band. Determine the laser wavelength of this material.

From Figure 5-27 we can estimate the bandgap energy to be 1.85 eV. If we assume that the laser transition will occur from the bottom of the conduction band to the top of the valence band, then the laser transition wavelength will correspond to the bandgap energy. Associating the bandgap energy with the wavelength, we therefore have

$$E = h\nu = h(c/\lambda)$$

and thus

$$\lambda = \frac{hc}{E} = \frac{(6.626 \times 10^{-34} \text{ J-s})(3 \times 10^8 \text{ m/s})}{(1.85 \text{ eV})(1.602 \times 10^{-19} \text{ J/eV})}$$

$$= 6.707 \times 10^{-7} \text{ m} = 670.7 \text{ nm}.$$

There will be some broadening of the emission, due to the finite distribution of electrons above the bottom of the conduction band and of holes below the top of the valence band, as indicated in Figure 5-22. The value just obtained (670.7 nm) represents the longest wavelength in the emission band produced by this laser.

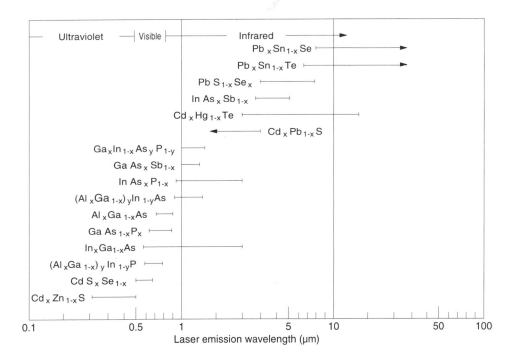

Figure 5-28. Possible materials available to make semiconductor lasers at a variety of wavelengths

A chart showing the possible materials that either have been used or might be used to make semiconductor lasers is shown as Figure 5-28. It can be seen that the possibilities extend from the near ultraviolet to the middle infrared spectral regions.

RECOMBINATION RADIATION TRANSITION PROBABILITY AND LINEWIDTH

When the electrons and holes arrive at the junction and recombine, the electrons revert to their location in the valence band, thereby giving up the excited energy they had while in the conduction band. As explained previously, this energy can either be given up to the crystalline lattice in the form of collisional relaxations (phonon interactions) or it can be radiated. The photon energy is at or near that of the bandgap, so that $E_G = hc/\lambda$. As in the other laser materials we have discussed, the electron–hole recombination time in semiconductors is the longest decay time of the system, equivalent to the radiative lifetime in a free atom. If collisions become significant, then the decay rate increases and the decay time is shortened to the typical relaxation time in a solid, approximately 10^{-13}–10^{-14} s. As was the case for organic dyes dissolved in solvents, radiative decay times for most semiconductor diode lasers are approximately 10^{-9} s, which corresponds to a radiative transition probability of $A = 10^9 \text{ s}^{-1}$. And just as with liquids and dielectric solids, the most desirable laser materials are generally those with a high radiation efficiency as opposed to phonon decay. Phonon relaxation, which leads to an undesirably high decay rate, is significantly enhanced by the presence of impurities. Thus it is essential to maintain a very clean system during the semiconductor growth process in order to minimize the presence of such impurities.

The spontaneous emission (or recombination radiation) linewidth is determined by the widths of the density-of-states distributions for both the conduction-band doped electrons and the valence-band doped holes, as shown in Figure 5-22; this linewidth is therefore sensitive to temperature. The width is reduced in a heterojunction laser and is even narrower for a quantum-well laser, as suggested in Figure 5-26. A typical linewidth at room temperature for a heterojunction laser is approximately 20 nm and for a quantum-well laser is approximately 5 nm. Of course, the actual laser output is significantly narrower than this value, which is reduced by gain narrowing and cavity effects that will be discussed in Chapter 10.

REFERENCES

F. J. Duarte and L. W. Hillman, eds. (1990), *Dye Laser Principles*. New York: Academic Press.

G. Herzberg (1945), *Infrared and Raman Spectra*. New York: Van Nostrand Reinhold.

G. Herzberg (1950), *Spectra of Diatomic Molecules*. New York: Van Nostrand Reinhold.

C. Kittel (1968), *Introduction to Solid State Physics,* 3rd ed. New York: Wiley.

W. Koechner (1992), *Solid State Laser Engineering,* 3rd ed. New York: Springer-Verlag.

B. G. Streetman (1990), *Solid State Electronic Devices,* 3rd ed. New York: Prentice-Hall.

PROBLEMS

1. Derive all of the allowed frequencies of radiation that might occur among the first four rotational energy levels of the hydrogen molecule H_2. Assume that H_2 has a permanent electric dipole moment (which it does not) so that it can radiate. Draw an energy-level diagram of those first four levels, and indicate the possible transitions and their frequencies.

2. Consider the first two vibrational energy levels of the CO molecule, which we will assume have an energy separation of 0.2 eV. Consider also the first four rotational levels of each of those translational levels. Derive all of the frequencies of radiation that can occur among all of these levels, including pure rotational transitions and rotational–vibrational transitions. Include an energy-level diagram for those levels, and indicate the allowed transitions and wavelengths. Assume that the separation between C and O is 0.111 nm.

3. Label the molecular electronic energy levels listed in the following table.

S	L
1	2
1	1
0	1
0	0
1	2

S	L
1/2	1
3/2	2
1/2	3
1	0
3/2	1
3/2	0
1/2	2

Identify which levels can have allowed transitions between them, indicating those transitions in the following form: $^2\Sigma \rightarrow {}^2\Pi$.

4. Which dyes of Figure 5-11 would make the best lasers at the following wavelengths:

(a) the red helium–neon laser wavelength;
(b) the $n = 3$ to $n = 2$ transition of atomic hydrogen;
(c) the $n = 4$ to $n = 2$ transition of atomic hydrogen.

5. For the dyes given in Figure 5-11, assume that the given range of emission wavelengths corresponds to the available laser wavelengths. Assume that the dyes radiate with a quantum yield of 100%. Compute the phase-interruption broadening time for each of the dyes. If their radiative decay time is 1 ns, what would be the natural linewidth for each of the dyes?

6. Compare the emission bandwidths of Nd:YAG and Nd:glass laser materials. Determine the collisional interaction time (either T_1 or T_2; see Section 4.3) that would be required to produce such broadening. Indicate whether this broadening is due to either T_1 or T_2.

7. Assume that both Nd:YAG and Nd:glass laser materials are doped to the same Nd concentration of $10^{26}/m^3$. Assume they are both pumped identically, so that they have approximately 0.1% of the Nd^{3+} ions in the upper laser level. Compare the radiated power of a cubic volume 10 mm on a side from each of these materials. Assume that both materials radiate at approximately the same wavelength of 1.06 μm. Compute the radiated power per unit frequency by assuming that all of the power radiated is averaged over the FWHM emission linewidth of each of the transitions.

8. Make a sketch of the emission spectrum for $Cr:BeAl_2O_4$, $Ti:Al_2O_3$, Cr:LiSAF, and Cr:LiCaF broadband laser materials. Compute the collision interruption time T_2 associated with phonon broadening for each of these materials by using the experimental emission linewidths. Compare these times to the upper laser level lifetimes.

9. Look up and compare the electrical conductivity of glass (insulator), silicon (semiconductor), and copper (conductor).

10. From the chart in Figure 5-27, determine the laser wavelength for each of the six different compounds listed in the chart: AlP, GaP, AlAs, GaAs, InP, and InAs.

11. Using the chart of Figure 5-28, determine which semiconductor laser materials could replace the He–Ne laser operating at 632.8 nm.

12. Assume that an electron is subjected to a potential energy function V of the form

$$V = \begin{cases} V_0 = 0 & \text{for } x = 0 \text{ to } x = L, \\ \infty & \text{for } x = -\infty \text{ to } x = 0, \\ \infty & \text{for } x = L \text{ to } x = +\infty. \end{cases}$$

Solve the Schrödinger equation to determine the energy levels for this system, which is called an *infinite square-well potential*. Assuming this is a quantum-well laser, how high would the first energy level occur above the bandgap?

RADIATION AND THERMAL EQUILIBRIUM
ABSORPTION AND STIMULATED EMISSION

6

SUMMARY Chapters 3 through 5 dealt with energy levels of gaseous, liquid, and solid materials, and the spontaneous radiative properties associated with transitions occurring between those levels. The rate of radiation occurring from energy levels was described in terms of transition probabilities and relative oscillator strengths of the transitions. So far, no attempt has been made to consider the collective properties of a large number of radiating atoms or molecules. Such considerations will be dealt with in this chapter, and will lead to the concept of radiation in thermodynamic equilibrium. Planck's radiation law evolved from the analysis of equilibrium radiation from dense bodies. From this law and other principles, Einstein was able to deduce the concept of stimulated emission, which is the underlying principle leading to the development of the laser.

6.1 EQUILIBRIUM

THERMAL EQUILIBRIUM

We will consider the processes that bring various masses into equilibrium when they are located near each other. In the study of thermodynamics, *thermal equilibrium* describes the case where all of the individual masses within a closed system have the same temperature. If a substance of small mass M1 and temperature T1 is placed near to or in contact with a much larger mass M2 at a higher temperature T2, the state of thermal equilibrium is achieved when the mass M1 reaches the same temperature as that of M2. The duration over which this occurs could be a period as short as picoseconds or as long as minutes or even hours, depending upon the situation. In fact, the final temperature will be less than T2, since M2 will also be cooled slightly as it transfers some of its energy to M1. The temperature decrease of M2 will depend upon the relative masses of M1 and M2, but the relevant factor, when a new equilibrium is reached, is that the total energy will be divided equally among all of the atoms of the combined system. Thermal equilibrium can be achieved between the two masses by one or more possible heat transfer processes: conduction, convection, and radiation.

THERMAL EQUILIBRIUM VIA CONDUCTION AND CONVECTION

If M1 is placed in direct contact with M2, as shown in Figure 6-1(a), this will generally bring the masses to equilibrium in the shortest period of

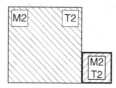

(a) Conduction

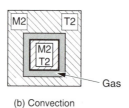

(b) Convection

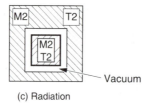

Figure 6-1. Examples of energy transfer via (a) conduction, (b) convection, and (c) radiation

(c) Radiation

time. In this case, the method of heat transfer is referred to as *conduction*. If M1 and M2 are composed of two metals, for example, the rapid heat transfer by the conduction electrons of the metals would bring the two masses quickly to an equilibrium temperature.

If the bodies are placed as shown in Figure 6-1(b), with M1 located in a large cavity inside M2, and if a gas (such as air) at atmospheric pressure fills the cavity between M1 and M2, then the molecules of the gas would provide the energy transport from M2 to M1 and so bring M1 to a final equilibrium temperature somewhere between T1 and T2. This case of energy transport leading to thermal equilibrium is referred to as *convection*. A small portion of the energy would also end up in the gas, which would arrive at the same final equilibrium temperature as that of the two masses. Since the mass density of gas at atmospheric pressure is approximately three orders of magnitude lower than that of solids, the convection process takes a much longer time to reach equilibrium than does the conduction process.

THERMAL EQUILIBRIUM VIA RADIATION

If all of the air is evacuated from the space between M1 and M2 as in Figure 6-1(c), so that there is no material contact between them, there is still another process that will bring the two masses into equilibrium – namely, *radiation*. For such a process to occur, two effects must take place: the masses must be radiating energy, and they must be capable of absorbing the radiation from the other body.

6.2 RADIATING BODIES

We will first consider the question of whether or not the masses are radiating. If a collection of atoms is at temperature T then, according to the Boltzmann equation, the probability distribution function f_i that any atom has a discrete energy E_i is given by

$$f_i(E_i) = C_1 g_i e^{-E_i/kT}. \tag{6.1}$$

In this equation, Boltzmann's constant k is 8.6164×10^{-5} eV/K and g_i is the statistical weight of level i. The C_1 term is a normalizing constant that is the same for all energy levels and is subject to the constraint that

$$\sum_i f_i = \sum_i C_1 g_i e^{-E_1/kT} = 1. \tag{6.2}$$

If N is the total number of atoms per unit volume of this species and N_i is the population density occupying a specific energy level i, then

$$\sum_i N_i = N, \tag{6.3}$$

where N_i could then be expressed as

$$N_i = f_i N = C_1 g_i e^{-E_i/kT} N. \tag{6.4}$$

For a high-density material such as a solid, the energy levels are usually continuously distributed (with some exceptions, including solid-state laser ions). Thus the distribution function would be expressed as a probability per unit energy $g(E)$ such that the probability of finding a fraction of that material excited to a specific energy E within an energy width dE would be given by

$$g(E)\,dE = C_2 e^{-E/kT}\,dE, \tag{6.5}$$

where we have ignored the statistical weights. This probability would also be subject to the normalizing constraint that

$$\int_0^\infty g(E)\,dE = \int_0^\infty C_2 e^{-E/kT}\,dE = 1. \tag{6.6}$$

From this equation we can readily show that $C_2 = 1/kT$, and thus $g(E)$ can be expressed as

$$g(E) = \frac{1}{kT} e^{-E/kT}. \tag{6.7}$$

Again, if N is the total number of atoms per unit volume in the solid and we refer to the number of atoms per unit volume within a specific energy range dE as $N(E)$, then the normalizing condition requires that

$$N = \int_0^\infty N(E)\, dE.$$ (6.8)

The number of atoms at energy E within a specific energy range dE can thus be given as

$$N(E)\, dE = \frac{N}{kT} e^{-E/kT}\, dE.$$ (6.9)

We can compute the ratio of the populations that exists at two specific energies, either for the case of discrete energy levels such as those for isolated atoms or for high-density materials such as solids. For discrete energy levels, according to (6.4) the ratio of the population densities N_2 and N_1 (number of particles per unit volume) of atoms with electrons occupying energy levels 2 and 1 (with corresponding energies E_2 and E_1) would be expressed as

$$\frac{N_2}{N_1} = \frac{g_2}{g_1} e^{-(E_2 - E_1)/kT} = \frac{g_2}{g_1} e^{-\Delta E_{21}/kT},$$ (6.10)

where it is assumed that E_2 is higher than E_1 and that $\Delta E_{21} = E_2 - E_1$. Similarly, the ratio of population densities of a dense material (such as a solid) at energies E_2 and E_1 within an energy interval dE can be expressed from (6.9) as

$$\frac{N(E_2)\, dE}{N(E_1)\, dE} = \frac{N(E_2)}{N(E_1)} = e^{-\Delta E_{21}/kT}.$$ (6.11)

It can be seen that these equations are identical except for the statistical weight factor, which we have ignored for the high-density material. Thus, when a collection of atoms – whether in the form of a gas, a liquid, or a solid – are assembled together and reach equilibrium, not just the kinetic energies related to their motion will be in thermal equilibrium; the distribution of their internal energies associated with the specific energy levels they occupy will also be in thermal equilibrium, according to (6.10) and (6.11).

EXAMPLE

Determine the temperature required to excite electrons of atoms within a solid to energies sufficient to produce radiation in the visible portion of the electromagnetic spectrum when the electrons decay from those excited levels.

We assume a typical solid material with a density of approximately $N = 5 \times 10^{28}$ atoms/m^3 in the ground state. For several different temperatures we will compute how many of those atoms would occupy energy levels high enough to radiate that energy as visible light. Visible radiation comprises wavelengths ranging from 700 nm in the red to 400 nm in the violet, or photons with energies ranging from approximately 1.7 to 3.1 eV. We would thus be interested in energy levels above 1.7 eV that are populated within the solid, since any energy higher than 1.7 eV will have the potential of radiating a visible photon. We will calculate the number of species that have an electron in an excited energy level that lies higher than 1.7 eV above the ground state for several different temperatures by taking the integral of $N(E)\,dE$ from (6.9) over the energy range from 1.7 eV to infinity. We must recognize that, for a typical solid, most of the atoms will decay nonradiatively from these excited levels, but a certain portion could emit visible radiation depending upon the radiation efficiency of the material.

At room temperature, $T \cong 300$ K and $kT = 0.026$ eV. Thus N_{vis} can be expressed as

$$N_{vis} \cong \frac{N}{kT} \int_{1.7\,eV}^{\infty} e^{-E/kT}\,dE = -N[e^{-E/kT}]_{1.7\,eV}^{\infty}$$

$$= (-5 \times 10^{28})[0 - 4 \times 10^{-29}] \cong \frac{0}{m^3}. \qquad (6.12)$$

Thus there are essentially no atoms in this energy range from which visible photons could radiate. This is, of course, why we can see nothing when we enter a room that has no illumination, even though the human eye is very sensitive and can detect as little as only a few photons. No thermal radiation in the visible spectrum could be emitted from the walls, floors, ceiling, or furniture when those various masses are at room temperature.

We next consider a temperature of 1,000 K or 0.086 eV. At this temperature we determine the population able to emit visible photons as

$$N_{vis} = \frac{N}{kT} \int_{1.7\,eV}^{\infty} e^{-e/kT}\,dE = -N[e^{-E/kT}]_{1.7\,eV}^{\infty}$$

$$= (-5 \times 10^{28})[0 - 2.6 \times 10^{-9}] = \frac{1.3 \times 10^{20}}{m^3}. \qquad (6.13)$$

Thus, increasing the temperature by little more than a factor of three, we have gone from essentially no atoms in those excited levels to an appreciable number in those levels. We can and do see such radiation: in the glowing coals of a campfire, in the glowing briquettes of a barbecue fire, or from the heating elements of an electric stove; all are at temperatures of approximately 1,000 K.

A temperature of 5,000 K or 0.43 eV (the temperature of the sun) can also be considered for comparison:

$$N_{vis} = \frac{N}{kT} \int_{1.7\,eV}^{\infty} e^{-e/kT}\,dE = -N[e^{-E/kT}]_{1.7\,eV}^{\infty}$$

$$= (-5 \times 10^{28})[0 - 1.9 \times 10^{-2}] = \frac{9.5 \times 10^{26}}{m^3}. \qquad (6.14)$$

EXAMPLE (cont.)

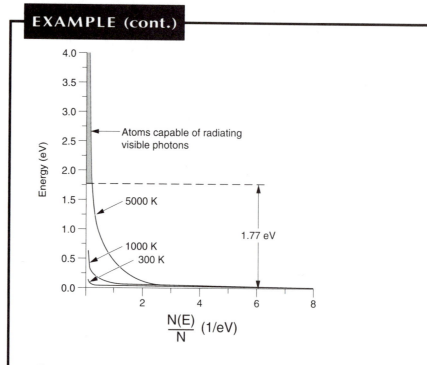

Figure 6-2. Population distribution of occupied states versus energy for temperatures of 300 K, 1,000 K, and 5,000 K

At this temperature nearly 10% of the atoms would be excited to an energy of 1.7 eV or higher, and the material would be radiating with an intensity that is too bright to look at.

A plot of $N(E)/N$, which is the distribution of population versus energy as taken from (6.9), is plotted in Figure 6-2 for 300 K, 1,000 K, and 5,000 K. The graph clearly shows the rapid increase in the population at higher energies as the temperature is increased. It can be seen that the energy levels that are high enough to produce visible radiation are populated only when the temperature is significantly above room temperature, as discussed in the preceding examples. Such radiation emitted from masses at those various temperatures would emit a continuous spectrum of frequencies over a certain frequency range. The radiation is referred to as *thermal radiation,* since it is emitted from objects in thermal equilibrium.

There are two effects we should consider in the observation of thermal radiation such as that of glowing coals. First, more energy is radiated from the object as the temperature is increased; this is described by the Stefan–Boltzmann law. Second, if the spectral content of the radiation from the glowing coals is analyzed, it will be found that the radiation increases with decreasing wavelength to a maximum value at a specific wavelength, and then decreases relatively rapidly at even shorter wavelengths. The wavelength at which the maximum value occurs can be obtained from Wien's law.

STEFAN–BOLTZMANN LAW

The Stefan–Boltzmann law is an empirical relationship obtained by Stefan and later derived theoretically by Boltzmann. It states that the total radiated intensity (W/m^2) emitted from a body at temperature T is proportional to the fourth power of the temperature, T^4. This can be written as

$$I = e_M \sigma T^4, \tag{6.15}$$

where $\sigma = 5.67 \times 10^{-8}$ W/m^2-K^4 and e_M is the emissivity that is specific for a given material. The emissivity, a dimensionless quantity that varies between zero and unity, represents the ability of a body to radiate efficiently and is also associated with its ability to absorb radiation, as will be discussed in more detail in Section 6.3. Equation (6.15) describes an extremely rapidly increasing function with temperature, and accounts for the tremendous flux increase we estimated to be radiating in the visible from our simple system at 300 K, 1,000 K, and 5,000 K. The total radiation from such a mass would increase in going from 300 K to 1,000 K by a factor of 123, and in going from 300 K to 5,000 K by a factor of over 77,000!

WIEN'S LAW

A graph of the spectrum of radiation emitted versus frequency from a heated mass is shown for three different temperatures in Figure 6-3. The

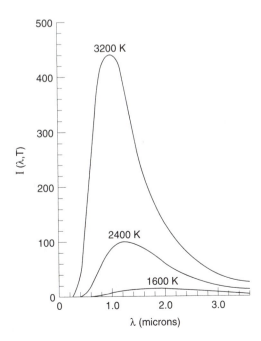

Figure 6-3. Spectrum of radiation versus wavelength of a heated mass (blackbody) for several temperatures

specific wavelength at which the radiation is a maximum was found to vary inversely with temperature. The wavelength at which the maximum emission occurs for any given temperature is described by Wien's law as follows:

$$\lambda_m T = 2.898 \times 10^{-3} \text{ m-K}. \tag{6.16}$$

IRRADIANCE AND RADIANCE

Recall the case depicted in Figure 6-1(c). Because M2 is initially at a higher temperature than M1, we surmise that M1 will eventually be raised to equilibrium with M2, at a temperature T3 such that T1 < T3 < T2, by absorbing the radiation from M2. However, when M1 reaches the same temperature as M2, *it must radiate as much energy as it absorbs*. Otherwise, it would continue to heat up to a higher temperature than T2, which defies the laws of thermodynamics. Thus the relationship – in total power per unit area – between the *irradiance* I_1 incident upon M1 and the *radiance* H_1 leaving M1 must be proportional:

$$H_1 = b_1 I_1 \quad \text{or} \quad I_1 = \frac{H_1}{b_1}, \tag{6.17}$$

where $b_1 \leq 1$ is the proportionality constant that represents the fraction of the power absorbed by M1. If instead of just M1 there were several masses (M1, M3, and M4) all inside the cavity of M2, we could express their radiative characteristics with respect to the incident flux in the same way as that of (6.17) for M1. We would obtain the relationships $H_1 = b_1 I_1$, $H_3 = b_3 I_3$, and $H_4 = b_4 I_4$. Because I (the power per unit area) arriving at each body would be the same, we can write

$$I = \frac{H_1}{b_1} = \frac{H_3}{b_3} = \frac{H_4}{b_4} = \cdots, \tag{6.18}$$

which indicates that the ratio of the power radiated from a body to the fraction of radiation absorbed is a constant, independent of the material. Thus strong absorbers are also strong radiators, and since b represents the fraction of radiation absorbed we can conclude that $b = e_M$. This relationship is known as *Kirchhoff's law*. Examples of emissivity are shown in Table 6-1. It can be seen that emissivities range from very small numbers for highly reflecting materials to values near unity for highly absorbing materials. Lampblack, the soot deposited from burning candles, is perhaps the blackest material we ever observe; it has an emissivity of $e_M = 0.95$.

The ideal case of a perfect absorber ($b = 1$) also describes a perfect emitter in that it would radiate as much energy as is incident upon it. Such a perfectly absorbing body is known as a *blackbody* and is the best emitter of thermal radiation, which in this case is referred to as *blackbody radiation*. It is also known as *cavity* radiation, since it meets the same requirements as those described previously for the mass M1 in a hollow cavity.

TABLE 6-1

TOTAL EMISSIVITY e_M AT LOW TEMPERATURES

Highly polished silver	0.02
Aluminum	0.08
Copper	0.15
Cast iron	0.25
Polished brass	0.60
Black gloss paint	0.90
Lampblack	0.95

6.3   CAVITY RADIATION

We have developed the idea that ideal blackbodies radiate with the same spectral power as that occurring within a cavity such as that depicted in Figure 6-1(c). It will be very useful for us to obtain the quantity and wavelength distribution of the radiated flux in order to later obtain the stimulated emission coefficient. Toward this end, we begin by considering the properties of radiation within a cavity. The boundary conditions – as deduced from electromagnetic theory – suggest that, in order for electromagnetic waves to be supported or enhanced (or at least not rapidly die away), the value of the electric vector must be zero at the boundary of the cavity. This involves the development of distinct "standing waves" within the cavity – that is, waves whose functions exhibit no time dependence. Such standing waves are referred to as *cavity modes*. The existence of each mode implies that the wave has an integral number of half-wavelengths occurring along the wave direction within the cavity. For example, three distinct cavity modes with a specific direction are shown in Figure 6-4: waves of one half-wavelength, two half-wavelengths (one complete cycle), and three half-wavelengths. If we can calculate the number of modes at each frequency or wavelength and multiply by the average energy of each mode at those wavelengths, we can obtain the frequency distribution of the emission

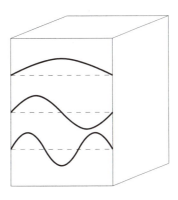

Figure 6-4. Several modes of electromagnetic radiation within a confined cavity

spectrum for cavity radiation. Hence we will now calculate the number of modes that exist within a cavity of a given size.

COUNTING THE NUMBER OF CAVITY MODES

Assume we have a cavity that is rectangular in shape with dimensions L_x, L_y, and L_z, as shown in Figure 6-1(c). Because we are considering standing waves within the cavity, in counting the number of modes we will include only the spatial dependence of the electromagnetic waves, which is described by an oscillatory function of the form $e^{-i(k_x L_x + k_y L_y + k_z L_z)}$. In order to satisfy the boundary condition that the field be periodic in L, the exponential phase factor must be an integral multiple of π. This can be achieved if we specify individual modes in the directions x, y, z such that

$$k_x L_x = n_x \pi, \quad k_y L_y = n_y \pi, \quad k_z L_z = n_z \pi,$$
$$n_x, n_y, n_z = 0, 1, 2, \dots . \tag{6.19}$$

Any mode in this cavity will have a specific value of k; its mode number can be identified by specifying mode numbers n_x, n_y, n_z since

$$k^2 = (k_x^2 + k_y^2 + k_z^2) = \left[\left(\frac{n_x \pi}{L_x} \right)^2 + \left(\frac{n_y \pi}{L_y} \right)^2 + \left(\frac{n_z \pi}{L_z} \right)^2 \right]. \tag{6.20}$$

The total number of modes in any volume $V = L_x L_y L_z$ for a given wavelength $\lambda = 2\pi/k$ can then be counted by using a three-dimensional space whose axes are defined as the number of modes. For example, the number of modes in the x direction would be the length of the cavity in that direction L_x divided by one half-wavelength or $n_x = L_x/(\lambda/2)$, which is equivalent to the mode number in that direction. Consequently, the number of modes along each axis of the volume for a specific wavelength λ is obtained as $n_x = 2L_x/\lambda$, $n_y = 2L_y/\lambda$, and $n_z = 2L_z/\lambda$. We will choose one octant of an ellipsoid to describe the mode volume, since this is a convenient way of counting the modes. (We choose only one octant with the foregoing dimensions for the major axes because all of the values of n are positive.) Each mode – say $n_x = 1$, $n_y = 2$, and $n_z = 1$, for example – is specified in our mode volume as the intersection of those specific mode numbers within that octant, as shown in Figure 6-5. Such a mode can be specified, for the purposes of this calculation, by assuming that it is represented by a cube of unit dimensions within that volume. This approximation is very good for volumes that are significantly larger than the wavelength λ, and since we are dealing with relatively small (short) wavelengths, the volume calculation for counting the number of modes is quite accurate.

The volume of the octant can be written as

$$\frac{1}{8} \cdot \frac{4\pi}{3} \cdot n_x n_y n_z = \frac{1}{8} \cdot \frac{4\pi}{3} \left(\frac{2L_x}{\lambda} \cdot \frac{2L_y}{\lambda} \cdot \frac{2L_z}{\lambda} \right); \tag{6.21}$$

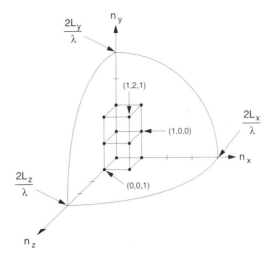

Figure 6-5.
Three-dimensional diagram of the cavity mode volume

likewise, since $\lambda\nu = c$, the number of modes M can be expressed as

$$M = \frac{1}{8} \cdot \frac{4\pi}{3} \left(\frac{2L_x 2L_y 2L_z}{\lambda^3} \right) = \frac{4}{3} \cdot \frac{\pi\nu^3}{c^3} V, \tag{6.22}$$

where $V = L_x L_y L_z$. This calculation defines the number of modes for *all frequencies up to and including the frequency ν within a volume V.* This number must be doubled to allow for the fact that two orthogonal polarizations of the electromagnetic waves must be represented for each spatial mode. Higher frequencies would be outside of this volume and thus would not be counted. For frequencies up to the value ν, the mode density ρ (number of modes per unit volume) is then given by

$$\rho(\nu) = \frac{2M}{V} = \frac{8\pi\nu^3}{3c^3}. \tag{6.23}$$

This equation can be differentiated to obtain the number of modes per unit volume within a given frequency interval between ν and $\nu + d\nu$,

$$\frac{d\rho(\nu)}{d\nu} = \frac{8\pi\nu^2}{c^3}. \tag{6.24}$$

We have used a rectangular cavity to derive the mode density, but it can be shown that (6.23) and (6.24) are generally applicable formulas for any shape of cavity.

RAYLEIGH–JEANS FORMULA

If we assume that the average energy per mode is of the order of kT, we can obtain the Rayleigh–Jeans formula for the energy density of radiation per unit volume within the frequency ν to $\nu + d\nu$. We do this by multiplying the mode density per unit frequency by the average energy kT per mode.

Therefore, using kT as the value of the energy per mode, for the energy density per unit frequency we have

$$u(v) = \frac{d\rho(v)}{dv} kT = \frac{8\pi v^2}{c^3} kT. \tag{6.25}$$

This result suggests that there is a continuous increase in energy density with frequency for a given temperature T. This expression is referred to as the Rayleigh–Jeans formula for the energy density of cavity radiation. It agrees with experiments for lower frequencies but does not predict the experimentally observed maximum value for a given temperature at higher frequencies. Instead, the Rayleigh–Jeans expression suggests that energy density approaches infinity as the frequency is increased.

PLANCK'S LAW FOR CAVITY RADIATION

Planck was disturbed by the invalidity of the Rayleigh–Jeans formula for the intensity of cavity radiation at higher frequencies, and therefore questioned the basic assumption of assigning a value of kT for the average energy per cavity mode. As an alternative, Planck explored the possibility of quantizing the mode energy, postulating that an oscillator of frequency v could have only discrete values mhv of energy, where $m = 0, 1, 2, 3, \ldots$. He referred to this unit of energy hv as a quantum that could not be further divided. We can apply this condition to obtain the energy of each cavity mode in thermal equilibrium. We will use the Boltzmann distribution function (eqn. 6.1) to describe the energy distribution of the modes; however, we will assign discrete values for the energy $E_m = mhv$ in the function instead of a continuous variable E, and we will ignore the statistical weight factor. Thus, for a given temperature T we can express the distribution function as

$$f_m = Ce^{-E_m/kT} = Ce^{-mhv/kT}. \tag{6.26}$$

The normalizing condition for the distribution function requires that

$$\sum_{m=0}^{\infty} f_m = \sum_{m=0}^{\infty} Ce^{-mhv/kT} = 1, \tag{6.27}$$

which can be solved for C to obtain

$$C = 1 - hv/kT. \tag{6.28}$$

We can then compute the average mode energy $\bar{E}$ of those oscillators in the usual manner:

$$\bar{E} = \frac{\sum_{m=0}^{\infty} E_m f_m}{\sum_{m=0}^{\infty} f_m} = \frac{C \sum_{m=0}^{\infty} (mhv)e^{-mhv/kT}}{C \sum_{m=0}^{\infty} e^{-mhv/kT}}$$

$$= \frac{[1 - (hv/kT)] \sum_{m=0}^{\infty} (mhv)e^{-mhv/kT}}{1}. \tag{6.29}$$

The value of the summation in the numerator on the right-hand side of (6.29) is

$$\sum_{m=0}^{\infty} (mh\nu)e^{-mh\nu/kT} = h\nu \sum_{m=0}^{\infty} me^{-mh\nu/kT} = \frac{h\nu e^{-h\nu/kT}}{(1-e^{-h\nu/kT})^2}. \qquad (6.30)$$

This leads to an expression for the average energy $\bar{E}$ of

$$\bar{E} = \frac{h\nu}{e^{h\nu/kT}-1}. \qquad (6.31)$$

Using this expression for the average energy per mode, instead of the kT value used in the Rayleigh–Jeans law to obtain (6.25), we arrive at the following relationship for the energy density per unit frequency:

$$u(\nu) = \frac{8\pi h\nu^3}{c^3(e^{h\nu/kT}-1)}. \qquad (6.32)$$

This relationship has a maximum at a specific frequency for a given temperature, and both the location of the maximum and the shape of the distribution agree very well with experimental observations. This expression became known as Planck's law for cavity radiation.

The expression $u(\nu)$ describes the energy density per unit frequency ν for radiation anywhere within an enclosed cavity at a temperature T. It consists of waves traveling in all directions within the cavity. If we wanted to compute the total energy density u emitted at all frequencies, we could of course simply integrate the energy density $u(\nu)\,d\nu$ over the frequencies:

$$u = \int_0^{\infty} u(\nu)\,d\nu. \qquad (6.33)$$

Carrying this integration out leads to the Stefan–Boltzmann law.

RELATIONSHIP BETWEEN CAVITY RADIATION AND BLACKBODY RADIATION

If we were able to make a small hole of unit area through mass M2 into the cavity within M2 of Figure 6-1(c), we would observe a small amount of radiation of intensity $I(\nu)$ at any specific frequency ν emerging from the hole; that intensity is related to the energy density $u(\nu)$ within the cavity. We could then calculate the total radiation flux emerging from that cavity at frequency ν, through the hole, traveling in all directions within a solid angle of 2π (a hemisphere). This value for the radiation flux at frequency ν would describe the irradiance of a blackbody, since we deduced earlier that the spectral density within a cavity has the same energy density as that at the surface of a blackbody. We will therefore refer to this radiation as the blackbody radiation intensity per unit frequency, or $I_{BB}(\nu)$.

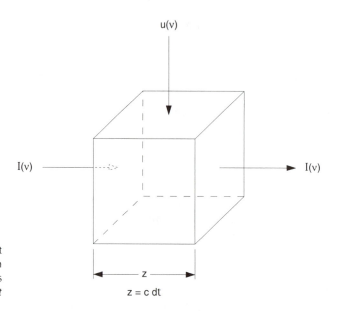

$u(v)$

$I(v)$ $I(v)$

z

$z = c\, dt$

Figure 6-6. Beam of light intensity $I(v)$ incident upon a volume of thickness $z = c\, dt$

Before performing this calculation, we will obtain a relationship between the energy density $u(v)$ (energy per unit volume-frequency) of electromagnetic radiation within a volume element and the flux $I(v)$ (energy per time-area-frequency) passing through a surface enclosing part of that volume element and traveling in a specific direction, as shown in Figure 6-6. Let us consider a beam of light of intensity per unit frequency $I(v)$ at frequency v, having a cross-sectional area dA and traveling in a direction z that passes through that area dA in a time dt. The energy density of radiation would be the product of the intensity of the beam, the cross-sectional area, and the time duration, or $I(v)\, dA\, dt$, which has units of energy per unit frequency. This would be equivalent to considering the energy density $u(v)$ of a beam existing within a volume $dV = dA \cdot z$ if an instantaneous photograph were taken of the beam within that volume dV of length $z = c\, dt$, where c is the velocity of the beam. This is described by the expression

$$I(v)\, dA\, dt = u(v)\, dV = u(v)\, dA \cdot z = u(v)\, dA \cdot c\, dt, \qquad (6.34)$$

which leads to the relationship

$$I(v) = u(v)c \qquad (6.35)$$

for a beam traveling in a specific direction.

We can now convert the expression for radiation within a cavity of temperature T to an expression for blackbody radiation emerging from a hole of unit surface area accessing that cavity. Radiation from any point within the cavity is traveling in all directions – that is, within a 4π solid angle. We consider the arrangement shown in Figure 6-7, where the fraction of radiation traveling within a solid angle $d\Omega/4\pi$ at a particular angle θ with respect to the normal to the plane of the hole is described by $I(v)\cos\theta\, d\Omega/4\pi$. In

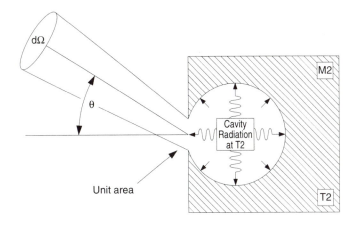

Figure 6-7. Blackbody radiation escaping from a confined cavity

this expression we have assumed that the intensity emitted from the source is independent of angle, but the factor $\cos\theta$ represents the decrease in the effective area of the hole with increasing angle from the normal as seen by the observer. Expressing the solid angle as $d\Omega = \sin\theta\, d\theta\, d\phi$, the radiation flux in that direction can be written as $I(\nu)\cos\theta\sin\theta\, d\theta\, d\phi/4\pi$. The total blackbody flux $I_{BB}(\nu)$ is obtained by integrating this component of the flux over the entire hemisphere (2π solid angle):

$$I_{BB}(\nu) = \int_0^{2\pi}\int_0^{\pi/2} \frac{I(\nu)\cos\theta\sin\theta\, d\theta\, d\phi}{4\pi}. \tag{6.36}$$

Replacing $I(\nu)$ by $u(\nu)$ from (6.35) gives the relationship between the blackbody radiation intensity per unit frequency, $I_{BB}(\nu)$, and the energy density per unit frequency, $u(\nu)$, of the cavity:

$$I_{BB}(\nu) = \int_0^{2\pi}\int_0^{\pi/2} u(\nu)c \cdot \frac{\cos\theta\sin\theta\, d\theta\, d\phi}{4\pi} = \frac{u(\nu)c}{4}. \tag{6.37}$$

Using (6.37) in conjunction with the value of $u(\nu)$ indicated in (6.23), we can obtain the spectral radiance of a blackbody as a function of ν and T as follows:

$$I_{BB}(\nu) = \frac{2\pi h\nu^3}{c^2}\frac{1}{e^{h\nu/kT}-1}. \tag{6.38}$$

WAVELENGTH DEPENDENCE OF BLACKBODY EMISSION

The total radiance of $I_{BB}(\nu)$ from a frequency interval $d\nu$ is given as

$$I_{BB}(\nu)\, d\nu = \frac{2\pi h\nu^3}{c^2}\frac{d\nu}{e^{h\nu/kT}-1}. \tag{6.39}$$

Because $\lambda\nu = c$ (i.e. $\nu = c/\lambda$), it follows that

$$|d\nu| = \frac{c}{\lambda^2}\, d\lambda. \tag{6.40}$$

We can therefore express the intensity per unit wavelength at temperature T as

$$I_{BB}(\lambda, T) = \frac{2\pi c^2 h}{\lambda^5 (e^{ch/\lambda kT} - 1)}. \tag{6.41}$$

The total radiance emitted from the blackbody surface within a specific wavelength interval $\Delta\lambda$ would then be expressed as

$$I_{BB}(\lambda, \Delta\lambda, T) = I_{BB}(\lambda, T) \Delta\lambda. \tag{6.42}$$

Specific values of the radiance $I_{BB}(\lambda, \Delta\lambda, T)$, in units of power per unit area (intensity) over the wavelength interval $\Delta\lambda$, can be obtained from the following explicit expression for $I_{BB}(\lambda, T)$:

$$I_{BB}(\lambda, T) = \frac{3.75 \times 10^{-22}}{\lambda^5 (e^{0.0144/\lambda T} + 1)} \ \text{W/m}^2\text{-}\mu\text{m}, \tag{6.43}$$

where λ must be given in meters, $\Delta\lambda$ in micrometers (μm), and T in degrees Kelvin. Alternatively,

$$I_{BB}(\lambda, T) = \frac{3.75 \times 10^{-25}}{\lambda^5 (e^{0.0144/\lambda T} - 1)} \ \text{W/m}^2\text{-nm} \tag{6.44}$$

for λ in meters, $\Delta\lambda$ in nanometers, and T in degrees Kelvin.

EXAMPLE

Compute the radiation flux or power in watts coming from a surface of temperature 300 K (near room temperature) and area 0.02 m^2 over a wavelength interval of 0.1 μm at a wavelength of 1.0 μm.

Use (6.42) and (6.43) to obtain the intensity (W/m^2) and multiply it by the area ΔA (m^2) to obtain the power P in watts coming from the surface:

$$P = I_{BB}(\lambda, T) \Delta\lambda \Delta A$$

$$= \frac{3.75 \times 10^{-22}}{(1 \times 10^{-6})^5 (e^{0.0144/(1 \times 10^{-6})300} - 1)} (0.1)(0.02) = 1.06 \times 10^{-15} \ \text{W}.$$

We can see that at room temperature the power radiated from a blackbody is almost too small to be measurable.

6.4 ABSORPTION AND STIMULATED EMISSION

We have examined the issues relating to the decay of electrons from a higher energy level to a lower level. We have also investigated the natural radiative decay process, which is inherent in all excited states of all materials and is referred to as *spontaneous emission*. However, we have seen in Chapter 4 that such emission is not always the dominant decay process. We have described how collisions with other particles (in the case of gases) or phonons (in the case of solids) can depopulate a level faster than the normal radiative process. Such collisions can also populate or *excite* energy levels.

Excitation or de-excitation can also occur by way of photons – "light particles" that have specific energies. The phenomenon of light producing excitation, referred to as *absorption,* has been known for well over 100 years. Such a process could also have been referred to as "stimulated" absorption, since it requires electromagnetic energy to stimulate the electron and thereby produce the excitation. There is no reason to expect that the inverse of that process would not also occur, but it was never seriously considered until Einstein suggested the concept of stimulated emission in 1917.

THE PRINCIPLE OF DETAILED BALANCE

Einstein was considering the recently developed Planck law for cavity radiation, the expression we just derived. He began questioning how the principle of detailed balance operated in the case of radiation in equilibrium, a situation associated with cavity radiation. This principle states that, *in equilibrium,* the total number of particles leaving a certain quantum state per unit time equals the number arriving in that state per unit time. It also states that in equilibrium the number leaving by a particular pathway equals the number arriving by that pathway.

We tacitly used this principle in the first part of this chapter when arguing that the mass M1 could not continue to receive excess energy from M2 once it had reached the new equilibrium temperature. The principle describes why the population of a specific energy state cannot increase indefinitely. This principle has also been called the principle of *microscopic reversibility,* and was originally applied to considerations of thermodynamic equilibrium.

The principle of detailed balance suggests that if a photon can stimulate an electron to move from a lower energy state l to higher energy state u by means of absorption, then a photon should also be able to stimulate an electron from the same upper state u to the lower state l. In the case of absorption, the photon disappears, with the energy being transferred to the absorbing species. In the case of stimulation, or stimulated emission, the species would have to radiate an additional photon to conserve energy. Such a stimulated emission process must occur in order to keep the population of the two energy levels in thermal equilibrium, if that equilibrium is determined by cavity radiation as described earlier in this chapter. Thus, the relationships between absorption and stimulated emission must be associated in some way with the Planck law for radiation in thermal equilibrium.

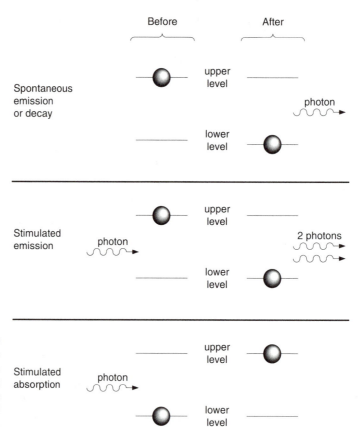

Figure 6-8. The three fundamental radiation processes associated with the interaction of light with matter: spontaneous emission, stimulated emission, and absorption

We have identified the three radiative processes producing interactions between two bound levels in a material: spontaneous emission, absorption (stimulated absorption), and stimulated emission. These processes are diagrammed in Figure 6-8. In the case of both of the stimulated processes (absorption and stimulated emission) occurring between bound (discrete) states, the energy (related to the frequency) of the light must correspond exactly to that of the energy difference between the two energy states. For stimulated emission, an additional photon is emitted at exactly the same energy (or frequency) as that of the incident photon (conservation of energy), and in exactly the same direction in phase with the incident photon (conservation of momentum).

ABSORPTION AND STIMULATED EMISSION COEFFICIENTS

We will now derive the absorption and stimulated emission coefficients associated with these processes by considering radiation in thermal equilibrium. We will consider a group of atoms having electrons occupying either energy levels u or l with population densities N_u and N_l (number of atoms per unit volume). We assume the atoms are in thermal equilibrium with each other and must therefore be related by the Boltzmann distribution function given in (6.10),

$$\frac{N_u}{N_l} = \frac{g_u}{g_l} e^{-(E_u - E_l)/kT} = \frac{g_u}{g_l} e^{-\Delta E_{ul}/kT}, \tag{6.45}$$

where g_u and g_l are the statistical weights of levels u and l and where $E_u - E_l = \Delta E_{ul}$.

We will consider photons interacting with such a collection of atoms. The photons will be assumed to have energies ΔE_{ul} such that $\Delta E_{ul} = h\nu_{ul}$, corresponding to the exact difference in energy between the levels u and l. We have defined A_{ul} as the spontaneous transition probability, the rate at which spontaneous transitions occur from level u to level l (number per unit time). Thus, the number of spontaneous transitions from u to l per unit time per unit volume is simply $N_u A_{ul}$.

We have also suggested that stimulated processes should occur. These processes would be proportional to the photon energy density $u(\nu)$ at frequency ν_{ul}, as well as to the population in the appropriate level. If we assume that the proportionality constant for such stimulated transitions is B, then the upward flux – the number of stimulated upward transitions per unit volume per unit time per unit frequency – would be $N_l B_{lu} u(\nu)$. Similarly, the downward flux would be $N_u B_{ul} u(\nu)$. The constants A_{ul}, B_{ul}, and B_{lu} are referred to as the Einstein A and B coefficients.

For the populations N_u and N_l to be in *radiative thermal equilibrium,* as described by (6.45), and for the principle of detailed balance to apply, the downward radiative flux should equal the upward radiative flux between the two levels:

$$N_u A_{ul} + N_u B_{ul} u(\nu) = N_l B_{lu} u(\nu). \tag{6.46}$$

From this equation we can solve for $u(\nu)$ as follows:

$$u(\nu) = \frac{N_u A_{ul}}{N_l B_{lu} - N_u B_{ul}}. \tag{6.47}$$

Dividing the top and bottom terms in the right-hand side of (6.47) by N_u, and using (6.45) for the ratio N_u/N_l, we are led to the expression

$$u(\nu) = \frac{A_{ul}}{B_{ul}} \left(\left[\frac{g_l B_{lu}}{g_u B_{ul}} \right] e^{h\nu_{ul}/kT} - 1 \right)^{-1}, \tag{6.48}$$

where we have used the relationship $\Delta E_{ul} = h\nu_{ul}$.

Equation (6.48) has a familiar form if we compare it to (6.32). Since both equations concern radiation in thermal equilibrium, if true then they must be equivalent. The equivalence follows if

$$\frac{g_l B_{lu}}{g_u B_{ul}} = 1 \quad \text{or} \quad g_l B_{lu} = g_u B_{ul} \tag{6.49}$$

and

$$\frac{A_{ul}}{B_{ul}} = \frac{8\pi h\nu^3}{c^3}. \tag{6.50}$$

We have thus derived the relationship between the stimulated emission and absorption coefficients B_{ul} and B_{lu} (respectively), along with their relationship to the spontaneous emission coefficient A_{ul}. We can rewrite (6.50) to obtain

$$B_{ul} = \frac{c^3}{8\pi h\nu^3} A_{ul}. \qquad (6.51)$$

We can now substitute the expression for A_{ul} in terms of $A_{ul}(\nu)$ from (4.60) to obtain

$$B_{ul} = \frac{c^3}{8\pi h\nu^3} \frac{(\nu - \nu_0)^2 + (\gamma_W^{ul}/4\pi)^2}{\gamma_W^{ul}/4\pi^2} A_{ul}(\nu). \qquad (6.52)$$

If we now define

$$B_{ul}(\nu) = \frac{\gamma_W^{ul}/4\pi^2}{(\nu - \nu_0)^2 + (\gamma_W^{ul}/4\pi)^2} B_{ul}, \qquad (6.53)$$

which describes the frequency dependence of B_{ul}, then (6.52) may be rewritten as

$$\frac{A_{ul}(\nu)}{B_{ul}(\nu)} = \frac{8\pi h\nu^3}{c^3} \equiv \frac{A_{ul}}{B_{ul}}. \qquad (6.54)$$

Hence the frequency-dependent expressions for $B_{ul}(\nu)$ and $B_{lu}(\nu)$ will satisfy (6.48) if

$$\frac{g_l B_{lu}(\nu)}{g_u B_{ul}(\nu)} = 1 \quad \text{or} \quad g_l B_{lu}(\nu) = g_u B_{ul}(\nu). \qquad (6.55)$$

We have thus derived the stimulated emission coefficients B_{ul} and B_{lu}, as well as their frequency-dependent counterparts $B_{ul}(\nu)$ and $B_{lu}(\nu)$, that define the way a photon beam interacts with a two-level system of atoms that was obtained by considering radiation in thermal equilibrium. These relationships provide the fundamental concepts that are necessary for producing a laser.

It is interesting to examine the ratio of stimulated to spontaneous emission rates from level u. This ratio can be obtained from (6.32) and (6.50) as

$$\frac{B_{ul}u(\nu)}{A_{ul}} = \frac{1}{e^{h\nu_{ul}/kT} - 1}. \qquad (6.56)$$

Thus, stimulated emission plays a significant role only for temperatures in which kT is of, or greater than, the order of the photon energy $h\nu_{ul}$. The ratio is unity when $h\nu_{ul}/kT = \ln 2 = 0.693$. For visible transitions in the green portion of the spectrum (photons of the order of 2.5 eV), such a relationship would be achieved for a temperature of 33,500 K. Thus, in the visible spectrum, the dominance of stimulated emission over spontaneous emission normally happens only in stars, in high-temperature and -density laboratory plasmas such as laser-produced plasmas, or in lasers. In low-pressure plasmas the radiation can readily escape, so there is no opportunity for the radiation density to build up to a value where the stimulated decay rate is comparable to the radiative decay rate. In lasers, the ratio of (6.56) can be significantly greater than unity.

EXAMPLE

A helium–neon laser operating at 632.8 nm has an output power of 1.0 mW with a 1-mm beam diameter. The beam passes through a mirror that has 99% reflectivity and 1% transmission at the laser wavelength. What is the ratio of $B_{ul}u(\nu)/A_{ul}$ for this laser? What is the effective blackbody temperature of the laser beam as it emerges from the laser output mirror? Assume the beam diameter is also 1 mm inside the laser cavity, and that the power is uniform over the beam cross section (this is only an approximation, as we will learn in Chapter 10). Assume also that the laser linewidth is approximately one tenth of the Doppler width for the transition.

The laser frequency is determined by

$$\nu = \frac{c}{\lambda} = \frac{3 \times 10^8 \text{ m/s}}{6.328 \times 10^{-7} \text{ m}} = 4.74 \times 10^{14} \text{ Hz}.$$

From (6.50) we can now compute the ratio of A_{ul}/B_{ul} as

$$\frac{A_{ul}}{B_{ul}} = \frac{8\pi h\nu^3}{c^3} = \frac{(8\pi)(6.63 \times 10^{-34} \text{ J-s})(4.74 \times 10^{14} \text{ Hz})^3}{(3 \times 10^8 \text{ m/s})^3}$$

$$= 6.57 \times 10^{-14} \text{ J-s/m}^3;$$

hence

$$\frac{B_{ul}}{A_{ul}} = 1.52 \times 10^{13} \text{ m}^3/\text{J-s}.$$

We must now compute the energy density $u(\nu)$, which from (6.35) is related to the intensity per unit frequency $I(\nu)$ as $u(\nu) = I(\nu)/c$. We can compute $I(\nu)$ by dividing the laser beam power in the cavity by the beam cross-sectional area and the frequency width of the beam. The Doppler width of the helium–neon 632.8-nm transition (see Table 4-1) is 1.5×10^9 Hz. Thus

$$u(\nu) = \frac{I(\nu)}{c}$$

$$= \frac{[(99 \times 1.0 \text{ mW})/(\pi \cdot (5 \times 10^{-4} \text{ m})^2)](0.1)(1.5 \times 10^9 \text{ Hz})}{3 \times 10^8 \text{ m/s}}$$

$$= 2.80 \times 10^{-12} \text{ J-s/m}^3,$$

so the ratio is

EXAMPLE (cont.)

$$\frac{B_{ul}u(\nu)}{A_{ul}} = (1.52 \times 10^{13} \text{ m}^3/\text{J-s})(2.80 \times 10^{-12} \text{ J-s/m}^3) = 42.6.$$

The stimulated emission rate is therefore almost 43 times the spontaneous emission rate on transitions from the upper to the lower laser level at 632.8 nm.

Using (6.56), the preceding ratio can be rewritten as

$$\frac{B_{ul}u(\nu)}{A_{ul}} = \frac{1}{e^{h\nu_{ul}/kT} - 1} = 42.6,$$

which yields

$$e^{h\nu_{ul}/kT} - 1 = 1/42.6 = 0.0235$$

or

$$h\nu_{ul}/kT = \ln(1.0235) = 2.32 \times 10^{-2}.$$

We can thus solve for T as follows:

$$T = \frac{h\nu_{ul}}{(2.32 \times 10^{-2})k} = \frac{(6.63 \times 10^{-34} \text{ J-s})(4.74 \times 10^{14} \text{ Hz})}{(2.32 \times 10^{-2})(1.38 \times 10^{-23} \text{ J/K})}$$

$$= 9.82 \times 10^5 \text{ K} = 982,000 \text{ K}.$$

This calculation indicates that the radiation intensity of the laser beam inside the laser cavity has a value equivalent to that of a 982,000-K blackbody, if we consider only the radiation emitted from the blackbody in the frequency (or wavelength interval) over which the laser operates.

REFERENCES

A. Corney (1977), *Atomic and Laser Spectroscopy.* Oxford: Clarendon Press, Chapter 9.
R. B. Leighton (1959), *Principles of Modern Physics.* New York: McGraw-Hill, Chapter 2.
R. Loudon (1973), *The Quantum Theory of Light.* Oxford: Clarendon Press, Chapter 1.

PROBLEMS

1. Calculate the number of radiation modes in a box 1 mm on a side for a spread of 0.001 nm centered at 514.5 nm and a spread of 0.01 μm centered at 10.6 μm.

2. Consider a 1-mm–diameter surface area of carbon (graphite). Calculate how many atoms would exist in energy levels from which they could emit radiation at wavelengths shorter than 700 nm (visible light and shorter wavelengths) for surface temperatures of 300 K, 1,000 K, and 5,000 K when the solid is in thermal equilibrium at those temperatures. Assume that only those atoms within a depth of 100 nm of the material surface can emit observable radiation.

3. In Problem 2, if the excited atoms that emit visible radiation decay in 10^{-13} s, and only 2% of them decay radiatively (quantum yield of 2%), how much power would be radiated from that surface at the aforementioned temperatures? Assume that half of the atoms that radiate emit into the 2π solid angle that would result in their leaving the surface of the material.

4. How much power is radiated from a 1-mm^2 surface of a body at temperature T when the peak measured wavelength is that of green light at 500 nm?

5. Determine the number of modes in a 1-cm^3 box for frequencies in the visible spectrum between 400 and 700 nm. Compare that value to the number of modes in a sodium streetlamp that emits over a wavelength interval of 3 nm at a center wavelength of 589 nm. Assume that the streetlamp is a cylinder of radius 0.5 cm and length 10 cm.

6. Estimate the number of photons in both the box and the streetlamp of Problem 5 for temperatures of 300 K, 1,000 K, and 5,000 K.

7. A 60-W incandescent lamp has a tungsten filament composed of a wire, 0.05 cm in diameter and 10 cm in length, that is coiled up to fit within the light bulb. Assume the filament is heated to a temperature of 3,000 K when the light bulb is turned on. How much power (watts) is emitted within the visible spectrum from the filament, assuming that it is emitting as a blackbody? As an approximation, you could divide the visible spectral region into several segments and compute the average contribution from each segment. Then simply add the averages together, instead of trying to integrate the blackbody function over the entire visible spectral range.

8. Show that Planck's radiation law of (6.38) will lead to the Stefan–Boltzmann relationship of (6.15) if the power radiated over all wavelengths is considered. Determine the coefficients of the Stefan–Boltzmann constant. *Hint:*

$$\int_0^\infty \frac{x^3 \, dx}{e^x - 1} = \frac{\pi^4}{15}.$$

9. An argon ion laser emits 2 W of power at 488.0 nm in a 2-mm diameter beam. What would be the effective blackbody temperature of the output beam of that laser radiating over the frequency width of the laser transition, given that the laser linewidth is approximately one fifth of the Doppler linewidth? Assume that the laser is operating at an argon gas temperature of 1,500 K and that the laser output is uniform over the width of the beam. Consider that the range of laser frequencies is just a small piece of the blackbody distribution spectrum as given in (6.43) or (6.44).

10. For the laser in Problem 9, how much power would be required for the stimulated emission rate to equal the radiative decay rate?

LASER AMPLIFIERS

CONDITIONS FOR PRODUCING A LASER

POPULATION INVERSIONS, GAIN, AND GAIN SATURATION

SUMMARY In Chapter 6 we developed the concept of stimulated emission and the relationships of the Einstein A and B coefficients. Those coefficients are associated with the interaction of radiation with two specific energy levels, where the radiation has the exact frequency corresponding to the energy separation between the two levels. We are now prepared to consider how gain (amplification) and absorption of radiation can occur in a medium containing populations in those two levels. We will derive the equation that predicts the amount of exponential growth or absorption of an incident light beam passing through such a medium based upon specific conditions of both the beam and the medium, including the beam frequency, the value of the stimulated emission cross section of the laser transition, the population densities of the upper and lower laser levels, and the length of the gain medium. The conditions associated with amplification will be shown to be *necessary* for making a laser but not *sufficient*. We will then obtain the sufficient conditions that suggest how much gain is necessary for the beam to reach the saturation intensity as it grows within the medium. We will consider increasing the gain length by putting either a mirror at one end of the medium or mirrors at both ends of the medium. Threshold conditions for laser operation will be obtained for mirrorless amplifiers as well as for amplifiers with mirrors. Finally, we will examine some of the effects that occur for a laser operating above the threshold gain.

7.1 ABSORPTION AND GAIN

ABSORPTION AND GAIN ON A HOMOGENEOUSLY BROADENED RADIATIVE TRANSITION (Lorentzian Frequency Distribution)

We will consider a beam of light having an intensity per unit frequency $I(\nu)$ at frequency ν and frequency width $\Delta\nu$, passing through a medium of thickness L and cross-sectional area A as indicated in Figure 7-1. We define the intensity I as $I = I(\nu)\Delta\nu$. We also let I_0 denote the value of I just before it enters the medium. Within the medium we assume there are atoms occupying at least two specific energy levels, designated as u for the upper (energetically higher) level and l for the lower level. Radiative transitions from u to l can occur over a narrow range of frequencies within the emission linewidth (with the center frequency at ν_0). We also assume that the frequency of the incident beam falls somewhere within that linewidth. Implicit in this assumption is that the energy difference ΔE_{ul} between levels u and l is nominally given by $\Delta E_{ul} = E_u - E_l = h\nu_{ul}$.

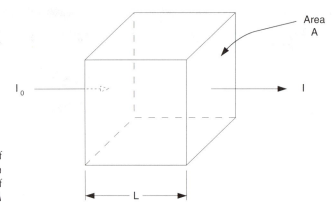

Figure 7-1. Light of intensity I_0 incident upon an absorptive material of length L and area A

We will now determine the effect of the medium on the beam as the beam passes through it. In other words, we will ask how I is changed after it has passed through the medium. To simplify the analysis, we will first assume that the dominant broadening of the transition under consideration is homogeneous, with a lineshape that is Lorentzian. We will later consider a similar situation for inhomogeneous broadening, where the emission lineshape is Gaussian.

For the case where the principal broadening is homogeneous, N_u and N_l represent the total number of atoms per unit volume in the upper and lower levels of the transition. The radiative transition probability for transitions occurring between energy levels u and l is given by A_{ul}. Recall from Section 4.3 that

$$A_{ul}(\nu) = \frac{\gamma_{ul}^T/4\pi^2}{(\nu - \nu_0)^2 + (\gamma_{ul}^T/4\pi)^2} A_{ul}. \tag{4.64}$$

Consider the three possible radiative interactions between the two levels u and l, as indicated in the diagram of Figure 7-2. The first one on the left is spontaneous emission downward from u to l at a spontaneous emission rate A_{ul}. The number of transitions downward per unit volume per unit time occurring by the spontaneous emission process at frequency ν from level u to level l is thus

$$N_u A_{ul}. \tag{7.1}$$

The other two processes shown in Figure 7-2 are the stimulated processes, which are proportional to the energy density $u(\nu)$ of the beam at frequency ν, as well as to the Einstein B coefficients derived in Chapter 6. We must use the frequency-dependent Einstein B coefficients multiplied by the frequency width of the beam $\Delta\nu$ to obtain the probability that a transition occurs at that particular frequency. We can therefore express the number of stimulated transitions (upward or downward) occurring between the two levels per unit volume per unit time as

$$N_l B_{lu}(\nu)\Delta\nu \cdot u(\nu) = N_l B_{lu}(\nu) I(\nu)\Delta\nu/c$$
$$= N_l B_{lu}(\nu) I/c \quad \text{(upward)} \tag{7.2}$$

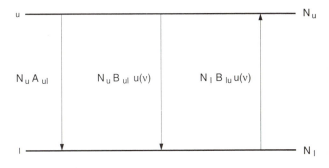

Figure 7-2. Energy levels of a two-level system, depicting the population flux transferred between the two levels via spontaneous emission, stimulated emission, and absorption

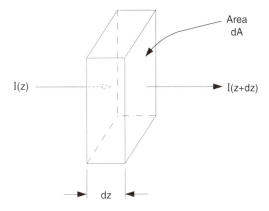

Figure 7-3. Light incident upon an incremental length dz of absorbing material

and

$$N_u B_{ul}(\nu)\Delta\nu \cdot u(\nu) = N_u B_{ul}(\nu)I(\nu)\Delta\nu/c$$
$$= N_u B_{ul}(\nu)I/c \quad \text{(downward)}, \tag{7.3}$$

where we have used the relationship $u(\nu)\Delta\nu = I(\nu)\Delta\nu/c = I/c$ in a manner similar to that of (6.35). Here we assume that the B coefficients per unit frequency, $B_{ul}(\nu)$ and $B_{lu}(\nu)$, have the same frequency dependence as $A_{ul}(\nu)$ of (4.64) for homogeneous broadening, as suggested by (6.53) and (6.55).

We now return to a situation similar to that of Figure 7-1, but will consider only a small length dz of the medium, as defined in Figure 7-3. In this case we will estimate the energy that – as a result of the stimulated processes given in (7.2) and (7.3) – can be added to and subtracted from the beam (of intensity I, frequency ν, and frequency width $\Delta\nu$) as it moves through a small distance dz. We will neglect the spontaneous emission contribution described in (7.1) for this calculation, since the spontaneous emission is radiated in all directions into a solid angle of 4π and thus contributes very little in the direction of the incident beam.

The amount of energy per unit time (intensity times cross-sectional area of the beam) added when the beam passes through a region of length dz and cross-sectional area dA within the medium can therefore be expressed as the difference between the number of transitions per unit time upward and the number of transitions per unit time downward within the volume, multiplied by the photon energy per transition:

$$[I(z+dz)-I(z)]\,dA$$

$$= [N_u B_{ul}(\nu)\Delta\nu\cdot u(\nu) - N_l B_{lu}(\nu)\Delta\nu\cdot u(\nu)]\,h\nu\,dA\,dz$$

$$= [N_u B_{ul}(\nu)I/c - N_l B_{ul}(\nu)I/c]\,h\nu\,dA\,dz$$

$$= [N_u B_{ul}(\nu) - N_l B_{lu}(\nu)]\frac{h\nu I\,dA\,dz}{c}. \tag{7.4}$$

Energy is added to or subtracted from the beam in discrete amounts $h\nu$ as a result of the two terms within the brackets on the right-hand side of (7.4), since those terms describe discrete events in which every stimulated transition is counted. The factor involving N_u leads to stimulated emission. Consequently, another photon is added to the beam and N_u is decreased by 1 every time the beam stimulates an electron from level u to level l. As this occurs, population within the medium is transferred from the upper level u to the lower level l. The factor containing N_l involves absorption of photons from the beam with population moving from the lower level l to the upper level u.

Using the definition $dI = I(z+dz)-I(z)$ and dividing through by dz and dA leads to the expression

$$\frac{dI}{dz} = [N_u B_{ul}(\nu) - N_l B_{lu}(\nu)]\frac{h\nu}{c}I. \tag{7.5}$$

Equation (7.5) is a differential equation of the form

$$\frac{dI}{dz} = CI, \tag{7.6}$$

where C is independent of z. The solution to this type of equation can be expressed as

$$I = I_0 e^{Cz}. \tag{7.7}$$

Thus, the solution to (7.5) can be expressed as

$$I = I_0 e^{g^H(\nu)z}, \tag{7.8}$$

where

$$g^H(\nu) = [N_u B_{ul}(\nu) - N_l B_{lu}(\nu)]\frac{h\nu}{c}; \tag{7.9}$$

the superscript H refers to homogeneous broadening. The term $g^H(\nu)$ is referred to as the *gain coefficient,* and has dimensions of $1/\text{length}$ or m^{-1} in the MKS system of units.

GAIN COEFFICIENT AND STIMULATED EMISSION CROSS SECTION FOR HOMOGENEOUS BROADENING

Using the relationships of the Einstein A and B coefficients from (6.54) and (6.55), we can rewrite the gain coefficient $g^H(\nu)$ (eqn. 7.9) as

$$g^H(\nu) = \left[N_u - \frac{g_u}{g_l}N_l\right]\frac{c^2}{8\pi\nu^2}A_{ul}(\nu). \tag{7.10}$$

Alternatively, by substituting the value for $A_{ul}(\nu)$ from (4.64) we have

$$g^H(\nu) = \left[N_u - \frac{g_u}{g_l}N_l\right]\frac{c^2}{8\pi\nu^2}\left[\frac{\gamma_W^{ul}/4\pi^2}{(\nu-\nu_0)^2+(\gamma_W^{ul}/4\pi)^2}\right]A_{ul}. \tag{7.11}$$

Let us define the first factor on the right-hand side of (7.11) as the population difference,

$$\Delta N_{ul} = \left[N_u - \frac{g_u}{g_l}N_l\right], \tag{7.12}$$

which has units of $1/\text{length}^3$ or number per unit volume. We define the second term on the right of (7.11) as

$$\sigma_{ul}^H(\nu) = \frac{c^2}{8\pi\nu^2}A_{ul}(\nu) = \frac{c^2}{8\pi\nu^2}\left[\frac{\gamma_W^{ul}/4\pi^2}{(\nu-\nu_0)^2+(\gamma_W^{ul}/4\pi)^2}\right]A_{ul}, \tag{7.13}$$

which has dimensions of length^2 or area; hence we refer to it as the *stimulated emission cross section*. Thus, (7.10) could be rewritten as

$$g^H(\nu) = \sigma_{ul}^H(\nu)\Delta N_{ul}. \tag{7.14}$$

We will generally be interested in the value of $g^H(\nu)$ at the center of the emission line ($\nu = \nu_0$), since it can be seen by examination of (7.11) that the gain coefficient is highest at that frequency. Therefore, for $\nu = \nu_0$ (center of the emission line), from (7.11) we obtain the gain coefficient $g^H(\nu_0)$ as

$$g^H(\nu = \nu_0) \equiv g^H(\nu_0) \equiv g_0^H = \frac{c^2}{2\pi\nu_0^2\gamma_{ul}}A_{ul}\left[N_u - \frac{g_u}{g_l}N_l\right], \tag{7.15}$$

where we have shortened the notation for the gain coefficient to g_0^H. From (7.14) and (7.15) we can also express the stimulated emission cross section $\sigma_{ul}^H(\nu_0)$ at the center frequency of the emission profile ($\nu = \nu_0$) as

$$\sigma_{ul}^H(\nu_0) = \frac{c^2 A_{ul}}{2\pi\nu_0^2\gamma_{ul}} = \frac{\lambda_{ul}^2 A_{ul}}{4\pi^2\Delta\nu_{ul}^H}, \tag{7.16}$$

where we have used the relationship $\gamma_{ul} = 2\pi\Delta\nu_{ul}^H$ from Chapter 4. For the special case in which natural broadening dominates (i.e., where $\gamma_{ul} = 2\pi\Delta\nu_{ul}^N$), we obtain the following expression for the stimulated emission cross section $\sigma_{ul}^N(\nu_0)$ at the center of the emission line,

$$\sigma_{ul}^N(\nu_0) = \frac{\lambda_{ul}^2}{2\pi}\frac{A_{ul}}{\sum_i A_{ui}+\sum_j A_{lj}}, \tag{7.17}$$

using the relationship (4.33) for $\Delta\nu_{ul}^N$.

The relationship for the intensity at a specific distance z into the medium for homogeneous broadening can now be expressed as

$$I = I_0 e^{g^H(\nu)z} = I_0 e^{\sigma_{ul}^H(\nu)[N_u - (g_u/g_l)N_l]z} = I_0 e^{\sigma_{ul}^H(\nu)\Delta N_{ul} z}. \tag{7.18}$$

Thus, since $g^H(\nu)$, $\sigma_{ul}^H(\nu)$, and z are positive quantities, if ΔN_{ul} is positive then the beam intensity will increase exponentially with distance z and so provide amplification; conversely, if ΔN_{ul} is negative then the beam will decrease in intensity, leading to absorption of the beam. The consequences of this relation will be explored in detail after we carry out the same analysis for Doppler broadening in the following section.

ABSORPTION AND GAIN ON AN INHOMOGENEOUSLY BROADENED RADIATIVE TRANSITION (Doppler Broadening with a Gaussian Distribution)

We will now consider the case for Doppler broadening in which the emission lineshape is Gaussian. Again we consider Figure 7-3 for a beam (of intensity I, frequency ν, and frequency width $\Delta\nu$) passing through a medium of thickness dz containing energy levels u and l and with the possibility that each of those levels is populated. For Doppler broadening, we have a population density per unit frequency of $N_u(\nu)$ and $N_l(\nu)$, as described by (4.69). Thus the amount of energy per unit time that is added to the beam when it passes through a volume of the medium having length dz and cross-sectional area dA is given by

$$[I(z+dz)-I(z)]\,dA$$
$$= [N_u(\nu)\Delta\nu \cdot B_{ul}u(\nu) - N_l(\nu)\Delta\nu \cdot B_{lu}u(\nu)]\,h\nu\,dA\,dz. \tag{7.19}$$

Again, as in the case of homogeneous broadening, photons are added to the beam via stimulated emission from the term involving $N_u(\nu)\Delta\nu$ and photons are removed from the beam via absorption from the term involving $N_l(\nu)\Delta\nu$.

Using (6.35) and the definition $dI = I(z+dz) - I(z)$ and dividing through by dz and dA leads to

$$\frac{dI}{dz} = [N_u(\nu)B_{ul} - N_l(\nu)B_{lu}]\frac{h\nu}{c}I, \tag{7.20}$$

where we have again used $u(\nu)\Delta\nu = I(\nu)\Delta\nu/c = I/c$. As in the previous case for homogeneous broadening, the solution to (7.20) can be written as

$$I = I_0 e^{g^D(\nu)z}, \tag{7.21}$$

where

$$g^D(\nu) = [N_u(\nu)B_{ul} - N_l(\nu)B_{lu}]\frac{h\nu}{c}. \tag{7.22}$$

Again, using the relationships of the Einstein A and B coefficients of (6.49) and (6.51), we can rewrite $g^D(\nu)$ (eqn. 7.22) as

$$g^D(\nu) = \left[N_u(\nu) - \frac{g_u}{g_l}N_l(\nu)\right]\frac{c^2 A_{ul}}{8\pi\nu^2}. \tag{7.23}$$

As shown in (4.69) for the case of Doppler broadening, we can express the population densities of the upper and lower laser levels per unit frequency as

$$N_{u,l}(\nu) = 2\sqrt{\frac{\ln 2}{\pi}}\frac{N_{u,l}}{\Delta\nu_D}\exp\left\{-\left[\frac{4\ln 2(\nu-\nu_0)^2}{\Delta\nu_D^2}\right]\right\}, \tag{7.24}$$

where $\exp\{x\}$ denotes e^x. Substituting values for $N_u(\nu)$ and $N_l(\nu)$ from (7.24) into (7.23) leads to

$$g^D(\nu) = \sqrt{\frac{\ln 2}{16\pi^3}}\frac{c^2 A_{ul}}{\nu_0^2\Delta\nu_D}\left[N_u - \frac{g_u}{g_l}N_l\right]\exp\left\{-\left[\frac{4\ln 2(\nu-\nu_0)^2}{\Delta\nu_D^2}\right]\right\}, \tag{7.25}$$

where we have replaced the variable ν in the denominator by the constant ν_0 because, for optical frequencies, the variation in frequency over the emission linewidth is negligible compared to the value of ν_0.

GAIN COEFFICIENT AND STIMULATED EMISSION CROSS SECTION FOR DOPPLER BROADENING

We are again interested in the value of the gain coefficient $g^D(\nu)$ at the center of the Doppler-broadened emission line ($\nu = \nu_0$), since it can be seen by examination of (7.25) that the gain is highest at that location. Thus we can write $g^D(\nu = \nu_0)$ as

$$g^D(\nu = \nu_0) \equiv g^D(\nu_0) \equiv g_0^D = \sqrt{\frac{\ln 2}{16\pi^3}}\frac{\lambda_{ul}^2 A_{ul}}{\Delta\nu_D}\left[N_u - \frac{g_u}{g_l}N_l\right], \tag{7.26}$$

where g_0^D is shorthand for the gain coefficient at the center of the emission line. The factor not associated with the population densities in (7.25) and (7.26) can again be separated from the population density difference and expressed as a stimulated emission cross section $\sigma_{ul}^D(\nu)$:

$$g^D(\nu) = \sigma_{ul}^D(\nu)\left[N_u - \frac{g_u}{g_l}N_l\right] = \sigma_{ul}^D(\nu)\Delta N_{ul}. \tag{7.27}$$

As for the case of homogeneous broadening, for Doppler broadening at the emission line center ($\nu = \nu_0$) we have the following definition of the stimulated emission cross section:

$$\sigma_{ul}^D(\nu_0) = \sqrt{\frac{\ln 2}{16\pi^3}}\frac{\lambda_{ul}^2 A_{ul}}{\Delta\nu_D}. \tag{7.28}$$

The stimulated emission cross section $\sigma_{ul}^D(\nu_0)$ for Doppler broadening at the center of the emission profile can also be expressed in a simplified numerical form as follows:

$$\sigma_{ul}^D(\nu_0) = (1.74 \times 10^{-4})\lambda_{ul}^3 A_{ul}\sqrt{M_N/T}, \tag{7.29}$$

where the Doppler width has been incorporated directly into the equation. In this formula $\sigma_{ul}^D(\nu_0)$ is in square meters, λ_{ul} is in meters, A_{ul} is in units of 1/seconds, T is degrees Kelvin, and M_N is the mass number of the laser species (the total number of protons and neutrons per atom or molecule; see the Appendix).

Thus, as in the case of homogeneous broadening expressed in (7.18), the intensity I of a beam passing through a medium at a frequency ν and width $\Delta\nu$ would, for the case of Doppler broadening, be modified according to the expression

$$I = I_0 e^{g^D(\nu)z} = I_0 e^{\sigma_{ul}^D(\nu)[N_u - (g_u/g_l)N_l]z} = I_0 e^{\sigma_{ul}^D(\nu)\Delta N_{ul}z} \tag{7.30}$$

at any specific distance z measured from the point the beam enters the medium.

RELATIONSHIP OF GAIN COEFFICIENT AND STIMULATED EMISSION CROSS SECTION TO ABSORPTION COEFFICIENT AND ABSORPTION CROSS SECTION

In (7.18) and (7.30) we have obtained two of the most important formulas for the understanding and development of lasers. These formulas – associated with the gain coefficient, or the product of the stimulated emission cross section and population difference – suggest how to achieve gain or amplification in a medium. Long before the concept of gain was proposed, such formulas were used by chemists and others who were primarily interested in measuring the absorption of light by various materials. The significant difference was their assumption that the population density N_u in level u was insignificant compared to N_l. This was based upon considerations of thermal equilibrium, which led to a very small ratio of N_u/N_l for excited-state energies that could lead to absorption and emission in or near the visible spectral region, as predicted by the Boltzmann distribution of (6.10). Thus, only the term involving N_l was considered when making absorption measurements. A general expression for the intensity variation of light passing through a medium of length z, for conditions associated only with absorption, can be written as

$$I = I_0 e^{-\alpha(\nu)z}, \tag{7.31}$$

where the absorption coefficient $\alpha(\nu)$ is related to the gain coefficient $g(\nu)$ by

$$\alpha(\nu) = \sigma_{ul}(\nu)\frac{g_u}{g_l}N_l = -g(\nu). \tag{7.32}$$

In (7.32), ΔN_{ul} is approximated as $\Delta N_{ul} = -(g_u/g_l)N_l$, since we are neglecting N_u. Equation (7.31) is known as *Beer's law*. Based upon (7.11) and (7.25), we can see that the frequency variations of both the gain and the absorption coefficients are the same as those of the spontaneous emission profile. We will use Beer's law in Chapter 9 when we discuss optical pumping of lasers, which involves the absorption of light by the gain medium during the pumping process.

Unless stated otherwise, any reference in this text to the coefficients of gain, absorption, or the stimulated emission cross section will be to the value *at the center of the emission line* as indicated by (7.16) and (7.28).

7.2 POPULATION INVERSION (Necessary Condition for a Laser)

We are now in a position to consider how amplification occurs in a medium with energy levels u and l and population densities N_u and N_l in those levels. We consider (7.18) and (7.30), but express them in general form by excluding reference to either homogeneous or Doppler broadening:

$$I = I_0 e^{\sigma_{ul}[N_u - (g_u/g_l)N_l]z}, \tag{7.33}$$

where I_0 represents the intensity of the beam (power per unit area) as it enters the medium and I represents the intensity at some distance z into the medium. It can be seen from (7.33) that if the value of the exponent is positive, the beam will increase in intensity or amplification will occur. If it is negative, the beam will decrease in intensity and absorption will occur. Since the values of σ_{ul} and z are always positive, we can see immediately that amplification will occur only if

$$N_u > \frac{g_u}{g_l}N_l \tag{7.34}$$

or, expressed in a different way,

$$\frac{g_l N_u}{g_u N_l} > 1. \tag{7.35}$$

The case of the upper level being more populated than the lower level (taking into account the statistical weights) is referred to as a *population inversion,* since the state is not normal under conditions of thermal equilibrium. The comparison of an inverted population with a population in thermal equilibrium at some temperature T (as described by eqn. 6.10) is shown in Figure 7-4. A population inversion is a *necessary* condition for amplification or laser action to occur, but it is not a *sufficient* condition. We

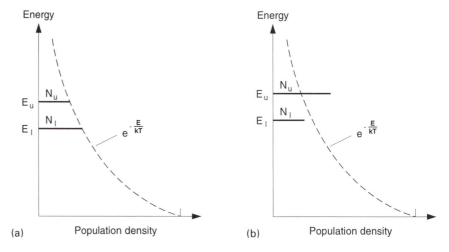

Figure 7-4. Population distribution between levels u and l for conditions of (a) thermal equilibrium and (b) a population inversion

will now address the requirements for sufficiency. The subject of how to produce population inversions will be explored in detail in the next two chapters.

7.3 SATURATION INTENSITY (Sufficient Condition for a Laser)

Equation (7.33) suggests the possibility of rapid growth of a beam as it passes through a medium of length z if the exponential coefficient is positive and of sufficient magnitude. We will next determine the requirements for the magnitude of that coefficient for various physical geometries that might be considered as possible laser configurations. Before doing so, we must consider the effect of saturation of the laser beam as it grows exponentially within a gain medium.

Assume there is a medium (such as that of Figure 7-1) in which a population inversion exists, and assume that the value of the gain – the exponent of (7.33) – is large enough to provide significant amplification of the beam. If we then increase the length of the medium to allow the intensity I to increase exponentially (via stimulated emission), (7.33) suggests that the beam could eventually reach an intensity (at some specific length z) such that the energy stored in the upper laser level is not sufficient to satisfy the exponential growth demands of the beam. Hence, there must be a limiting expression to estimate the intensity at which this *saturation* process occurs. If we consider what happens on a microscopic scale as the beam transits the medium, we will be able to derive a simple relationship that provides the intensity at which the beam stops growing exponentially. The length z of the medium at which that saturation effect occurs can be expressed as the *saturation length* L_s. From our previous analysis, we can anticipate that the specific length L_s would be associated with the specific value of the product $\sigma_{ul}\Delta N_{ul}$ for a given medium. The intensity achieved by the beam when $z = L_s$ will be referred to as the *saturation intensity*, I_{sat}, in units of power per unit area or energy per unit time per unit area.

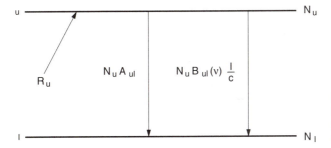

Figure 7-5. Terms associated with the flux input and decay from energy level u

We will perform a simple calculation to estimate I_{sat} by considering levels u and l in a case where a steady-state population density N_u exists in the upper level. We are assuming that a population inversion exists between levels u and l, so we can surmise that the steady state is not one of thermal equilibrium. We will assume, as indicated in Figure 7-5, that there is a pumping flux R_u that is populating level u which has a lifetime τ_u and would therefore normally decay at a rate of $1/\tau_u$. We also assume that the population in level u can be reduced by stimulated emission resulting from a beam of intensity $I(\nu)$ and frequency ν_{ul} transiting the medium such that $\Delta E_{ul} = h\nu_{ul}$. We assume this beam has a frequency width $\Delta\nu$, so that the frequency width of the beam is less than or equal to the natural linewidth of the relevant energy levels of the medium: $\Delta\nu \leq \Delta\nu^N$. We are thus assuming that frequencies outside of the natural linewidth have no significant effect in stimulating population from level u to level l. In summary, we are considering the effect of a beam of intensity $I = I(\nu)\Delta\nu$ as it interacts with the two-level system of the medium. The stimulated emission rate from level u to level l, $B_{ul}(\nu)\Delta\nu \cdot u(\nu) = B_{ul}(\nu)I(\nu)\Delta\nu/c = B_{ul}(\nu)I/c$, is also shown in Figure 7-5. Here we have again used the relationship $u(\nu) = I(\nu)/c$ to relate the energy density per unit frequency to the intensity per unit frequency.

We have assumed that level u is in a steady-state equilibrium, so we can write a rate equation taking into account all of the population changes affecting level u and equate it to zero, since "steady state" implies that there are no net changes in N_u. This equation can be expressed as

$$\frac{dN_u}{dt} = R_u - N_u\left[\left(\frac{1}{\tau_u}\right) + \frac{B_{ul}(\nu)I}{c}\right] = 0. \tag{7.36}$$

This analysis is carried out for the case where homogeneous broadening dominates. For Doppler broadening, a similar conclusion would result if $N_u(\nu)$ were used. Equation (7.36) suggests that the steady-state solution for N_u is

$$N_u = \frac{R_u}{\dfrac{1}{\tau_u} + \dfrac{B_{ul}(\nu)I}{c}}. \tag{7.37}$$

When $I = 0$, the value of the population density N_u is given as

$$N_u = R_u\tau_u. \tag{7.38}$$

In previous chapters we considered values of τ_u for many different lasers, and found that they vary over many orders of magnitude. For upper laser levels of most visible laser levels the values of τ_u range from 10^{-3} to 10^{-9} s. We can examine (7.37) and see that as I increases (according to eqn. 7.33) as the beam traverses a long gain medium, a value of z ($z = L_s$) would eventually be reached such that the intensity I, and consequently the stimulated emission term in the denominator of that equation, could eventually become as large as the term associated with the level lifetime. Since the value of $1/\tau_u$ can be quite large (ranging from 10^3/s to 10^9/s), a significant intensity I would be required for that to happen. We will *define* that intensity as I_{sat}. When saturation occurs, the population of level u would decrease by a factor of two to a value of $R_u \tau_u/2$. According to (7.37), the exponential growth factor $\sigma_{ul}(\nu)\Delta N_{ul}z$ of (7.33) for $z = L_s$ (the gain) would also decrease by approximately a factor of two because N_u would decrease by two. Further increase in I would further decrease the gain in regions where the beam propagates further into the medium as suggested by (7.37). Thus, for our analysis we will arbitrarily define I_{sat} as that intensity at which the stimulated rate downward equals the normal decay rate:

$$\frac{I_{sat} B_{ul}(\nu)}{c} = \frac{1}{\tau_u}. \tag{7.39}$$

This leads to a value of I_{sat} of

$$I_{sat} = \frac{c}{B_{ul}(\nu)\tau_u}. \tag{7.40}$$

Using the Einstein relationship of (6.54) between $B_{ul}(\nu)$ and $A_{ul}(\nu)$, we find that

$$I_{sat} = \frac{c 8\pi h\nu_{ul}^3}{c^3 A_{ul}(\nu)\tau_u} = \frac{8\pi \nu_{ul}^2}{c^2 A_{ul}(\nu)} \frac{h\nu_{ul}}{\tau_u}. \tag{7.41}$$

Using the expression for $\sigma_{ul}^H(\nu)$ of (7.13), we can write

$$I_{sat} = \frac{h\nu_{ul}}{\sigma_{ul}^H(\nu)\tau_u}. \tag{7.42}$$

For situations in which the stimulating beam has a pulse duration $\Delta\tau_p$ that is shorter than τ_u ($\Delta\tau_p < \tau_u$), we can define a saturation energy E_{sat} (energy per unit area) instead of a saturation intensity (power per unit area) as

$$E_{sat} = I_{sat}\Delta\tau_p = \frac{h\nu_{ul}}{\sigma_{ul}^H(\nu)}\left(\frac{\Delta\tau_p}{\tau_u}\right). \tag{7.43}$$

This expression for E_{sat} would be particularly applicable to solid-state lasers where τ_u is relatively long.

7.4 DEVELOPMENT AND GROWTH OF A LASER BEAM

GROWTH OF BEAM FOR A GAIN MEDIUM WITH HOMOGENEOUS BROADENING

We now have a formula for the intensity I_{sat} above which the laser beam can no longer grow exponentially according to (7.33). It would be useful to obtain a value of the gain or the exponent of (7.33), $\sigma_{ul}(\nu)\Delta N_{ul}z$, at which the beam would reach the saturation intensity. We will consider a beam that starts by spontaneous emission and is subsequently amplified over a frequency width approximately equal to the homogeneous linewidth $\Delta\nu^H$ centered at the center of the emission frequency ν_{ul}. We will not be able to derive an exact value of the gain required to reach I_{sat} that can be applied to every situation. We will, however, obtain an approximate *range* of values that are dependent on the length and width of the gain medium.

Consider a cylindrical gain medium, as shown in Figure 7-6, that has a length L, a cross-sectional area A, and a diameter d_a. Within that gain medium we assume that a population inversion exists. Let the population density in the upper laser level be N_u, and suppose that the radiative rate from level u to level l at frequency ν_{ul} is A_{ul}. Because this calculation is only an approximation, we will assume that the population inversion is large enough that we can neglect the population density N_l in the lower level when using the exponential growth formula given in (7.33). In other words, we are assuming that $\Delta N_{ul} \cong N_u$. This would essentially define an upper limit on the gain exponent, $\sigma_{ul}^H(\nu)\Delta N_{ul}L$.

For simplicity we will consider the beam as starting at one end of the medium in a region of length l_g, as shown in Figure 7-6. We define l_g as one

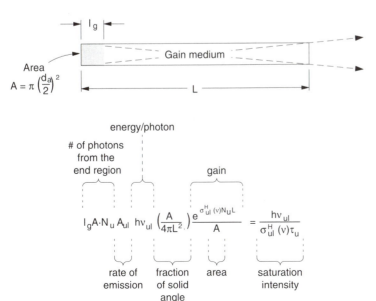

Figure 7-6. Growth and development of a laser beam from an elongated gain medium

gain length such that $\sigma_{ul}^H(\nu)\Delta N_{ul}l_g \cong \sigma_{ul}^H(\nu)N_u l_g = 1$; implicit in this definition is that $l_g < L$. We will assume that the atoms in level u within that region are radiating at a rate A_{ul} and with an energy per photon of $h\nu_{ul}$. Some of these photons are emitted in the elongated direction of the amplifier and would therefore be enhanced by stimulated emission as they transit through the length L of the medium. Thus, a beam would evolve as the radiation propagates down the length of the medium and the intensity grows exponentially. We will calculate the conditions required for the beam to reach the saturation intensity when it arrives at the opposite end of the amplifier. Of course, a similar situation would also happen at the other end of the medium for a beam traveling in the opposite direction; the calculation would be the same for both directions. We could also consider atoms in the next volume element to the right of the region of length l_g, but the beam initiated by such a volume element would not quite reach the saturation intensity since it would not have traversed as much length as the beam originating from the end region. We will therefore consider this calculation as an approximation by neglecting effects from other volume elements to keep it simple. The result, in fact, provides a reasonably good approximation even when compared to that of a more rigorous calculation.

We thus consider the energy radiated per unit time into a 4π solid angle from within the volume $A \cdot l_g$ as $N_u(A \cdot l_g)A_{ul}h\nu_{ul}$. This is multiplied by the fractional portion of the energy radiating within a solid angle $d\Omega$ that would reach the opposite end of the medium: $d\Omega/4\pi$. This fraction of the total solid angle can be expressed as $(A/L^2)(1/4\pi)$, or simply $A/4\pi L^2$. We assume that the energy radiated from that volume element per unit time is amplified by an amount $e^{\sigma_{ul}^H(\nu)N_u L}$ by the time it reaches the other end of the medium. We divide that energy per unit time by the area A to obtain an intensity. We then equate that intensity to the saturation intensity I_{sat}. This entire process is described as follows:

$$(N_u \cdot A \cdot l_g)A_{ul}h\nu_{ul}\frac{A}{4\pi L^2}\frac{e^{\sigma_{ul}^H(\nu)N_u L}}{A} = I_{sat} = \frac{h\nu_{ul}}{\sigma_{ul}^H(\nu)\tau_u}. \tag{7.44}$$

For the simplest case – where the only decay process from level u is via radiative decay to level l – we can express the decay time as $\tau_u = 1/A_{ul}$, so that (7.44) can be rewritten as

$$(N_u \cdot A \cdot l_g)A_{ul}h\nu_{ul}\frac{A}{4\pi L^2}\frac{e^{\sigma_{ul}^H(\nu)N_u L}}{A} = \frac{h\nu_{ul}A_{ul}}{\sigma_{ul}^H(\nu)}, \tag{7.45}$$

which reduces to

$$N_u \cdot \left[\pi\left(\frac{d_a}{2}\right)^2\right] \cdot \left(\frac{1}{\sigma_{ul}^H(\nu)N_u}\right) \cdot \frac{1}{4\pi L^2}e^{\sigma_{ul}^H(\nu)N_u L} = \frac{1}{\sigma_{ul}^H(\nu)}, \tag{7.46}$$

where the area A has been rewritten as $A = \pi(d_a/2)^2$ and l_g has been replaced by $1/\sigma_{ul}^H(\nu)N_u$. Further simplification leads to

$$e^{\sigma_{ul}^H(\nu)N_u L} = 16(L/d_a)^2. \tag{7.47}$$

This equation can be expressed more simply as

$$e^x = 16(L/d_a)^2, \tag{7.48}$$

where $x = \sigma_{ul}^H(\nu)N_u L \cong \sigma_{ul}^H(\nu)\Delta N_{ul}L$. Choosing a length l_g leads to a simplified result. If a very much shorter region than l_g were chosen then a significant amount of the energy that might eventually be amplified would be left out of the calculation. If a significantly longer region than l_g were chosen, a much shorter exponential growth length than L would have to be used and the beam would not gain as much energy through amplification. However, minor changes do not alter the results significantly. A problem at the end of this chapter considers a region of length d_a (the diameter of the amplifier) rather than l_g, and it will be seen that such a length change for the initial radiating region does not make a significant difference in the final result, that is, in the amount of gain required to reach I_{sat}.

SHAPE OR GEOMETRY OF AMPLIFYING MEDIUM

The solution to (7.48) can be graphed in the form of L/d_a versus x, or L/d_a versus $\sigma_{ul}^H\Delta N_{ul}L$ as shown in Figure 7-7. This graph has many implications. First: The ratio of the length of the amplifier to its diameter is an important factor in how much gain is needed to reach saturation. In most cases, it is difficult to generate a large gain factor (such as $gl = \sigma_{ul}^H\Delta N_{ul}L = 12$) in an amplifying medium of any reasonable length. One might therefore think it simpler to use an amplifier with a low value of L/d_a, since according to Figure 7-7 that amplifier would require less gain; however, such reasoning turns out to be erroneous. Let us consider two extreme situations that will clarify this issue.

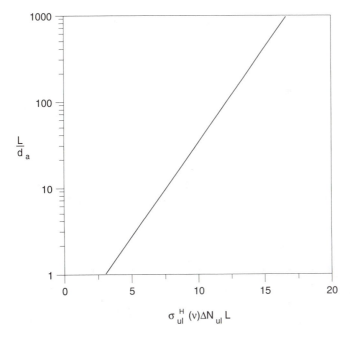

Figure 7-7. Graph of the ratio of gain length L to diameter d versus the exponential gain coefficient $\sigma \cdot \Delta N \cdot L$

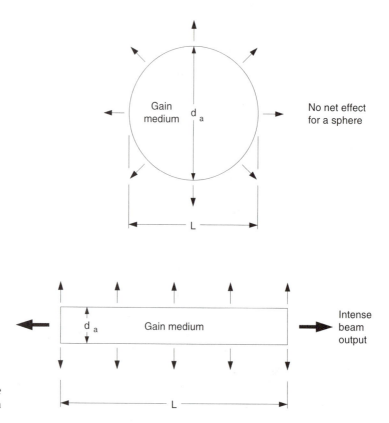

Figure 7-8. Two possible types of gain media

We will consider two differently shaped gain media, as shown in Figure 7-8. One is a long cylinder, similar to that of Figure 7-6; the other is a sphere of diameter d. For the long cylinder, say $L/d_a = 100$, spontaneous emission will originate at one end of the medium and will emerge at the other end in an elongated shape, as shown by the dashed lines exiting from the cylinder. This could certainly be considered a beam. We can see from Figure 7-7 that the value of $\sigma_{ul}^H(\nu)N_u L$ would be approximately 12 and thus the beam would have grown by an amount $e^{12} = 1.6 \times 10^5$! This is an extremely large increase resulting in a very intense beam with an extremely low divergence.

Now consider the case of the sphere, where $L/d_a = 1$. For this case the use of (7-48) is not completely accurate, but some conclusions can nevertheless be drawn from the simplified analysis. First, Figure 7-7 suggests that the value of $\sigma_{ul}^H(\nu)N_u L$ is approximately 2.7 for $L/d_a = 1$. This would represent an exponential growth factor of $e^{2.7} = 15$, which is significantly lower than the value of 1.6×10^5 obtained for the elongated medium. While such a "laser" would be just as intense at the surface of the sphere as that from the elongated medium (since it too would reach I_{sat}), the intensity would be rapidly reduced as one moves away from the sphere. In other words, the radiation originating from different locations within the sphere would cause the beam to diverge very rapidly (rather than remain concentrated in a specific direction), in the same manner as radiation emitting from a spherically shaped incoherent source.

Second: The radiation emitted from the entire sphere will be emitted equally in all directions, since the sphere is completely symmetrical. Therefore, given a certain population density N_u in level u within the medium, the maximum amount of energy E_{ul} that could be radiated on that transition from u to l is $E_{ul} = N_u V h \nu_{ul}$ within a time of $1/A_{ul}$, where V represents the volume of the sphere. This is also the same amount that would be radiated by spontaneous emission at a rate A_{ul}. Because E_{ul} represents all of the energy stored in level u during the level lifetime, there is no additional energy available to be radiated by stimulated emission. Thus the same amount of energy would emerge from the sphere under the presence of gain as would occur with no gain in the medium! This means that, for a spherical medium, the only effect caused by the existence of gain is that the energy might emerge from the sphere at a slightly higher rate for the case of stimulated emission, if the rate thereof exceeds that for spontaneous emission.

Thus, it makes no sense to construct a laser that radiates in all directions, since the purpose of having a laser – that of concentrating the available energy in both direction and frequency – is thereby defeated. In Chapter 1 it was mentioned that all a laser does is take energy that would normally be radiated in all directions and concentrate it into a specific direction. We have begun to show how this concentration is achieved, using analysis of stimulated emission and the considerations of beam growth necessary to reach the saturation intensity I_{sat}.

Figure 7-9 shows how a beam would grow as length is added to the amplifier: it grows exponentially over the length L_{sat}. Then, as it reaches the saturation intensity, it can no longer grow at that rate. It then begins to extract most of the energy from the medium to the right of $L = L_s$ and grows at an approximately linear rate. When the beam reaches that location in

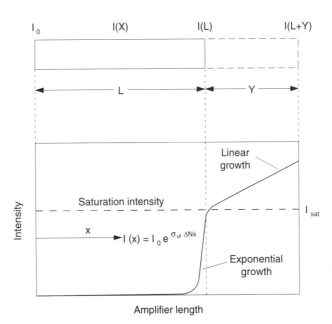

Figure 7-9. Exponential growth and saturation of a laser beam as a function of amplifier length

the amplifier, with the beam intensity above I_{sat}, the stimulated emission rate will exceed the spontaneous emission rate so that most of the available energy, in the form of excited-state population in level u, will be transferred to the beam, permitting significant enhancement of the beam energy.

We have seen that attempting to make a spherically shaped laser essentially defeats the purpose of making a laser, that of concentrating the beam energy into a specific direction. We return to Figure 7-7, where we see it would be useful to make L/d_a reasonably large so that the beam can develop a well-defined direction before I_{sat} is reached. Typical lasers require an L/d_a ratio ranging from about 10 to 1,000, which suggests a desirable range of gain values from 7 to 17 in order to reach saturation:

$$\sigma_{ul}^{H}(\nu)N_{ul}L \cong 12 \pm 5. \tag{7.49}$$

GROWTH OF BEAM FOR DOPPLER BROADENING

The analysis for the growth and development of a laser beam was carried out for homogeneous broadening in equations (7.44)–(7.49) by estimating the amount of gain ($\sigma_{ul}^{H}(\nu)\Delta N_{ul}L$) required for the beam to reach I_{sat}. We can perform a similar analysis for Doppler broadening by considering a similar beam-growth equation to that of (7.44). We still need to reach the same value of I_{sat} on the right-hand side of the equation, since that value is associated only with the beam intensity needed to produce a certain rate of stimulated emission and not with the details concerning the frequency distribution of the spontaneous emission and the gain profile.

We must use only a portion of the population residing in N_u to generate the spontaneous emission factor that initiates the beam at one end of the medium. We do this by using the portion of the N_u population that is within the homogeneous linewidth of a given group of the radiating atoms, $(\Delta\nu^{H}/\Delta\nu^{D})N_u$. We must also replace the stimulated emission cross section $\sigma_{ul}^{H}(\nu)$ with the value $\sigma_{ul}^{D}(\nu)$ for Doppler broadening in the exponential growth factor of (7.44). We will simplify the process by considering the ratio of $\sigma_{ul}^{H}(\nu)/\sigma_{ul}^{D}(\nu)$ at the center frequency ν_0, since that frequency value has the highest gain and is therefore the frequency at which I_{sat} will first be reached. The ratio of those cross sections is given by

$$\frac{\sigma_{ul}^{H}(\nu_0)}{\sigma_{ul}^{D}(\nu_0)} = \frac{\dfrac{\lambda_{ul}^{2}A_{ul}}{4\pi^{2}\Delta\nu^{H}}}{\sqrt{\dfrac{\ln 2}{16\pi^{3}}}\dfrac{\lambda_{ul}^{2}A_{ul}}{\Delta\nu^{D}}} = \frac{1}{\sqrt{\pi\ln 2}}\frac{\Delta\nu^{D}}{\Delta\nu^{H}}, \tag{7.50}$$

and thus we have

$$\sigma_{ul}^{H}(\nu_0) = \frac{1}{\sqrt{\pi\ln 2}}\frac{\Delta\nu^{D}}{\Delta\nu^{H}}\sigma_{ul}^{D}(\nu_0) = 0.68\frac{\Delta\nu^{D}}{\Delta\nu^{H}}\sigma_{ul}^{D}(\nu_0). \tag{7.51}$$

Equation (7.44) can then be written as

$$\left[N_u\left(\frac{\Delta\nu^H}{\Delta\nu^D}\right)\cdot A\cdot l_g\right]A_{ul}h\nu_{ul}\frac{A}{4\pi L^2}\frac{e^{\sigma_{ul}^D(\nu_0)N_uL}}{A}=\frac{h\nu_{ul}A_{ul}}{\sigma_{ul}^H(\nu_0)}, \qquad (7.52)$$

where we have again assumed that the dominant decay mode from level u is radiative decay to level l such that $1/\tau_u=A_{ul}$. For this case $l_g=1/\sigma_{ul}^D(\nu_0)N_u$. We did not have to adjust the value for N_u in the exponent because this was already taken into account by using the lower stimulated emission cross section for Doppler broadening, as demonstrated previously in equations (7.23)–(7.25). We can now re-arrange (7.52) as

$$N_u\left(\frac{\Delta\nu^H}{\Delta\nu^D}\right)\sigma_{ul}^H(\nu_0)\left(\frac{1}{\sigma_{ul}^D(\nu_0)N_u}\right)\left[\pi\left(\frac{d_a^2}{4}\right)\right]\frac{1}{4\pi L^2}e^{\sigma_{ul}^D(\nu_0)N_uL}=1. \qquad (7.53)$$

Thus, for the case of Doppler broadening, if we substitute the expression for $\Delta\nu^H/\Delta\nu^D$ from (7.51) we obtain

$$\left(\frac{1}{\sqrt{\pi\ln 2}}\right)e^{\sigma_{ul}^D(\nu_0)N_uL}=16\left(\frac{L}{d_a}\right)^2. \qquad (7.54)$$

We have thus derived an expression similar to that of (7.48), with similar requirements for the gain. If we neglect the factor $1/\sqrt{\pi\ln 2}$, which is nearly unity, we arrive at the same conclusion for the requirements of the exponential growth factor for Doppler broadening,

$$\boxed{\sigma_{ul}^D(\nu_0)\Delta N_{ul}L\cong 12\pm 5. \qquad (7.55)}$$

7.5 EXPONENTIAL GROWTH FACTOR (Gain)

It is appropriate at this time to examine the various components of the gain in order to understand the important factors associated with producing a gain of 12. From the analysis we will see that mirrors are required to enable most lasers to reach I_{sat}.

STIMULATED EMISSION CROSS SECTION   If we look at the stimulated emission cross section σ_{ul}, as summarized by the formulas in Table 7-1, we can see that none of the coefficients can be easily adjusted. The transition probability is fixed for a given laser transition. We could make the wavelength longer, since σ_{ul} depends upon the square of the wavelength, but we may need a laser to be at a specific wavelength – ruling out any such adjustments. We could try to reduce the linewidth. For Doppler broadening this would involve reducing the temperature, but since the linewidth depends upon the square root of the temperature, that is not a very sensitive factor. In other words, there is generally not much we can do about increasing the stimulated emission cross section; we must think of it as a constant for a specific laser transition in a specific gain medium. Typical values of σ_{ul} for various lasers are shown in Figure 7-10. We can see why, given the strong wavelength dependence of σ_{ul} shown in Table 7-1, it is difficult to make short-wavelength lasers.

TABLE 7-1

EXPRESSIONS FOR STIMULATED EMISSION CROSS SECTION FOR THE GAIN MEDIUM

FORMULA FOR σ MAXIMUM (usually at the center of the emission band)

Homogeneous Broadening

$$\sigma_{ul}^{H}(\nu_0) = \frac{1}{4\pi^2}\left(\frac{\lambda_{ul}^2 A_{ul}}{\eta^2 \Delta\nu^H}\right)$$

Doppler Broadening

$$\sigma_{ul}^{D}(\nu_0) = \frac{1}{4\pi}\left(\frac{\ln 2}{\pi}\right)^{1/2}\left(\frac{\lambda_{ul}^2 A_{ul}}{\eta^2 \Delta\nu^D}\right)$$

λ_{ul} = wavelength
η = index of refraction
A_{ul} = radiative transition probability from level u to l
$\Delta\nu$ = bandwidth of emission

Laser	λ (nm)	σ_{ul} (m^2)
He-Ne	632.8	3.0×10^{-17}
Argon	488.0	2.5×10^{-16}
He-Cd	441.6	9.0×10^{-18}
Copper (CVL)	510.5	8.6×10^{-18}
CO$_2$	10,600.0	3.0×10^{-22}
Excimer	248.0	2.6×10^{-20}
Dye (Rh6G)	577.0	2.5×10^{-20}
Semiconductor	800.0	1.0×10^{-22}
Nd:YAG	1064.1	6.5×10^{-23}
Nd:Glass	1062.3	3.0×10^{-24}
Ti:Al$_2$O$_3$	800.0	3.4×10^{-22}
Cr:LiSrAlF	850.0	4.8×10^{-24}

Figure 7-10. Stimulated emission cross section for a variety of lasers

POPULATION DIFFERENCE The second component of gain is the population difference factor ΔN_{ul}, or $[N_u - (g_u/g_l)N_l]$. There are two important aspects of this function. One is the requirement that N_u be greater than $(g_u/g_l)N_l$. Satisfying this requirement will be addressed in Chapter 8, when we talk about the types of level systems that can be used to develop population inversions. Making N_u greater than N_l generally is determined by the populating and decay rates of both the upper and lower laser levels during the excitation process.

The second important aspect is the necessity for a relatively large population density difference ΔN_{ul}. Since the gain must be of the order of 12, and since σ_{ul} is typically of the order of 10^{-16} m^2, it would be desirable to have the population density difference be of the order of $12/\sigma_{ul}$, or in this case approximately 1.2×10^{17} m^{-3} for a laser with $L \cong 1$ m. Such a density is not easy to achieve. This issue will be addressed in more detail in Chapter 9, when we deal with excitation mechanisms.

GAIN LENGTH We now examine the third component of gain: requirements of the length L of the gain medium. Lasers have been made with gain media lengths ranging from less than 1 millimeter to longer than 10 meters. Long lasers, for the most part, have not turned out to be very useful (except in some very special cases, such as the high-power NOVA laser at Lawrence Livermore National Laboratories). Long lasers are too cumbersome, too difficult to set up and operate, and hence not very practical. People using lasers naturally prefer L to be as small as possible. However, this requirement sometimes creates problems.

In many cases there are limitations on making ΔN_{ul} larger to make up for having a shorter L. For example, consider the well-known helium–neon (He–Ne) laser emitting at 632.8 nm. From Figure 7-10, the stimulated emission cross section for that laser is $\sigma_{ul}^D = 3 \times 10^{-17}$ m^2 and ΔN_{ul} is of the order of 5×10^{15} m^{-3}. Thus it would take a length of the order of $L = 80$ m to make a laser! We know that making a laser with such a gain length is not realistic, so there must be another technique available to overcome this limitation. That technique is the use of mirrors to effectively increase the length of the medium.

7.6 THRESHOLD REQUIREMENTS FOR A LASER

Most lasers have limitations on the population density that can be achieved in any particular energy level, which in turn places limits on the maximum ΔN_{ul} that can be obtained for a specific laser transition. Also, the stimulated emission cross section is essentially a constant for a specific laser transition. Therefore, the only factor of the gain that can be increased is the *effective* length of the amplifier, since it is not practical to make long lasers. This increased effective length can be realized with the use of mirrors. We will determine the threshold requirements for producing a laser with no mirrors, and also with either one or two mirrors located at the ends of the gain medium. The threshold conditions are defined as the necessary requirements for the beam to grow to the point at which it reaches the saturation intensity I_{sat}.

LASER WITH NO MIRRORS

Consider an amplifier similar to that shown in Figure 7-6. A diagram of that amplifier is shown in Figure 7-11(a), along with an outline of the beam envelope that would emerge if the length L were sufficient for the beam to

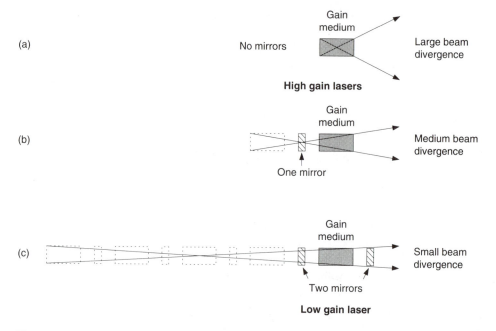

Figure 7-11. Laser beam divergence for an amplifier with (a) no mirrors, (b) one mirror, and (c) two mirrors

reach I_{sat} at the end of the amplifier. The beam would then meet the gain requirement that

$$\sigma_{ul} \Delta N_{ul} L \cong 12 \pm 5. \qquad (7.56)$$

EXAMPLE

A laser gain medium has a gain coefficient of 100 m^{-1} at the center of the emission profile and an amplifier length of 0.14 m. For what diameter d of the gain medium would the saturation intensity be reached in a single pass through the amplifier? What would be the divergence half-angle of the beam exiting from the medium? (See Figure 7-11(a).)

Since $g(\nu_0) = 100$ m^{-1} and $L = 0.14$ m, we have an exponential growth factor of 100 m^{-1}·0.14 m = 14. Thus, spontaneous emission originating from the end of the amplifier would grow according to $I = I_0 e^{14} = (1.2 \times 10^6) I_0$. Using (7.48) we can determine that, for a gain of 14, $L/d_a = 274$ and thus $d_a = 0.51$ mm. The half-angle of the emerging beam would have an angular divergence determined by the length divided by the diameter, since spontaneous emission originating from any radial position at the end of the amplifier would emit radiation (toward the other end of the amplifier) that would reach the saturation intensity when those rays were amplified as they traveled the entire length L of the amplifier. This half-angle divergence d_a/L would thus be $1/274 = 3.6 \times 10^{-3}$ rad (3.6 mrad).

LASER WITH ONE MIRROR

If the length L were not sufficient to reach saturation, we could add a mirror behind the amplifier as shown in Figure 7-11(b). Adding a mirror can be thought of as adding a second amplifier behind the first one, as shown in the figure, since the mirror effectively produces an image of the existing amplifier located behind that amplifier. For this simplified discussion, we assume that the mirror reflectivity is 100% at the laser wavelength and that the mirror is placed directly behind the amplifier. In this case, we can again indicate the beam envelope that would occur, assuming that the beam just reaches I_{sat} after two passes through the amplifier. We can see that the beam is narrower, as it emerges from the end of the amplifier, when compared to the case with no mirrors shown in Figure 7-11(a). For this arrangement involving one mirror, the beam would meet the gain requirement if

$$\sigma_{ul}\Delta N_{ul}(2L) \cong 12 \pm 5. \tag{7.57}$$

EXAMPLE

A laser amplifier with a length of 0.12 m and a gain of 60 m^{-1} has a 100% reflecting mirror at the laser wavelength coated on one end of the laser rod. For a rod diameter of 5 mm, what would be the half-angle of the diverging beam exiting from the amplifier? (See Figure 7-11(b).) Would the beam reach the saturation intensity as it emerges from the rod after having made a double pass through the rod?

With one mirror on the end of the rod, this laser has an effective length $L_{eff} = 2L$, as shown in Figure 7-11(b). Thus the effective L/d_a ratio for this laser is $L_{eff}/d_a = (2 \times 0.12 \text{ m})/(0.005 \text{ m}) = 48$. The half-angle would be determined by the factor $d_a/2L = (0.005 \text{ m})/2(0.12 \text{ m}) = 2.1 \times 10^{-2}$ rad (21 mrad).

Equation (7.48) can be expressed as

$$e^x = 16(L_{eff}/d_a)^2 = 16[(2 \times 0.12 \text{ m})/(0.005 \text{ m})]^2$$
$$= 16(48)^2 = 3.7 \times 10^4,$$

where $x = g(\nu_0)L_{eff} = g(\nu_0)2L$. The beam would thus reach the saturation intensity for an amplifier length of $L = x/2g(\nu_0)$. We can solve for x in the equation just displayed to find that $x = 10.5$ and $L = x/2g(\nu_0) = 10.4/2(60)$ m $= 0.087$ m. Thus the 0.12-m rod will be more than sufficient to reach the saturation intensity. If the rod had not been coated (i.e., no mirror) then the beam would have grown by a factor of only $e^{5.25} = 191$ as it made a single pass through the amplifier, and would therefore not have become a very intense beam. It would also have had twice the divergence angle as that determined previously for the double-pass laser.

LASER WITH TWO MIRRORS

If the length L were still not sufficient for the beam to reach I_{sat} after two passes through the amplifier, we could add a second mirror in front of the amplifier. We will allow a slight amount of light to "leak" out of the end by using a mirror with only 99.9% reflectivity, so that a portion of the beam can escape and provide an observable signal. Placing a mirror at each end of the amplifier effectively adds an infinite series of amplifiers behind the original amplifier, as shown in Figure 7-11(c); this essentially allows the amplifier to have as much length as necessary to reach I_{sat}. For such an arrangement, the beam emerges with a very narrow angular divergence. This results in a very low-divergence laser beam, and is therefore the arrangement used for most lasers. The details of such a two-mirror laser and the associated beam that develops are much more complex than suggested in this simple analysis. These details will be covered extensively in Chapter 10, when we discuss the cavity effects associated with having two mirrors surrounding an amplifier. However, the analysis here provides a simple visualization of how a low-divergence beam evolves.

Two factors determine if a two-mirror laser can reach I_{sat}.

(1) *Net gain per round trip* – There must be a net gain per round trip of the beam through the amplifier. In other words, all of the losses (including transmission losses, scattering losses, and absorption losses) must be lower than the gain or increase per round trip through the amplifier.

(2) *Sufficient gain duration* – The gain duration is the amount of time that the gain exists in the amplifier. (This duration is relevant to pulsed lasers only.) The gain duration t_s must last long enough to allow a sufficient number of passes of the beam through the amplifier in order for the beam to reach I_{sat}.

NET GAIN PER ROUND TRIP

● SIMPLE CASE   Requirement (1) can be satisfied as follows. Consider a round-trip pass of the beam through the amplifier, and assume that the gain is uniform over the amplifier length (not changing in time). In this case the beam will experience an exponential growth of $e^{g(\nu_0)2L}$ for a round trip through the amplifier. It will also experience a loss at each mirror of $1-R$, where R is the mirror reflectivity. For the simple situation in which both mirrors have the same reflectivity R, the minimum round-trip steady-state requirement for the *threshold* of lasing is that the gain exactly equal the loss. In this case the beam will remain unchanged after it makes one round trip through the medium. Thus, any increase in gain beyond the threshold will cause the beam to grow. This threshold can be expressed as

$$R^2 e^{g(\nu_0)2L} = 1 \tag{7.58}$$

or

$$g(\nu_0) = \frac{1}{2L} \ln \frac{1}{R^2}. \qquad (7.59)$$

● GENERAL CASE   A more general situation would be that in which the mirrors have different reflectivities R_1 and R_2. We will allow for fractional losses a_1 and a_2 at the "Brewster windows" (see Section 10.2) or at any other region in the path of the beam other than inside the amplifier, and we also include a possible distributed loss α within the gain medium; α does not involve the particular laser levels u and l but occurs at the same frequency as that of the laser beam. For this general situation, the threshold equation for a round-trip pass becomes

$$R_1 R_2 (1-a_1)(1-a_2) e^{(g(\nu_0)-\alpha)2L} = 1. \qquad (7.60)$$

The solution for the threshold gain is then

$$g(\nu_0) = \frac{1}{2L} \left[\frac{1}{R_1 R_2 (1-a_1)(1-a_2)} \right] + \alpha. \qquad (7.61)$$

SUFFICIENT GAIN DURATION   Requirement (2) concerns the minimum gain duration t_s for the beam to reach I_{sat}, and can be obtained by using the term m to describe the number of passes through the amplifier before I_{sat} is reached. Thus, the effective gain length becomes mL. The threshold condition for producing a laser can then be given as

$$g(\nu_0)mL \cong 12, \qquad (7.62)$$

where we have taken an average value of the gain in (7.49). An expression for m can then be obtained as

$$m \cong \frac{12}{g(\nu_0)L}. \qquad (7.63)$$

If mirror reflectivity is significantly less than 100%, the analysis is more complicated and will not be carried out here. We will therefore consider (7.63) as indicating a minimum number of passes to reach the saturation intensity.

   Knowledge of m can be used to determine t_s. We will use a laser mirror separation d such that $d \geq L$. For this general situation, the time t_s is simply

$$t_s = \frac{\text{distance}}{\text{velocity}} = \frac{md}{c/\eta}. \qquad (7.64)$$

The velocity of light is here c/η, where c is the velocity of light in a vacuum and η is the index of refraction of the medium in which the developing beam travels. Taking into account the possible difference in refractive index for the beam when it is within the gain medium versus when it is between the gain medium and the mirrors, we can express (7.64) in a more general form as

$$t_s = m[\eta_C(d-L)+\eta_L L]/c, \tag{7.65}$$

where η_C is the index of refraction of the medium between the mirrors but outside of the laser amplifier (which is usually air, $\eta_C \cong 1$), $d-L$ is the distance between the mirrors that does not include the amplifier, and η_L is the index of refraction within the gain medium.

EXAMPLE

For a He–Ne laser, assuming a mirror reflectivity of 99.9% for each mirror and a gain-medium length of 0.2 m, how many passes must occur through the amplifier before I_{sat} is reached?

We will solve this simplified example by assuming that $R = 100\%$. From Table 7-2 we find that $g_{ul} = 0.15$ m^{-1}. For these conditions we can use (7.63) to obtain a value of $m \approx 12/g_{ul}L = 12/(0.15$ m$^{-1} \times 0.2$ m) $= 400$, or approximately 400 passes through the amplifier to reach the saturation intensity.

For the case of the He–Ne laser just described but with $d = 0.3$ m and $\eta \cong 1$, $t_s = 0.5$ μs. The minimum gain duration can be useful for determining several types of restrictions. For a continuously operating laser, referred to as a *cw laser* (continuous wave), t_s will indicate how long it will take for the laser beam to grow to an equilibrium situation, provided that the gain

TABLE 7-2

AMPLIFIER PARAMETERS FOR VARIOUS LASERS

Laser	g_{ul} (m^{-1})	L (m)	m	d (m)	η	t_s (s)
He–Ne	0.15	0.2	500	0.3	1.0	5.0×10^{-7}
Argon	0.5	1.0	30	1.1	1.0	1.1×10^{-7}
He–Cd	0.3	0.5	100	0.6	1.0	2.0×10^{-7}
Copper (CVL)	5	1.0	3	1.1	1.0	1.1×10^{-8}
CO_2	0.9	1.0	19	1.1	1.0	7.0×10^{-8}
Excimer	2.6	1.0	6	1.1	1.0	2.2×10^{-8}
Dye (Rh6G)	500	0.02	1.5	0.04	1.5	3.0×10^{-10}
GaAs	100,000	0.0001	1.5	0.0001	3.4	1.7×10^{-12}
Nd:YAG	10	0.1	17	0.2	1.82	2.0×10^{-8}
Nd:Glass	3	0.1	50	0.2	1.55	5.2×10^{-8}

medium itself stabilizes immediately. For a pulsed laser, it will indicate how long the gain must last in order for the beam to reach I_{sat}. If the medium is pulsed, and the gain duration lasts for a duration less than t_s, the laser beam will not develop fully. Thus, either g_{ul} would have to be increased or d would have to be decreased, subject to the constraints mentioned previously, for the full laser output to occur.

VALUES OF m AND t_s FOR SEVERAL LASERS Values of m and t_s are estimated in Table 7-2 for several well-known lasers. It must be remembered that these are only estimates rather than exact calculations, and are intended only as guidelines. For some lasers, the gain lasts for only about 10 ns. Assuming that $n = 1$, $L = 0.2$ m, and $d = L$, if $t_s = 10$ ns then we find that the maximum number of passes would be $m = 15$ before the gain would cease. For $d = 1$ m, we find that $m = 3$. These estimates are overestimates because they do not allow for at least one extra pass, after I_{sat} is reached, for the beam to extract energy from the amplifier, as suggested in Figure 7-9.

7.7 LASER OSCILLATION ABOVE THRESHOLD

RATE EQUATIONS OF THE LASER LEVELS THAT INCLUDE STIMULATED EMISSION

In Section 7.3 we obtained an expression for the saturation intensity I_{sat} by equating the radiative decay rate from the upper laser level u to the stimulated emission rate from that level. We were able to do this because we know that the intensity of the beam I in a medium with net gain will continue to grow exponentially until those two rates are in fact equal. However, we did not consider the effect of this intense beam on the population densities N_u and N_l of levels u and l.

Following the development of an expression for L_{sat} initiated with (7.36), we can write steady-state rate equations for the change in the population densities of levels u and l in the presence of a beam of intensity I as

$$\frac{dN_u}{dt} = R_u + N_l \frac{B_{lu}(\nu)I}{c} - N_u\left(A_{ul} + \frac{B_{ul}(\nu)I}{c}\right) = 0; \qquad (7.66)$$

$$\frac{dN_l}{dt} = R_l - N_l\left(A_l + \frac{B_{lu}(\nu)I}{c}\right) + N_u\left(A_{ul} + \frac{B_{ul}(\nu)I}{c}\right) = 0. \qquad (7.67)$$

As before, R_u and R_l represent the excitation flux from any source other than that from one of the two levels u and l to the other. We have also assumed for simplicity that the only pathway for radiative decay from level u is to level l at a rate of A_{ul} and A_l from level l, and that there is no collisional decay. Both (7.66) and (7.67) are equated to zero in order to obtain steady-state solutions. All of the processes in these equations are depicted in Figure 7-12. The equations can be solved for population densities N_u and N_l in terms of the laser intensity I within the gain medium.

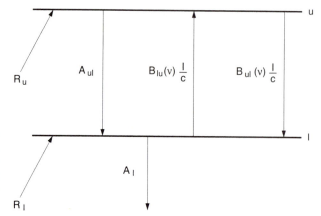

Figure 7-12. Relevant excitation and decay processes for a two-level system in which stimulated emission is a significant process

SMALL SIGNAL GAIN COEFFICIENT

Adding (7.66) and (7.67), we can eliminate N_u and obtain an expression for N_l as

$$N_l = \frac{R_u + R_l}{A_l} = N_l^0. \tag{7.68}$$

We can see that N_l is independent of I and is therefore identically equal to N_l^0, the population in the lower laser level when $I = 0$. The small signal gain $g^0(\nu_0)$ – that is, the gain at the center of the emission when no beam is present – is thus defined as $g_{ul}^0 = \sigma_{ul}(N_u^0 - (g_u/g_l)N_l^0) = \sigma_{ul}\Delta N_{ul}^0$.

From (7.66) we can express N_u as

$$
\begin{aligned}
N_u &= \frac{R_u + \dfrac{B_{lu}(\nu)I}{c}N_l}{A_{ul} + \dfrac{B_{ul}(\nu)I}{c}} = \frac{R_u + \dfrac{(g_u/g_l)B_{ul}(\nu)I}{c}N_l}{\dfrac{B_{ul}(\nu)I_{\text{sat}}}{c} + \dfrac{B_{ul}(\nu)I}{c}} \\
&= \frac{R_u A_{ul} + (g_u/g_l)(I/I_{\text{sat}})N_l}{1 + I/I_{\text{sat}}} \\
&= \frac{R_u/A_{ul} + (g_u/g_l)(I/I_{\text{sat}})N_l^0}{1 + I/I_{\text{sat}}} = \frac{N_u^0 + (g_u/g_l)(I/I_{\text{sat}})N_l^0}{1 + I/I_{\text{sat}}}. \tag{7.69}
\end{aligned}
$$

In this formula we have used the relationship (7.39) that $B_{ul}(\nu)I_{\text{sat}}/c = 1/\tau_u \cong A_{ul}$, that $B_{lu}(\nu) = (g_u/g_l)B_{ul}(\nu)$ as obtained from (6.55), and that $N_u^0 = R_u/A_{ul}$, which represents the steady-state population in level u when there is no laser beam present. Examining this expression, we can see that the population in level u is significantly changed (reduced) by the presence of the field of intensity I, especially when I becomes of or greater than the order of I_{sat}.

SATURATION OF THE LASER GAIN

We can now investigate the effect that a beam of intensity I has on the population difference $\Delta N_{ul} = N_u - (g_u/g_l)N_l$ associated with the laser gain. Using (7.68) and (7.69) for N_u and N_l, we express the population difference as

$$\Delta N_{ul} = N_u - (g_u/g_l)N_l = \frac{N_u^0 + (g_u/g_l)(I/I_{sat})N_l^0}{1 + (I/I_{sat})} - (g_u/g_l)N_l^0$$

$$= \frac{N_u^0 - (g_u/g_l)N_l^0}{1 + (I/I_{sat})} = \frac{\Delta N_{ul}^0}{1 + (I/I_{sat})}. \tag{7.70}$$

In this formula we have written $\Delta N_{ul}^0 \equiv N_u^0 - (g_u/g_l)N_l^0$ as the population difference when there is no laser beam present. This would be the value used in determining the small signal gain, or the gain when the beam is not strong enough to alter significantly the population of level u. We saw previously that the beam does not alter the population in level l even when the intensity I is very high.

Because the gain coefficient $g(\nu)$ is expressed as $g(\nu) = \sigma_{ul}(\nu)\Delta N_{ul}$, as obtained from either (7.14) or (7.27), we can write a more explicit expression for $g(\nu)$ in terms of the laser beam intensity I within the gain medium as

$$g(\nu) = \frac{g^0(\nu)}{1 + (I/I_{sat})} = \frac{\sigma_{ul}(\nu)\Delta N_{ul}^0}{1 + (I/I_{sat})}. \tag{7.71}$$

Thus, for a specific value of the small signal gain provided within the medium, the beam will grow until the intensity I causes the gain to be reduced as described in (7.71). It will grow further until the gain coefficient is reduced to the same value as described for the threshold gain in (7.59) or (7.61); this value is equivalent to the losses within the cavity, as can be seen in Figure 7-13. At that point the beam can no longer continue to grow and so the gain – and consequently the population in the upper laser level – are reduced no further. Figure 7-13 also shows that, with homogeneous broadening, the entire gain profile is reduced by this effect, since every atom in the upper laser level contributing to the gain is affected in the same way as any other atom. The case of an inhomogeneously broadened gain profile will be described in detail in Section 10.3, when we discuss the effect of "hole burning" associated with laser cavity modes.

At first it may seem odd that the laser intensity within the gain medium does not alter the population density N_l in level l, as indicated in (7.68). The only way this can happen is if the additional population flux into level l from level u due to stimulated emission is equal to the population flux passing via absorption from level l to level u. This would require that

$$N_u \frac{B_{ul}(\nu)I}{c} = N_l \frac{B_{lu}(\nu)I}{c} \tag{7.72}$$

or

$$N_u B_{ul}(\nu) = N_l B_{lu}(\nu). \tag{7.73}$$

Using the relationship from (6.55) that $g_l B_{lu}(\nu) = g_u B_{ul}(\nu)$, we can rewrite (7.73) as

$$N_u = \left(\frac{g_u}{g_l}\right)N_l \quad \text{or} \quad \left(N_u - \frac{g_u}{g_l}N_l\right) = \Delta N_{ul} = 0. \tag{7.74}$$

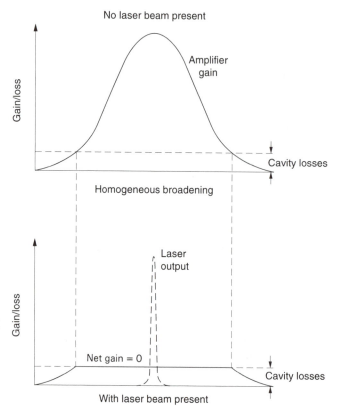

Figure 7-13. Laser output for a homogeneously broadened gain medium that has no additional frequency-selective elements within the laser cavity

This suggests that the gain is reduced to zero, since (7.74) suggests that $\Delta N_{ul} = 0$. This would be true for the ideal situation in which there are no other losses to decrease the intensity I of the beam. However, for a real situation the *net* gain will go to zero, and thus N_u of (7.74) will always be slightly greater than $(g_u/g_l)N_l$, adjusting to a steady-state value in which there is just enough increase in the beam per pass through the amplifier to make up for the mirror transmission losses and the absorption and scattering losses.

RELAXATION OSCILLATIONS

Many solid-state lasers tend to have relatively high gain and also a long upper laser level lifetime. This combination can lead to a phenomenon known as relaxation oscillations in the laser output intensity. For pulsed (non–Q-switched) lasers in which the gain lasts for many microseconds or milliseconds, these oscillations occur in the form of a regularly repeated spiked laser output superimposed on a lower steady-state value. For cw lasers the relaxation takes the form of a sinusoidal oscillation of the output. The phenomenon is caused by an oscillation of the gain due to the interchange of pumped energy between the upper laser level and the laser field in the cavity. This effect can be controlled by using an active feedback mechanism, in the form of an intensity-dependent loss, in the laser cavity.

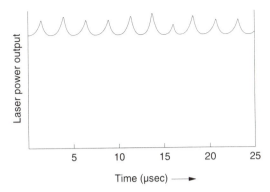

Figure 7-14. Graph of relaxation oscillations in the output intensity of a ruby laser

An example of the spiked output resulting from relaxation oscillations is shown in Figure 7-14.

7.8 LASER AMPLIFIERS

We have defined lasers as having a gain or amplifying medium that has both a population inversion and sufficient gain for the developing beam to reach the saturation intensity I_{sat}. We have also suggested that, when the laser reaches I_{sat}, the beam will continue to grow until the gain has been reduced to the value of the losses within the medium. In many cases this value of laser intensity, or the quality of the beam that develops, may not be sufficient for a specific application. Overcoming these problems may require the use of additional amplifiers to enhance the beam further. Such amplifiers could therefore be placed at a location beyond the laser output and would add more photons to the beam in replica of the input beam, since stimulated emission occurs in the same direction as that of the stimulating photons.

An amplifier is usually (though not always) composed of the same gain medium as that of the input laser beam. In order to extract energy efficiently from the amplifier, it is desirable to have the input intensity be above I_{sat} for the amplifier. Then the beam will transfer all of the stored energy, in the form of population in the upper laser level, to stimulated emission. Of course, this will occur only in the regions of the amplifier accessed by the input beam. It is thus very desirable to ensure that the amplifier is designed properly, so as to use the maximum volume possible. For amplifiers in which many stages are used to build the beam up to many orders of magnitude in intensity (or energy) greater than that of the input beam, the beam cross-sectional area is increased from one amplifier to the next to take advantage of more stored energy within the larger amplifier. In the earlier amplifier stages the use of a large cross-sectional area is not possible because the beam intensity (power per unit area) is insufficient for the beam to be above I_{sat} for such a large area. But as the beam grows in intensity, I_{sat} can be maintained over a much larger cross-sectional area and thus larger-diameter amplifiers can be accommodated. See for example the Nd:glass amplifier discussed in Section 14.3.

In some situations the input beam to the amplifier cannot be sufficiently reduced in cross-sectional area to have a value greater than or equal to I_{sat} of the amplifier. There is, however, a minimum value it must have in order to be effective. The intensity of the input beam must be significantly greater (at least by a factor of 10) over the same frequency bandwidth as that produced by the spontaneous emission of the amplifier itself propagating within the same amplifier volume (same solid angle) as the input beam, as described in (7.44). Otherwise the amplified spontaneous emission (ASE) from the amplifier will appear as part of the emerging beam and will not generally have the desired characteristics of the input beam that is being amplified. Equation (7.44) described spontaneous emission that began at one end of an amplifying medium and grew to a much larger beam at the other end. In that analysis we determined how high the gain would have to be within the medium for the spontaneous emission to reach I_{sat} by the time it reached the other end of the amplifier. We depended upon the spontaneous emission to initiate the laser beam. In the present case we do not want that spontaneous emission to contribute to the amplified beam. In cases where the input beam is weak, it may be necessary to estimate the contribution from ASE in the same manner as was done in (7.44) to determine how intense the input beam must be for a given gain in the amplifier. If the ASE appears to be too great, then either the input beam must be increased or the amplifier gain reduced until the ASE is below the tolerable limit. This is usually a problem only in high-gain amplifiers.

REFERENCES

A. Corney (1977), *Atomic and Laser Spectroscopy.* Oxford: Clarendon Press.

A. C. G. Mitchell and M. W. Zemansky (1961), *Resonance Radiation and Excited Atoms.* Cambridge University Press.

A. E. Siegman (1986), *Lasers.* Mill Valley, CA: University Science.

PROBLEMS

1. Derive the frequency distribution for $A_{ul}(\nu)$, $B_{ul}(\nu)$, and $B_{lu}(\nu)$ for the case of natural broadening. Assume that the total decay rate for level u (when taking into account all frequencies associated with the decay from level u to level l) is A_{ul}, and that the total B coefficients are B_{ul} and B_{lu} when integrated over all frequencies associated with the emission of the specific transition from level u to level l.

2. Derive the frequency distributions of the population densities $N_u(\nu)$ and $N_l(\nu)$, as described by (7.24), for the case of Doppler broadening. Assume that the total population densities in levels u and l are N_u and N_l.

3. Calculate the stimulated emission cross section for the He–Ne laser operating at 632.8 nm, and for the He–Cd laser at 325.0 nm, using the necessary coefficients for those transitions given in earlier chapters. It will be necessary to determine which broadening mechanism is dominant in order to use the appropriate expression for the stimulated emission cross section.

4. Calculate the stimulated emission cross section for the Nd:YAG laser using (7.16) together with the necessary values from Table 14-4 for the Nd:YAG laser.

5. Obtain the saturation intensities of the following lasers, using the expression derived in this chapter.

(a) A helium–neon laser operating at 632.8 nm.
(b) An argon ion laser at 488.0 nm and at 514.5 nm.
(c) A helium–cadmium laser operating at 441.6 nm.
(d) A copper vapor laser at 510.5 nm and 578.2 nm.

Relevant parameters are available from Chapter 4 or Chapter 13.

6. What is the ratio of the stimulated emission cross sections at the peak of the emission linewidth for two identical transitions (same A_{ul} and $\Delta\nu_{ul}$), where one is dominated by natural broadening and the other by Doppler broadening?

7. In Figure 7-6, assume an incremental length of d_a (equal to the diameter of the amplifier) at the left end of the amplifier instead of l_g. Obtain an expression similar to (7.47) in terms of x and L/d_a. Make a plot of L/d versus $\sigma_{ul}^H(\nu)\Delta N_{ul}L$, comparing it to the graph shown in Figure 7-7. You will realize that the incremental length chosen for the end region that initiates the amplification process does not make a significant change in the implications of the graph of Figure 7-7.

8. For a cavity with mirrors having reflectivities of 98%, what would the minimum gain need to be in order for the laser to reach threshold if the amplifier length is 0.2 m and the mirror separation is 0.4 m? Assume no losses in the cavity other than the mirror transmission losses.

9. (a) If you were to design a solid-state laser that reached I_{sat} when pumped, and if the gain medium (a solid-state crystal) were limited to a length of 10 cm, what would the minimum single-pass small signal exponential gain $(\sigma_{ul}\Delta N_{ul}L)$ have to be at line center if the gain duration lasts only 10 ns? Assume a rod diameter of 6 mm and that the mirrors are 100% reflecting and coated onto the ends of the laser rod.

(b) Assume the crystal has an index of refraction of 2.0, the linewidth of the homogeneously broadening medium is 100 nm, the laser wavelength at line center is 700 nm, and the radiative transition probability of the laser transition is 10^4 s^{-1}. What population density difference ΔN_{ul} would be required to produce the gain value of part (a)?

10. If a copper vapor laser is operated in the double-pass mode (only one mirror), what would the gain and population difference ΔN_{ul} have to be at 510.5 nm in order to reach I_{sat} at the line center in a double pass through the discharge? The gain region of the tube has a length of 0.1 m and a bore size of 25 mm. Assume the single mirror is installed immediately next to the gain medium at one end of the tube. Use the conditions and information of Problem 5(d) if needed. The copper vapor laser operates at a discharge temperature of approximately 1,600°C.

11. A He–Ne laser has a gain region of length 0.2 m, mirrors of reflectivity 99%, and scattering losses per pass of 0.5%. What is the threshold gain necessary to make a laser? What would be the population difference ΔN_{ul} required to reach that threshold? If the population difference were increased by a factor of 10 from those you determined for the threshold conditions, what intensity would be required in the cavity (compared to I_{sat}) to reduce the gain to the point that it would equal the losses in the cavity (the steady-state situation)?

8

REQUIREMENTS FOR OBTAINING POPULATION INVERSIONS

SUMMARY In Chapter 7 we showed that a population inversion between the upper and lower laser levels was essential in order to have amplification in a medium. We also showed that such an inversion was a necessary but not a sufficient condition for laser development. It was later shown how much gain is needed to make a laser, in terms of the value of the exponential growth factor $g(\nu)L$ or $\sigma_{ul}(\nu)\Delta N_{ul}L$ necessary to reach I_{sat}. This chapter will address the issue of obtaining population inversions in a more detailed manner. It will describe how inversions are created even though they seem to defy the concept of thermal equilibrium, a principle that we generally accept as an everyday fact of life. We will first show that it is not possible to

have an inversion between two levels that are the only two levels of the system. We will then describe how steady-state inversions are created in two types of three-level systems and also in the more general four-level system. Then we will consider transient inversions and show that their requirements are different from those of a steady-state system. At the end of the chapter, we will describe three types of processes that can prevent or destroy inversions. The first involves radiation trapping, the second involves collisional thermalization of the upper and lower laser levels, and the third is absorption within the gain medium on transitions other than those involving levels u and l.

8.1 INVERSIONS AND TWO-LEVEL SYSTEMS

Consider a hypothetical atom having just two levels u and l with an energy difference $\Delta E_{ul} = E_u - E_l = h\Delta\nu_{ul}$ separating the levels, where $\Delta\nu_{ul}$ is the frequency emitted from a radiative transition occurring between the levels. Assume that ν_{ul} is in or near the visible spectral region so that thermal excitation of the upper level u from the lower level l via collisions with other particles is negligible at room temperature. The question is whether or not an inversion could be created between those two levels. We assume a cell of dimension L containing atoms of total density N at room temperature, and that the cell is square shaped with optical-quality walls so that a beam could be transmitted through the cell without being distorted by the cell walls. The total decay rate from the upper level to the lower level is given by γ_{ul}, and the radiative decay from level u to level l occurs at a rate A_{ul}. Thus $\gamma_{ul} \geq A_{ul}$, depending upon whether collisions increase the decay rate from level u to level l above the rate of radiative decay. We will use the general form of $\sigma_{ul}(\nu)$ without specifying the dominance of either homogeneous or Doppler broadening, since the discussion is applicable to both types. We also shorten the expression for the stimulated emission cross section to σ_{ul}.

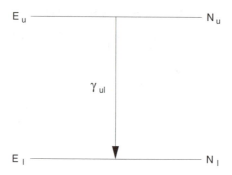

$$\Delta E_{ul} = E_u - E_l = h\nu_{ul}$$

Figure 8-1. Energy levels and decay rate of a two-level system

Figure 8-1 shows the two levels, with population moving from level u to level l at a rate γ_{ul}. Atoms in the cell are at room temperature so that – according to the Boltzmann distribution – nearly all of the atoms would be in the lower level l (the ground state) of this hypothetical atom, which is another way of saying that collisional or thermalizing excitation from level l to level u can be neglected. Thus, initially $N_l \cong N$. We will attempt to "pump" those atoms from level l to level u by shining light of intensity I_0 and of frequency ν_{ul} into the cell, and then examine the intensity I emerging from the opposite side after having passed through the cell. In this case the light would be absorbed by the atoms in level l, and for every absorbed photon, an atom would be promoted to level u. This process would occur according to the following equation which was derived in Chapter 7 (see eqns. 7.18 and 7.30) for two different types of broadening:

$$I = I_0 e^{\sigma_{ul}\Delta N_{ul}z} = I_0 e^{\sigma_{ul}(N_u - N_l)L}. \tag{8.1}$$

In the right-side of (8.1) we have assumed that $g_u/g_l = 1$ for simplicity, and have replaced the variable length z by the cell length L.

Because we assumed that the total population density of the combined two levels is N and that all of the population is initially in level l such that $N_l \cong N$, (8.1) can be written as

$$I = I_0 e^{-\sigma_{ul}NL}. \tag{8.2}$$

We can thus see from (8.2) that, as we increase the input light intensity I_0, the energy absorbed in the medium would increase and therefore, for very high intensities I_0, we might expect all of the population to be transferred from level l to level u by absorption of photons from the beam.

This is not the case, however, as is suggested by examining (8.1). As soon as population is pumped up to level u, N_u becomes greater than zero. Since $N_u + N_l = N$ (or $N_u = N - N_l$), we can rewrite (8.1) as

$$I = I_0 e^{\sigma_{ul}[1 - 2(N_l/N)]NL}. \tag{8.3}$$

Thus from (8.3) we can see that as the population leaves level l, the ratio N_l/N begins to drop from a value of unity, and when it reaches 0.5, no more energy will be absorbed since the value of the exponent in (8.3) will be reduced to zero. If no more energy can be absorbed, then there is no mechanism to increase the population in level u. Also, when $N_l/N = 0.5$, $N_u = N_l$ and since no further absorption can occur, N_u can never exceed N_l. Radiative decay (or whatever other decay process is dominant) will continually reduce the population in level u, which will also increase the population in level l and therefore lead to more absorption. But N_l/N will never decrease to a value lower than 0.5, so there will never be a population inversion.

This is a simple-minded argument. To analyze the situation more thoroughly, we could write out a set of rate equations by considering the flux of population associated with excitation and de-excitation of level u from level l. Not only would transitions from level l occur via absorption of the radiation at frequency ν_{ul}, but also transitions from level u to level l would occur via stimulated emission. However, as long as $N_u < N_l$, the net effect would be absorption of the beam in the cell.

A similar argument can be made if energetic electrons are present in the cell that could provide excitation from level l to level u. For the case of electrons, N_l/N will never be reduced to the value of 0.5 ($N_u/N_l = 1$), because the electrons will try to create a thermal distribution of the two populations according to the Boltzmann equation $N_u/N_l = e^{-\Delta E_{ul}/kT_e}$, where T_e is the electron temperature. It can be seen that this ratio can never have a value of unity except for an infinite electron temperature T_e. Although we are not addressing the issue of excitation techniques in this chapter, we can point out that it is the principle of detailed balance that leads to this Boltzmann distribution of N_u/N_l, with electrons colliding with atoms in either level l or level u. The more atoms that are excited to level u, the more collisions electrons make with those atoms in level u, returning them to level l. We have not calculated the additional effect of the radiative decay rate A_{ul}, which would only further decrease the ratio N_u/N_l. This in turn increases the ratio of N_l/N to a value somewhat greater than 0.5, such that – according to (8.3) – there would never be any gain.

We can thus conclude from this simple analysis that it is impossible to have a population inversion between levels u and l for a system with only those two energy levels. In the next section we will therefore add another level and consider inversions in two different types of three-level systems.

8.2 STEADY-STATE INVERSIONS IN THREE- AND FOUR-LEVEL SYSTEMS

We now consider two types of three-level systems and also a general four-level system, as diagrammed in Figure 8-2, that can provide steady-state population inversions. Steady-state conditions are those in which the pumping flux is constant and the populations are not changing with time, even though there is population moving in and out of the relevant levels.

The first three-level system we investigate will be one in which the upper laser level is the intermediate of the three levels; this is the arrangement in the ruby laser, the first laser. We will then consider the three-level scheme

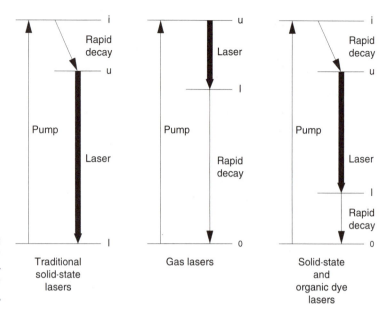

Figure 8-2. Energy-level arrangements and pumping and decay processes for three- and four-level lasers

in which the upper of the three levels is the upper laser level. This arrangement is used in many gas lasers, where the pumping requirements necessary to achieve sufficient gain to reach I_{sat} are much less stringent than for solid-state lasers. Population density requirements for N_u can be as low as 10^{14} per cubic meter for gas lasers but as high as 10^{25} per cubic meter for solid-state lasers, owing to their much smaller stimulated emission cross section. Although this requirement for higher N_u may at first seem a disadvantage for solid-state lasers, the much higher energy per unit volume stored in the upper laser level for such lasers makes up for the greater requirements on N_u, thus leading to *compact* high-power lasers.

THREE-LEVEL LASER WITH THE INTERMEDIATE LEVEL AS THE UPPER LASER LEVEL

For the traditional three-level arrangement, shown in Figure 8-3, we have levels i, u, l with populations N_i, N_u, N_l and $E_i > E_u > E_l$. We assume that the laser transition occurs from level u to level l (in this case the ground state). We also assume that the gain medium is in thermal equilibrium before we begin pumping from level l to level i at a rate Γ_{li}. In such a situation we have decay from level i to level u at a rate γ_{iu}, decay from level i to level l at a rate γ_{il}, and decay from the upper laser level u to the lower laser level l at a rate γ_{ul}, which can be made up of both the radiative decay rate and a collisional decay rate. We assume that decay rates γ_{iu} and γ_{il} both occur by thermalizing (collisional) processes in the solid-state gain medium. We assume no thermalizing excitation from the ground-state level l to levels u and i, since we assume that the energies of u and i are sufficiently high that such processes are very small. This can be shown by using the principle of detailed balance, in which

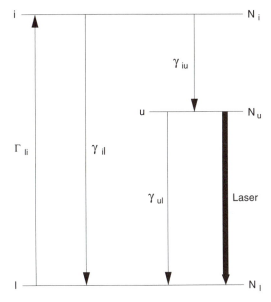

Figure 8-3. Energy levels and relevant excitation and decay processes for the traditional three-level laser

$$N_u \gamma_{ul} = N_l \gamma_{lu} \tag{8.4}$$

for the conditions of thermal equilibrium as described in Chapter 6. Using the Boltzmann relationship $N_u/N_l = e^{-\Delta E_{ul}/kT}$, again assuming $g_u = g_l$, we have

$$\frac{\gamma_{lu}}{\gamma_{ul}} = e^{-(E_u - E_l)/kT}, \tag{8.5}$$

which at room temperature indicates that γ_{lu} is very small for any value of E_u associated with visible and near-infrared lasers. We must realize that (8.5) is applicable in all conditions, and not just for the case of thermal equilibrium; we only assumed the conditions of thermal equilibrium to obtain the relationship.

We will now consider the rate equations for the population densities of levels i, u, and l by considering the flux entering and leaving each level per unit time:

$$\frac{dN_l}{dt} = -\Gamma_{li}N_l + \gamma_{ul}N_u + \gamma_{il}N_i = 0, \tag{8.6}$$

$$\frac{dN_u}{dt} = -\gamma_{ul}N_u + \gamma_{iu}N_i = 0, \tag{8.7}$$

$$\frac{dN_i}{dt} = \Gamma_{li}N_l - (\gamma_{il} + \gamma_{iu})N_i = 0. \tag{8.8}$$

We wish to consider steady-state solutions of these equations and so have equated them to zero. We use Γ to denote an externally applied pumping or excitation rate and γ to describe an excitation or decay process inherently associated with the medium. We have neglected γ_{lu}, γ_{li}, and γ_{ui} because of the implications of (8.5) and because the energy separations are much greater than kT. We have also assumed that no significant external pumping process occurs from level l to level u, which is typically true for this type of laser.

We can write the solutions for the populations of the upper and lower laser levels N_u and N_l in terms of the total population N, since $N_i + N_u + N_l = N$:

$$N_l = \frac{\gamma_{ul}(\gamma_{iu} + \gamma_{il})}{\gamma_{ul}(\gamma_{il} + \gamma_{iu}) + (\gamma_{ul} + \gamma_{iu})\Gamma_{li}} N, \tag{8.9}$$

$$N_u = \frac{\gamma_{iu}\Gamma_{li}}{\gamma_{ul}(\gamma_{il} + \gamma_{iu}) + (\gamma_{ul} + \gamma_{iu})\Gamma_{li}} N. \tag{8.10}$$

If we then consider whether or not a population inversion can occur, we investigate the ratio N_u/N_l (assuming that $g_u \cong g_l$) such that

$$\frac{N_u}{N_l} = \frac{\gamma_{iu}\Gamma_{li}}{\gamma_{ul}(\gamma_{iu} + \gamma_{il})} \overset{?}{>} 1; \tag{8.11}$$

we would thus have a population inversion if

$$\Gamma_{li} > \gamma_{ul}\left(1 + \frac{\gamma_{il}}{\gamma_{iu}}\right). \tag{8.12}$$

We can see that it is desirable to have the ratio γ_{il}/γ_{iu} be as small as possible in order to minimize the requirements on the pumping rate Γ_{li}. In fact, this is the case for most solid-state lasers, since collisional depopulation occurs much more frequently between nearby energy levels than between levels separated by a large energy. If γ_{il}/γ_{iu} is small, and the decay from level u to level l is primarily by radiative decay at a rate A_{ul}, then an inversion is produced if

$$\Gamma_{li} > A_{ul}. \tag{8.13}$$

Hence for this type of three-level laser, if the pumping rate Γ_{li} exceeds the radiative rate A_{ul} then the population will be transferred to level i and then to level u by rapid collisional decay, which eventually produces an inversion between level u and level l. Thus, for extremely high pumping rates, most of the population will reside in level u and thereby allow the production of large gains.

EXAMPLE

The first laser ever developed used the three-level pumping scheme. The laser was produced in a ruby crystal and was pumped by a flashlamp. A diagram of this system is shown in Figure 8-4. The energy levels for this laser were also discussed in Chapter 5. Pumping is achieved by absorption of the pump light on transitions from the ground level to the $^4F_1$ excited level (also known as $^4T_1$) in the blue spectral region

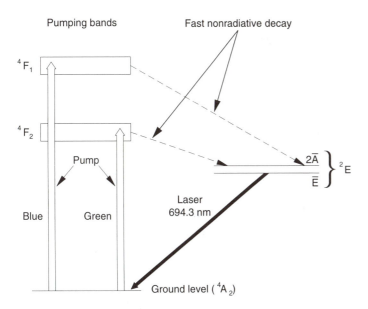

Figure 8-4. Energy-level diagram of the three-level ruby laser

and to the $^4F_2$ excited level (also known as $^4T_2$) in the green. Both of these transitions have a broad absorption spectrum, allowing a broad-spectrum pumping lamp to be used with reasonable efficiency. The population pumped to these excited states rapidly decays to the ^2E state, where a population inversion builds up on the $\bar{E}$ upper laser level with respect to the ground state $^4A_2$, and laser action occurs on the radiative transition between those two states.

We will refer to the broad excited levels of Figure 8-4 as level i in (8.11). In the ruby laser crystal $\gamma_{iu} \gg \gamma_{il}$, since the collisional re-laxation or phonon relaxation occurs preferentially to the energet-ically nearby upper laser level rather than to the ground state. Also, the decay from the upper laser level is primarily due to radiative de-cay; we can therefore infer that $\gamma_{ul} \cong A_{ul} \cong 1/\tau_u$, where τ_u is the life-time of the upper laser level. For ruby, $\tau_u \cong 3$ ms and thus from (8.13) we have the threshold pumping condition that $\Gamma_{li} > A_{ul} = 1/\tau_u = 1/(3 \times 10^{-3}$ s$) = 3.33 \times 10^2$ s^{-1}. For a population density of N_l initially in the ground state, we can see from (8.6) that the pumping flux (pho-tons per cubic meter per second) is just $N_0\Gamma_{li}$. For a doping concen-tration of 0.05% by weight (see Section 5.3), $N_0 \approx 1.6 \times 10^{25}$/m^3 and thus $N_0\Gamma_{li} > 5.33 \times 10^{27}$/m^3-s is the pumping flux necessary to reach a population inversion *threshold*. For significant laser output to occur, the inversion must of course be higher than that. This is an extremely high pumping flux. Problem 2 at the end of this chapter involves cal-culating how many pump photons and how much pump power are required to reach the laser pumping threshold in a typical ruby laser rod over a duration of the upper laser level lifetime.

The ruby laser has been a very useful laser. However, in recent years it has been overshadowed by more efficient solid-state lasers using four-level systems, since the pumping requirements for those systems are much lower. The four-level system will be described after we consider the other type of three-level system.

THREE-LEVEL LASER WITH THE UPPER LASER LEVEL AS THE HIGHEST LEVEL

We have seen how to create a population inversion in a three-level system when the upper laser level is not the highest level in that system. We will now investigate such an inversion when the upper laser level is the highest in the three-level arrangement. Consider the energy-level diagram shown in Figure 8-5, with the highest level as the upper laser level u, the intermediate level as the lower laser level l, and the lowest level (or ground state) of that species as level 0. We allow a general pumping process that provides flux to both levels u and l at rates Γ_{0u} and Γ_{0l} from the ground state 0, as shown in the figure. We assume population densities N_u, N_l, and N_0, and assume that N_0 is nearly equal to the total population N of the system before pumping of the higher levels is initiated. This is approximately true for atomic systems (gas lasers), since ΔE_{l0} and ΔE_{u0} are generally much greater than kT for the system.

The pumping flux from level 0 to levels u and l is given by $N_0\Gamma_{0u}$ and $N_0\Gamma_{0l}$, respectively. We have allowed levels u and l to decay by rates γ_{ul}, γ_{u0}, and γ_{l0}. We also assume no significant thermal excitation, and so γ_{0u}, γ_{0l}, and γ_{lu} are neglected.

With these additional conditions, we can write the rate equations for the flux entering and leaving levels u and l as

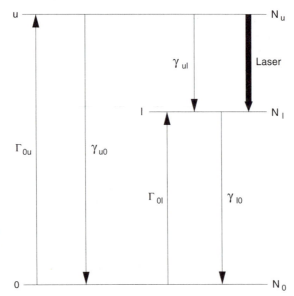

Figure 8-5. Energy-level diagram and relevant excitation and decay processes of an atomic three-level laser system

$$\frac{dN_u}{dt} = N_0\Gamma_{0u} - N_u(\gamma_{ul} + \gamma_{u0}) = 0, \tag{8.14}$$

$$\frac{dN_l}{dt} = N_0\Gamma_{0l} + N_u\gamma_{ul} - N_l\gamma_{l0} = 0. \tag{8.15}$$

For the steady-state solution of N_u and N_l, we equate the right-hand side of both equations to zero to obtain

$$N_u = \frac{N_0\Gamma_{0u}}{\gamma_{ul} + \gamma_{u0}}, \tag{8.16}$$

$$N_l = \frac{N_0\left[\Gamma_{0l} + \dfrac{\Gamma_{0u}\gamma_{ul}}{\gamma_{ul} + \gamma_{u0}}\right]}{\gamma_{l0}}. \tag{8.17}$$

Again, we investigate the possibility of obtaining a population inversion between level u and l by examining the ratio N_u/N_l and assuming that $g_u = g_l$ for simplicity. From (8.16) and (8.17) we obtain

$$\frac{N_u}{N_l} = \frac{\Gamma_{0u}\gamma_{l0}}{\gamma_{u0}\Gamma_{0l} + \gamma_{ul}[\Gamma_{0l} + \Gamma_{0u}]} \overset{?}{>} 1. \tag{8.18}$$

We consider a situation that often occurs in atomic systems, where $\gamma_{ul} = A_{ul}$, $\gamma_{u0} = A_{u0}$, and $\gamma_{l0} = A_{l0}$. This is equivalent to saying that collisional decay processes are negligible compared to radiative decay. Equation (8.18) could then be expressed as

$$\frac{N_u}{N_l} = \frac{\Gamma_{0u}A_{l0}}{A_{ul}(\Gamma_{0l} + \Gamma_{0u}) + A_{u0}\Gamma_{0l}} \overset{?}{>} 1. \tag{8.19}$$

It is evident that a population inversion can be obtained if the decay from level l is significantly greater than the decay from level u, provided that the pumping to level l is not highly favored over that to level u. For an atomic system in which A_{l0} is large, A_{u0} would be very small because levels u and 0 would have to be of the same parity (see the selection rules listed in Section 4.4). We can therefore rewrite (8.19) in an approximate form as

$$\frac{N_u}{N_l} \cong \frac{1}{(1 + \Gamma_{0l}/\Gamma_{0u})} \frac{A_{l0}}{A_{ul}} \overset{?}{>} 1. \tag{8.20}$$

Thus we can see that, for inversions to occur, the ratios of A_{l0}/A_{ul} and Γ_{0l}/Γ_{0u} must be favorable in some combination as indicated here. For example, if $\Gamma_{0l}/\Gamma_{0u} = 1$ then A_{l0} must be greater than $2A_{ul}$. In any case it is most desirable to have a fast decay out of the lower laser level, and a higher pumping flux to the upper laser level.

EXAMPLE

Let us now apply (8.20) to the 441.6-nm He–Cd laser transition in the Cd^+ ion. Figure 8-6 shows the three relevant energy levels of that ion with the radiative decay rates listed for each transition. The relevant transition probabilities are $A_{ul} = 1.4 \times 10^6$, $A_{u0} \approx 0$, and $A_{l0} = 2.8 \times 10^8$ per second. Thus, using (8.20), the ratio N_u/N_l becomes

$$\frac{N_u}{N_l} = \frac{200}{1 + \Gamma_{0l}/\Gamma_{0u}} \stackrel{?}{>} 1. \tag{8.21}$$

Thus, an inversion can be obtained unless the pumping flux from the ground level 0 to level l exceeds 199 times the pumping flux to level u, that is, unless $\Gamma_{0l} > 199\Gamma_{0u}$. One means of exciting the He–Cd laser has been shown to occur by a process that could be modeled after this three-level system, where electron collisions provide the Γ_{0u} pumping flux to the upper laser level from the Cd^+ ion ground state (see Section 13.3).

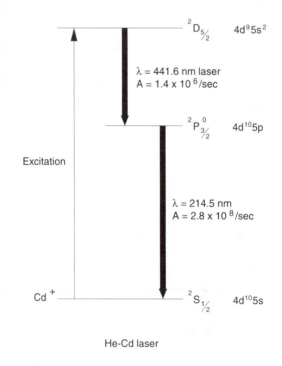

Figure 8-6. Energy-level diagram of a three-level He–Cd laser

FOUR-LEVEL LASER

We now consider an arrangement similar to that of the traditional three-level system (Figure 8-3) but with the level 0 added below the lower laser

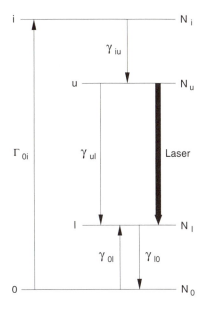

Figure 8-7. Energy-level diagram and relevant excitation and decay processes of a four-level laser system

level l as shown in Figure 8-7. This arrangement is typical of many solid-state lasers. The level 0 is the ground state, and the majority of the atoms are initially in that level before pumping occurs. In this scheme, we neglect the upward rates γ_{0u}, γ_{0i}, γ_{lu}, γ_{li}, and γ_{ui} because of the large energy separation and the consequences of (8.5). The specific downward rates γ_{il}, γ_{i0}, and γ_{u0} are also neglected since they are very small in solid-state laser crystals of this type, owing to the large energy separations of the specific levels. We will assume that the dominant decay rate from level u to level l will most likely be radiative, but we must keep in mind that $\gamma_{ul} = A_{ul} + 1/T_1$ (as discussed in Section 4.3) so if T_1 is short then this situation could be significantly altered. We described in Section 4.1 how the lifetime of the upper laser level of dielectric solid-state lasers can be shortened by collisions that are detrimental to producing inversions. We have also allowed for thermal excitation and de-excitation of level l from level 0 by including the rates γ_{0l} and γ_{l0}. Although these rates may not be very large, they can still play a role in the excitation process, as will be seen from the example at the end of this section.

As we did for the three-level system described previously, we can now write the rate equations as follows:

$$\frac{dN_l}{dt} = \gamma_{0l} N_0 - \gamma_{l0} N_l + \gamma_{ul} N_u = 0, \tag{8.22}$$

$$\frac{dN_u}{dt} = -\gamma_{ul} N_u + \gamma_{iu} N_i = 0, \tag{8.23}$$

$$\frac{dN_i}{dt} = \Gamma_{0i} N_0 - \gamma_{iu} N_i = 0. \tag{8.24}$$

Because $N_0 + N_l + N_u + N_i = N$, where N is the total number of laser species (dopant ions) per unit volume and is therefore constant, we can easily show by differentiation that

$$\frac{dN_0}{dt} = -\frac{dN_l}{dt} - \frac{dN_u}{dt} - \frac{dN_i}{dt}. \tag{8.25}$$

From these equations we can solve for N_u and N_l in terms of N and obtain

$$N_u = \frac{\gamma_{iu}\Gamma_{0i}}{\gamma_{ul}\gamma_{iu}} N_0 = \frac{\Gamma_{0i}}{\gamma_{ul}} N_0, \tag{8.26}$$

$$N_l = \left[\frac{\gamma_{0l}}{\gamma_{l0}} + \frac{\Gamma_{0i}}{\gamma_{l0}}\right] N_0 = \left[\frac{(\gamma_{0l} + \Gamma_{0i})}{\gamma_{l0}}\right] N_0. \tag{8.27}$$

Again we consider the conditions for a population inversion by taking the ratio N_u/N_l, assuming again for simplicity that $g_u = g_l$. We are thus led to the ratio

$$\boxed{\frac{N_u}{N_l} = \frac{\gamma_{l0}\Gamma_{0i}}{\gamma_{ul}[\gamma_{0l} + \Gamma_{0i}]} \overset{?}{\gtrless} 1.} \tag{8.28}$$

Therefore a population inversion will occur for a pumping flux Γ_{0i} if

$$\boxed{\Gamma_{0i} > \frac{\gamma_{0l}\gamma_{ul}}{(\gamma_{l0} - \gamma_{ul})}.} \tag{8.29}$$

Because γ_{l0} is typically much greater than γ_{ul} (owing to the energy-level separation), we can approximate (8.29) as

$$\boxed{\Gamma_{0i} > \frac{\gamma_{0l}\gamma_{ul}}{\gamma_{l0}} = e^{-\Delta E_{l0}/kT}\gamma_{ul},} \tag{8.30}$$

where we have used an argument similar to that used for (8.5) to obtain the Boltzmann relationship for the ratio γ_{0l}/γ_{l0}. If we compare this result to (8.12) and (8.13) we can see that the pumping requirements of the four-level system are significantly reduced, by the factor $e^{-\Delta E_{l0}/kT}$ compared to the traditional three-level system, since ΔE_{l0} is typically much greater than kT and thus $e^{-\Delta E_{l0}/kT} \ll 1$ for most solid-state laser crystals.

EXAMPLE

We will compare the threshold pumping flux required for the four-level Nd:YAG laser with that of the ruby laser, which uses the three-level pumping scheme. The energy-level diagram for Nd:YAG is shown in Figure 8-8. Excitation occurs via optical pumping from the $^4I_{9/2}$ ground state to a band of excited states that we will refer to as level i. Nonradiative decay then occurs very rapidly to the $^4F_{3/2}$ upper laser level u with $\gamma_{iu} \approx 10^{12}$ to 10^{14} per second. The upper laser level decays primarily radiatively and has a lifetime of 230 μs such that $\gamma_{ul} \cong A_{ul} = 1/\tau_u = 1/(2.3 \times 10^{-4}\text{ s}) = 4.35 \times 10^3 \text{ s}^{-1}$. Using (8.30) for the threshold condition, we need values for the energy separation between the lower

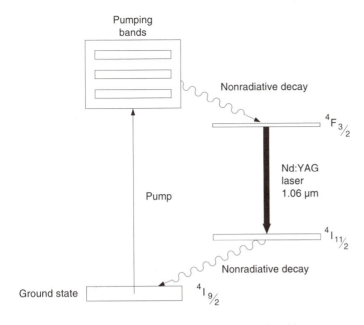

Figure 8-8. Energy-level diagram of a Nd:YAG four-level laser

laser level and the ground state as well as the temperature of the rod. For this case $\Delta E_{l0} = 0.25$ eV, and we will assume the laser crystal is at room temperature (approximately 300 K).

Putting these values into (8.30), we find that

$$\Gamma_{0i} > e^{-\Delta E_{l0}/kT}\gamma_{ul} = e^{-0.25/(8.6 \times 10^{-5} \times 300)}(4.35 \times 10^3 \text{ s}^{-1})$$
$$= e^{-9.7}(4.35 \times 10^3 \text{ s}^{-1}) \cong 0.265.$$

If we compare this pumping rate to that obtained for ruby in the previous example, we see that the pumping rate is reduced by a factor of $333/0.265 = 1,257$! Thus, even though the transition probability of the Nd:YAG is significantly higher than that of the ruby laser and would therefore increase the pumping threshold, the reduction due to the exponential factor far outweighs this increase. Hence we have a much lower threshold pumping rate for a four-level system than for a three-level system.

8.3 TRANSIENT POPULATION INVERSIONS

We have so far considered only steady-state solutions to the populations of two-, three-, and four-level systems. We will now investigate the possibility of a transient inversion in a three-level system similar to that shown in Figure 8-5, where the upper level is the upper laser level u and the intermediate level is the lower laser level l. We will allow pumping to both the upper and lower laser levels from level 0. We have seen how rapid decay

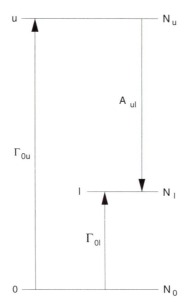

Figure 8-9. Energy-level diagram and relevant excitation and decay rates for a transient three-level laser

from the lower laser level promotes a population inversion under steady-state conditions in both three- and four-level lasers. For the transient situation, we will choose the extreme case in which there is no decay from the lower laser level. Although we realize that all levels above the ground state eventually decay, we are assuming that the decay from level l is so slow that it can be neglected for the time period under consideration. We also assume that there is no decay from level u to level 0. This approximates a type of gas laser in which the decay from level u to level 0 is prevented by a process known as *radiation trapping*. We will discuss radiative trapping in Section 8.4, but for now we will simply assume its existence and hence $\gamma_{u0} = 0$. We can therefore assume that $\tau_u = 1/\gamma_u \cong 1/\gamma_{ul} = 1/A_{ul}$, which suggests that there is no collisional decay and that the only radiative decay is from level u to level l at a rate A_{ul}.

We will use the energy-level diagram of Figure 8-9 as a model for this analysis. The rate equation for the change in population density of levels u and l, when taking into account the constraints mentioned previously, can be written as

$$\frac{dN_u}{dt} = N_0 \Gamma_{0u} - N_u A_{ul}, \tag{8.31}$$

$$\frac{dN_l}{dt} = N_0 \Gamma_{0l} + N_u A_{ul}. \tag{8.32}$$

Since we are not considering steady-state conditions, we cannot equate these expressions to zero as in the other situations. The solutions for the populations N_u and N_l are given as

$$N_u = \frac{N_0 \Gamma_{0u}}{A_{ul}} (1 - e^{-A_{ul}t}), \tag{8.33}$$

$$N_l = \frac{N_0 \Gamma_{0u}}{A_{ul}} \left[\left\{ \left(\frac{\Gamma_{0l}}{\Gamma_{0u}} \right) + 1 \right\} A_{ul} t - (1 - e^{-A_{ul}t}) \right]. \tag{8.34}$$

It can be seen from (8.33) that the value of N_u approaches $N_u = (\Gamma_{0u}/A_{ul})N_0$ when $t \gg 1/A_{ul}$, which is just the steady-state solution of (8.31). The value of N_l, on the other hand, just continues to increase with time (eqn. 8.34) because there is no decay from level l.

We again consider the conditions necessary to produce a population inversion between levels u and l by taking the ratio of N_u/N_l. We have assumed that $g_u = g_l$ for simplicity, and have (again) used the approximation $\tau_u \cong 1/A_{ul}$ to obtain

$$\frac{N_u}{N_l} = \frac{1 - e^{-t/\tau_u}}{\left[\left(\frac{\Gamma_{0l}}{\Gamma_{0u}} \right) + 1 \right] \frac{t}{\tau_u} - (1 - e^{-t/\tau_u})} \overset{?}{>} 1. \tag{8.35}$$

We plot the ratio N_u/N_l in units of τ_u in Figure 8-10, since the lifetime of the upper laser level τ_u is the relevant time factor for this situation. In plotting the ratio, we have chosen three different values of Γ_{0l}, the pumping rate into level l in terms of Γ_{0u}, such that $\Gamma_{0l} = 0$, $\Gamma_{0l} = \Gamma_{0u}/2$, and $\Gamma_{0l} = \Gamma_{0u}$. We can see that there is indeed a temporary inversion for the first two situations. This inversion lasts longer for $\Gamma_{0l} = 0$, as might be anticipated since it would take longer for level l to fill up and thereby terminate the gain when there is no pumping into that level from the ground state (level 0). In that case, the only population flowing into level l comes from level u. Of course, when laser action occurs, and the gain is high enough to exceed the saturation intensity on the transition ν_{ul}, then the stimulated rate $B_{ul} I(\nu)/c$ from level u to level l will exceed the spontaneous emission rate A_{ul}. The

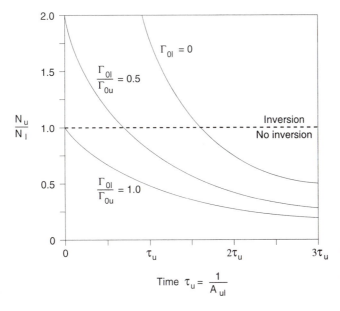

Figure 8-10. Population density ratio N_u/N_l versus reduced time τ_u for a three-level transient inversion

gain can therefore be extinguished faster via stimulated emission than as shown in Figure 8-10 where stimulated emission is not included.

EXAMPLE

Relevant energy levels of three lasers that can be described by the preceding equations are shown in Figure 8-11. These are the copper vapor

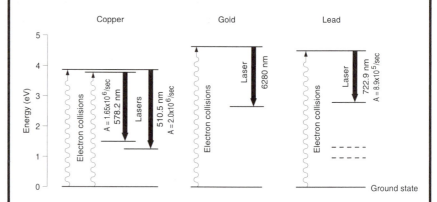

Figure 8-11. Examples of transient three-level lasers

laser (CVL), the gold vapor laser, and the lead vapor laser. For these lasers, decay to the other lower levels (other than the lower laser level), such as to the ground state or to the two dashed levels in the case of the lead laser, is prevented by radiation trapping effects enabled by the conditions under which these lasers typically operate. (Radiation trapping is described in detail in Section 8.4.) There may appear to be other possible routes via which the upper laser level can decay radiatively, but such processes do not occur: the upper laser levels effectively decay only to the lower laser levels. In the case of the copper vapor laser there are two laser transitions, making the analysis slightly more complex than that of equations (8.31)–(8.34), but in principle all three of these lasers work the same way. The effective decay times for their upper laser levels, ignoring decay to levels other than the lower laser level, are simply the reciprocal of the radiative transition probabilities, shown in Figure 8-11 for the CVL and the lead laser. Thus, for example, for the 510.5-nm CVL transition $\tau_u = 1/(2 \times 10^6 \text{ s}^{-1}) = 5 \times 10^{-7}$ s.

8.4 PROCESSES THAT INHIBIT OR DESTROY INVERSIONS

We will now describe three phenomena that can prevent population inversion from occurring. The first prevents the lower laser level from undergoing its normal decay process in a gas laser; this is known as *radiation trapping*. The second, also applicable to gas or plasma lasers, tends to reduce the population of the upper laser level and move it to the lower laser level via collisions; this is referred to as *electron collisional thermalization* of the laser levels. The third is *absorption within the gain medium* by various processes present within the medium that are not associated with levels u or l.

RADIATION TRAPPING IN ATOMS AND IONS

Radiation trapping is associated with the inhibition of radiative decay of the lower laser level. Referring to the analysis of the three-level system shown in Figure 8-5, we indicated in equations (8.18)–(8.20) that the inversion between levels u and l is produced primarily when there is a high decay rate A_{l0} from level l. Radiation trapping becomes a relevant factor when the population density in level 0 becomes high enough that every photon radiated due to a transition from level l to level 0 is immediately reabsorbed by a nearby atom in level 0, thereby effectively preventing the decay of population in level l. This phenomenon is diagrammed in Figure 8-12. For the case of Figure 8-12(a), two identical atoms are shown near each other. One has an electron in level l and the other in level 0. When the electron in level l decays to level 0 and emits a photon, that photon leaves the atom and, in the first case, escapes from the gain medium, thereby effectively reducing the population in level l. In the second case of Figure 8-12(b), where N_0 is much higher and consequently the absorption coefficient α_{0l} (for the transition from level 0 to level l) is also much higher, the photon is

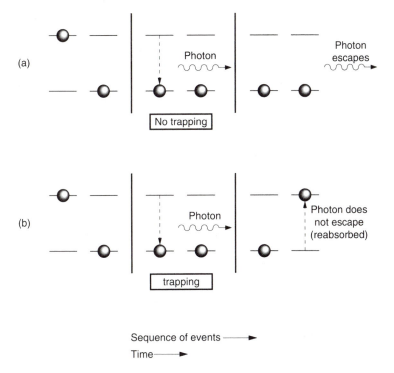

Figure 8-12.
Radiation trapping

reabsorbed by a nearby atom and so leaves that atom in level l, effectively allowing no decay to take place. This reabsorption effect is dependent upon the density N_0 of level 0 and it is therefore desirable to keep the total density of level 0 as low as possible. We must remember that level 0 is often the source of pumping to level u, so we cannot always reduce it arbitrarily without reducing the population in level u. The expression for the absorption coefficient was originally derived as (7.31) and (7.32), but we have rewritten it here with the levels u and l replaced by levels l and 0 (since we are dealing with the transition from the lower laser level l to level 0, rather than the transition from level u to level l):

$$I = I_0 e^{\sigma_{l0}[N_l - (g_l/g_0)N_0]z}. \tag{8.36}$$

For most situations in which radiation trapping is applicable, $N_l/N_0 \ll 1$, as suggested by the Boltzmann expression of (6.10), so we can neglect N_l in this formula. Therefore, in considering (7.31) and (7.32), we are finally left with the following equation for the absorption at frequency ν_{l0} in the medium:

$$I \cong I_0 e^{-\sigma_{l0}(g_l/g_0)N_0 z} = I_0 e^{-\sigma_{0l}N_0 z} = I_0 e^{-\alpha_{0l}z}. \tag{8.37}$$

We recall from Section 7.1 that this equation is known as Beer's law and is used extensively by chemists and others to measure quantities such as the concentration of N_0 in various materials.

We would like to estimate when this effect of radiation trapping is applicable. As shown in Figure 8-12, if the photon can escape the medium then the decay will have taken place and the population of level l will have been reduced, allowing an inversion between levels u and l. If the photon does not escape the medium, effectively causing the retention of population in level l, then the population of level l remains high and no inversion occurs. The critical factor is whether or not the absorption occurs before the photon escapes.

We thus consider the shortest path over which the photon might have an opportunity to escape from the gain medium. For a cylindrical gain medium, which is usually the case for gaseous or plasma lasers (for which this effect is applicable), the shortest distance over which the photon can travel to escape from the center of the medium is the radius b of the cylinder. Thus we can substitute b for z in (8.37) and so can estimate when significant absorption occurs by setting the exponent equal to unity ($\sigma_{0l}N_0 b = 1$), thereby suggesting that

$$b \approx 1/\sigma_{0l}N_0. \tag{8.38}$$

We will obtain a more accurate expression for $\sigma_{0l}N_0 b$ in what follows. But first, in considering (8.38) we can see that the effect of radiation trapping is not only dependent upon the absorption coefficient σ_{0l}, which has a specific value for a given transition, but is also (and strongly) dependent upon the value of the density N_0 of the level 0 to which the lower laser level decays. If this level 0 is the ground state of the neutral atom, then N_0 is related to the gas pressure. For a gas at a pressure of 1 torr (1/760 of an atmosphere), the density at room temperature is approximately 3.3×10^{22} per cubic meter; thus, if the level 0 is such a ground state then N_0 would have that density.

Radiation trapping can be incorporated into the rate equations that determine the population densities of the levels associated with the laser transition. If the rate equation for the change in population of the lower laser level is written as in (8.15), but we assume that only radiative decay is relevant, then

$$\frac{dN_l}{dt} = N_0 \Gamma_{0l} + N_u A_{ul} - N_l A_{l0}. \tag{8.39}$$

We will now rewrite this equation to incorporate a *trapping factor* $F_{l0} \leq 1$. This factor will effectively reduce the decay rate A_{l0} from level l for values of $F_{l0} < 1$, as shown in the following revised equation for the population entering and leaving level l:

$$\frac{dN_l}{dt} = N_0 \Gamma_{0l} + N_u A_{ul} - N_l F_{l0} A_{l0}. \tag{8.40}$$

It has been shown that, for a cylindrically shaped gain medium of radius b, F_{l0} can be defined as

$$F_{l0} = \frac{1.60}{\sigma_{0l} N_0 b (\pi \ln[\sigma_{0l} N_0 b])^{1/2}}. \tag{8.41}$$

The relationship between σ_{l0} and σ_{0l} for a specific transition was given in (8.37), as well as in (7.32). For Doppler broadening, the type of broadening usually associated with radiation trapping, the value of σ_{0l} is given by

$$\sigma_{0l} = \left(\frac{g_l}{g_0}\right) \sigma_{l0} = \left(\frac{g_l}{g_0}\right) \sqrt{\frac{\ln 2}{16\pi^3}} \frac{\lambda_{l0}^2 A_{l0}}{\Delta \nu^D}, \tag{8.42}$$

as obtained from (7.28) and (7.32) by substituting level l for level u and level 0 for level l (since those expressions are valid for any transition and not just laser transitions). Thus the trapping factor F_{l0} can be estimated for a specific radiative transition, gain medium of known dimensions, and gas density. This will allow determination of the maximum density of the species N_0 before the inversion density is reduced or eliminated. Quite often, this level 0 is the ground state of the laser species, whether it be a neutral or an ionic species. In some situations the maximum density N_0 that is possible without encountering radiation trapping might be low enough ($\approx 10^{-2}$ torr) to prevent the operation of the gas discharge. Thus, another gas might have to be added to the medium to operate the discharge. In this case, one should ensure the additional gas does not have an absorbing transition at the lasing wavelength λ_{ul} of the laser species, or at the wavelength λ_{l0} associated with the decay of the lower laser level. (However, the latter is highly unlikely for gaseous lasers, since there are not many transitions in a typical spectrum and the emission linewidths are relatively narrow.)

Equation (8.41) is not an easy formula to work with. For example, for values of $\sigma_{0l}N_0b < 1$, F_{l0} becomes negative and therefore meaningless in this context. A relevant expression for predicting the onset of radiation trapping is the value of $\sigma_{0l}N_0b$ at which F_{l0} is equal to unity, since we have ranged the values of F_{l0} as $0 < F_{l0} < 1$. Therefore, any value of F_{l0} above unity is meaningless. That minimum value of $\sigma_{0l}N_0b$ at which F_{l0} is unity – and therefore at which radiation trapping begins to increase N_l – can be shown to be

$$\sigma_{0l}N_0b = 1.46. \tag{8.43}$$

Thus, any value for $\sigma_{0l}N_0b$ that exceeds 1.46 will suggest that trapping has begun to reduce F_{l0} and hence also the decay rate of the lower laser level l. When $\sigma_{0l}N_0b$ is estimated, (8.41) can be used to determine the trapping factor F_{l0}; then F_{l0} can be used in equations of the form of (8.40) to see if the inversion has been reduced or eliminated.

EXAMPLE

For hydrogen atoms, determine the ground-state density N_0 at which radiation trapping would begin to occur on the $2 \rightarrow 1$ transition to prevent the development of an inversion on the $3 \rightarrow 2$ transition.

If we assume that the laser transition is from $n = 3$ to $n = 2$ then (from our three-level model) level l is the $n = 2$ level and level 0 is the $n = 1$ level, or the ground state of atomic hydrogen; see Figure 8-5. We are interested in estimating the density N_0 at which radiation trapping would slow down the decay of level l to level 0. We are given the transition probability A_{l0}, which is necessary in order to calculate σ_{0l}. We will consider the effect for discharge tubes with bores of radius 1 mm (0.001 m) and 5 mm (0.005 m). We assume a temperature in the gas discharge of 500 K. Using these values, σ_{0l} can be determined from (8.42) to be 6.6×10^{-18} m^2. Using this value in (8.43), we find that F_{l0} becomes less than unity for $N_0 < 2.2 \times 10^{20}$ m^{-3} for the 1-mm–bore radius and $N_0 < 4.4 \times 10^{19}$ m^{-3} for the 5-mm–bore radius.

Thus, to make a laser on the $n = 3 \rightarrow n = 2$ transition of hydrogen at 656.3 nm, we would want to operate a discharge with a partial pressure of hydrogen of no greater than 0.007 torr for the 1-mm–bore radius discharge tube and no greater than 0.0013 torr for the 5-mm–bore tube; these values would yield the greatest possibility of an inversion. Greater pressures would reduce the inversion ratio below that dictated by the natural decay rates of the levels.

ELECTRON COLLISIONAL THERMALIZATION OF THE LASER LEVELS IN ATOMS AND IONS

In gaseous discharges or plasmas in which population inversions are produced, free electrons in the discharge are the most common form of excitation of the laser levels. The pumping electrons (discharge electrons) are referred to as *free electrons,* since they are not attached to the atoms. These free electrons collide with atoms and excite the atoms to higher energy levels. Details of electron excitation processes will be summarized in the next chapter. For now, we will show that the presence of too many pumping electrons in the laser gain medium is potentially detrimental to making an inversion, since the electrons can also de-excite levels. Thus, too many collisions between electrons and atoms within a specific time period will lead to *thermalization* of the populations of the upper and lower laser levels.

Assume that a population inversion exists between levels u and l. For a given electron temperature T_e, which in a gas discharge is one or two orders of magnitude above room temperature, there is a certain electron density n_e (number of free electrons per unit volume) at which electrons colliding with population in the upper laser level u will significantly de-excite or reduce the population density N_u of that level (collisional decay), transferring that population to the lower laser level l in an attempt to produce a thermalizing Boltzmann distribution of the form $N_u/N_l = e^{-\Delta E_{ul}/kT_e}$ between those levels. Such a distribution is far from that of a population inversion and is therefore undesirable.

We can make an estimate of that detrimental collisional process by considering the electron collisional decay rate from level u to level l and comparing it to the radiative decay rate. Then, just as we did in deriving the saturation intensity I_{sat} (eqn. 7.41), we can determine at what electron density the collisional decay rate would be equal to the radiative decay rate between levels u and l; this density is denoted n_e^{max}. We must then restrict the electron density to values at or below n_e^{max} if a population inversion is desired.

Consider the energy-level diagram of Figure 8-13. It includes the radiative decay rate A_{ul} from level u to level l, as well as the electron collisional decay rate $n_e k_{ul}$, which also has units of 1/time. The k_{ul} term is a

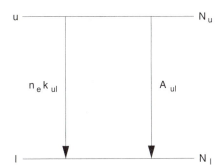

Figure 8-13. Relevant decay rates from level u when electron collisional depopulation of the level competes with radiative decay

rate constant associated with the collisional process. Thus, since the units of n_e are $1/\text{length}^3$, the units of k_{ul} are $\text{length}^3/\text{time}$. The flux downward from u to l by radiative decay would thus be

$$N_u A_{ul}, \tag{8.44}$$

and the flux downward due to electron collisions is given by

$$N_u n_e k_{ul}. \tag{8.45}$$

Hence we define $n_e^{\max}$ by equating the flux caused by radiative decay to the flux caused by electron collisional decay, in a similar way to the derivation of I_{sat}. That is,

$$N_u A_{ul} = N_u n_e^{\max} k_{ul}. \tag{8.46}$$

Any value of n_e higher than $n_e^{\max}$ would tend to destroy the inversion, by reducing N_u and increasing N_l. From (8.46) we obtain

$$n_e^{\max} = \frac{A_{ul}}{k_{ul}}. \tag{8.47}$$

We must therefore obtain an expression for k_{ul}. A formula, developed by Seaton (1962), provides an expression for the electron collisional excitation rate from any lower level l to an upper level u for hydrogen-like ions (ions that have energy levels similar to that of hydrogen):

$$k_{lu} = (7.8 \times 10^6) \frac{\lambda_{ul}^3 A_{ul}}{T_e^{1/2}} \left(\frac{g_u}{g_l} \right) e^{-\Delta E_{ul}/kT_e} \ \text{m}^3/\text{s}, \tag{8.48}$$

where T is in degrees Kelvin, λ_{ul} is in meters, and A_{ul} has units of $1/\text{sec}$onds. We will consider this formula in Section 9.4, when we discuss electron collisional excitation as a pumping or excitation mechanism for producing population inversions. For now, we need an expression not for k_{lu} but rather for k_{ul}.

Using the principle of detailed balance for conditions of thermal equilibrium, $N_u k_{ul} = N_l k_{lu}$ or $N_u/N_l = k_{lu}/k_{ul}$. For thermal equilibrium we also obtain the ratio of N_u/N_l as

$$\frac{N_u}{N_l} = \left(\frac{g_u}{g_l} \right) e^{-\Delta E_{ul}/kT} = \frac{k_{lu}}{k_{ul}}. \tag{8.49}$$

From this formula, involving the ratio of the upward rate k_{lu} to the downward rate k_{ul}, we obtain

$$k_{ul} = \left(\frac{g_l}{g_u} \right) \frac{k_{lu}}{e^{-\Delta E_{ul}/kT}}. \tag{8.50}$$

Using the expression in (8.48) for k_{lu} leads to

$$k_{ul} = 7.8 \frac{\lambda_{ul}^3 A_{ul}}{T^{1/2}}. \tag{8.51}$$

We can thus use this value for k_{ul} in (8.47) to obtain an expression for $n_e^{\max}$ of

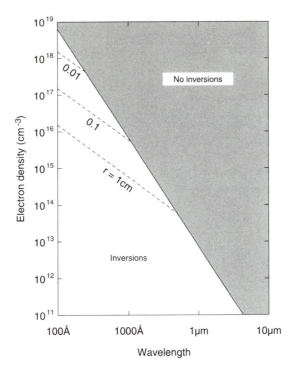

Figure 8-14. Graph of electron density versus laser wavelength, depicting the region where population inversions are possible

$$n_e^{\max} = \frac{0.13\sqrt{T_e}}{\lambda_{ul}^3} \; \text{m}^{-3}, \tag{8.52}$$

where T_e is in degrees Kelvin and λ_{ul} is the transition wavelength in meters.

A graph of the wavelength regions where inversions are allowed at various values of $n_e^{\max}$ is shown in Figure 8-14. The dashed lines show even more stringent requirements that are added by considerations of radiation trapping, where b is the minimum dimension of the gain region as outlined in the previous section.

EXAMPLE

As a specific example, we can consider the maximum electron density for the argon ion laser operating at 488.0 nm. The electron temperature in such a discharge is of the order of 100,000 K (the electron temperature can be much higher than the ion temperature in a gas discharge). Thus the maximum electron density can be determined from (8.52) as

$$n_e^{\max} = \frac{0.13\sqrt{T_e}}{\lambda_{ul}^3} = \frac{0.13\sqrt{100,000}}{(4.88 \times 10^{-7})^3} \cong 3.5 \times 10^{20} \; \text{m}^{-3}.$$

This is very close to the experimentally determined value based on measurements of argon ion laser plasmas. Figure 8-15 shows the actual n_e for a wide range of plasma lasers covering wavelengths from

EXAMPLE (cont.)

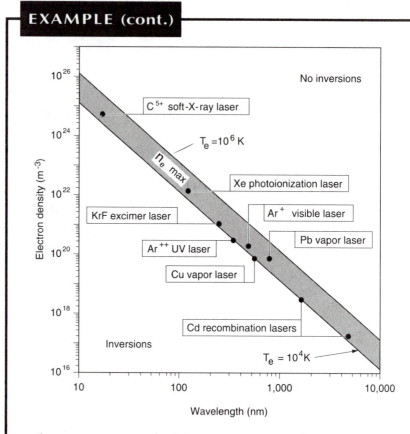

Figure 8-15. Experimental indication of the electron density limit for a variety of lasers

the visible region to the soft–X-ray region. Also shown is n_e^{max} versus wavelength for electron temperatures of 10,000 K and 1,000,000 K. It can be seen that all of these lasers operate at an electron density very near to that of n_e^{max}.

ABSORPTION WITHIN THE GAIN MEDIUM

There are several different mechanisms that can lead to absorption within the gain medium for various types of lasers. These absorptions, represented by the coefficient α in (7.60) and (7.61), serve to reduce or sometimes eliminate the effective gain that might otherwise be obtained due to a population inversion between levels u and l. In all cases, this reduction in gain occurs as a result of absorption on transitions within the gain medium other than the laser transition from level u to level l. In every case the absorption is inherent in the specific gain medium and therefore cannot be eliminated.

GROUND-STATE ABSORPTION IN DYE LASERS   Organic dye lasers have an overlap between the absorption (pumping) spectrum and the emission

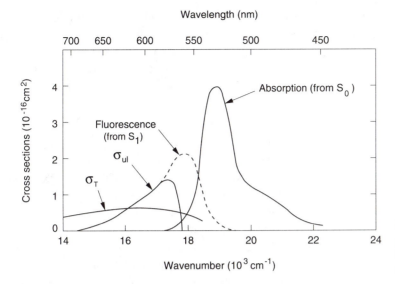

Figure 8-16. Absorption processes associated with absorption from S_0 and absorption from T_1 (σ_T)

spectrum, as shown in Figure 5-9 for the dye RhB. This overlap is determined by the relationship of the singlet ground-state S_0 and excited-state S_1 energy levels shown in Figure 5-11. In Chapter 7 we showed that in general the gain spectrum for a laser transition from level u to level l follows the spontaneous emission spectrum, as indicated in (7.11) for homogeneous broadening and in (7.25) for Doppler broadening. In the case of dye lasers, since there are always ground-state dye molecules present even under intense pumping conditions, there will always be absorption present within the medium at wavelengths associated with the singlet absorption from the ground state S_0 to the excited state S_1. Thus the gain is either reduced or eliminated in the wavelength region where emission and absorption overlap, with greater absorption and consequently reduced gain occurring at shorter wavelengths (where the gain is lower and the absorption is higher). Figure 8-16 shows the singlet absorption cross section, the stimulated emission cross section, and also the triplet absorption cross section (to be discussed in the next section). The dashed lines display the spontaneous emission spectrum, which is generally referred to as the *fluorescence* spectrum. Thus, the reduced gain due to ground-state absorption can be seen in Figure 8-16 as compared to the spontaneous emission spectrum. In dyes, this reduced gain occurs primarily on the short-wavelength side of the maximum of the singlet emission spectrum. For example, it can be seen that there is no gain below 568 nm even though the emission spectrum extends to wavelengths as short as 520 nm.

TRIPLET ABSORPTION IN DYE LASERS   Dyes also have a separate absorption within the medium, owing to the triplet state as discussed in Section 5.2. In this case some of the excited-state singlet atoms from level S_1 decay to the triplet state T_1. Absorption from T_1 to T_2 then can occur and has its own spectrum, which is also shown in Figure 8-16. This absorption is a function of the T_1 population and thus it is desirable to minimize the number of molecules in the triplet state. This is done by "flowing the dye" – in effect, continually moving a new gain medium into the pumping region as the laser is operated. Thus it is possible to either minimize or eliminate the triplet absorption, but not the singlet ground-state absorption described previously.

EXCITED-STATE ABSORPTION IN EXCIMER LASERS When excimer lasers are pumped to the upper laser level, it is possible for absorption to occur to higher excited molecular states. This effectively reduces the gain of the medium, since every photon interacting with the upper laser level species can either stimulate it to make a transition downward, thereby contributing to the laser emission, or stimulate it upward and thereby contribute to absorption. It is therefore possible to create a laser only if the stimulated emission cross section for that level exceeds the excited-state absorption cross section, since both processes begin from the upper laser level u. This effect is thus inherent in the medium and cannot be adjusted. It is also possible to create other undesired molecular species within the excimer medium that absorbs the laser photons. Such species are sometimes a result of the presence of "impurity gases" resulting from reactions of the laser medium with the laser chamber (fluorine species can be highly reactive). All of these potential absorptions are wavelength dependent, and would be treated as the α in (7.60) and (7.61). We must remember that these absorptions can be very wavelength sensitive and with a different spectral variation than that of the gain; they should consequently be indicated as $\alpha(\nu)$ instead of α to remind us of the varying frequency response.

EXCITED-STATE ABSORPTION IN SOLID-STATE LASERS Many of the transition metal solid-state lasers also suffer from excited-state absorption (ESA) similar to that just described for excimer lasers. (Chromium is an exception in that the ESA is significantly reduced compared to other transition metal lasers.) The alexandrite laser uses a $Cr:BeAl_2O_4$ medium. The ESA cross section in alexandrite is less than 10% of the stimulated emission cross section at the center of the gain spectrum at 750 nm.

In sapphire (Al_2O_3) hosts the Ti^{3+} ion has no appreciable ESA, thus allowing the $Ti:Al_2O_3$ laser to be tuned over the very broad wavelength spectrum that corresponds to the spontaneous emission spectral range shown in Table 5-1. The energy-level structure is relatively unique for transition metal laser ions in that there are no d-state energy levels above the upper laser level to which excited-state absorption from the laser level could occur.

ABSORPTION IN SEMICONDUCTOR LASERS In semiconductor lasers the valence band is full or nearly full when no electric current is flowing in the diode. For a population inversion to occur, a significant number of holes must be created in that band. For example, at any given energy within the valence band, half of the possible sites must be occupied by holes before there can be inversion. Below that number of holes, there would be absorption on a recombination transition that terminates at that energy. Thus, an extremely high current must be initiated simply to overcome the initial absorption before gain exists. Although not the same, this absorption can be compared to that of the ruby laser, in which more than half of the ground-state population must be pumped to the excited state before gain can occur. There is typically 0.2 m^{-1} of inherent absorption in a semiconductor laser, which must be overcome by current flowing in the diode and adding holes within the junction region (see Sections 5.4 and 14.9).

REFERENCES

A. Corney (1977), *Atomic and Laser Spectroscopy*. Oxford: Clarendon Press.

T. Holstein (1947), "Imprisonment of resonance radiation in gases," *Physical Review* 72: 1212–33.

T. Holstein (1951), "Imprisonment of resonance radiation in gases, II," *Physical Review* 83: 1159–68.

M. J. Seaton (1962), "Electron collisional excitation," in *Atomic and Molecular Processes* (D. R. Bates, ed.). New York: Academic Press.

A. E. Siegman (1986), *Lasers*. Mill Valley, CA: University Science, Chapter 4.

PROBLEMS

1. Using the equations (8.6), (8.7), and (8.8) that were developed to describe changes in the populations of a three-level system, obtain expressions for the populations in the upper and lower laser levels, the inversion ratio, and the minimum pumping requirements given in (8.9), (8.10), (8.11), and (8.12).

2. Referring to the example in Section 8.2 on the threshold pumping requirements for a ruby laser, estimate the number of pumping photons and the power required to pump the laser rod to the threshold value for a 6-mm–diameter, 0.1-m-long ruby laser rod doped to the concentration given in the example during a pumping cycle that lasts for the duration of the upper laser level lifetime of 3 ms. Assume that the laser is pumped by photons of wavelength 550 nm. Remember that $N_0\Gamma_{li}$ represents the number of ground-state Cr^{3+} ions pumped per cubic meter every second into the upper laser level.

3. Beginning with the rate equations given in (8.14) and (8.15), obtain the threshold inversion requirements for a three-level atomic laser as given in (8.18).

4. Refer to the energy-level diagram, Figure 9-5(b), of the argon ion with the relevant energy levels for the 488.0-nm laser transition. Assuming that it operates as a three-level laser when considering only the single-ion energy levels, what would be the maximum ratio of excitation from the ground state to the lower laser level when compared to the excitation rate from the ground state to the upper laser level at threshold?

5. Given the rate equations (8.22), (8.23), (8.24), and (8.25), obtain the expressions for N_u, N_l, and Γ_{0i} for a four-level laser system given in (8.26), (8.27), and (8.30).

6. Determine the single-pass gain of a 0.1-m–long Nd:YAG laser rod operating at 1.06 μm at room temperature. Assume the following:

(a) $A_{ul} = 4 \times 10^3$ s^{-1}, no significant collisional or phonon broadening occurs on that transition, and there exist no radiative decay routes from level u other than to level l;

(b) the pumping level i decays primarily to the upper laser level u, and the lower laser level l decays to the ground state 0, at rates of 10^{12} s^{-1};

(c) the lower laser level l is 0.27 eV above the ground state;

(d) the pumping rate to the intermediate level i is 100 times the minimum value given in (8.30); and

(e) the doping concentration of the Nd:YAG rod is 10^{26} m^{-3}.

7. Beginning with the differential rate equations (8.31) and (8.32), solve for the expressions for N_u and N_l as a function of time as given in (8.33) and (8.34).

8. Referring to Figure 4-15 of the lead vapor laser of Problem 8 in Chapter 4, graph the time dependence of the population inversion (in real time) of the $^3P_1^o \rightarrow {}^1D_2$ 722.9-nm laser transition, assuming that the pumping flux to the upper laser level is twice that to the lower laser level, and that the pumping is applied at a constant rate beginning at $t = 0$. Assume that radiation trapping prevents any decay to the three levels ($^3P_2$, $^3P_1$, $^3P_0$) below the lower laser level, and assume there is essentially no decay to the higher lying $^1S_0$ level as shown in the figure.

9. At what electron density would radiation trapping begin to slow down the decay of the lower laser level of the He–Cd laser transition at 441.6 nm? Assume a plasma tube radius of 1 mm. Assume that the gas temperature is 350°C, and that half of the ions in the discharge are Cd$^+$ ions and the other half are He$^+$ ions. Assume also that there are no other ions in the discharge, and that charge neutrality exists: the total electron charge of the free electrons that provide the discharge current equals the total ion charge in the discharge. Use the energy-level diagram and the associated coefficients of Figure 8-6 for the relevant Cd$^+$ energy levels. *Hint:* the Cd$^+$ ion ground state is the state that causes radiation trapping on the decay of the lower laser level, and its population accounts for half of the free electrons that make up the electron density in the discharge (the other half are He$^+$ ions).

10. At what population density of the argon ion ground state would radiation trapping begin to reduce the population inversion for the laser transition of Problem 3?

11. For the H-like $n = 3 \rightarrow n = 2$ transition of the B^{4+} ion, what is the maximum electron density at which a population inversion could occur for an electron and ion temperature of 300 eV? What would the gain be on that transition under those conditions, assuming that the pumping flux to the lower laser levels, other than radiative decay from the $n = 3$ level, is negligible? Assume that the population in the upper laser level is 10^{-5} n_e. Note that the transition probabilities for H-like ions scale as Z^4 compared to hydrogen, and that $A_{32} = 4.4 \times 10^7$ s^{-1} for hydrogen.

9 LASER PUMPING REQUIREMENTS AND TECHNIQUES

SUMMARY It was shown in Chapter 7 that a population inversion was a necessary but not a sufficient condition for laser action. Sufficiency is realized by having enough gain in the amplifier to reach the saturation intensity within the laser cavity. Providing sufficient gain involves optimizing the product $\sigma_{ul} \Delta N_{ul} L$ in the laser amplifier. Since σ_{ul} is essentially a constant for most laser transitions, the burden of providing enough gain falls on maximizing either the population density difference ΔN_{ul} or the length L. However, efforts in the laser R & D community generally aim at *decreasing* L, for a variety of reasons including cost, compactness, and availability of amplifier material. Therefore the burden of obtaining sufficient gain falls mostly upon producing sufficient ΔN_{ul}, which in turn requires the generation of adequate population density N_u in level u. Such efforts involve optimization of excitation or pumping mechanisms to populate the laser levels. This chapter deals with such pumping techniques.

Two principal types of pumping or excitation are used to produce lasers. One type involves optical pumping, generally with either flashlamps or other lasers. The other type involves particle pumping, in the form of particles within a gaseous or plasma discharge or particle beams interacting with a potential gain medium. Particle pumping is usually done with electrons, but can also employ metastable atoms or ions. Other pumping methods, not as commonly used, include chemical reactions or the flux from nuclear reactors. This chapter will summarize these various techniques and describe some of the methods used to implement them. Before these methods are explained, we will outline some general excitation processes that are applicable to many of the pumping techniques.

9.1 EXCITATION OR PUMPING THRESHOLD REQUIREMENTS

In this chapter we will be concerned with the applied excitation flux, which was defined in Chapter 8 as the product of the density of the pumping state N and the rate of excitation Γ, or $N\Gamma$. Other factors associated with producing population inversions, such as thermalizing collisions (both upward and downward) by phonons or electrons, as well as radiative decay, were already dealt with sufficiently in Chapter 8. Excitation processes, and the rates associated with them, can in general be considered in terms of a relatively simple concept: the amount of population flux transferred to the upper laser level within the average lifetime of the population residing in that level. In other words, any amount of population that is pumped into level u will decay within the lifetime associated with that level. Thus, the pumping process to sufficiently fill level u effectively starts over for each lifetime duration τ_u of that level. It does not matter which process determines that lifetime; it could be radiative decay or phonon-related collisional decay (T_1). Equation (8.16) can be modified and used to indicate the significance of this concept. The steady-state solution for N_u in (8.16) is

$$N_u = \frac{N_j \Gamma_{ju}}{\gamma_u} = N_j \Gamma_{ju} \tau_u. \tag{9.1}$$

In this equation, N_j is the density in the state j of the species from which the energy is to be transferred. (In many cases this would be the ground state 0 and the rate would be Γ_{0u}.) Γ_{ju} is the pumping rate to the upper laser level u, and γ_u is the decay rate of level u, from which we have $\tau_u = 1/\gamma_u$ ($\gamma_{ul} + \gamma_{u0}$ was replaced by its equivalent γ_u). Because the population density N_u is directly proportional to τ_u, it can be seen that the longer the lifetime τ_u of the level, the more population will build up in that level.

If we express the exponential gain factor as $\sigma_{ul}[N_u - (g_u/g_l)N_l]L$, we know that the portion of that factor leading to gain is the product $\sigma_{ul}N_u L$, whereas the factor $-\sigma_{ul}(g_u/g_l)N_l$ associated with the population density N_l in the lower laser level l is detrimental to producing gain. Thus, to determine the *minimum* pumping flux required, we will deal only with N_u. Of course, more pumping flux would be required if N_l becomes significant compared to N_u. We can therefore express the minimum threshold condition for making a laser by substituting N_u for ΔN_{ul} in (7.56) and (7.59), and by using the value for N_u from (9.1) to obtain an expression involving the minimum pumping flux Γ_{ju} as

$$\sigma_{ul}N_u L = \sigma_{ul}N_j \Gamma_{ju} \tau_u L \approx 12 \pm 5 \quad \text{with no mirrors,} \tag{9-2a}$$

$$\sigma_{ul}N_u L = \sigma_{ul}N_j \Gamma_{ju} \tau_u L = \frac{1}{2}\ln\left(\frac{1}{R_1 R_2}\right) \quad \text{with two mirrors.} \tag{9.2b}$$

Equations (9.2) determine the *threshold value* of the amplifier gain necessary to make a laser, in terms of the factors associated with providing enough population in level u to make such a laser. This threshold value represents a minimum value for the factors shown. If level l has a significant population, then one or more of the pumping factors of (9.2) would have to be increased in order to make up for the absorption loss due to the population in level l.

The first term in (9.2) that is associated with pumping is the stimulated emission cross section, which is essentially fixed for a particular laser transition. We have seen how to calculate its value, but we cannot make any significant changes to it. One should use the value σ_{ul}^H for the case of homogeneous broadening or σ_{ul}^D for Doppler broadening.

The second term is N_j, the population density in the level j that is being transferred to the upper laser level. In many cases this is the ground state of the laser species. For a solid-state laser this would be the dopant concentration (number of dopant ions per cubic meter) or the number of laser species per unit volume. For an organic dye laser, N_j would be the dye concentration mixed into the solvent; for a gas laser, it might be the density of the gas. It could also be a separate species (within the laser gain medium) that can accumulate and store population for later transfer to the upper laser level, as in the case of N_2 in the CO_2 laser.

The third term in (9.2) is the pumping rate Γ_{ju}, which has units of 1/seconds. Γ_{ju} can be provided by electromagnetic waves or light (optical pumping), or by particles such as electrons (particle pumping). It represents the process by which we inject energy into the gain medium to populate the upper laser level. Optical pumping can be provided either by lamps or other incoherent sources or by lasers (coherent sources), and is generally used for solid-state and dye lasers. We will later use (9.2) to determine how much lamp power or laser power is needed to pump a laser. Particle pumping is generally accomplished with electrons, and is primarily confined to gaseous or plasma lasers. We will later determine how much electron current is needed to pump a gas laser.

The requirements defined in (9.2) represent the threshold requirements for laser action. It must be kept in mind that, in order to obtain significant power from a laser, the pumping must significantly exceed these threshold requirements. In a steady-state laser the gain will be reduced to zero as the power is extracted, according to (7.71), owing to the fact that the stimulated emission rate will increase the population decay of level u until the net gain is zero. Thus, the stimulated emission rate is an automatic adjusting parameter that keeps the net gain at a zero value. If the net gain were not yet zero, the beam intensity would increase until the stimulated emission rate brought the net gain to zero (adding more power to the beam as it makes this adjustment), since we are assuming conditions of steady state.

For pulsed lasers, the threshold pumping requirements can be different than those just summarized. The population for N_u was obtained from a steady-state solution of the rate equations. For pulsed conditions, the duration and time dependence of the excitation pulse must be considered in determining the desired value of N_u.

9.2 PUMPING PATHWAYS

Two types of excitation pathways can be used to produce pumping of the upper laser level: direct pumping and pump and transfer. It is useful to visualize these two processes separately to help sort out the important pumping parameters before we become involved in estimating specific pumping requirements and thresholds.

EXCITATION BY DIRECT PUMPING

In the direct pumping process, the excitation flux is sent directly to the upper laser level u from a source or target state j in which the source state is the highly populated ground state 0 of the laser species. This direct pumping

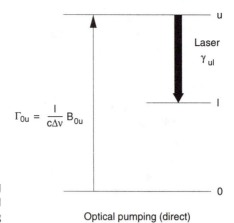

Figure 9-1. Optical pumping rate associated with direct pumping

$\Gamma_{0u} = \dfrac{I}{c\Delta\nu} B_{0u}$

Optical pumping (direct)

process is diagrammed in Figure 9-1. The pumping rate Γ_{0u} can be described in general terms for either optical pumping or particle pumping.

OPTICAL PUMPING For optical pumping, the excitation involves absorption of the pumping light, so we can write Γ_{0u} as

$$\Gamma_{0u} = \frac{I}{c \cdot \Delta\nu} B_{0u}, \tag{9.3}$$

as was first outlined in Chapters 6 and 7. As before, the optical pumping intensity is the intensity I that is within the absorption linewidth $\Delta\nu$ of the absorbing species. This can be converted to an energy density by replacing I/c by the energy density u according to (6.35). The B_{0u} coefficient is related to the optical transition probability A_{u0} of the absorbing transition (the absorption band), as described by (6.49) and (6.51).

PARTICLE PUMPING For particle pumping, Γ_{0u} can be written as

$$\Gamma_{0u} = N_p k_{0u}, \tag{9.4}$$

where N_p is the pumping particle density (equivalent to the intensity factor I in eqn. 9.3 for optical pumping) and k_{0u} is the reaction probability for causing a transition from level 0 to level u when a particle p collides with the laser species in level 0; k_{0u} has dimensions of volume per time. This probability k_{0u} is often broken down into the more useful factors σ_{0u}, the cross section or probability for the transfer of energy from level 0 to level u, and $\bar{v}_{p0}$, the average relative velocity between the colliding species p and the target species 0. The relation between these terms can be expressed as

$$k_{0u} = \bar{v}_{p0} \sigma_{0u}, \tag{9.5}$$

where k is given in (8.48). Thus the population flux rate into level u from level 0 can be expressed as

$$\Gamma_{0u} N_0 = N_p \bar{v}_{p0} \sigma_{0u} N_0. \tag{9.6}$$

In the case of a gaseous discharge, where the particles labeled p are electrons, (9.6) can be rewritten as

$$\Gamma_{0u} = n_e \bar{v}_e \sigma_{0u}^e. \tag{9.7}$$

Here n_e is the electron density (number per volume), $\bar{v}_e$ is the average electron velocity, and σ_{0u}^e is the velocity-averaged electron excitation cross section from level 0 to level u.

DISADVANTAGES OF DIRECT PUMPING   At first examination, direct pumping might seem to be a very straightforward and simple technique. However, many drawbacks are revealed when it is examined in more detail. Several effects can prevent direct pumping from being an effective excitation process for many potential lasers. These effects are listed as follows.

1. There may be no efficient direct route from the ground state 0 to the laser state u. For optical pumping, that would mean that B_{0u} is too small to produce enough gain; for particle pumping, it would mean that σ_{0u} is too small.

2. There may be a good direct route from 0 to u, but there may also be a better route from 0 to l (the lower laser level) by the same process. In this case Γ_{0l}/Γ_{0u} may be too large, according to (8.20), to allow an inversion.

3. Even though there may be a good probability for excitation, via absorption either of the pump light associated with B_{0u} for optical excitation or of σ_{0u}^e for particle excitation, there may not be a good source of pumping flux available. That is, there may be insufficient intensity I for optical pumping, or insufficient density N_p for particle pumping, at the specific energies necessary for pumping population from level 0 to level u.

This third effect can be described more explicitly by reconsidering the flux rate into level u. Instead of writing the population flux rate or number per unit volume per second (eqn. 9.6), we write the energy required per unit volume per second or the power per unit volume by multiplying the population flux rate by ΔE_{u0}, the energy separation between levels u and 0, to obtain the minimum pumping power per unit volume:

$$P_{0u} = \Gamma_{0u} N_0 \Delta E_{u0}. \tag{9.8}$$

To obtain the pumping power needed, one must refer to (9.2) for the required values of Γ_{0u} and N_0; ΔE_{u0} is obtained from energy-level diagrams. Equation (9.8) formulates a minimum pumping power, since in many cases (for both optical and particle pumping) the pump energy exceeds ΔE_{u0}. For optical pumping, this can happen by pumping to higher bands (see e.g. Figure 8-8 for the Nd:YAG laser), which will be described in the next section as one of the processes involving pump and transfer. For particles and, more specifically, for electrons, the probability cross section σ_{0u}^e is generally largest for electron energies that are several times ΔE_{u0}.

Thus, examining (9.8), it can be seen that the pumping power required to produce an inversion increases with energy-level separation ΔE_{u0}. Producing an inversion where the upper laser level u is not very high above

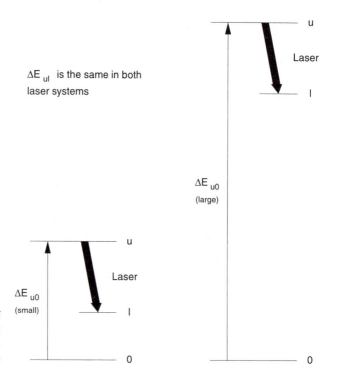

ΔE_{ul} is the same in both laser systems

Figure 9-2. Comparison of pumping energy required for low-lying versus high-lying upper laser levels

level 0 (ΔE_{u0} small) would therefore be much easier, in terms of pumping power requirements, than producing such an inversion where level u is very high above level 0 (ΔE_{u0} large), even though the laser transition might occur at the same wavelength in both cases. This effect is displayed in Figure 9-2.

EXCITATION BY INDIRECT PUMPING (Pump and Transfer)

Because of the restrictions on direct pumping outlined previously, multiple-step pumping processes – referred to here as *indirect pumping* or *pump and transfer* – have evolved over the years. These processes provide alternate routes to obtaining sufficient population inversion ΔN_{ul} for laser action in addition to that of direct pumping. Indirect pumping processes all involve an intermediate level q, and can be considered in three general categories as diagrammed in Figure 9-3: transfer from below, transfer across, and transfer from above. Although these are simplified descriptions, they can be used to categorize, in a reasonably systematic way, almost all of the laser excitation processes other than direct pumping.

For all three cases, the flux transfer rate from a level q to the upper laser level u can be written as

$$N_p \bar{v}_p \sigma_{qu} N_q \tag{9.9}$$

for particle transfer, such as electrons or heavy particles, and

$$\frac{I}{c \cdot \Delta \nu} B_{qu} N_q \tag{9.10}$$

for transfer by photons.

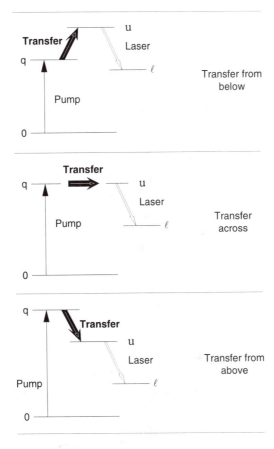

Figure 9-3. Three types of indirect pumping

ADVANTAGES OF INDIRECT PUMPING Before we investigate these processes individually, there are a number of general comments that can be made about the advantages of such pump-and-transfer schemes. First, they all involve an intermediate level q that receives the pumping flux before it arrives at the upper laser level. In some cases populating the intermediate level q is a multiple-step process, but in most cases it involves only a single step. In any event, *level q is always much more closely located in energy to the upper laser level u* than is the initial level 0. This makes the transfer of energy from level q to the upper laser level u a much easier task than pumping directly from level 0 to level u. This intermediate level q is used in different ways to overcome some of the possible problems outlined previously for direct pumping.

1. In some cases, the intermediate level q has a lifetime τ_q that is much longer than the lifetime τ_u of level u ($\tau_q \gg \tau_u$). Level q might therefore accumulate more population than level u. This can be seen by rewriting (9.2) to address the pumping of level q instead of level u. Doing so yields

$$N_q = \Gamma_{0q} N_0 \tau_q. \tag{9.11}$$

Thus, for the case $\Gamma_{0q} = \Gamma_{0u}$, N_q will still be much larger than N_u of (9.2) if $\tau_q \gg \tau_u$. Hence level q can serve as a reservoir of population that is energetically near level u, with the possibility of transfer from q to u being much simpler than direct transfer from 0 to u (owing to the much smaller energy separation).

2. In some cases the pumping probability (cross section) for pumping from level 0 to level q is much greater than that from 0 directly to u. This can significantly lower the pumping requirements.

3. Transfer from q to u can be quite selective in many cases, which implies that it occurs much more favorably to the upper laser level than to the lower laser level. This can often happen in a situation where direct pumping from 0 to u would not be selective, or might even be detrimental to the generation of an inversion.

4. Level q can belong to a different species of material than that associated with level u. This could allow the use of a material that can be efficiently excited to a storage level q. The stored energy is then transferred over to the laser level, either by a collisional or a radiative process, from one species to another. In this case the transfer process usually involves a collision of the species in the excited level q with the laser species in its ground state. The energy is then exchanged from level q in one species to level u in the laser species, and level q reverts to its own ground state. One of the most common processes of this type is the excitation and storage of energy in helium metastable levels (via an electrical discharge), which is then transferred to the neon laser levels to produce the well-known helium–neon laser; see Figure 9-4. In a broader sense, optical pumping can be thought of as occurring in this same manner: the "other species" is the flashlamp or laser pumping medium, and the transfer takes place via optical transfer from either spontaneous emission within the lamp or stimulated emission from the laser.

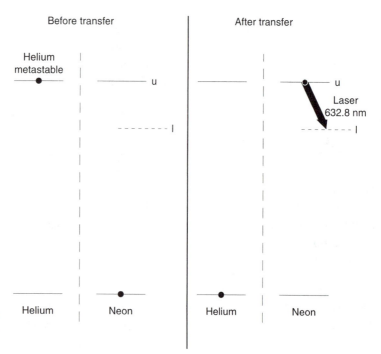

Figure 9-4. Collisional transfer of energy to upper laser level

5. Level q can have a very broad width and thus accept pumping flux of intensity I (eqn. 9.10) over a broad range of energies, in contrast with the upper laser level u, which might be quite narrow in order to provide a high stimulated emission cross section. This q-level capability is particularly advantageous for solid-state lasers in which the pump bands q are broad enough to collect the flux from a flashlamp having a broad spectral output. This is also applicable to atoms when pumping occurs to highlying levels of a particular atom or ion (say $n > 4$), which effectively act as one broad level q, with subsequent decay to an upper laser level u such as $n = 4$ or $n = 3$ (using the Bohr model).

SPECIFIC PUMP-AND-TRANSFER PROCESSES

We will now examine each of the pump-and-transfer processes separately, and decribe the many types of lasers that fall into each category.

TRANSFER FROM BELOW Transfer from below generally occurs for gas lasers in which the level q is an excited state with a long lifetime. The level q thus accumulates energy according to (9.11) and, because of its long lifetime, serves as a storage state. One example is the argon ion laser, where level q is the ground state of Ar^+ as shown in Figure 9-5(a). Excitation of the other noble-gas ion lasers (krypton and xenon) and one excitation process suggested for the He–Cd laser also occur in this way.

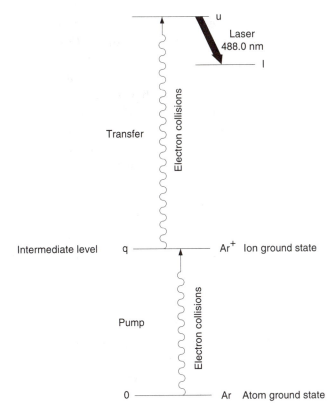

Figure 9-5(a). Example of transfer from below in the argon ion laser, where the transfer level q is the ion ground state

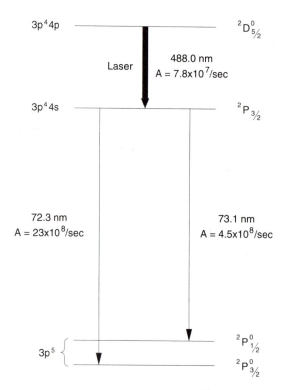

Figure 9-5(b). Decay rates of the argon ion laser (detail)

Two other lasers that should be mentioned include the Ar^{++} laser, in which two stages of ionization are involved in the pumping portion of the excitation, and the Se^{24+} soft–X-ray laser, in which 24 stages of ionization are involved in the pumping portion (Figure 9-6). For both lasers, as more energy is pumped into the discharge or plasma, a higher ion stage (level q) is produced with most of the population residing in that specific ion ground state, rather than having the population distributed among all ion stages. It is therefore possible to place essentially all of the population originally located in either the Ar or Se neutral ground state into the ground state (level q) of the specific ion stage associated with the laser transition. This makes for efficient transfer of that population to an upper laser level u by making N_q of (9.9) very large.

For transfer from below, N_p and $\bar{v}_p$ in (9.9) refer to electrons of density n_e ($N_p = n_e$) and velocity $\bar{v}_e$ ($\bar{v}_p = \bar{v}_e$), and $\sigma_{qu} \equiv \sigma^e_{qu}$ is the electron excitation cross section from the ion ground state (level q) of the specific ion species to the upper laser level u. (The bar above $\bar{v}_p$ and $\bar{v}_e$ denotes average velocity.) Once the appropriate ionization stage is reached, the inversion analysis for all three lasers (Ar^+, Ar^{++}, Se^{24+}) is the three-level analysis diagrammed in Figure 8-5, with inversion conditions described by (8.19) and (8.20).

TRANSFER ACROSS   These indirect pumping techniques generally occur in gaseous media. By "transfer across" we mean that the intermediate level q has the same energy as the upper laser level u. In gases, energy can be transferred via a collision between two atoms or molecules, but the process

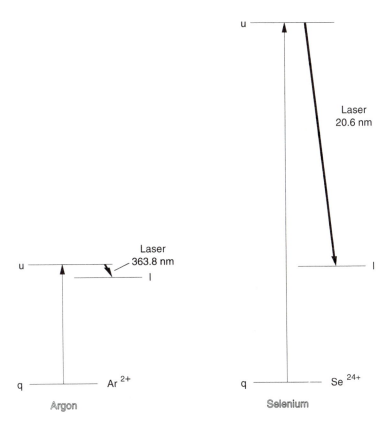

Figure 9-6. Examples of transfer from below, where the transfer level q is a highly ionized ion

must conserve both energy and momentum of the combined two-body system. For transfer to occur, the energies of the initial state of the first particle and the final state of the second particle must be equal, or nearly equal. Two well-known lasers that operate in this fashion are the helium–neon laser and the CO_2 laser.

The helium–neon laser was already discussed briefly as an example of how energy could be transferred from one type of particle (helium) to another (neon). In this section we will emphasize the importance of the energy coincidence of the energy levels associated with these particles. Figure 9-7 shows the relevant energy levels for both helium and neon. Helium has two energy-storage states, the $2s^3S_1$ and $2s^1S_0$ metastable levels, that accumulate significant population when excited in a gas discharge. When small concentrations of neon are placed within the discharge, energy can be transferred directly from the helium metastable atoms to those neon levels that are in direct or near-energy coincidence with the He metastable levels. Figure 9-8 shows a close-up of the energy levels in neon that serve as upper laser levels (the levels that are solid lines) and their energetic relationship with the helium metastable levels.

Figure 9-9 shows a similar relationship between a metastable nitrogen level and the CO_2 upper laser level. In this case, the nitrogen metastable level accumulates significant population during the pumping process and – owing to the energy coincidence – transfers it to the CO_2 molecule.

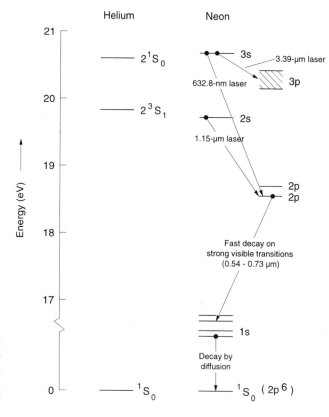

Figure 9-7. Relevant energy levels of the helium–neon laser in which the transfer-across process is used

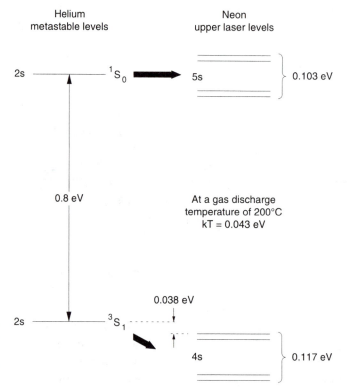

Figure 9-8. Collisional transfer of energy illustrated by relationship to the helium metastable levels in a helium–neon laser (detail)

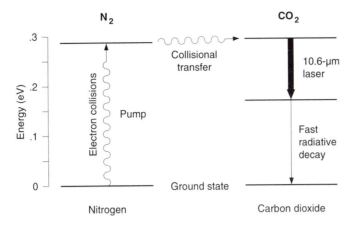

Figure 9-9. Relationship of the nitrogen vibrational level and the CO_2 upper laser level

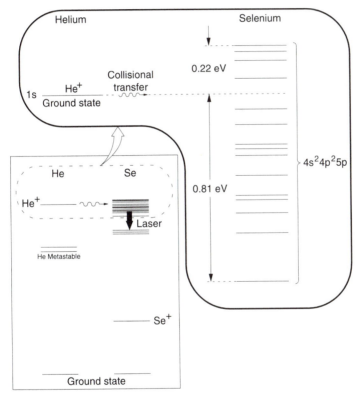

Figure 9-10. Charge-transfer pumping process in the helium–selenium laser

Another laser with such a coincidence is the helium–selenium laser in Se^+, where the intermediate state q is the helium ion ground state. In a gaseous discharge of helium, the helium single-ion ground state can accumulate as much, and sometimes more, population than the helium metastable atoms referred to earlier. This population can be transferred to the selenium ion by a process known as a *charge exchange* or *charge transfer reaction*. In this reaction, a helium single ion (He^+) collides with a selenium atom in its neutral ground state, and the energy is transferred from helium ion ground-state species to selenium ion excited states, as shown in Figure 9-10. Again, as in Figure 9-7, it can be seen that those levels in Se^+

in near-energy coincidence with He^+ serve most effectively as upper laser levels.

Thus, the process of transfer across can effectively serve as a pumping scheme for cases where considerations of conservation of energy and momentum allow efficient transfer from one excited energy state to another state, but only if those two states have equal or nearly equal energy.

TRANSFER DOWN This form of indirect pumping is probably the most widely used excitation process for lasers, when considered in all of its variations. It is especially effective in producing inversions for the following reasons.

1. In most situations the pump energy can occur over a wide range of excitation energies. Thus, although we refer to level q as a single level, it most frequently represents a band of levels or simply an energy band. This band of levels, for the purposes of laser excitation, behaves as a single level. For optical pumping, the pumping band represents the range of wavelengths that can be absorbed by the gain medium to produce population in level q. For dye lasers, solid-state lasers, and optically pumped semiconductor lasers, this spectral range of pump wavelengths can be many tens of nanometers if the pumping is in the visible spectral region. For gas lasers, the energy range for the pump electrons that provide excitation to various excited states can be several times the energy of the level being excited.

2. Because the intermediate level q lies above the upper laser level, the population preferentially decays to the upper laser level as opposed to the lower laser level, since typical decay probabilities, or cross sections for collisional decay, favor levels that are energetically near that of the transfer level. In other words, $k_{qu} \gg k_{ql}$. As an example of this favorable transfer, the collisional transfer rate k_{ul} from level u to a lower level l can be seen from k_{ul} in (8.51) to vary as the inverse cube of the energy separation between the levels, since $\lambda \propto 1/\Delta E$.

3. In most cases the energy moves to the upper laser level from level q "automatically," or without additional stimulus of any kind, at a very fast rate. The transfer is produced by phonon collisions for solids, or collisions with the surrounding atoms in the case of solids, liquids, or high-density plasmas. These transfer processes usually occur at rates of 10^{12} per second or greater for liquids and solids, and can approach that rate in high-density gases. This decay is a thermalization process, where the downward rate is controlled by thermal equilibrium forces that try to produce a Boltzmann relationship between the upper laser level and the higher level q (or group of levels q). When decay from level q begins, the ratio of populations in the upper laser level u to the higher level (or band) q is in a high state of inversion, since there is typically no pumping directly to level u from the ground-state level 0. Thus the population shift is rapid, bringing the population ratio N_q/N_u into thermal equilibrium: $N_q/N_u = e^{-\Delta E_{qu}/kT}$.

EXAMPLES

We will now briefly summarize the well-known lasers that use this excitation process. These lasers include solid-state lasers in dielectric materials, semiconductor lasers, organic dye lasers, excimer lasers, and gaseous lasers (referred to as *recombination* lasers). A few examples of the first category, the solid-state lasers in dielectric materials, are diagrammed in Figure 9-11. The intermediate level q in these examples always consists of a band of levels. For semiconductor lasers, the energy band of the n-type doping lies higher than that of the p-type doping. These different doping materials are physically separated, and the process that causes the energy to flow from the n-type doped material (level q) to the p-doped material (upper laser level u) is the electric field applied across the semiconductor junction. This is diagrammed in Figure 9-12.

The relevant energy levels for dye lasers are shown in Figure 9-13. In this case, the intermediate level q and the upper laser level u are both part of the same energy band. For organic dye molecules dissolved in a solvent, when pumped by a broad-spectrum lamp, the pumping flux goes directly to producing population over the entire range of the energy band referred to as q in Figure 9-13. The population rapidly decays to the bottom of the band (which we define as the upper laser level u) by collisions, which rapidly drive the population within the band to a Boltzmann distribution. Since the temperature is relatively low (a dye is typically at or near room temperature), most of that population quickly decays to the bottom of the band, which serves as the upper laser level. The population moves into this level in times of the order of 10^{-12} s, whereas the population at the bottom of the band (level u) decays in a much slower time (10^{-9} s) to level l. Thus the population

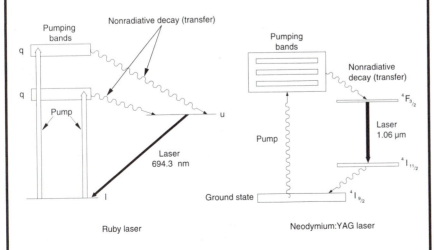

Figure 9-11. The transfer-down process in a solid-state laser

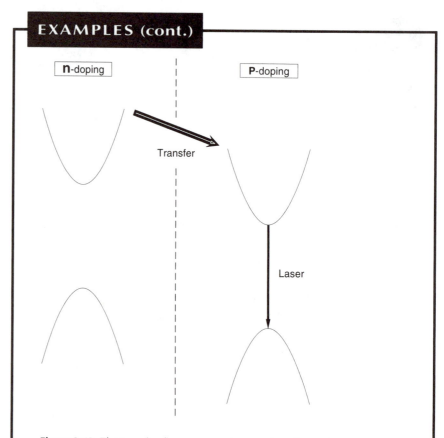

Figure 9-12. The transfer-down process in a semiconductor laser

accumulates at level u, and an inversion develops between u and the lower laser level l (which in this case is also a band of energies).

For the case of excimer lasers, there are typically two intermediate steps or levels. Typically, a gas mixture including the two laser species is subjected to a high electric field. This field ionizes the noble-gas atoms such as argon, krypton, or xenon (to produce Ar^+, Kr^+, or Xe^+), and at the same time produces negative ions of the halogen molecules (F^- and Cl^-) in very large quantities. These intermediate species are then attracted to each other, since they have opposite charges, and combine to form excited states of rare-gas halogen molecules ArF^*, KrF^*, XeF^*, $XeCl^*$, et cetera. When these rare-gas halogen molecules are formed, they form in either highly excited states or ionized states, which then decay rapidly via collisions with electrons to the lowest energy state of that band, or the upper laser level u. These steps are diagrammed in Figure 9-14.

The atomic (or molecular) recombination laser, diagrammed in Figure 9-15, also belongs to this general excitation category. In this type of laser, the intermediate level q is the next higher ion stage above the stage that contains the upper laser level u. The objective in making

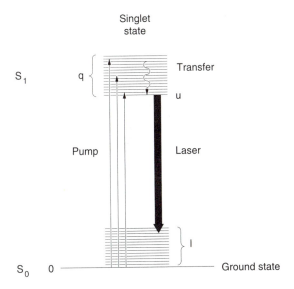

Figure 9-13. The transfer-down process in a dye laser

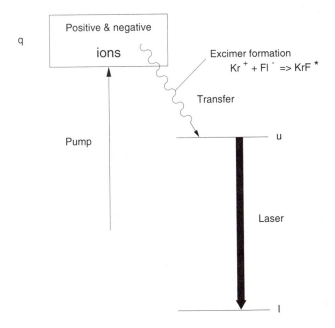

Figure 9-14. The transfer-down process in an excimer laser

this laser operate effectively is to produce a very large population in that next higher ion stage. This population, consisting of positively charged ions, attracts electrons in a process called *recombination*. When the electrons recombine with the ions, they proceed to decay and move downward in energy within that next lower ion stage until they reach the upper laser level *u*, where the population begins to accumulate. This accumulation is due to the fact that a large energy gap

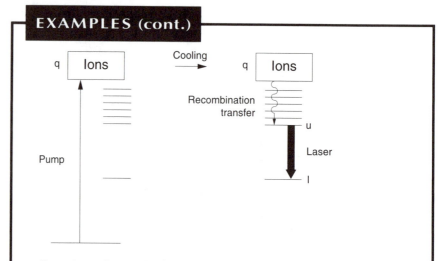

Figure 9-15. The transfer-down process in a plasma recombination laser

is reached, at which point the collisional decay rate is significantly reduced and the relevant decay rate to lower levels from this level becomes radiative instead of collisional. A population therefore builds up in level u, and an inversion is established between that level and a lower-lying level l, the lower laser level.

9.3 SPECIFIC EXCITATION PARAMETERS ASSOCIATED WITH OPTICAL PUMPING

PUMPING GEOMETRIES

Optical pumping can be accomplished by many different light sources, including flashlamps, lasers, solar flux, flashcubes, and laser-produced plasmas. In all cases, the same pumping considerations apply. In order for pumping to occur, the light from the pumping source must be absorbed by the gain medium. This can be accomplished by a number of techniques, summarized in Figure 9-16. The first example, Figure 9-16(a), shows a lamp located next to a cylindrical gain medium; a small portion of the flux from the lamp is directly intercepted by the gain medium. To increase the amount of flux intercepted by the laser rod from the flashlamp, an elliptically shaped pumping cavity is used, a cross section of which is shown in Figure 9-16(b). An elliptical shape is used because any optical ray emitted from one focus of the ellipse is refocused at the other focus of the ellipse. This elliptical arrangement is configured to take advantage of the common linear type of flashlamp that is available as a reliable, long-lasting pumping lamp. This pumping configuration can be improved further by using a double elliptical cavity, shown in Figure 9-16(c), where two flashlamps are used to increase the total pumping flux to the rod. These improvements help to overcome

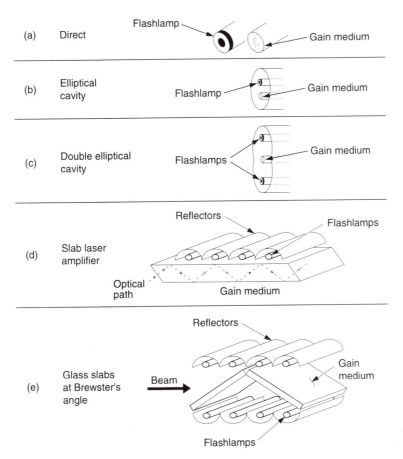

(a) Direct

(b) Elliptical cavity

(c) Double elliptical cavity

(d) Slab laser amplifier

(e) Glass slabs at Brewster's angle

Figure 9-16. Summary of optical pumping processes (continued on next page)

some of the problems (summarized later in this chapter) associated with optical pumping of a cylindrical laser rod. An alternative geometry for a solid-state laser is the slab geometry shown in Figure 9-16(d), where the laser beam undergoes a zigzag path through the slab. This geometry, and the associated optical pathway within the gain medium, have been shown to eliminate both stress-induced birefringence and thermal and stress-induced focusing. Pumping is accomplished by flashlamps partially enclosed by reflectors located adjacent to the slab, as shown in the figure. This geometry is used mostly for high-power Nd:glass amplifiers. Another geometry used for very large-aperture glass amplifiers is that shown in Figure 9-16(e). It involves locating a series of glass slabs at Brewster's angle with the pumping lamps, as well as associated reflectors located above and below the slabs so as not to interfere with the optical pathway of the beam. This allows the possibility of very large apertures (up to 0.4 m), and is a configuration used for some of the highest-power lasers such as those used for laser fusion.

The use of another laser as the pumping source is shown in the next three diagrams of Figure 9-16. In Figure 9-16(f) the process is known as *transverse pumping,* wherein a (typically) cylindrical lens focuses the laser pumping flux to a linear region within the gain medium, at the surface of that gain medium. In this case the pumping flux arrives in the gain medium

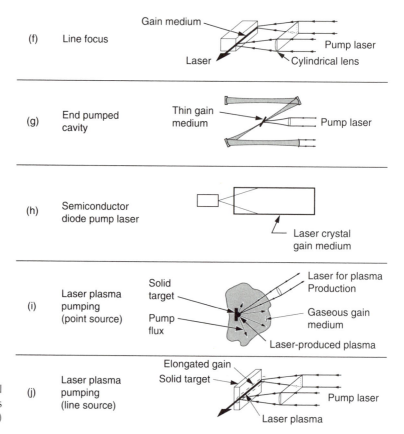

(f) Line focus

(g) End pumped cavity

(h) Semiconductor diode pump laser

(i) Laser plasma pumping (point source)

(j) Laser plasma pumping (line source)

Figure 9-16. Optical pumping processes (continued)

in a direction transverse to that of the developing laser beam. It is desirable to absorb most of the pumping flux within a short depth into the gain medium, in order to define a narrow gain medium and to increase the pumping efficiency. This technique is often used for pulsed dye lasers. In the other two cases, Figures 9-16(g) and 9-16(h), the process is known as *end pumping*. The pumping laser is directed more in the longitudinal direction of the laser gain medium, and thereby enters the end region or regions of the gain medium. In Figure 9-16(g) the pumping is within a thin gain medium, either a dye laser jet stream or a thin laser crystal, designed to obtain an extremely high pumping intensity. In Figure 9-16(h), a semiconductor pumping laser (typically with a highly diverging beam) is closely coupled to a laser rod.

Laser-produced plasmas have also been involved in pumping of lasers. They have been used to provide electron collisional excitation of ions, such as that described in Section 9.2 for the Se^{24+} soft–X-ray laser. Plasmas have also been used as intense light sources for optical pumping of other lasers. In this case, as shown in Figures 9-16(i) and 9-16(j), the flux from the laser-produced plasma pumps a gain medium directly adjacent to the laser plasma source. This medium could be a solid, a liquid, or a gas. In the point-source arrangement of Figure 9-16(i), a region surrounding the laser-produced plasma is pumped relatively uniformly in all directions that are accessible to the flux. This technique is not particularly amenable to producing the desirable elongated gain medium (as suggested in Section 7.4). The linear laser-produced plasma pumping arrangement was developed to refine this pumping process further, as shown in Figure 9-16(j). In this case the flux is concentrated in a linear shape that would be more suitable to producing an elongated gain region.

PUMPING REQUIREMENTS

Considerations for optical pumping can be described in terms of how the gain medium can best be designed to take advantage of the pumping flux. With such considerations in mind, we provide a simple analysis for determining how to utilize that flux optimally. We examine the simplified model shown in Figure 9-17, and assume that the pumping flux is incident upon the gain medium with an intensity I_0. It would be desirable to use as much of the pumping flux as possible (subject to other constraints), in order to make the pumping process efficient. We assume that the gain medium has a uniform distribution of laser species in the ground level 0 with a population density N_0. From this starting point, population is moved either directly to the upper laser level u, or indirectly via an intermediate level q by virtue of the absorption of the pump flux. That absorption process is described by an equation of the form of Beer's law as given in (8.37). The change in intensity within the rod thus varies as

$$I = I_0 e^{-\sigma_{0u} N_0 x} = I_0 e^{-\alpha_{0u} x} \quad \text{(direct pumping)}, \tag{9.12}$$

$$I = I_0 e^{-\sigma_{0q} N_0 x} = I_0 e^{-\alpha_{0q} x} \quad \text{(indirect pumping)}. \tag{9.13}$$

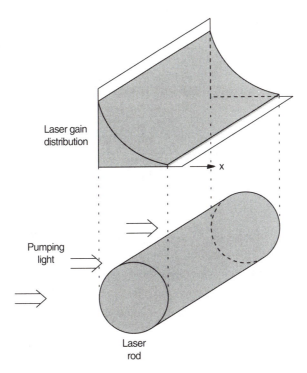

Laser gain distribution

x

Pumping light

Laser rod

Figure 9-17. Gain distribution across a laser rod due to the non-uniform absorption of the pumping light

The direction of the pump beam is designated as x, and the direction of the laser beam has been specified previously as z. However, it is important to emphasize that the pump beam could be in any direction, including the z direction. As the pumping flux penetrates the gain medium and populates the upper laser level u (either directly, or indirectly through an intermediate level q), the population in the upper laser level u at that time would have a distribution in the x direction that follows Beer's law. This occurs because the energy absorbed in the pumping process is proportional to the pumping beam intensity at any given value of x, as shown in Figure 9-17, provided that N_0 is distributed uniformly with the gain medium (which is usually the case).

In order to use (9.12) or (9.13) effectively, we must know how to determine the value of the intensity I_0 required to produce the necessary gain in the amplifier. We can express the desired gain coefficient g_0 within the medium we are attempting to pump optically as

$$g_0(\nu) = \sigma_{ul} \Delta N_{ul} \cong \sigma_{ul} N_u, \tag{9.14}$$

where we have approximated ΔN_{ul} by assuming that $\Delta N_{ul} \cong N_u$. This would provide the absolute minimum pumping flux. For cases where N_l is large enough to affect the inversion, we must realize that the threshold condition for the pumping intensity I must be increased above the minimum obtained here. In determining the value of N_u, we can use (9.1) and (9.3) to obtain

$$N_u = N_0 \Gamma_{0u} \tau_u = N_0 \frac{B_{0u} I \tau_u}{c \Delta \nu_{0u}} = N_0 \frac{B_{0u} I}{c A_{ul} \Delta \nu_{0u}}, \tag{9.15}$$

where we have assumed that $\tau_u \approx 1/A_{ul}$. Using the relationship $B_{0u} = (g_u/g_0)(A_{u0} c^3/8\pi h\nu_{0u}^3)$ from (6.49) and (6.51), we can now write

$$N_u = \frac{g_u}{g_0} \frac{N_0 A_{u0} c^3 I}{8\pi h\nu_{u0}^3 A_{ul} c \Delta \nu_{u0}}$$

$$= \frac{g_u}{g_0} \left[\frac{A_{u0} \lambda_{u0}^2}{4\pi^2 \Delta \nu_{u0}} \right] \frac{\pi N_0 I}{2(h\nu_{u0}) A_{ul}} = \frac{g_0(\nu)}{\sigma_{ul}}, \tag{9.16}$$

where we have used (9.14) to obtain $N_u = g_0(\nu)/\sigma_{ul}$. The term in brackets we also recognize as the stimulated emission cross section σ_{u0} on the pumping transition from level u to level 0 as described by (7.16). Recall that all transitions – not only laser transitions – have a stimulated emission cross section defined by (7.16).

We can thus rewrite (9.16) as

$$N_u = \frac{g_u}{g_0} \sigma_{u0} N_0 \frac{\pi I}{2(h\nu_{u0}) A_{ul}} = \alpha_{0u} \frac{\pi I}{2(h\nu_{u0}) A_{ul}} = \frac{g_0(\nu)}{\sigma_{ul}}, \tag{9.17}$$

and solve for the pump intensity I that will produce a specific gain coefficient $g_0(\nu)$ to be used in either (9.12) or (9.13):

$$I = \frac{2(h\nu_{u0}) A_{ul}}{\pi \alpha_{0u} \sigma_{ul}} g_0(\nu_0), \tag{9.18}$$

where $\alpha_{0u} = (g_u/g_0)\sigma_{u0}N_0$ from (7.32) has been used to incorporate the absorption coefficient α_{0u} (in place of the stimulated emission cross section σ_{u0}) for the pumping transition, since it is the absorption coefficient that is generally quoted for a specific doping concentration. Thus, the pumping intensity within the absorption bandwidth of the absorbing transition is determined by the photon energy $h\nu_{u0}$ of the pump beam, the laser transition probability A_{ul}, the desired gain coefficient $g_0(\nu_0)$ at the center of the desired laser transition, the absorption coefficient α_{0u} at the center of the pumping transition, and the stimulated emission cross section σ_{ul} of the laser transition. We can see from (9.18) that the gain will vary in proportion to the pumping intensity. It will thus vary exponentially within the medium in the x direction during the pumping process, according to (9.12) or (9.13).

We will next consider two different pumping directions oriented with respect to the laser beam direction. First we will examine the factors relating to transverse pumping, where the pump beam direction x is perpendicular to the laser beam axis z within the gain medium. We will then consider the consequences of longitudinal pumping, where the pump axis x is parallel (or nearly parallel) to the laser beam axis z.

TRANSVERSE PUMPING

The transverse optical pumping geometry provides a way to pump a large amount of energy into a laser gain medium. The gain medium can be made as long as desired, and since the energy input per unit length is a constant, more length generally provides more laser energy. However, transverse pumping leads to problems of non-uniformity in the gain medium in the transverse direction to the laser beam axis. It is desirable in most situations to have a uniform gain profile in a direction transverse to the laser beam axis z. For transverse pumping from a single lamp, in an arrangement such as that of Figure 9-16(a), the transverse gain distribution would exhibit a profile similar to that of Figure 9-17, which is far from uniform. The laser output from an amplifier with such a distribution would be highly non-uniform, especially when one considers that the exponential beam growth factor (the gain) is proportional to the value of gain at each transverse location in the gain profile. Pumping from a single direction is also not an efficient use of the optical flux from a flashlamp, because most of the flashlamp energy is emitted in directions that would never be coupled into the laser gain medium. Hence elliptical cavities as in Figures 9-16(b)–(c) were developed to collect and use the pumping light more efficiently than with transverse pumping into the medium from only one direction. The single elliptical cavity can also significantly improve the gain symmetry around the central gain axis in the transverse direction of the gain medium, as indicated in Figure 9-18(a) for a weakly absorbing gain medium and in Figure 9-18(b) for a strongly absorbing gain medium. The double elliptical cavity of Figure 9-18(c) is shown to be even more effective by providing a

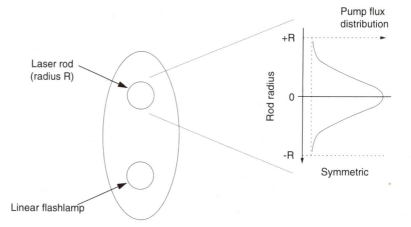

Figure 9-18(a). Radial gain profile of a weakly absorbing rod in a single elliptical cavity

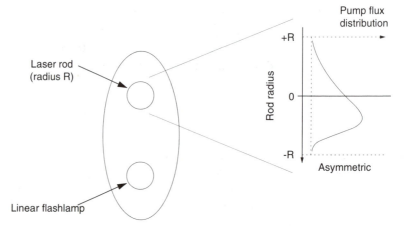

Figure 9-18(b). Radial gain profile of a strongly absorbing rod in a single elliptical cavity

more *uniform* gain distribution in the transverse direction of the gain medium than that obtained from the single elliptical cavity.

For transverse pumping, it is generally desirable to absorb the pump energy within the width (or diameter) d_a of the gain medium. The uniformity is improved – but the efficiency is reduced – if the absorption length is longer than d_a, thereby allowing part of the light to escape. Conversely, the uniformity is decreased if the pump light is absorbed in a distance shorter than d_a. Thus, if most of the energy is to be absorbed in a distance d_a, we can equate the exponent of (9.12) to unity:

$$\sigma_{0u} N_0 d_a = 1 \quad \text{or} \quad \sigma_{0q} N_0 d_a = 1. \tag{9.19}$$

We can solve for the value of N_0 that would be necessary to meet this requirement as follows:

$$N_0 = \frac{1}{\sigma_{0u} d_a} \quad \text{or} \quad N_0 = \frac{1}{\sigma_{0q} d_a}. \tag{9.20}$$

Thus, for a solid-state laser rod, the density of laser species (such as Nd^{3+} or Cr^{3+} ions) can in most cases be made to the specifications of (9.20). In

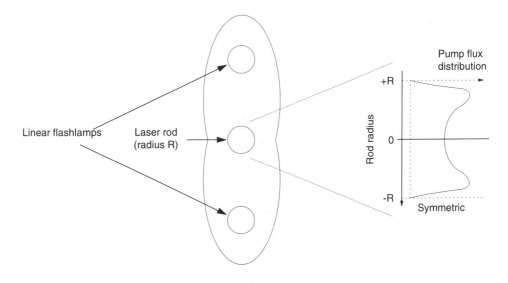

Figure 9-18(c). Radial gain profile of a strongly absorbing rod in a double elliptical cavity

some instances, the density prescribed by (9.20) cannot be achieved in a crystal, owing to limitations upon the growth process of the crystal or to detrimental interactions of the dopant ions if they are located too close to each other (a process known as *quenching*). For an elliptical cavity this reduced density can result in a gain profile that is low around the periphery of the laser rod and higher in the center, as shown in Figure 9-18(a). This non-uniformity can cause a variation in the index of refraction of the crystal during the pumping process. This will cause focusing effects and thus distortion of the laser beam within the gain medium (self-focusing), thereby limiting the amount of pumping flux that can be used; see Section 15.4.

END PUMPING

One of the most useful arrangements for pumping dye lasers or tunable solid-state lasers is the end pumping (or longitudinal pumping) arrangement shown in Figure 9-16(g). In making a cw dye laser, in order to get rid of the excited-state absorption from unwanted triplet states (see Section 5.2), it is necessary to flow the dye rapidly to physically remove those states from the gain region. A clever way to do this is to flow a very thin stream of dye solution through the mode volume of the laser. In this arrangement, a laser pump beam is focused into the region where the laser gain is desired, such that an amplified beam can develop within the optical cavity. This "jet stream" of the dye is very thin, so a high density of dye molecules in the solution is necessary for efficient absorption to occur. The dye will then be absorbed according to (9.12), but in this case of end pumping the gain profile in the direction perpendicular to that of the dye laser beam will be uniform. The arrangement shown in Figure 9-19 involves a cavity in which the astigmatism due to the off-axis imaging of two curved mirrors is

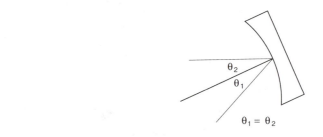

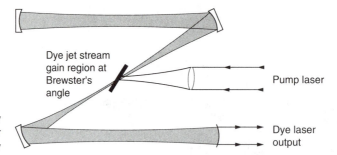

Figure 9-19. Astigmatically compensated four-mirror laser cavity

compensated by that of the gain cell, as described in more detail in Chapter 12. Such pumping techniques are also used for solid-state lasers such as the Ti:Al$_2$O$_3$ laser, in which the dye jet stream is replaced by a thin Ti:Al$_2$O$_3$ crystal.

The advantages of longitudinal pumping primarily include the production of a uniform gain over a region normal or perpendicular to the laser beam direction z, thus providing a uniform gain region in which the beam can develop. It also allows the possibility of pumping very thin regions that could not be accessed from the transverse direction. Pumping can be produced specifically only within the mode volume of the laser gain medium, and thus pump energy is not wasted by pumping regions that are not used. Longitudinal pumping is generally done with another laser beam as the pump source.

End pumping is also a very useful arrangement for laser–diode pumping of solid-state lasers. Such an arrangement is shown in Figure 9-20. In this situation, the diode pumping flux can be matched directly into a solid-state laser crystal, in a closely coupled cavity. In this case, the pumping (diode) laser and the solid-state laser are adjacent to each other, which makes for a very compact laser. The diode laser serves as an efficient pumping source that matches well with the pumping bands of Nd:YAG and several other laser systems.

CHARACTERIZATION OF GAIN MEDIUM WITH OPTICAL PUMPING – SLOPE EFFICIENCY

A standard criterion for measuring the efficiency of an optically pumped gain medium is referred to as the *slope efficiency:* the slope of a plot of the

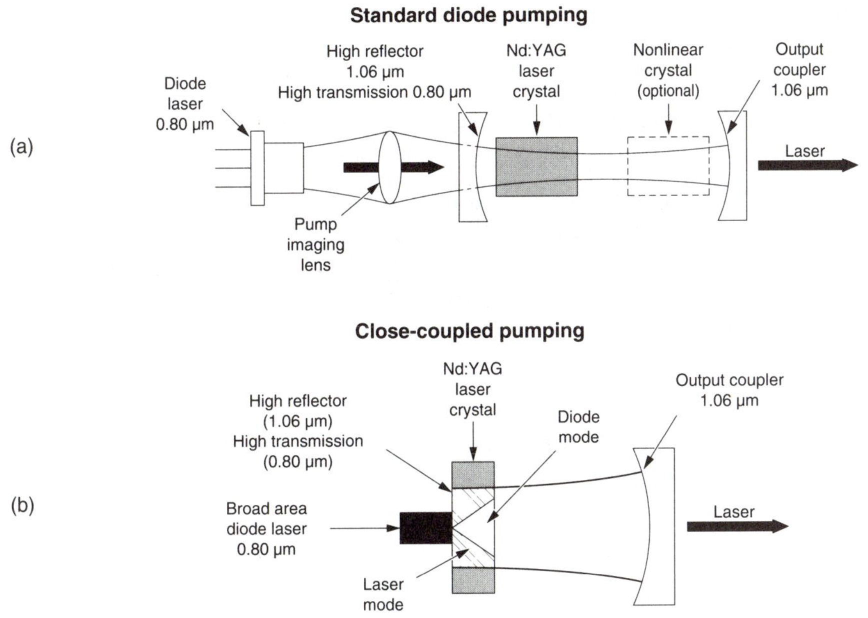

Standard diode pumping

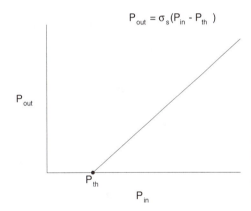

Figure 9-20. Diode pumping cavities: (a) standard cavity; (b) closely coupled cavity, in which the pumping laser is located adjacent to the laser crystal

Figure 9-21. Graph of laser power output versus optical pumping power input defining the slope efficiency σ_s

laser output power versus the input pumping power. Both solid-state lasers and dye laser gain media are characterized in this fashion. We can write an expression for the laser output power as P_{out} and the laser input power as P_{in}, where P_{in} is the pumping power either of another laser or of a flash-lamp or some other optical source. Since there is also a threshold pumping power P_{th} at which the gain exceeds the losses within the cavity, the laser power will increase linearly above P_{th} according to the expression

$$P_{out} = \sigma_s(P_{in} - P_{th}), \tag{9.21}$$

where σ_s is a dimensionless proportionality constant relating the output power to the input power. A typical example is shown in Figure 9-21. The proportionality constant σ_s associated with the slope of the curve is called the *slope efficiency*. Both σ_s and P_{th} are related to pumping parameters and cavity losses of the laser, as well as to the effectiveness of the gain medium itself. The highest theoretical slope efficiency would be for a 45° slope, with the pump wavelength equal to the laser wavelength; in this case the slope efficiency would be 100%. Since the pump wavelength is generally shorter than the laser wavelength, the theoretical maximum slope efficiency is $100\% \times [(h\nu)_{laser}/(h\nu)_{pump}] = (h\nu)_{laser}/(h\nu)_{pump}$. In practice, slope efficiencies are more typically in the range of 5% or less, except for diode pumping when they can be as high as 30%. These reduced efficiencies are attributable to the losses associated with the pumping process, as described earlier in this section.

9.4 SPECIFIC EXCITATION PARAMETERS ASSOCIATED WITH PARTICLE PUMPING

There are two general processes involved in particle pumping. The first uses electrons, and the second uses atomic or molecular particles. These processes will be summarized separately.

ELECTRON COLLISIONAL PUMPING

Electron collisional excitation processes and ionization pumping processes are generally produced by applying an electric field within a gaseous medium; the field produces and accelerates the free electrons within that medium to velocities of 10^6 to 10^7 m/s. These electrons then collide with atoms or molecules in their ground state within the discharge, and produce transitions within those atoms or molecules to excited energy states. The process is regulated in the sense that the electrons gain energy by the electric field, but lose energy by collisions with the atoms or molecules. At very low electron energies the principal collisions are elastic ones, since the electron energies are so low that no excitation occurs. Because the electron is significantly less massive than the atom or ion, very little energy per collision is transferred from the electron to the atom (or ion) when elastic collisions occur.

As the electron energy or velocity is increased, inelastic collisions begin to dominate. These inelastic collisions involve raising the atom or molecule to an excited energy level. The collisions can result in either an *excitation* process, which raises the bound electron to an excited energy state, or an *ionization* process, which frees the electron from the atom. In both of these collision processes, the electron gives up part of its energy to the atom, and is then accelerated again by the field until it suffers another collision.

The free electrons are the principal contributing factor to the current that flows when the electric field is applied to the gaseous medium. Those electrons are characterized by a density n_e and a temperature T_e. The current density (A/m^2) is related to n_e by the relationship

$$j_e = n_e e \bar{v}_D, \tag{9.22}$$

where e is the charge of the electron (1.6×10^{-19} C) and $\bar{v}_D$ is the average electron drift velocity, or the net velocity in the direction of the field. This velocity $\bar{v}_D$ is usually an order of magnitude or more lower than the thermal or average velocity $\bar{v}_e$ of the electrons. The relationship of $\bar{v}_D$ to the applied electric field E in the discharge is given by

$$\bar{v}_D = \mu E, \tag{9.23}$$

where μ is the mobility of the electrons.

Assigning an electron temperature T_e assumes that a Boltzmann distribution of electron energies exists. Although this is not completely accurate, it is a good approximation. Thus we can refer to the electron temperature either in degrees Kelvin (K) or in electron volts (eV), where it is understood that the temperature in energy units such as eV involves multiplying T_e by Boltzmann's constant. From (4.29), the average velocity of a particle of mass M and temperature T in thermal equilibrium is given by

$$\bar{v} = \sqrt{\frac{8kT}{M\pi}}, \tag{9.24}$$

so we can write the average electron velocity in terms of the electron temperature T_e as

$$\bar{v}_e = \sqrt{\frac{8kT_e}{m_e\pi}}. \tag{9.25}$$

As mentioned previously, electrons are more than three orders of magnitude *less* massive than the atomic or molecular particles with which they collide in a gas discharge, so the electrons transfer only a small portion of their energy to those particles. As a result, they have an effective temperature T_e that is significantly higher than that of the ambient gas. Thus, while a gas is at a temperature of 500–1,500 K, the electron temperature can be at 5,000–100,000 K. Table 9-1 provides some examples of atom and electron velocities obtained from (9.24) and (9.25) to show their striking differences. From the table it can be seen that, owing to the difference in mass, electron velocities are very much higher than the velocities of heavy particles, even when both are at the same temperature. Moreover, the difference increases significantly as the electrons are accelerated in the gas discharge.

The electron velocities are determined by acceleration due to the applied electric field E, and by deceleration due to occasional collisions with atoms of pressure p within the discharge. The frequency of collisions of the electrons with the much more massive atoms would therefore be expected to have an effect upon the electron temperature T_e. This frequency is determined by the density of gas N in the discharge. This gas could be primarily the laser species of density N_0, or it could be a separate gas of density N that is provided in the discharge for reasons other than for use as a lasing species. The gas pressure is proportional to the gas density according to the ideal gas law $p = NkT$, where k is Boltzmann's constant.

TABLE 9-1

ATOM AND ELECTRON AVERAGE VELOCITIES IN A GAS DISCHARGE

Average Atom and Molecule Velocities

Laser Species	Laser Discharge Temperature (K)	Average Velocity (m/s)
Neon (Ne)	400	650
Argon$^+$ (Ar$^+$)	1,500	890
Copper (Cu)	1,800	775
Cadmium (Cd)	600	330
Selenium X-ray (Se^{23+})	6,000,000	4,000
Carbon Dioxide (CO$_2$)	400	440
Krypton Fluoride (KrF)	400	290

Average Electron Velocities

Electron Temperature in Discharge	Average Electron Velocity (m/s)
58,000 K ($kT_e = 5$ eV)	1.5×10^6
116,000 K ($kT_e = 10$ eV)	2.1×10^6
174,000 K ($kT_e = 15$ eV)	2.6×10^6

Thus it is not surprising that the average electron velocity and the associated electron temperature T_e can be related to the quantity E/p, since the field E accelerates the electrons and the pressure p of the gas retards the electrons via collisions. Thus, as the gas pressure p is reduced or the field E is increased, the average electron velocity $\bar{v}_e$ or temperature T_e will increase. Likewise, $\bar{v}_e$ and T_e will be reduced by increasing the pressure or reducing the field.

The primary consideration for understanding the interaction of the various parameters in a gas discharge is estimating the flux rate into the upper laser level u. This flux, obtained from (9.2a) and (9.2b), is given by an equation of the form

$$\Gamma_{0u}N_0 = n_e v_e \sigma_{0u}^e N_0 \quad \text{or} \quad \Gamma_{qu}N_q = n_e v_e \sigma_{qu}^e N_q, \tag{9.26}$$

depending upon whether the transfer of energy to the upper laser level u occurs from the ground state 0 of density N_0 or from the intermediate state q having density N_q. The electron collisional excitation cross sections expressed in (9.26) are described for a specific electron energy, as determined by the value of the electron velocity. The excitation cross section is also related to the specific transition wavelength (energy), as given in (8.48) for either λ_{u0} (ΔE_{u0}) or λ_{uq} (ΔE_{uq}). Although (8.48) gives the expression for k_{lu}, the expression is the same for any electric-dipole–allowed transition and can therefore be used also to describe k_{0u} or k_{qu}. Since k_{0u} and k_{qu} are related to σ_{0u} and σ_{qu} by the relationships $k_{0u} = n_e \sigma_{0u}$ and $k_{qu} = n_e \sigma_{qu}$, we can easily obtain the energy-dependent electron excitation cross sections from (8.48). Often it is convenient to use the *average* electron energy, which then allows us to use the concept of electron temperature T_e as described in (9.25). Thus, both the electron density n_e and the electron velocity v_e (related to the electron temperature T_e) are quantities we need to know. Both can be obtained from (9.22) and (9.25) if the discharge current and T_e are known.

The remaining factor is the electron excitation cross section σ_{0u}^e or σ_{qu}^e. Such cross sections are known in many cases (and can be looked up in tables), but it is useful to give a range for these values. For electron collisions that cause transitions to occur between energy levels with a high radiative transition probability or a high oscillator strength, the cross section would be of the order of 10^{-21} m^2. A cross section for an unallowed transition, according to the selection rules of Chapter 4, would range from 10^{-23} to 10^{-24} m^2. Equation (8.48) is an expression involving electron collisions transferring population from level l to level u where $k_{lu} = v_e \sigma_{lu}$.

Electrons can also be accelerated to very high energies in either an electron accelerator or an electron beam (e-beam) machine. These are devices designed to transfer excitation energy, in the form of high-energy electrons, from the high-vacuum chamber in which they are produced to a separate gaseous gain medium where pumping takes place. In such an accelerator, electrons are ejected from a solid material by application of a very intense electric field, and accelerated to very high energies – many thousands of electron volts (KeV) or even millions of electron volts (MeV). It is essential to use an evacuated chamber so that the electrons will not lose energy via collisions during their acceleration process; hence the need for a separate accelerating chamber. The electrons are injected into the gaseous gain medium through a very thin but strong metallic window, and the energy is absorbed within the gas to pump the laser levels. This technique is particularly useful for providing excitation in high-pressure gases, where discharge excitation is difficult to produce since collisions are so frequent that the electric field cannot accelerate the electrons to sufficiently high velocities. A high-pressure gas is essential for e-beam excitation, in order to absorb sufficient pump energy transferred from the beam to the gas. This is due to the fact that the cross section σ_{0u}^e of (9.26) for very high-energy electrons is very low, and thus the gas density N_0 must be very high in order for sufficient transfer to take place.

HEAVY PARTICLE PUMPING

In the context of pumping, the term "heavy particles" refers to any particles other than electrons. These could include particles in metastable atomic levels (such as the neutral helium singlet and triplet metastable atoms), ions, nuclear reactor fragments, or high-energy ions produced by accelerators. If any such species have a density N_s and an average velocity V_s then they can produce excitation from level 0 to laser level u of the laser species of density N_0, by providing a flux into level u determined by

$$\Gamma_{0u} N_0 = N_s V_s \sigma_{0u}^s N_0 \tag{9.27}$$

in a similar fashion to that for electrons (eqn. 9.26). In this case the excitation cross section can be approximately estimated by the size of the colliding particle. For atoms with dimensions typically of the order of 1 Å (10^{-10} m), their cross-sectional area that would be "seen" by another colliding particle would be of the order of $(1 \text{ Å})^2$ or 10^{-20} m^2. Such cross sections are known as *gas kinetic* cross sections. Exact values have been measured and can be obtained from the research literature for many particles. Cross sections for transfer of energy from helium metastable atoms to other atoms typically range from 10^{-19} to 10^{-20} m^2.

REFERENCES

FLASHLAMP PUMPING

J. M. Eggleston, T. J. Kane, K. Kuhn, J. Unternahrer, and R. L. Byer (1984), "The slab geometry laser – Part I: Theory," *IEEE Journal of Quantum Electronics* 20: 289–300.

W. Koechner (1992), *Solid State Laser Engineering,* 3rd ed. New York: Springer-Verlag.

LASER PUMPING

F. J. Duarte (1990), "Technology of Pulsed Dye Lasers," in *Dye Laser Principles* (F. J. Duarte and Lloyd W. Hillman, eds.). New York: Academic Press, Chapter 6.

LASER-PLASMA PUMPING

W. T. Silfvast and O. R. Wood II (1987), "Photoionization lasers pumped by broadband soft–X-ray flux from laser-produced plasmas," *Journal of the Optical Society of America B* 4: 609–18.

PARTICLE PUMPING

S. C. Brown (1966), *Introduction to Electrical Discharges in Gases.* New York: Wiley.

Y. B. Raizer (1991), *Gas Discharge Physics.* New York: Springer-Verlag.

C. S. Willett (1974), *Introduction to Gas Lasers.* New York: Pergamon.

PROBLEMS

1. If a laser-pumping light source emits a power of 10 W at a wavelength of 500 nm, and if we assume that 10% of the photons are absorbed by the laser gain material within the useful mode volume of the laser that is being pumped, how many photons per second are absorbed in the material?

2. In Problem 1, how many upper laser level species would accumulate within an upper laser level lifetime of 5 ns (dye laser), 3.8 μs (titanium:sapphire laser), 230 μs (Nd:YAG laser), and 3 ms (ruby laser)? Assume that one upper laser level species is produced for every pump photon absorbed within the material.

3. The flux from the sun arriving at the surface of the earth is approximately 1 kW/m^2 on average during the daytime. If 10% of that flux falls within the pumping band of a Nd:YAG laser crystal, how big would a collecting lens have to be (in diameter) to collect enough power to pump the Nd:YAG laser to reach laser threshold if the laser crystal is 0.1 m long and 6 mm in diameter? Assume that all of the sunlight passing through the lens is concentrated within the laser rod such that all of the flux within the pump absorption bandwidth is absorbed and converted to upper laser level species. Assume that the flux accumulates for a duration of the lifetime of the upper laser level and that the average photon energy of the solar flux within the absorption pumping band of the Nd:YAG rod is 2 eV. Also assume that both of the laser mirrors have a reflectivity of 95%.

4. A 5-W argon ion laser operating at 514.5 nm is used as a pumping beam for a Rh6G dye laser. The dye gain medium consists of a rapidly flowing Rh6G dye jet stream of thickness 0.5 mm located within a cavity similar to that shown in Figure 9-16(g). The dye concentration is of sufficient density such that 50% of the pump beam is absorbed in the dye. It is desirable that the pump beam spot size be the same as the dye laser spot size, to provide efficient absorption of the energy. What spot size (diameter) would the pump laser and the dye laser need within the dye jet stream in order to produce a single-pass exponential gain of 1 (i.e., $\sigma_{ul}^H N_u L = 1$)? Assume the upper laser level lifetime of the dye is 5 ns and every pump photon that is absorbed is converted to an upper laser level species within the lifetime of that level. Assume that the lower laser level population is negligible since it rapidly decays to the bottom of the singlet ground state. If the spot size were increased to 1 mm *diameter,* by what factor would the gain be reduced?

5. An electrical current is applied to a copper vapor laser discharge tube filled with Cu atoms at a density of 10^{23} atoms per cubic meter. The electron temperature obtained within the discharge is approximately 15,000 K. Assume that the velocity-averaged electron excitation cross section from the Cu atom ground state to the upper laser level is $\sigma_{0u}^e \cong 10^{-22}$ m^2. At what electron density would the pumping flux be sufficient to produce a single-pass gain coefficient of 10 m^{-1}? Assume that the lifetime of the upper laser level is 5×10^{-7} s. *Hint:* Determine the average velocity of the electrons for the given temperature.

6. A flashlamp is available that radiates a power of 25 W within the pumping band of a Nd:YAG laser at an average wavelength of 700 nm. Approximately 15% of that power is collected by the laser rod and absorbed *uniformly* within the rod by the use of a multi-elliptical cavity. The rod is doped with Nd ions such that the absorption length of the pump radiation is equal to the diameter of the rod. For a 0.1-m-long, 6-mm–diameter rod, what is the single-pass gain obtained with this pumping flux? The absorption cross section at 700 nm (to level q) is equal to the stimulated emission cross section from level u to level l. Assume also that $A_{ul} = \gamma_u = 4.2 \times 10^3$ s^{-1} and $\gamma_{l0} = \gamma_{qu} \approx 10^{12}$ s^{-1}, and that the laser operates at room temperature.

7. What pump intensity at 514.5 nm would be required to pump a Ti:Al$_2$O$_3$ laser rod optically to achieve a gain of 3/m if the rod is doped to a concentration of 10^{25} per cubic meter and the cross section associated with absorption of the pump flux at the pump wavelength exceeds the stimulated emission cross section of the laser by a factor of three? Assume that the stimulated emission cross section of Ti:Al$_2$O$_3$ at the operating laser wavelength is 3.5×10^{-23} m^2.

8. A 40-W cw argon ion laser operating at 514.5 nm is used to end pump a Rh6G dye jet stream of 500 μm thickness (effective gain length). The pump radiation is concentrated on the jet in a 0.1-mm–diameter circular spot. What molar concentration (moles/liter) of the dye in the solvent will produce a single-pass gain of 1% at 577 nm? Assume that the profile of the absorption cross section for the dye has the wavelength dependence shown in Figure 5-10. The peak absorption cross section is $\sigma_{abs} = 3.8 \times 10^{-20}$ m^2. The molecular weight of Rh6G is 479. The stimulated emission cross section is given in Figure 7-10. The radiative lifetime was given in Problem 4. Assume that the population in the lower laser level is negligible. *Hint:* The molecular weight in grams of a substance will provide Avogodro's number of molecules (6.023×10^{23} molecules) of that substance. If that amount is dissolved in one liter of solvent (10^{-3} m^3) it will provide 6.023×10^{26} dye molecules per cubic meter.

9. Assume the He–Cd upper laser level is populated by collisions with helium metastable atoms in a discharge tube. Assume there are 10^{18}/m^3 He triplet metastable atoms in the discharge and that they are the primary source of excitation of the He–Cd laser; assume the Cd density in the discharge is 5×10^{20}/m^3. For a gas discharge temperature of 540 K, what must the population be in the upper laser level, and what would be the single-pass gain at 441.6 nm for a 0.3-m–long gain region? Assume Doppler broadening and that there is a single even isotope of Cd^{114} in the laser tube, so that there is no isotope broadening. Assume that the Penning ionization cross section from He metastables to Cd neutral ground-state atoms that populates the upper laser level is $\sigma_P = 10^{-19}$ m^2. Assume also that there is no collisional depopulation of the upper laser level u and that radiative decay occurs from level u only to level l (i.e., there are no other decay pathways).

10. An argon ion laser operates at a gas temperature of 1,200 K and pressure of 0.1 torr and an electron temperature of 80,000 K. Assume the laser discharge has a length of 0.5 m and a bore diameter of 2 mm. Also assume that both mirrors have a transmission at the laser wavelength (488.0 nm) of 1% and that there are no other losses within the cavity. What must the electron collisional excitation cross section be in order for the laser to operate at threshold with a discharge current of 10 A? Assume the electron drift velocity is 10^5 m/s. Assume that the gas is completely ionized and composed of singly ionized argon ions and electrons, and that excitation of the upper laser level occurs by electron collisions with the argon ion ground state. An energy-level diagram of the relevant argon ion laser levels is shown in Figure 9-5(b).

LASER RESONATORS

LASER CAVITY MODES ▮10

SUMMARY Chapters 3 through 9 dealt with various aspects of the laser amplifier or gain medium. Those chapters covered the concepts of energy levels, emission linewidth, stimulated emission, population inversions, sufficient gain, and gain saturation. They also considered the requirements for pumping into the upper laser level. Brief mention was made in Chapter 7 of the use of mirrors at both ends of the laser amplifier to obtain a longer effective gain length than is available from either a single or double pass of a beam through the laser amplifier. In that analysis, the mirrors were considered only from the standpoint of providing an extension of the gain length. The concept of cavity modes of electromagnetic radiation in a completely enclosed cavity were also discussed briefly in Chapter 6 and related to radiation in thermal equilibrium or blackbody radiation.

In this chapter we consider the properties associated with the optical cavity of a laser that has mirrors located at each end of the laser gain medium. These properties, which will be related to cavity modes, play a significant role in determining the output characteristics of the laser beam. We will first discuss the Fabry–Perot optical cavity (resonator) and thereby develop the concept of longitudinal modes. We will describe how these modes are associated with a laser resonator. We will then proceed with the analysis of a cavity with mirrors of finite size at the ends of the amplifier, and of the diffraction losses associated with such a cavity. This will lead to the development of transverse modes in the laser cavity. It will be shown that these laser modes can be easily characterized as Gaussian-shaped beams.

10.1 LONGITUDINAL LASER CAVITY MODES

When mirrors are placed at the ends of an amplifying medium, they not only effectively increase the length of the gain medium (as discussed in Chapter 7) but also place boundary conditions upon the electromagnetic field (the laser beam) that develops between the two mirrors. From our discussions in Chapter 6 concerning modes that develop within a cavity in radiative thermal equilibrium, we might expect that similar modes develop also within the laser cavity, with similar boundary conditions such that the electric field must be zero at the reflecting surfaces of the mirrors. We will therefore begin our analysis by considering what happens when a beam of light is incident upon a two-mirrored cavity, known as a *Fabry-Perot resonator,* with no optical elements between the mirrors. We will then consider the effect of placing an amplifying medium between the mirrors.

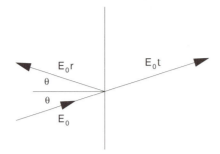

Figure 10-1. Transmitted and reflected rays when an electromagnetic wave arrives at a reflecting surface

E_0 = Amplitude

r = Coefficient of reflection

t = Coefficient of transmission

FABRY–PEROT RESONATOR

First we consider a beam of light interacting with a single reflecting surface. Assume that a beam of light of amplitude E_0 is incident upon a partially reflecting surface at an angle θ to the normal of that surface, as shown in Figure 10-1. Assume that the light is a plane wave of electromagnetic radiation of infinite extent and that the reflecting surface is also of infinite extent, to eliminate the effects of diffraction associated with the edges of a finite reflecting surface. A portion of the light incident upon the surface will be reflected and a portion will be transmitted. The reflected portion will be $E_0 r$ and the transmitted portion will be $E_0 t$, where r and t are the amplitude coefficients of reflection and transmission (respectively), with values ranging from 0 to 1. The time dependence of E_0 has been neglected for this analysis.

We now add another mirror, so that we have a Fabry–Perot resonator consisting of two parallel partially reflecting surfaces separated by a distance d. Initially we will assume that an index of refraction of unity exists in all regions of the analysis; later we will allow for a material with a different index of refraction in the region between the mirrors. The beam arrives at the first surface, where a portion is reflected and the remaining portion is transmitted, as indicated in Figure 10-1. The transmitted beam then continues until it reaches the second surface, and again part is reflected and part is transmitted. The transmitted portion has an amplitude of $E_0 t^2$ and the reflected portion is $E_0 tr$. The reflected beam then continues back to the first surface, where part is reflected ($E_0 tr^2$) and this beam again arrives at the second surface. The transmitted portion is $E_0 t^2 r^2$ and the reflected portion is $E_0 tr^3$. This continues indefinitely, as shown in Figure 10-2.

Let us consider the extra path length that the reflected ray from the second surface undergoes in order to return to the wavefront location indicated by the dashed line of Figure 10-3, exactly in phase with the next cycle

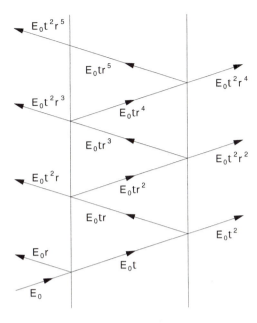

Figure 10-2. Multiple reflections from two reflective surfaces

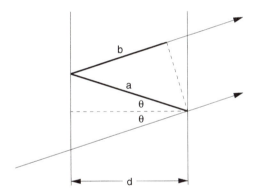

Additional path length = a + b

$$a = \frac{d}{\cos \theta} \qquad b = a \cos 2\theta$$

$$a + b = \frac{d}{\cos \theta} [1 + 2 \cos^2 \theta - 1] = 2 \, d \cos \theta$$

Figure 10-3. The extra path length of a ray reflected from two surfaces

of the original ray such that the two rays might add together in phase. From the figure it can be seen that the extra path length of the reflected ray when it returns to this wavefront location is

$$2d \cos \theta. \qquad (10.1)$$

Because we are considering an electromagnetic plane wave of the form e^{ikz}, where z represents the path of the wave, we can describe the exponential portion of this wave as $e^{i\phi}$, since the exponent of an imaginary exponential function is simply a phase factor (see Section 2.2). Because the expression in (10.1) is just the additional path length z that the reflected wave travels to return to the same wavefront location, we can express the phase factor as

$$kz = \phi = 2kd\cos\theta \tag{10.2}$$

or

$$\phi = \frac{4\pi}{\lambda}d\cos\theta, \tag{10.3}$$

where k has been replaced by the relationship $k = 2\pi/\lambda$.

We can now sum up all of the transmitted amplitudes as

$$E_t = E_0t^2 + E_0t^2r^2e^{i\phi} + E_0t^2r^4e^{2i\phi} + \cdots \tag{10.4}$$

or

$$E_t = E_0t^2(1 + r^2e^{i\phi} + r^4e^{2i\phi} + \cdots) = E_0t^2\sum_{n=0}^{\infty}r^{2n}e^{in\phi}. \tag{10.5}$$

This is just a geometric series with a ratio of successive terms of $r^2e^{i\phi}$, which can be expressed as

$$\sum_{n=0}^{\infty}r^{2n}e^{in\phi} = \frac{1}{1-r^2e^{i\phi}}. \tag{10.6}$$

Thus, the total transmitted wavefront can be rewritten as

$$E_t = \frac{E_0t^2}{1-r^2e^{i\phi}}. \tag{10.7}$$

Since the transmitted intensity I_t is expressed as $I_t = |E_t|^2$, we have

$$I_t = E_0^2\frac{|t|^4}{|1-r^2e^{i\phi}|^2} = I_0\frac{|t|^4}{|1-r^2e^{i\phi}|^2}, \tag{10.8}$$

where $I_0 = E_0^2$. We can allow for a phase change upon reflection such that

$$r = |r|e^{i\phi_r/2}, \tag{10.9}$$

where $\phi_r/2$ is the phase change for one reflection. For a dielectric interface, the phase change $\phi_r/2$ is either 0 (when going from a more dense to a less dense medium) or π (for the reverse). For a metal, the phase change could be any value.

If we define the reflectivity R associated with the intensity as

$$R = |r|^2 \tag{10.10}$$

and define transmission T as

$$T = |t|^2, \tag{10.11}$$

then we can write I_t as

$$I_t = I_0\frac{T^2}{|1-Re^{i\Phi}|^2}, \tag{10.12}$$

where

$$\Phi = \phi + \phi_r. \qquad (10.13)$$

The expression in the denominator of (10.12) can be rewritten as

$$
\begin{aligned}
|1 - Re^{i\Phi}|^2 &= (1 - Re^{i\Phi})(1 - Re^{-i\Phi}) \\
&= 1 - Re^{i\Phi} - Re^{-i\Phi} + R^2 \\
&= 1 - 2R\cos\Phi + R^2 \\
&= (1 - R)^2 \left[1 + \frac{4R}{(1-R)^2} \sin^2\frac{\Phi}{2} \right]. \qquad (10.14)
\end{aligned}
$$

If we set

$$F' = \frac{4R}{(1-R)^2}, \qquad (10.15)$$

then we can rewrite I_t as

$$I_t = I_0 \frac{T^2}{(1-R)^2} \frac{1}{1 + F'\sin^2(\Phi/2)}. \qquad (10.16)$$

The expression

$$\frac{1}{1 + F'\sin^2(\Phi/2)} \qquad (10.17)$$

is referred to as the *Airy function*.

If we assume that there is no absorption and thus $R + T = 1$ (or $T = 1 - R$), then the ratio of the transmitted intensity to the incident intensity can be written as

$$\frac{I_t}{I_0} = \frac{1}{1 + F'\sin^2(\Phi/2)}, \qquad (10.18)$$

which is just the Airy function. A plot of I_t/I_0 versus $\Phi/2$ is shown in Figure 10-4 for several values of R. The function has a maximum value

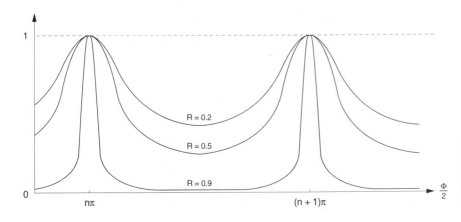

Figure 10-4. Plot of transmitted intensity (from a Fabry–Perot interferometer) versus phase change

of unity for $\sin(\Phi/2) = 0$ or $\Phi/2 = n\pi$, $n = 0, 1, 2, \ldots$, and a minimum for $\sin(\Phi/2) = 1$ or $\Phi/2 = (2n+1)\pi/2$. The minimum depends upon F' but can be very small for large values of R. The maximum values of I_t/I_0 can be seen to occur at

$$\Phi/2 = n\pi, \tag{10.19}$$

where $n = 0, 1, 2, \ldots$. We will refer to these values of Φ as Φ_{max}. Using (10.13), we have

$$\Phi_{max} = 2n\pi = \frac{4\pi}{\lambda_0} d \cos\theta + \phi_r. \tag{10.20}$$

Each of the peaks of the Airy function shown in Figure 10-4 is identical in shape. Thus we can obtain the FWHM (full width half maximum) of the Airy function by considering the width of the peak for $n = 0$. For larger values of R, say $R > 0.6$, we can approximate $\sin(\Phi/2)$ by $\Phi/2$. The value of $\Phi/2$ at which the Airy function then reduces to half its maximum value will be referred to as Φ'. This is obtained by setting

$$\frac{1}{1 + F'(\Phi/2)^2} = \frac{1}{2} \tag{10.21}$$

or

$$F'(\Phi'/2)^2 = 1, \tag{10.22}$$

which leads to

$$\Phi' = \frac{2}{\sqrt{F'}}. \tag{10.23}$$

The FWHM is just twice this value:

$$\boxed{\text{FWHM} = 2\Phi' = \frac{4}{\sqrt{F'}}.} \tag{10.24}$$

Since the separation between the peak values of the function is $\Delta\Phi = 2\pi$, we can obtain an expression F for the ratio of the separation between peaks to the FWHM as follows:

$$\boxed{F = \frac{\Delta\Phi}{\text{FWHM}} = \frac{2\pi}{4/\sqrt{F'}} = \frac{\pi\sqrt{F'}}{2} = = \frac{\pi}{2}\frac{2\sqrt{R}}{(1-R)} = \frac{\pi\sqrt{R}}{1-R},} \tag{10.25}$$

where F is referred to as the *finesse*. For cavities in which the two mirrors have different reflectivities R_1 and R_2, F is given by

$$\boxed{F = \frac{\pi(R_1 R_2)^{1/4}}{1 - (R_1 R_2)^{1/2}}.} \tag{10.26}$$

Let us now consider the simple case in which $\theta = 0$ and $\phi_r = 0$ (for a dielectric). From (10.20) we can then obtain the following relationship between the integer n, the wavelength λ, and mirror separation d:

$$n = \frac{2d}{\lambda}. \tag{10.27}$$

The wavelength at which the maxima occur in Figure 10-4 can be referred to as λ_n^{max}, where

$$\lambda_n^{max} = \frac{2d}{n}. \tag{10.28}$$

These maxima occur at an infinite sequence of wavelengths, decreasing in separation with increasing n. It is more convenient to write them in terms of the frequencies ν_n^{max} at which the maxima occur by using the relationship $\lambda \nu = v = c/\eta$. Thus

$$\nu_n^{max} = \frac{nc}{2\eta d}. \tag{10.29}$$

We now take the difference between two successive frequencies $\nu_{n+1}^{max} - \nu_n^{max}$ as

$$\nu_{n+1}^{max} - \nu_n^{max} = \frac{c}{2\eta d}[(n+1) - n] = \frac{c}{2\eta d}. \tag{10.30}$$

We can see that this frequency difference is independent of n. We can thus refer to the frequency difference as

$$\Delta\nu = \frac{c}{2\eta d}, \tag{10.31}$$

where ηd is the optical path length separating the mirror surfaces. For $\eta = 1$, $\Delta\nu$ can be expressed as

$$\Delta\nu = \frac{c}{2d}. \tag{10.32}$$

These maxima (or "enhancements") therefore occur at equal frequency spacings that are independent of the specific value of the frequency or wavelength.

We have thus derived the expression necessary to understand the response of a two-mirrored cavity or resonator to a beam of light of intensity I_0. Such a cavity is referred to as a Fabry-Perot resonator or *interferometer* for most optical applications, and as a Fabry-Perot *etalon* if the spacing d between the mirrors is firmly fixed. For laser applications it is generally referred to as a two-mirror optical cavity or resonator. The theory is applicable for any value of d, including large values associated with many lasers, and the intensity of the light within the cavity is enhanced at frequency spacings given by (10.31) or (10.32) for frequencies within the bandwidth of the laser emission.

Determine the frequency difference between successive maxima for a Fabry–Perot etalon (see Section 12.5) in which the mirrors are separated by 0.01 m and the mirrors are located in air.

Because the mirrors are located in air, the index of refraction is essentially unity and we can thus use (10.32) to compute the frequency difference $\Delta\nu$. Thus we find that

$$\Delta\nu = \frac{c}{2d} = [(3\times10^8 \text{ m/s})/2](0.01 \text{ m}) = 1.5\times10^{10} \text{ Hz}.$$

Hence the spacing is about 10 times the typical emission linewidth of a Doppler-broadened gas laser emission linewidth as indicated in Table 4-1.

The separation between maxima in terms of frequency is given by (10.31), so we can obtain $\Delta\nu_{\text{FWHM}}$ for each of these maxima by using (10.25) and (10.31):

$$\Delta\nu_{\text{FWHM}} = \frac{\Delta\nu}{F} = \left(\frac{1}{F}\right)\frac{c}{2\eta d} = \frac{c(1-R)}{2\pi\eta d\sqrt{R}} \tag{10.33}$$

for mirrors of equal reflectivity; for mirrors of unequal reflectivity R_1 and R_2, we have

$$\Delta\nu_{\text{FWHM}} = \frac{c[1-(R_1R_2)^{1/2}]}{2\pi\eta d(R_1R_2)^{1/4}}. \tag{10.34}$$

The quality factor Q of a cavity can be given as $\nu_0/\Delta\nu$ and, like the finesse F, is a measure of the sharpness of the frequency transmission under consideration. Using (10.33), we can write

$$Q = \frac{\nu_0}{\Delta\nu_{\text{FWHM}}} = \frac{2\pi\eta d\sqrt{R}\,\nu_0}{c(1-R)} \tag{10.35}$$

or, for mirrors of different reflectivities R_1 and R_2,

$$Q = \frac{2\pi\eta d(R_1R_2)^{1/4}\nu_0}{c[1-(R_1R_2)^{1/2}]} \tag{10.36}$$

FABRY–PEROT CAVITY MODES

We can rewrite (10.28) as

$$d = n\left(\frac{\lambda^{\text{max}}}{2}\right), \tag{10.37}$$

where we have assumed that $\eta = 1$ for simplicity.

Equation (10.37) indicates that each enhancement occurs when an integral multiple of half-wavelengths fit into the cavity spacing of length d such that the electric vector of the electromagnetic wave is zero at the reflecting surfaces. Each of these waves that is enhanced by virtue of its exactly "fitting into the spacing d" or exactly "fitting into the cavity" is referred to as a *mode,* in a similar way to that of the cavity modes discussed in Section 6.3. Such modes result from the interference effects that occur when light interacts with two parallel reflecting surfaces, as shown earlier in this section.

Rewriting (10.29) for $\eta = 1$ yields

$$\nu = \frac{nc}{2d}. \tag{10.38}$$

We can thus see that there are essentially an infinite number of frequencies that would fit within such a cavity. If a wide range of frequencies were to be considered, the reflectivity of the mirrors would have to be kept high over that entire range in order to obtain sharp discrete enhancements similar to those shown in Figure 10-4 over that frequency range. For the visible spectral region, ν is typically of the order of 5×10^{14} s^{-1} or greater. For a Fabry–Perot cavity of separation $d = 0.01$ m, n would thus be of the order of 35,000, meaning that the waves would have somewhere in the range of 30,000 to 40,000 half-cycles within the cavity.

In order for the resonance condition of (10.31) and (10.37) to be applicable over a large mirror surface, the mirror quality – or the *variation* in d at different portions of the reflecting surface – would have to be less than $(\lambda/2)/2$ or $\lambda/4$. In actual practice the surface variation, often referred to as the *surface quality* or *surface figure,* must be better than $\lambda/10$ to obtain the conditions of high finesse associated with lasers.

If, for example, we consider a frequency at which $I_t/I_0 = 1$ and $T = 0.01$ ($R = 0.99$), then we know that the intensity inside the cavity at that frequency must be 99 times greater ([1−0.01]/0.01) than that outside of the cavity. A Fabry–Perot cavity can therefore serve as an energy storage device for those wavelengths or frequencies at which (10.28) and (10.29) are satisfied. Thus, when the beam is initially fed into the cavity, the intensity builds up within the cavity until the transmitted beam is equal to the input beam for the frequencies at which the *resonances* (locations of maximum transmission) occur. The beam intensity cannot increase beyond the input value without violating the law of energy conservation.

LONGITUDINAL LASER CAVITY MODES

When a laser gain medium is inserted within a Fabry–Perot cavity with mirrors located at the ends of the medium, a similar set of enhancements or modes in the form of standing-wave patterns equally spaced in frequency will build up within the cavity. When the gain medium is initially turned on, the amplifier begins emitting spontaneous emission at the laser frequency in all directions. However, the rays that are directed toward the mirrors are reflected and return through the amplifier, and are thus enhanced by the gain during each transit through the medium. A highly directional beam eventually evolves in the axial direction and reaches an intensity that exceeds I_{sat} if the gain duration is sufficiently long (see Section 7.6). At this point the beam begins to approximate a plane wave in that it has very low divergence – the case for which we obtained the modes within the Fabry–Perot cavity. According to (10.19), the boundary condition at the mirrors is $\Phi/2 = n\pi$, which is equivalent to requiring the electric field to be zero at the mirrors (as was required for the modes of the cavity radiation discussed in Section 6.3). This condition establishes a standing-wave pattern of the laser beam within the cavity. The various standing waves, each of a different frequency according to (10.38), are referred to as *longitudinal modes* because they are associated with the longitudinal direction of the electromagnetic waves within the cavity. The modes thus occur at wavelengths or frequencies at which the electric field has an integral multiple of half-wavelengths with zero magnitude at the mirror surfaces.

LONGITUDINAL MODE NUMBER

We have seen that one or more longitudinal laser mode frequencies can occur when a two-mirror cavity is placed around the laser gain medium and sufficient time (typically 10 ns to 1 μs) is allowed for such modes to develop. The total number of modes present is determined by the separation d between the mirrors, as well as by the laser bandwidth and type of broadening (homogeneous or inhomogeneous) that is present. The actual laser mode frequencies can be obtained by recalling (10.38), which we now generalize as

$$\nu = n\left(\frac{c}{2\eta d}\right) \tag{10.39}$$

for lasers in which the index of refraction η is the same throughout the pathway of the laser beam within the optical cavity. Remember that n is a positive integer so as to satisfy the standing-wave conditions of the Fabry-Perot resonator. This expression is valid for almost all gas lasers, as well as for solid-state and dye lasers in which the mirrors are placed immediately at the ends of the gain medium. If a laser has a space of length $d - L$ between the gain medium and the mirrors, and if that cavity space has a different value for the index of refraction η_C than the index η_L of the laser gain medium, then a specific laser mode frequency associated with a mode number n can be expressed as

$$\nu = n\left(\frac{c}{2}\left[\frac{1}{\eta_C(d-L)+\eta_L L}\right]\right). \tag{10.40}$$

> **EXAMPLE**
>
> Determine the mode numbers at the extreme ends of the optical frequency region for the following examples: a wavelength of 700 nm in the red and a spacing of $d = 0.35$ m; and a wavelength of 400 nm in the blue and a spacing of $d = 0.4$ m.
>
> The mode numbers can easily be determined by using (10.39). In both cases we assume that $\eta = 1$, yielding
>
> red light: $\quad n = \dfrac{2\eta d\nu}{c} = \dfrac{2\eta d}{\lambda} = \dfrac{2 \cdot 1 \cdot 0.35 \text{ m}}{700 \times 10^{-9} \text{ m}} = 1{,}000{,}000;$
>
> blue light: $\quad n = \dfrac{2\eta d}{\lambda} = \dfrac{2 \cdot 1 \cdot 0.4 \text{ m}}{400 \times 10^{-9} \text{ m}} = 2{,}000{,}000.$

REQUIREMENTS FOR THE DEVELOPMENT OF LONGITUDINAL LASER MODES

Longitudinal modes may develop within any frequency or wavelength region in which:

(1) the gain within the laser amplifier at that frequency exceeds the losses according to (7.59) and (7.61), and
(2) there exists an integral value of n such that that specific frequency can be satisfied according to (10.39) or (10.40).

Thus, for example, the total frequency width over which all of the longitudinal modes of a specific laser transition occur could be somewhat smaller than the gain bandwidth of the gain medium for that transition, if the losses within the laser cavity are high. A laser could thus have just one longitudinal mode, or many thousands of longitudinal modes for gain media such as dye lasers or broadband solid-state lasers that have extremely wide gain bandwidths. A diagram indicating two such longitudinal modes within a laser cavity is shown in Figure 10-5.

It is possible that not all of the potential longitudinal modes will actually appear when observing the laser output. For the case of homogeneous broadening, the mode at the highest value of gain will develop first after the laser amplifier is turned on. As that mode develops and removes the population from the upper laser level (by converting the energy to photons of the laser beam), the entire gain spectrum will be reduced since all of the upper laser level population contributes equally at any wavelength over the emission spectrum for homogeneous broadening, as described in (4.65). Thus a typical homogeneously broadened laser might have only one longitudinal mode. This mode can be tuned by inserting additional tuning elements (variable frequency losses), as described in Section 12.5.

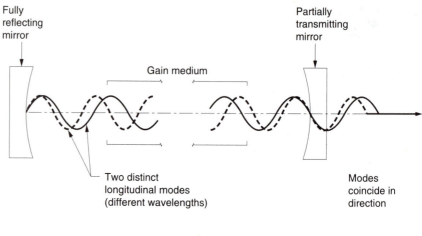

Figure 10-5. Diagram of two longitudinal laser modes

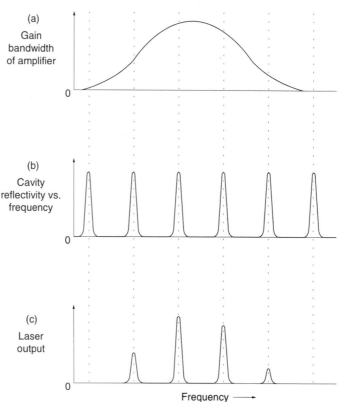

Figure 10-6. Resulting laser cavity modes when a gain bandwidth of a laser amplifier is combined with resonances of a two-mirror laser cavity

For an inhomogeneously broadened laser, such as for Doppler broadening, all of the modes that meet requirements (1) and (2) will be present, provided that the natural linewidth is narrower than the separation between modes as dictated by (10.31). The presence of many longitudinal modes will lead to an unusual gain distribution and "frequency hole burning," which will be described in Section 10.2.

A description of how the cavity modes appear within the gain bandwidth of an inhomogeneously broadened medium is shown in Figure 10-6. The gain bandwidth of the amplifier is shown in Figure 10-6(a). Figure 10-6(b)

shows the possible equally spaced Fabry–Perot resonances (possible cavity modes), as described by either (10.39) or (10.40). When these two effects are combined and the possible modes occur where there is net gain (i.e., where gain − losses > 0), we are left with the laser output spectrum shown in Figure 10-6(c). In this case, the laser has four longitudinal modes operating within the cavity. An example of the number of modes within a helium–neon laser cavity is described next.

EXAMPLE

A helium–neon laser operating at 632.8 nm has a mirror separation of 0.5 m. Determine the number of longitudinal modes operating within the laser cavity if the region of the spontaneous emission spectrum (gain spectrum) that exceeds the cavity losses occurs between the two frequencies at which the gain is half of the maximum.

We can first compute the frequency separation $\Delta\nu$ between individual longitudinal modes based upon the separation between mirrors of $d = 0.5$ m. From (10.38) we have

$$\Delta\nu = \frac{c}{2d} = \frac{3 \times 10^8 \text{ m/s}}{0.5 \text{ m}} = 6.0 \times 10^8 \text{ Hz} = 0.6 \text{ GHz}. \qquad (10.41)$$

The gain bandwidth (FWHM) of the helium–neon laser is determined by the Doppler width of the spontaneous emission, since the broadening is dominated by Doppler broadening (see Table 4-1) and (7.25). From Table 4-1 we can see that the FWHM gain bandwidth is therefore 1.5 GHz. Thus the number of modes of spacing 0.6 GHz that would fit within the 1.5-GHz gain bandwidth would be 3, since three modes would cover a frequency width of $2\Delta\nu = 1.2$ GHz. Four modes would require a frequency width of $3\Delta\nu = 1.8$ GHz, which is wider than the effective gain bandwidth as determined by taking into account the cavity losses as discussed previously.

10.2 TRANSVERSE LASER CAVITY MODES

In the previous section we analyzed the effects of two parallel reflecting surfaces (of infinite extent) on a plane wave arriving from a direction normal to, or nearly normal to, the reflecting surfaces. We then described how this would lead to discrete longitudinal modes within a laser cavity, since the laser beam was similar to that of a plane wave. However, the beam is *not* a plane wave because a plane wave must be of infinite lateral extent, and this is not possible within a laser amplifier. We will therefore consider the consequences of having a beam that is "almost" a plane wave. The finite lateral size of the beam, due to either the finite size of the mirrors or some other "limiting aperture" within the amplifier (such as the amplifier diameter), will cause diffraction of the beam to occur. This leads to losses within the laser cavity that were not considered in the preceding analysis of the plane wave.

Thus we will consider two modifications to our previous analysis. We will now assume that the laser mirrors are of finite extent and of circular shape, and we will also assume that the source of light originates from the laser amplifier region between the mirrors, thus altering our previous conditions of having plane waves incident upon the mirrors. This simulates the conditions in which a laser beam begins to develop, when the laser gain medium located between the mirrors is first activated or pumped. We will do the exact geometrical analysis assuming that the mirrors are flat or "plane parallel," so that the description of the optical rays is simplified. We will first obtain the expression for the transverse profile of the beam that builds up within the cavity after having undergone many reflections as the beam oscillates back and forth between the mirrors. We will then compare these results for plane-parallel mirrors with those derived elsewhere for slightly curved mirrors. It will be shown that mirrors of certain specific curvatures (in relation to the mirror separation distance d) exhibit much lower diffraction losses for each reflection of the developing laser beam than do plane-parallel mirrors, and are therefore generally more desirable for a laser cavity.

FRESNEL–KIRCHHOFF DIFFRACTION INTEGRAL FORMULA

We will use a technique first developed by Huygens in the seventeenth century for the analysis of a number of optical interference effects, including diffraction. In his analysis, light could emanate from a source S, expanding spherically with a $1/r^2$ intensity dependence, until it reached an aperture A such as a circular aperture shown in Figure 10-7. At this point, secondary "wavelets" would emanate from this aperture region to arrive at any point P on the other side of the aperture. The distance from S to a specific point on A is represented by the position vector $\mathbf{r}'$, and the distance from P to the same point on A is represented by the vector $\mathbf{r}$.

The Fresnel–Kirchhoff integral formula,

$$U_P = -\frac{ik}{4\pi} \int\int_A U_0 \frac{e^{ikr'}}{r'} \frac{e^{ikr}}{r} [\cos(\mathbf{n}, \mathbf{r}) - \cos(\mathbf{n}, \mathbf{r}')] \, dA, \qquad (10.42)$$

can then be used to obtain the value of the wave function U_P arriving at P, resulting from the source S, after having passed through aperture A. In

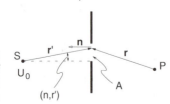

Figure 10-7. Symbols used in the Fresnel–Kirchhoff diffraction integral formula

this formula, U_P represents the wave function describing the wave *amplitude* at P from all secondary wavelets originating from the aperture region of A, each of which originated from a point S of amplitude U_0. The integral occurs only over the aperture A, and it is assumed that points outside of that aperture provide no contribution to the integral. In (10.42) the factors $e^{ikr'}/r'$ and e^{ikr}/r represent the amplitude functions of expanding spherical waves leaving the source point S and the aperture A, respectively. In these expressions, r and r' represent the magnitudes of the vectors $\mathbf{r}$ and $\mathbf{r'}$. The factors $(\mathbf{n}, \mathbf{r})$ and $(\mathbf{n}, \mathbf{r'})$ are the angles that the vectors $\mathbf{r}$ and $\mathbf{r'}$ make with the normal to the surface of integration, where the normal $\mathbf{n}$ is always defined as positive with respect to the point P. Equation (10.42) is obtained from the use of Green's function, which is designed to evaluate scalar point functions, such as U, that satisfy conditions of integrability and continuity and also satisfy the wave equation.

In our evaluation of diffraction effects due to electromagnetic waves traversing between circular laser mirrors, we will consider a special case of the Fresnel–Kirchhoff integral formula in which the source S is located symmetrically with respect to the aperture A of Figure 10-7. We will take the surface of integration for the source function at the aperture to be a sphere of radius r' equal to the distance from the source to the edge of the aperture. In this case the magnitude of the vector $\mathbf{r'}$ is a constant over the sphere of integration, and the normal vector $\mathbf{n}$ always points in the opposite direction of $\mathbf{r'}$. For this case we can thus express the Fresnel–Kirchhoff integral formula as

$$U_P = -\frac{ik}{4\pi} \iint_A U_A \frac{e^{ikr}}{r} [\cos(\mathbf{n}, \mathbf{r}) + 1]\, dA. \tag{10.43}$$

Here, $\cos(\mathbf{n}, \mathbf{r'}) = -1$ and $U_A = U_0(e^{ikr'}/r')$.

DEVELOPMENT OF TRANSVERSE MODES IN A CAVITY WITH PLANE-PARALLEL MIRRORS

We will now apply the Fresnel-Kirchhoff integral formula of (10.43) to the case of a laser amplifier as described at the beginning of this section. We will consider two parallel circular mirrors separated by a distance d, as shown in Figure 10-8. We will consider a distribution of light beginning at various points on the primed mirror, radiating toward the unprimed mirror, and then reflecting back to a point on the primed mirror. For a steady-state "mode" to develop, the amplitude distribution of the light on the primed mirror must have the same transverse spatial dependence as that on the unprimed mirror. We therefore consider a source point function $U(x, y)$ at point (x, y) on the unprimed mirror, which is the sum of the contributions of radiation from all points leaving the primed mirror that arrive at the location (x, y) on the unprimed mirror. This source point $U(x, y)$ then radiates back to the primed mirror to arrive at various points (x', y') with an amplitude function $U'(x', y')$ on that mirror, after having traveled a distance r. In this situation, we define r as

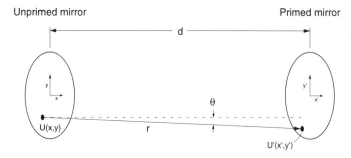

Figure 10-8. Two parallel circular mirrors considered as apertures when applying the Fresnel–Kirchhoff integral formula to a laser cavity

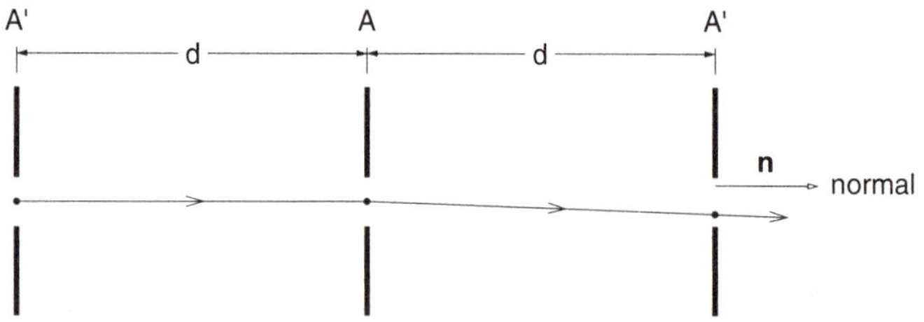

$$\cos\theta = \frac{d}{r}$$

$$r = [d^2 + (x' - x)^2 + (y' - y)^2]^{\frac{1}{2}}$$

Figure 10-9. Equivalent aperture description of a two-mirror reflective laser cavity

$$r = (d^2 + (x'-x)^2 + (y'-y)^2)^{1/2}, \tag{10.44}$$

and θ is defined as the angle between d and r.

We now wish to determine the distribution of radiation $U'(x', y')$ on the primed mirror as a result of the radiation of distribution $U(x, y)$ at the unprimed mirror (which originated previously from radiation from the primed mirror). It will be easier to picture the application of the Fresnel–Kirchhoff integral formula if the two mirrors in Figure 10-8 are transposed into a series of successive apertures. This is possible since, for plane mirrors, sequential images through two mirrors appear as a successive row of virtual images of the mirror apertures, each spaced a distance d from the preceding mirror aperture as shown in Figure 10-9.

Thus we ask: What is the distribution $U'(x', y')$ at the primed aperture resulting from a distribution $U(x, y)$ at the unprimed aperture? We can apply (10.43) to this situation and obtain

$$U'(x', y') = -\frac{ik}{4\pi} \int\!\!\int_A U(x, y) \frac{e^{ikr}}{r} (\cos\theta + 1)\, dx\, dy, \tag{10.45}$$

where $U'(x', y')$ is the amplitude function at the primed mirror of Figure 10-8 resulting from radiation from the unprimed mirror of the figure, which in turn originated from the primed mirror of the figure. In this case, by using (10.45), the source is again assumed to be symmetrical with respect to the primed aperture, and $\cos(\mathbf{n}, \mathbf{r}) = \cos\theta$.

We are seeking solutions for the case in which the light has bounced back and forth between the mirrors many times. We therefore seek functions describing the light that originates at the left mirror (unprimed) of Figure 10-8 and passes to the right mirror with *no change in the transverse shape* of the amplitude of the function. This is equivalent in Figure 10-9 to the light leaving the left aperture A', passing through the unprimed aperture A, and arriving at the right aperture A' with no change in shape (although reduced by the factor γ). This could thus be designated as a stable situation that would perpetuate itself within a laser amplifier, since only functions that repeat themselves would lead to a steady-state beam within the amplifier. We must include the possibility of a decrease in the amplitude at all values of x' and y' by a constant factor of γ, which is determined by diffraction. This diffraction loss is in addition to the losses described in Section 7.6 for the threshold conditions for a laser with two mirrors and included in the threshold calculations of (7.59) and (7.61). This decrease in amplitude, as well as the mirror reflection losses and absorption and scattering losses, must be made up by the gain in the laser amplifier if a laser beam is to develop.

We thus need solutions such that the two functions U' and U are proportional ($U' \propto U$) for every point (x', y') on the primed mirror. This can be expressed by using the constant factor γ such that

$$U'(x', y') = \gamma U(x', y') = \int\int U(x, y) K(x, y, x', y') \, dx \, dy, \qquad (10.46)$$

where

$$K(x, y, x', y') = -\frac{ik}{4\pi}(\cos\theta + 1)\frac{e^{ikr}}{r}. \qquad (10.47)$$

Equation (10.46) is an integral equation in U; K is known as the *kernel* of that equation and γ is the *eigenvalue*. The γ in this analysis has no relation to the γ used in previous chapters to denote a decay rate. We use γ here to be consistent with the original theories published in this field. It will not be carried through to any resulting formulas, and will therefore not lead to a potential conflict with the γ associated with decay rates.

There are an infinite number of solutions U_n and γ_n to (10.46); each set is associated with a specific value of n, where n can take on values of $n = 1, 2, 3, \ldots$. These solutions correspond to the normal modes of this resonator. They are referred to as *transverse modes* because they represent amplitude distributions of the electromagnetic field in directions transverse to the propagation of the laser beam within the resonator.

We will describe γ_n by allowing both an amplitude and a phase factor as follows,

$$\gamma_n = |\gamma_n| e^{i\phi_n}, \qquad (10.48)$$

where $|\gamma_n|$ is the ratio of the amplitudes of the distributions after two successive passes from one mirror to the next, and ϕ_n represents the possible phase shift. The energy loss per transit is then given as

$$\text{loss/transit} = 1 - |\gamma_n|^2, \tag{10.49}$$

where $|\gamma_n|^2$ represents the square of the field amplitude, or the intensity of the radiation.

A solution of (10.45) can then be obtained by making a simple approximation for its kernel:

$$K(x, y, x', y') = Ce^{-ik_1(xx'+yy')}, \tag{10.50}$$

where C and k_1 are constants. Now we can rewrite (10.46) as

$$\gamma U(x', y') = C \int\int_A U(x, y) e^{ik_1(xx'+yy')} \, dx \, dy. \tag{10.51}$$

In this equation it can be seen that $U(x, y)$ is its own Fourier transform. The simplest solution of such a function is the Gaussian function

$$U(x, y) = e^{-\rho^2/w^2} = e^{-(x^2+y^2)/w^2}, \tag{10.52}$$

where w is a scaling constant and $\rho^2 = x^2 + y^2$ is radial distance to any location on the mirror from the central point on the mirror. This function $U(x, y)$ gives the variation of the distribution of the electric field amplitude over the mirror at various locations (x, y). The Gaussian expression of (10.52) is a symmetrical distribution that decreases with distance ρ from the center of the mirror ($\rho = 0$). The term w in this expression represents the value of ρ at which the electric field amplitude decreases to a value of $1/e$ from the center of the mirror, or the value of ρ at which the intensity of the beam decreases to $1/e^2 = 0.86$ of the value at the center of the mirror (since the intensity is the absolute value of the square of the field amplitude).

Functions that are their own Fourier transforms can be written more generally as the products of Hermite polynomials and the Gaussian distribution function of (10.52):

$$U_{pq}(x, y) = H_p\left(\frac{\sqrt{2}x}{w}\right) H_q\left(\frac{\sqrt{2}y}{w}\right) e^{-(x^2+y^2)/w^2}. \tag{10.53}$$

In this solution, p and q are integers that designate the order of the Hermite polynomials. Thus every set of (p, q) represents a specific stable distribution of wave amplitude at one of the mirrors, or a specific transverse mode of the open-walled cavity. We can list several Hermite polynomials as follows:

$$H_0(u) = 1, \qquad H_1(u) = 2u,$$

$$H_2(u) = 2(2u^2 - 1),$$

$$H_m(u) = (-1)^m e^{u^2} \frac{d^m(e^{-u^2})}{du^m}, \tag{10.54}$$

where u denotes one or the other of the terms $\sqrt{2}x/w$ or $\sqrt{2}y/w$ from (10.53). We can designate these transverse mode distributions as TEM$_{pq}$, where TEM stands for "transverse electromagnetic." The lowest-order mode is designated TEM_{00}. Such modes could also be written as TEM_{npq}, where n would designate the longitudinal mode number of (10.39) or (10.40). Since this longitudinal mode number is generally very large for optical frequencies, it is not normally written. For example, for a gas laser with a mirror separation $d = 0.5$ m operating at a green wavelength of 500 nm, from (10.39) we have $n = 2d\nu/c = 2d/\lambda = (2 \cdot 0.5$ m$)/(5 \times 10^{-7}$ m$) = 2 \times 10^{6}$, where we have assumed $\eta \cong 1$ for a gas laser. It would clearly be too cumbersome to include such longitudinal mode numbers in the TEM designation.

TRANSVERSE MODES USING CURVED MIRRORS

The derivation leading to transverse modes was obtained by considering plane-parallel mirrors located at the ends of the laser amplifier. However, the use of slightly curved, spherically shaped mirrors was considered in the early development of laser cavities. An analysis similar to that of the previous section was made for a laser beam as it developed between such curved mirrors. Using curved mirrors, it was thought that if the beam were focused slightly after each reflection then the value of the beam amplitude near the edges of the mirror would be reduced, thus reducing the amount of beam amplitude subject to diffraction losses. These considerations turned out to be justifiable.

An analysis was made to compare the diffraction losses for both plane-parallel mirrors and curved mirrors. The *Fresnel number* or *confocal parameter* $N = a^{2}/\lambda d$ was used as a variable in which a denotes the radius of the mirror (or limiting aperture at the mirror), λ the laser wavelength, and d the separation between mirrors. For the cavity with plane-parallel mirrors, the diffraction losses, described as the fractional power loss per pass through the resonator, are shown in Figure 10-10 plotted against the Fresnel

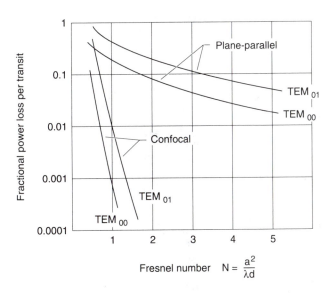

Figure 10-10. Fractional power loss per transit versus Fresnel number for a laser cavity

number N for TEM$_{00}$ and TEM$_{01}$ modes. Also shown is a plot of the fractional losses per pass for a specific type of curved mirror cavity, known as a *confocal* cavity, in which the radius of curvature of the mirrors is equal to the separation d between the mirrors. In this case the diffraction losses can be seen to be significantly lower than those for a plane-parallel resonator, especially for a cavity at larger Fresnel numbers. As a consequence, most of the laser mirrors presently used are slightly curved rather than plane. Moreover, the alignment tolerances for curved mirrors are much less stringent than those for plane-parallel mirrors. Different types of curved mirror cavities will be considered in more detail in Chapter 11.

TRANSVERSE MODE SPATIAL DISTRIBUTIONS

We have obtained the transverse electric field profile of various modes at the laser mirrors, as determined by the product of the Hermite polynomials and the basic Gaussian function described in (10.53). If there are no radially nonsymmetric losses within the laser cavity, such as those induced by Brewster angle windows (to be described shortly), then the laser beam that develops within the cavity will have circular symmetry and the evolving modes will have circularly shaped patterns. The transverse mode distribution for curved mirrors is similar to that for plane-parallel mirrors, and we can therefore use the distributions obtained in (10.53) and (10.54) for either case.

EXAMPLE

Write out the mode distributions at the mirrors for the TEM$_{00}$, the TEM$_{01}$ and the TEM$_{11}$ modes in terms of the transverse variables x, y and $\rho = \sqrt{x^2 + y^2}$.

From (10.53) we can write out the distribution of the electric field on the mirrors for the TEM$_{00}$ mode as follows:

$$U_{00}(x, y) = \text{TEM}_{00} = H_0\left(\frac{\sqrt{2}x}{w}\right) H_0\left(\frac{\sqrt{2}y}{w}\right) e^{-(x^2+y^2)/w^2}$$

$$= e^{-(x^2+y^2)/w^2} = e^{-\rho^2/w^2},$$

given that H_0 for both x and y is unity. For the TEM$_{01}$ mode we have

$$U_{01}(x, y) = \text{TEM}_{01} = 1 \cdot H_1\left(\frac{\sqrt{2}y}{w}\right) e^{-(x^2+y^2)/w^2} = \frac{2\sqrt{2}y}{w} e^{-(\rho^2)/w^2}.$$

In this case, the value of H_p for the x direction is unity; H_q for the y direction is given from (10.54) as $H_1 = 2\sqrt{2}\,y/w$. For the TEM$_{11}$ mode, the value of H is the same for both the x and y directions, so we have the product of $H_1 \times H_1 = H_1^2$:

$$U_{11}(x, y) = \text{TEM}_{11} = \left(\frac{2\sqrt{2}x}{w}\right)\left(\frac{2\sqrt{2}y}{w}\right) e^{-(x^2+y^2)/w^2}$$

$$= \frac{8xy}{w^2} e^{-(x^2+y^2)/w^2}.$$

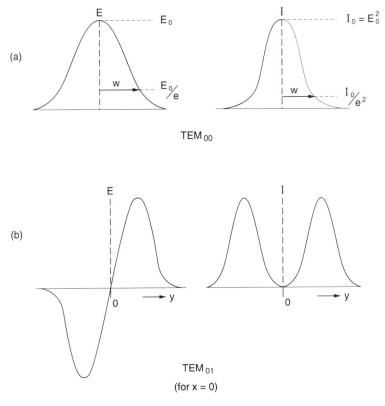

(a)

E E_0

I $I_0 = E_0^2$

w E_0/e

w I_0/e^2

TEM$_{00}$

(b)

E

I

0 y

0 y

TEM$_{01}$
(for $x = 0$)

Figure 10-11. Electric field and intensity distributions at the laser mirrors for (a) a TEM$_{00}$ laser mode and (b) a TEM$_{01}$ laser mode

The TEM$_{00}$ mode has a Gaussian distribution of the form $E_0 e^{-\rho^2/w^2}$ for the electric field as shown in Figure 10-11(a), where E_0 is the value of the electric field at the center of the mirror ($\rho = 0$). Adjacent to that diagram is the intensity distribution, which is determined by taking the square of the electric field at each point on the mirror such that $I_0 = E_0^2$. The radial distribution of the intensity from the center of the mirror is then $I_0 e^{-2\rho^2/w^2}$. The values of the electric field distribution and intensity for the TEM$_{01}$ mode on the mirrors are also shown in Figure 10-11(b).

TRANSVERSE MODE FREQUENCIES

Earlier in this chapter we described longitudinal modes as modes having the same optical path within the cavity, but with slightly different frequencies depending upon the mode number n. Different *transverse* modes are modes having either the same or different values of n (the same longitudinal mode number), but with different values of p and/or q. Transverse modes with the same n but with different values of p and q would have slightly different optical path lengths d owing to slightly different angular

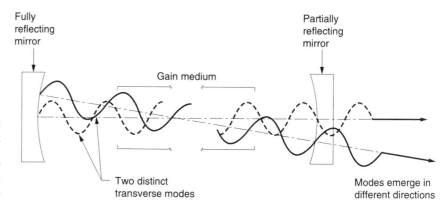

Fully
reflecting
mirror

Partially
reflecting
mirror

Gain medium

Two distinct
transverse modes

Modes emerge in
different directions

Figure 10-12. Simplified description of two distinct transverse laser modes, showing the larger effective path length for an off-axis mode

distributions of the amplitude functions within the cavity, as suggested by the transverse mode distributions of (10.53). Such modes consequently have slightly different frequencies, according to (10.39) or (10.40), because of the different value of the distance d. Generally the frequency difference between longitudinal modes is significantly greater than the frequency difference between transverse modes. However, in a cavity with two mirrors, a standing-wave pattern is always produced that leads to longitudinal modes. Hence every transverse mode, if examined in the frequency domain, will be found to consist of one or more longitudinal modes, depending on the separation d between mirrors as well as on the gain bandwidth of the laser. A simplified example of two different transverse modes having different frequencies is shown in Figure 10-12, where the two transverse modes have slightly different effective path lengths within the cavity. The off-axis mode is not realistic in that transverse modes do not necessarily follow a straight-line path between the mirrors as shown in the figure. For example, the TEM_{11} mode distribution actually follows a slightly curved path traveling between the mirrors.

GAUSSIAN-SHAPED TRANSVERSE MODES WITHIN AND BEYOND THE LASER CAVITY

In the foregoing analysis for transverse modes, we were primarily concerned with the profile and distribution of the electric field and the intensity at the laser mirrors, such that the same beam profile was duplicated after many passes of the beam through the amplifier. We did not address the issue of what the beam profile looks like, within the amplifier or elsewhere, other than at the mirrors. In Chapter 11 we will investigate the mode shape and size in other locations within the cavity and beyond the cavity. Since we did not specify the exact separation of the mirrors in demonstrating the existence of the Hermite–Gaussian mode distributions, we can argue that the mirrors could have been moved either closer together or farther apart and we still would have obtained the Hermite-Gaussian distribution. The only difference we would have found is that, in the case of curved mirrors, the parameter w of (10.52) and (10.53) would have been smaller if the mirrors were closer together and larger if they were farther apart. In effect, the mode retains the same transverse shape as it propagates back and forth between the mirrors. The TEM_{00} mode would have been found to have a Gaussian shape everywhere, including between the mirrors and also beyond the mirrors (for the part of the beam that is transmitted through the mirrors). Consequently, because transverse laser modes involve Gaussian functions, we will be able to develop equations and techniques for describing such beams at any spatial location. This topic is left for the next chapter. We must remind ourselves, however, that each of these transverse modes consists of one or more longitudinal (standing-wave) modes, as described earlier in the chapter. Transverse modes can also be described as spatial modes, whereas longitudinal modes are more associated with frequencies.

Images of several mode distributions are shown in Figure 10-13(a) as they might be observed if a laser beam were projected onto a screen. These mode patterns occur for a completely symmetric laser medium in the directions transverse to the direction of travel of the laser beam. The patterns comprise waves of all polarizations, since there are no polarization discriminating optical elements within the laser cavity.

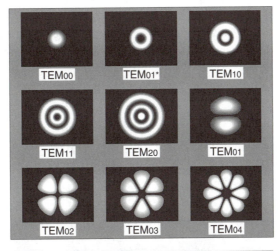

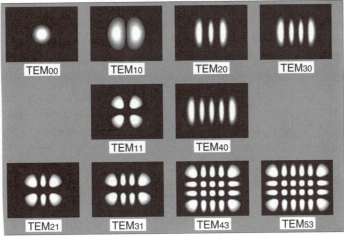

Figure 10-13. Mode patterns for various transverse laser modes: pure modes in (a) circular symmetry and (b) (x, y) symmetry

● BREWSTER ANGLE WINDOWS WITHIN THE LASER CAVITY   In Figure 10-13(b), the mode patterns have a definite preferred orientation rather than a cylindrical symmetry. This is determined entirely by a radially nonsymmetric loss within the cavity that is lower for a beam propagating with the electric field in one transverse direction than the other. In Figure 10-13(b), the presence of Brewster angle windows within the laser cavity introduces such a radial variation of the beam loss.

The Brewster angle window arrangement is used to minimize the losses resulting when the laser beam passes from the amplifier to the region between the gain medium and the laser mirrors. It is particularly useful when the laser mirrors are mounted external to the laser gain medium rather than directly at the ends of the gain medium, or at the ends of a gas laser discharge tube in the case of some gas lasers. At Brewster's angle there is essentially no loss, as shown in Figure 10-14(a), for the component of the laser beam polarized in the transverse direction that lies in a plane normal to the plane of the window and perpendicular to the direction of propagation of the beam, as indicated by the vertical arrows in Figure 10-14(b). The significance in minimizing such losses as the beam propagates back and forth within the cavity was considered in (7.61), and will also affect the expression in (11.82) for estimating laser power output. A maximum loss (15% for a glass reflector) during transmission through the Brewster angle window occurs for the electric field vector, polarized in a transverse

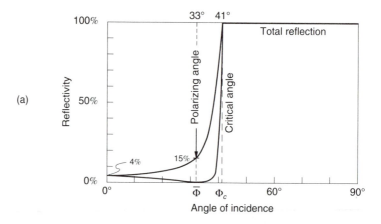

(a)

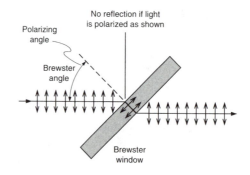

Figure 10-14. (a) Reflected intensity versus angle for light reflected from a dielectric interface; (b) Brewster angle laser window providing very low reflection loss for light polarized in the plane of the figure

(b)

direction that is in the plane of the window and perpendicular to the beam propagation direction. This also is shown in the graph of Figure 10-14(a).

10.3 PROPERTIES OF LASER MODES

MODE CHARACTERISTICS

At this point we can summarize the characteristics of laser modes. It has been pointed out earlier in this chapter that it is possible to have more than one laser mode, either transverse or longitudinal, oscillating simultaneously within the laser cavity.

SPATIAL DEPENDENCE Each mode with its associated mode number (n, l, m) in a typical two-mirror, standing-wave laser cavity is in fact a distinct standing wave, and has a value of the electric field of zero at the mirrors. Thus each transverse mode must be associated with a specific standing wave and thus to a specific longitudinal mode number n as defined in (10.38), as well as to unique transverse mode numbers l and m. This distinct standing-wave pattern for each mode is a distinct spatial distribution of laser intensity between the mirrors that is slightly different from that for any other mode, as indicated in Figure 10-13. Thus, although all operating modes typically use mostly the same gain medium, they can also access different spatial regions of the gain medium owing to their different shapes within the laser cavity.

FREQUENCY DEPENDENCE Each mode also has a slightly different frequency. Two transverse modes could have the same mode number n and yet have different frequencies, since they may have slightly different optical path lengths d within the cavity as defined in (10.38) and depicted in Figure 10-12. However, adjacent longitudinal modes (where n differs by 1) typically have a greater frequency difference (as described by eqn. 10.31) than do two transverse modes with the same longitudinal mode number n but different values of p and q (as defined in eqn. 10.53).

MODE COMPETITION For the case of homogeneous broadening, the waves associated with different modes within the same gain region are all competing for the same upper laser level species after the laser amplifier is turned on and the various modes begin to develop. Each mode is attempting to grow toward reaching its saturation intensity by stimulating more emission into its mode. The mode at the center of the gain profile, where the gain is highest, will reach its saturation intensity first; then the entire gain curve will begin to decrease, since every upper laser species is affected by that saturation according to (7.70). Thus it will be difficult for more than one mode to lase unless the weaker mode can "feed on" a spatial region of gain that is distinct from that of the strong mode. Hence it is common, in the case of homogeneous broadening, for more than one transverse mode but only a single longitudinal mode to lase, since the transverse modes have distinctly different spatial regions.

In the case of inhomogeneous broadening, different longitudinal modes can operate somewhat independently if their natural linewidths do not overlap, since they are not competing for the same upper laser level species. However, even if the distinct longitudinal modes are sufficiently separated in frequency so as not to interfere with each other, transverse modes with the same longitudinal mode number n can be close enough in frequency that they may compete for the same upper laser level species. They must therefore seek their additional gain in different spatial regions from that of the strongest mode, which is typically the fundamental TEM_{00} mode. Thus close proximity, either spatial or in frequency, can lead to competition between modes.

EFFECT OF MODES ON THE GAIN MEDIUM PROFILE

Laser modes that reach saturation intensity have a significant effect on the gain profile within the laser amplifier, as pointed out in (7.70). Each stimulated photon results in a decrease in the population of the upper laser level by one and an increase in the population of the lower laser level by one, for a net change of two. Thus when a laser beam that might consist of 10^{10} photons bounces back and forth within the cavity, the populations of the upper and lower laser levels, and thus the gain, can be altered dramatically from that which existed before the beam developed, as indicated in Figure 7-13. Two effects of mode competition that can have a significant impact upon this gain profile are spectral and spatial hole burning.

SPECTRAL HOLE BURNING   It was mentioned previously (and described in Section 7.7; see Figure 7-13) that for a gain spectrum that is homogeneously broadened, as a longitudinal laser mode develops, the stimulated emission process will reduce the gain profile to a value at which it equals the losses within the laser cavity. These losses include mirror transmission and absorption losses and scattering losses as discussed in Section 7.6, as well as absorption losses within the gain medium as described in Section 8.4. In contrast, for the inhomogeneous Doppler-broadened medium typical of most gas lasers, the population in the upper laser level will be reduced only at the frequencies where the modes are developing, since different populations within the upper laser level contribute to different frequency components of the gain spectrum as described by (4.69). The gain will thus be depleted only at the locations of those modes over a frequency width equivalent to the natural linewidth of each of those radiating atoms. Thus, if the natural emission linewidth of the transition is significantly narrower than the Doppler width (which it is for most visible gas lasers), then the gain spectrum while the laser is operating will have a periodic variation

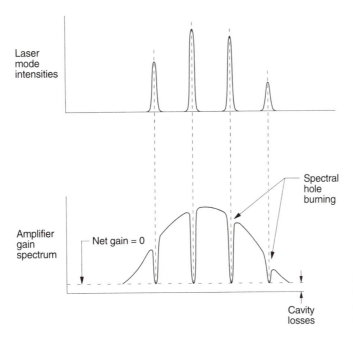

Laser
mode
intensities

Spectral
hole
burning

Amplifier
gain
spectrum

Net gain = 0

Cavity
losses

Figure 10-15. Laser gain
distribution within a laser
amplifier due to spectral
hole burning

spaced according to the longitudinal mode spacing shown in Figure 10-15. The gain will be reduced to the value of the cavity losses at the frequencies given by either (10.39) or (10.40), as shown in the figure. This effect is known as *spectral hole burning*. If the natural linewidth is nearly that of the Doppler width, then the gain medium becomes more like a homogeneously broadened medium. The modes then compete with each other, and there is a tendency to have fewer longitudinal modes lasing.

SPATIAL HOLE BURNING When the standing-wave pattern of a single longitudinal mode develops within the homogeneously broadened laser cavity that is typical of most solid-state lasers, the laser intensity pattern along the length of the gain medium is regularly spaced as shown in Figure 10-16. The spacings occur at half-wavelength intervals, with the intensity being zero at the cavity mirrors, and at every half-wavelength interval between the mirrors. As long as the cavity length remains the same and the amplifier gain remains constant, this pattern remains stable. Within the amplifier, at those null points of the intensity pattern where the electric field is zero, there is no stimulated emission and thus no reduction in the gain according to (7.71). The gain profile thus has a periodic spatial variation within the gain medium that is 90° ($\pi/2$) out of phase with the laser intensity profile within the gain medium, as shown in Figure 10-16. Such a spatial distribution leads to wasted gain. It is possible that higher-order transverse modes if present will use part of this gain, but otherwise it is wasted energy. Ring lasers have been developed (see Section 12.4) to take advantage of this extra gain by eliminating the standing-wave mode pattern and thereby making a more effective use of the gain medium.

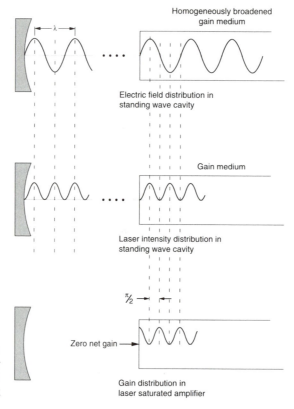

Figure 10-16. Laser gain distribution within a laser amplifier due to spatial hole burning

REFERENCES

LONGITUDINAL MODES

M. Born and E. Wolf (1980), *Principles of Optics,* 6th ed. New York: Pergamon.
A. E. Siegman (1986), *Lasers.* Mill Valley, CA: University Science.

TRANSVERSE MODES

G. D. Boyd and J. P. Gordon (1961), "Confocal multimode resonator for millimeter through optical wavelength masers," *Bell System Technical Journal* 40: 489–508.
A. G. Fox and T. Li (1961), "Resonant modes in a laser interferometer," *Bell System Technical Journal* 40: 453–88.
H. Kogelnik and T. Li (1966), "Laser beams and resonators," *Proceedings of the IEEE* 5: 1550–67.

PROBLEMS

1. Show that the two expressions for I_t of (10.12) and (10.16) are equivalent; that is, show that

$$I_0 \frac{T^2}{|1 - Re^{i\Phi}|^2} = I_0 \frac{T^2}{(1-R)^2} \frac{1}{1 + F' \sin^2(\Phi/2)}.$$

2. Compute the FWHM linewidth for a plane-parallel Fabry–Perot inter-ferometer with two identical mirrors, separated by a distance of 1 cm, with air between the mirrors. Assume mirror reflectivities of 99.99%, 95%, and 70%. De-termine the cavity Q for these three cases if the output of a GaAs diode laser op-erating at 800 nm is transmitted through the interferometer.

3. A very narrow-frequency 10-mW cw single-mode He–Ne laser beam is trans-mitted through an air-spaced Fabry–Perot interferometer with the laser wavelength exactly corresponding to one of the 100% transmission peaks of the interferome-ter, as indicated in Figure 10-4. The reflecting surfaces of the interferometer, sep-arated by a distance of 0.1 m, are dielectric coatings with a reflectivity at the laser wavelength of 99% and no absorption losses. What is the intensity of the beam inside of the Fabry–Perot cavity? Assume that the laser is instantly shut off in a time of the order of 1 ns or less. How long will it then take for the beam inside the Fabry–Perot cavity to decay to $1/e$ of its steady-state value prior to the shut-off?

4. A 0.1-m–long Nd:YAG laser rod is coated with a 99% reflector on one end and a 95% reflector on the other end. Determine the longitudinal mode spacing. How many modes would be operating if the gain bandwidth is a Lorentzian shape with a FWHM of 0.45 nm and a peak at 1.06 μm? Assume that the maximum gain is 10% at the peak of the emission spectrum and that the cavity losses (in addition to mirror losses) are 2% per pass. The index of refraction of Nd:YAG at 1.06 μm is 1.82. *Note:* Since the transmission losses are different at each mirror, you must consider the total round trip gain and loss.

5. A helium–cadmium laser is operating at 441.6 nm with a natural mixture of Cd in the discharge. Assume the laser has a gain of 12% for the isotope 114 at its maximum emission frequency. The laser mirrors are separated by a distance of 0.4 m and the mirrors have reflectivities of 96% and 99.9% at the laser wave-length. Assume that there are scattering losses of 1% per pass through the cavity. Also remember that the index of refraction within a gas laser is 1. How many longitudinal modes would be lasing in this system? Refer to Figure 4-14 for infor-mation about isotopes of Cd.

6. Show that the solution for a stable mode passing back and forth between two plane-parallel mirrors within a limiting aperture of radius a, as given by (10.51), is the Gaussian function $U(x, y) = e^{-(x^2+y^2)/w^2}$ as expressed in (10.52).

7. Show that, for a TEM$_{00}$ mode passing between two laser mirrors, the transverse displacement – from the center axis, where (according to eqn. 10.53) the intensity reaches a maximum, to the location where the intensity of the mode has a value of $1/e^2$ of the maximum – corresponds to the value w in the (10.53) (for a Gaussian mode distribution).

8. Write out the expressions for the transverse spatial distribution at the mir-rors of the TEM$_{11}$, TEM$_{20}$, and TEM$_{22}$ modes.

9. For a confocal cavity with a Fresnel number of 1, compare the losses per pass with those of a plane-parallel resonator for both TEM$_{00}$ and TEM$_{01}$ modes. What are the implications of having a Fresnel number of 1 for a visible laser oper-ating at 500 nm and an infrared laser operating at 10 μm?

10. Assume an argon ion laser is operating at 488.0 nm in a confocal cavity in the TEM$_{00}$ mode with a mirror separation of 0.6 m. Assume that the gain medium is 0.4 m long. What must the gain coefficient be for the laser to operate at threshold with mirror reflectivities of 99.99% at the laser wavelength? Assume that the beam diameter (the effective limiting aperture) for the laser beam is 0.5 mm at each mirror. Assume also that the only losses in the cavity are the diffraction losses at the mirrors. (The mirrors are mounted at the ends of the laser discharge tube such that there are no Brewster angle windows.)

STABLE LASER RESONATORS AND GAUSSIAN BEAMS

11

SUMMARY In Chapter 10 we described the longitudinal and transverse laser cavity modes that are produced when a laser amplifier is placed within an optical cavity. The properties of those modes were shown to be associated with the cavity properties and the laser wavelength. We also indicated that curved mirrors gave much lower diffraction losses than plane-parallel mirrors. In this chapter we will first consider several possible types of curved mirror cavities. We will then determine which of those cavities provide stable modes by doing an analysis using *ABCD* matrices. Stability will be determined by considering which laser cavities would allow a laser beam to operate on a steady-state basis with a uniform stable power output. The properties of a hypothetical Gaussian-shaped beam that originates within the laser cavity will be considered, and expressions for determining how such a beam would propagate will be provided. We will then examine real laser beams and indicate how they propagate as compared to ideal Gaussian beams. A simple way to determine the propagation of Gaussian beams through various optical elements using *ABCD* matrices will then be discussed. Finally, we will consider how to determine the optimization of the output coupling mirror of a laser in order to obtain the maximum power output from the laser.

11.1 STABLE CURVED MIRROR CAVITIES

CURVED MIRROR CAVITIES

We suggested in Chapter 10 that curved mirrors have lower diffraction losses than plane-parallel mirrors. An analysis of the reduced losses for curved mirrors was shown in Figure 10-10 for a confocal curved mirror cavity, where the radius of curvature of the mirrors is equal to the spacing between the mirrors. There are a number of different types of curved mirror laser cavities, as shown in Figure 11-1. At first glance they all look like they would make reasonable laser cavities. As indicated in the figure, the cavities are distinguished from each other in terms of the relative value of the radius of curvature of the mirrors in comparison to the separation distance d between the mirrors. However, using these definitions does not tell us whether or not the cavities will function as useful, stable laser cavities. We deduced in Chapter 10 that, in order for a beam to evolve into a stable steady-state beam, the beam profile at the mirrors must duplicate itself after successive passes through the amplifier, except for an amplitude factor γ that is constant over the profile as suggested in (10.46).

Our approach will therefore be to analyze the round-trip propagation of a beam from one mirror to the other many successive times, and determine

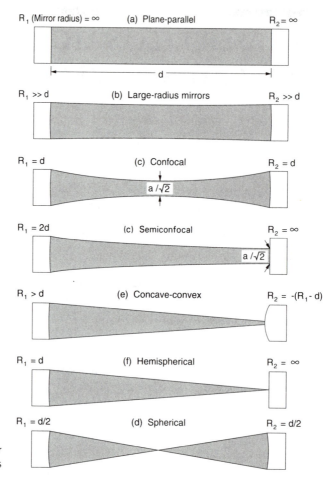

Figure 11-1. Two-mirror laser cavities

the conditions for which the beam remains concentrated within the cavity as opposed to diverging out of the cavity. Those conditions for which it converges will then be designated as the stable conditions. We will use ray matrix theory – incorporating what are often called *ABCD* matrices – to aid us in developing the stability criteria for curved laser resonators, since it offers a convenient way of mathematically allowing a large number of round-trip passes of the beam to be made within the resonator.

ABCD MATRICES

Ray matrices, or *ABCD* matrices, offer a convenient form for describing the propagation of optical rays through various optical elements. Because they are matrix representations, a sequence of events can be combined by matrix multiplication to yield a final result, such as an image of an object in the form of a single matrix. It is thus possible, for example, to propagate radiation from an object to a lens, refract it through the lens, propagate it to another lens, refract through that lens and then propagate it to the image location, and to describe all these events with a single matrix equation.

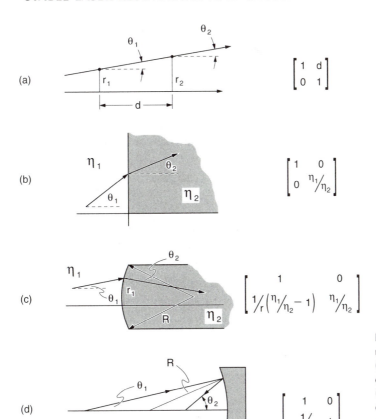

(a)

$$\begin{bmatrix} 1 & d \\ 0 & 1 \end{bmatrix}$$

(b)

$$\begin{bmatrix} 1 & 0 \\ 0 & \eta_1/\eta_2 \end{bmatrix}$$

(c)

$$\begin{bmatrix} 1 & 0 \\ 1/r\left(\eta_1/\eta_2 - 1\right) & \eta_1/\eta_2 \end{bmatrix}$$

(d)

$$\begin{bmatrix} 1 & 0 \\ -1/f & 1 \end{bmatrix}$$

Figure 11-2. *ABCD* matrices associated with (a) translation, (b) index-of-refraction change, (c) passing through a curved boundary with an index change, and (d) reflecting from a curved mirror

We will first consider the propagation of a beam from point 1 to point 2 that translates the beam a distance d in the direction of the optical propagation axis, as shown in Figure 11-2(a). In this example r_1 represents the initial displacement of point 1 above the axis and θ_1 represents the angular direction of propagation with respect to the axis. Thus, as the beam arrives at point 2, it has been displaced a distance d along the axis and it is still propagating at an angular direction $\theta_2 = \theta_1$. This can be expressed mathematically as

$$r_2 = r_1 + d\theta_1,$$
$$\theta_2 = \qquad \theta_1. \tag{11.1}$$

This set of equations can be written in matrix form as follows:

$$\begin{bmatrix} r_2 \\ \theta_2 \end{bmatrix} = \begin{bmatrix} 1 & d \\ 0 & 1 \end{bmatrix} \begin{bmatrix} r_1 \\ \theta_1 \end{bmatrix}. \tag{11.2}$$

Thus the matrix for translation over a distance d can be expressed as

$$\begin{bmatrix} 1 & d \\ 0 & 1 \end{bmatrix}. \tag{11.3}$$

This matrix is termed an *ABCD matrix,* in that each of the coefficients in the matrix is designated A, B, C, D as follows:

$$\begin{bmatrix} A & B \\ C & D \end{bmatrix};$$

(11.4)

for (11.3), $A = 1$, $B = d$, $C = 0$, and $D = 1$.

Similarly, the ray matrix for refraction of a beam (no translation) at a flat interface normal to the optical axis, going from a medium of index η_1 to a medium of index η_2, is shown in Figure 11-2(b) along with its associated *ABCD* matrix.

A ray passing through a curved boundary of radius R, at a dielectric interface going from index η_1 to index η_2, is shown in Figure 11-2(c). For a beam reflected from a curved mirror of focal length f such that $f = R/2$, the appropriate equations and the *ABCD* matrix are shown in Figure 11-2(d). This same expression is also valid for propagation through a thin lens, where f is the focal length of the lens.

An example of combining two separate processes to obtain a single matrix will now be shown by considering two thin lenses of focal length f_1 and f_2 placed adjacent to each other so that, in the first approximation, there is no effective distance between them. The final result of propagation through both lenses can then be described as

$$\begin{bmatrix} r_2 \\ \theta_2 \end{bmatrix} = \begin{bmatrix} 1 & 0 \\ -1/f_2 & 1 \end{bmatrix} \begin{bmatrix} 1 & 0 \\ -1/f_1 & 0 \end{bmatrix} \begin{bmatrix} r_1 \\ \theta_1 \end{bmatrix}.$$

(11.5)

These two matrices, when combined and simplified, become a single matrix leading to

$$\begin{bmatrix} r_2 \\ \theta_2 \end{bmatrix} = \begin{bmatrix} 1 & 0 \\ -(1/f_1 + 1/f_2) & 1 \end{bmatrix} \begin{bmatrix} r_1 \\ \theta_1 \end{bmatrix};$$

(11.6)

here the *ABCD* designations are $A = 1$, $B = 0$, $C = -(1/f_1 + 1/f_2)$, and $D = 1$. For further information regarding the use of *ABCD* matrices, see the references listed at the end of the chapter.

CAVITY STABILITY CRITERIA

We are now in a position to derive the stability of a laser cavity. We consider a cavity composed of two mirrors of equal curvature R and focal length $f = R/2$, separated by a distance d on the axis as shown in Figure 11-3(a). It is easier to visualize the equivalent situation in which lenses of equivalent focal length replace the mirrors, as shown in Figure 11-3(b). Propagation of a ray over a distance of one pass through the cavity of Figure 11-3(a) and then reflected by the mirror is equivalent to an axial displacement d in the lens diagram of Figure 11-3(b) and then a refraction due to the lens. This can be expressed in ray matrix form as

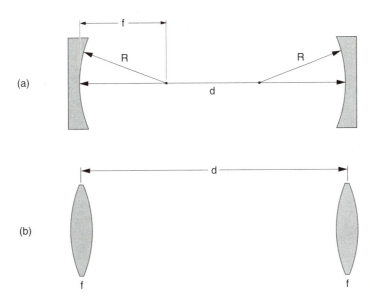

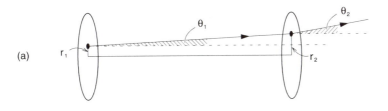

Figure 11-3. Parameters associated with the stability analysis of a two-mirror laser cavity

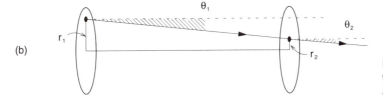

Figure 11-4. Laser beam tending to (a) instability, and (b) stability

$$\begin{bmatrix} r_2 \\ \theta_2 \end{bmatrix} = \begin{bmatrix} 1 & 0 \\ -1/f & 1 \end{bmatrix} \begin{bmatrix} 1 & d \\ 0 & 1 \end{bmatrix} \begin{bmatrix} r_1 \\ \theta_1 \end{bmatrix}$$

$$= \begin{bmatrix} 1 & d \\ -1/f & 1-d/f \end{bmatrix} \begin{bmatrix} r_1 \\ \theta_1 \end{bmatrix}, \tag{11.7}$$

where the propagation matrix operates on r_1 and θ_1 first and is then followed by the refraction matrix, since that is the sequence in which the light
encounters these two processes.

Now, in order to determine stability, consider the situations depicted in
Figure 11-4(a) and 11-4(b). If a ray leaves lens 1 (mirror 1), propagates to
lens 2 (mirror 2), and is refracted by lens 2 (reflected from mirror 2), we can
ask whether r_2 is greater than or less than r_1 at that point, and whether θ_2 is

greater or less than θ_1. If $r_2 > r_1$ and $\theta_2 > \theta_1$ then the beam will be on a diverging path that would lead to instability after many passes, as seen in Figure 11-4(a), since the beam would sooner or later walk its way out of the cavity. However, if $r_2 < r_1$ and $\theta_2 < \theta_1$ then we could conclude that the beam would tend toward stability, since it would always be attempting to converge to the optic axis as indicated in Figure 11-4(b).

We will now attempt to solve (11.7) by asking whether solutions exist in which the ray (r_2, θ_2) will differ from the ray (r_1, θ_1) by only a constant factor λ:

$$\begin{bmatrix} r_2 \\ \theta_2 \end{bmatrix} = \lambda \begin{bmatrix} r_1 \\ \theta_1 \end{bmatrix}. \tag{11.8}$$

For such a solution the ray would be diverging for $\lambda > 1$ since in that case $r_2 > r_1$ and $\theta_2 > \theta_1$, which produces a diverging ray as seen in Figure 11-4(a). It would likewise be converging for $\lambda < 1$ since then $r_2 < r_1$ and $\theta_2 < \theta_1$. We will develop the solution of (11.8) by using $ABCD$ matrices to solve for the possible values of λ. The stability criteria will then be obtained after considering a large (say, N) number of passes back and forth within the cavity.

Writing the relationship of (11.8) with the relevant $ABCD$ matrix leads to

$$\begin{bmatrix} r_2 \\ \theta_2 \end{bmatrix} = \begin{bmatrix} A & B \\ C & D \end{bmatrix} \begin{bmatrix} r_1 \\ \theta_1 \end{bmatrix} = \lambda \begin{bmatrix} r_1 \\ \theta_1 \end{bmatrix}. \tag{11.9}$$

The two right-hand parts of this equality can be combined to give

$$\begin{bmatrix} A-\lambda & B \\ C & D-\lambda \end{bmatrix} \begin{bmatrix} r_1 \\ \theta_1 \end{bmatrix} = 0. \tag{11.10}$$

Because λ is a constant, the above equation is a characteristic eigenvalue equation that will be satisfied only if the determinant of the coefficients of the matrix is zero:

$$\begin{vmatrix} A-\lambda & B \\ C & D-\lambda \end{vmatrix} = 0. \tag{11.11}$$

Using the relevant $ABCD$ values for the laser cavity with two curved mirrors as given in (11.7) leads to the following determinant,

$$\begin{vmatrix} 1-\lambda & d \\ -1/f & 1-d/f-\lambda \end{vmatrix} = 0, \tag{11.12}$$

which must be solved for the eigenvalues or characteristic values. Solving the determinant leads to the eigenvalue equation

$$\lambda^2 - 2\lambda \left(1 - \frac{d}{2f} \right) + 1 = 0, \tag{11.13}$$

which is an equation of the form

$$x^2 - 2\alpha x + 1 = 0, \tag{11.14}$$

where $x = \lambda$ and $\alpha = 1 - d/2f$. This equation has both real and imaginary solutions. The real solution for x occurs for $|\alpha| > 1$ and can be written as

$$x = \lambda = \alpha \pm \sqrt{\alpha^2 - 1} = e^{\pm\phi}, \quad |\alpha| > 1, \tag{11.15}$$

where the solution for λ is expressed also as an exponential in the form of $\lambda = e^{\pm\phi}$. The imaginary solution for x is

$$x = \lambda = \alpha \pm i\sqrt{1 - \alpha^2} = e^{\pm i\phi}, \quad |\alpha| < 1, \tag{11.16}$$

where $i = \sqrt{-1}$ and we have expressed the imaginary solution in the form of $\lambda = e^{\pm i\phi}$. In both cases ϕ is a real number.

Returning now to the question of stability, we ask what happens to the ray after it makes N passes through the cavity. The answer requires N successive applications of (11.7). We consider the matrix representation of this as

$$\begin{bmatrix} r^N \\ \theta^N \end{bmatrix} = \lambda^N \begin{bmatrix} r_1 \\ \theta_1 \end{bmatrix}. \tag{11.17}$$

Thus, for the solution of (11.15) for $|\alpha| > 1$, for N passes we would have

$$\begin{bmatrix} r^N \\ \theta^N \end{bmatrix} = \lambda^N \begin{bmatrix} r_1 \\ \theta_1 \end{bmatrix} = e^{\pm N\phi} \begin{bmatrix} r_1 \\ \theta_1 \end{bmatrix}, \tag{11.18}$$

which would clearly diverge for large N, leading to an unstable cavity situation.

For the solution of (11.16) for $|\alpha| < 1$, for N passes we would have

$$\begin{bmatrix} r^N \\ \theta^N \end{bmatrix} = \lambda^N \begin{bmatrix} r_1 \\ \theta_1 \end{bmatrix} = e^{\pm iN\phi} \begin{bmatrix} r_1 \\ \theta_1 \end{bmatrix}, \tag{11.19}$$

from which we conclude that the trajectory would clearly converge, since $|e^{-iN\phi}| \le 1$. Thus the beam would always remain confined to the region of the axis of the resonator. The requirement for stability would therefore be $|\alpha| < 1$, or

$$\lambda = \left(1 - \frac{d}{2f}\right) \pm i\left[1 - \left(1 - \frac{d}{2f}\right)^2\right]^{1/2}$$

$$= 1 - \frac{d}{2f} \pm i\left[\frac{d}{f}\left(1 - \frac{d}{4f}\right)\right]^{1/2}. \tag{11.20}$$

The value of λ will remain imaginary, leading to stability, only if

$$1 > \frac{d}{4f} \quad \text{or} \quad 0 < d < 4f \tag{11.21}$$

or, for spherical mirrors of radius R such that $R = 2f$,

$$0 < d < 2R. \tag{11.22}$$

For two mirrors of unequal curvature ($f_1 \ne f_2$) separated by a distance d, we have the relationship

$$0 < \alpha_1\alpha_2 < 1. \tag{11.23}$$

In this case, the solutions for α_1 and α_2 are

$$\alpha_1 = 1 - \frac{d}{2f_1} = 1 - \frac{d}{R_1} = g_1,$$

(11.24)

$$\alpha_2 = 1 - \frac{d}{2f_2} = 1 - \frac{d}{R_2} = g_2.$$

We have redefined α_1 and α_2 as g_1 and g_2 to be consistent with the laser literature. These gs are not to be confused with the statistical weights or the gain of a laser.

For stability, we thus have the requirement that

$$0 < \left(1 - \frac{d}{R_1}\right)\left(1 - \frac{d}{R_2}\right) < 1 \quad \text{or} \quad 0 < g_1 g_2 < 1.$$

(11.25)

This condition can be expressed in the form of a stability diagram, as shown in Figure 11-5. The clear regions are the regions where (11.25) is not satisfied and $g_1 g_2 > 1$. For such conditions the relationships between R_1, R_2, and d will therefore not lead to stability; in other words, the cavity is unstable. For the shaded regions, (11.25) is satisfied, $g_1 g_2 < 1$, and the cavity is stable. See the inside back cover for some general stability conditions.

Three particular points in Figure 11-5 are of special interest. They represent cavities depicted in Figure 11-1 that can be seen from Figure 11-5 to be on the verge of instability:

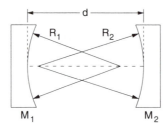

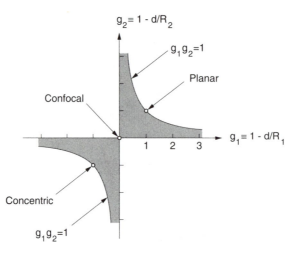

Figure 11-5. Stability diagram for two mirrors with radii of curvature R_1 and R_2

$R_1 = R_2 = d/2$ (symmetric concentric),

$R_1 = R_2 = d$ (confocal),   (11.26)

$R_1 = R_2 = \infty$ (plane-parallel).

All three of these cavities are on the edge of stability in the diagram, and can become extremely "lossy" for slight deviations into the shaded regions. Thus it would be wise to purposely design those cavities so that the g_1, g_2 parameters move slightly into the stable zones indicated in Figure 11-5.

11.2 PROPERTIES OF GAUSSIAN BEAMS

In a gain medium located within an optical resonator, the TEM_{00} Gaussian mode that develops when the single-pass gain exceeds the cavity losses (including diffraction losses) was shown in Chapter 10 to have a Gaussian profile at the mirrors in the direction transverse to the direction of propagation of the beam. A Gaussian beam has the following property: If the beam has a Gaussian transverse profile at one location then it will have a Gaussian transverse profile at all other locations, unless optical elements introduce a distortion that is non-uniform across the wavefront of the beam. Such a Gaussian beam can be characterized completely at *any* spatial location by defining both its "beam waist" and its wavefront curvature at a *specific* location of the beam. Moreover, an unaltered Gaussian beam always has a minimum beam waist w_0 at one location in space. The coordinate axis z that is used to define the propagation direction of the beam can be defined to have a value of $z = 0$ at the location of the minimum beam waist. Thus a beam having a Gaussian spatial distribution at the mirrors implies that a Gaussian beam profile exists at all other locations of that beam as it propagates between and beyond the mirrors. We can therefore conclude that laser resonators generate Gaussian beams that can be well characterized at any spatial location, as we will proceed to describe in more detail.

As mentioned in Section 10.2, the transverse distributions of the intensity of a simple Gaussian beam is of the form

$$I = I_0 e^{-2r^2/w^2}, \tag{11.27}$$

where I_0 is the maximum intensity and w is the beam radius (or waist) inside of which 86.5% of the energy is concentrated, as shown in Figure 11-6.

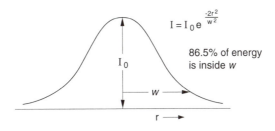

$$I = I_0 e^{\frac{-2r^2}{w^2}}$$

86.5% of energy is inside w

Figure 11-6. Radial distribution of Gaussian beam

The Gaussian beam minimum waist w_0 for a typical laser resonator mode occurs either at a point of focus after having passed through a lens, or in the region between the two mirrors of an optical resonator. For example, the minimum beam waist w_0 in a confocal optical resonator ($R_1 = R_2 = d$) occurs halfway between the two mirrors. The beam then expands and diverges from that location, such that the beam waist at a distance of $\pm z$ from the minimum beam waist w_0 can be described as

$$w(z) = w_0 \left[1 + \left(\frac{\lambda z}{\pi w_0^2} \right)^2 \right]^{1/2} \tag{11.28}$$

or, if the minimum beam waist w_0 occurs at a value of z_0 such that $z_0 \neq 0$, then

$$w(z) = w_0 \left[1 + \left(\frac{\lambda(z - z_0)}{\pi w_0^2} \right)^2 \right]^{1/2}. \tag{11.29}$$

Equation (11.28) can also be expressed as

$$w(z) = w_0 \left(1 + \frac{z^2}{z_R^2} \right)^{1/2}, \tag{11.30}$$

where

$$z_R = \frac{\pi w_0^2}{\lambda}. \tag{11.31}$$

The term z_R is known as the *Rayleigh range*. It is also referred to as the *depth of focus* when focusing a Gaussian beam. An alternative term $b = 2z_R$ is referred to as the *confocal parameter,* a term that is commonly used to characterize Gaussian beams. Both of these terms are indicated in Figure 11-7.

The beam wavefront curvature of a Gaussian beam at a location z, in terms of the minimum beam waist w_0 and the wavelength λ, is given by

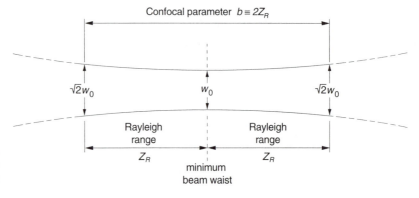

Figure 11-7. Gaussian beam parameters associated with the minimum beam waist

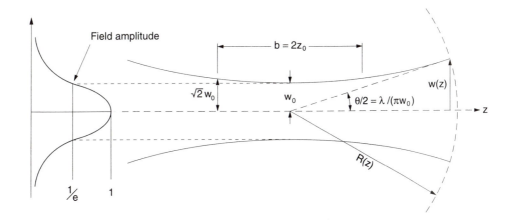

Figure 11-8. Gaussian beam parameters associated with angular divergence

$$R(z) = z\left[1 + \left(\frac{\pi w_0^2}{\lambda z}\right)^2\right]. \tag{11.32}$$

The angular spread of a Gaussian beam for a value of $z > z_R$ is given as

$$\theta(z) = \lim_{z \to \infty} \frac{2w(z)}{z} = \frac{2\lambda}{\pi w_0} = 0.64\frac{\lambda}{w_0}, \tag{11.33}$$

as shown in Figure 11-8. The $\theta(z)$ term is the full angle, at a given location z, over which the beam reduces to half of its maximum intensity at the center of the beam.

For a *symmetrical cavity* formed by two mirrors, each with radius of curvature R and separated by a distance d, the minimum beam waist w_0 is given by

$$w_0^2 = \frac{\lambda}{2\pi}[d(2R - d)]^{1/2} \tag{11.34}$$

and the radius of curvature r_c of the wavefront is

$$r_c = z + \frac{d(2R - d)}{4z}. \tag{11.35}$$

For a confocal resonator in which $R = d$, w_0 is given by

$$w_0 = \sqrt{\frac{\lambda d}{2\pi}} \tag{11.36}$$

and the beam waist (spot size) at each mirror located a distance $d/2$ from the minimum is

$$w = \sqrt{\frac{\lambda d}{\pi}}. \tag{11.37}$$

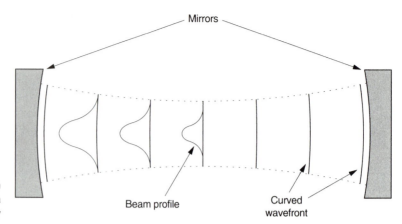

Figure 11-9. Gaussian beam profiles within a stable cavity

Equation (11.32) describes the wavefront curvature of a Gaussian beam expanding from the beam minimum where the wavefront is planar and the waist is w_0. A *stable laser cavity* is a cavity in which the mirrors have *curvatures identical to that predicted for the Gaussian beam at that location,* as shown in Figure 11-9.

11.3   PROPERTIES OF REAL LASER BEAMS

The previous section described the parameters associated with the dimensions and propagation of an ideal Gaussian-shaped beam (TEM$_{00}$ mode), which is often referred to as a *diffraction-limited* beam since it can be focused to the smallest (diffraction-limited) spot size. However, higher-order modes are not Gaussian-shaped, as can be seen from (10.53). Therefore real laser beams, operating multi-mode, are not perfect Gaussian beams and the formulas in Section 11.2 associated with the propagation of Gaussian beams are thus not applicable. However, a parameter has been developed that describes how close a real laser beam is to that of an ideal Gaussian beam. A real laser beam has both a waist and divergence that are larger than that of an ideal Gaussian beam.

For a diffraction-limited Gaussian beam, let us consider the product of the beam waist and the angular spread using (11.33):

$$w(z) \cdot \theta(z) = w(z) \cdot \frac{2\lambda}{\pi w_0}. \tag{11.38}$$

The minimum possible product would occur for $w(z) = w_0$ as follows:

$$w_0 \cdot \theta(z) = w_0 \cdot \frac{2\lambda}{\pi w_0} = \frac{2\lambda}{\pi}. \tag{11.39}$$

Real beams can be defined via multiples of the diffraction-limited beam. For example, let us define the divergence of a real beam as

$$\Theta = M\theta \qquad\qquad (11.40)$$

and the minimum waist of a real beam as W_0 such that

$$W_0 = Mw_0, \qquad\qquad (11.41)$$

where M is a positive integer (greater than or equal to unity). Thus, substituting these values into (11.39) gives the minimum possible product of the waist and the angular spread (for a real beam) as

$$\Theta \cdot W_0 = M\frac{2\lambda}{\pi w_0} \cdot Mw_0 = M^2 \frac{2\lambda}{\pi}, \qquad\qquad (11.42)$$

which is just M^2 times the product for a Gaussian beam. Thus the factor M^2 describes the relationship of the real beam to that of an ideal Gaussian beam. This M^2 is referred to as a *propagation constant* or a *propagation factor*. The constant M^2 can also be related to the minimum beam diameter $2W_0$ as

$$M^2 = \frac{\pi\Theta W_0}{2\lambda} = \frac{\pi\Theta(2W_0)}{4\lambda}, \qquad\qquad (11.43)$$

where $2W_0$ denotes the minimum beam diameter *measured* at the $1/e^2$ values of the beam intensity. The advantage of using M^2 is that commercial instruments are available to measure its value.

By substituting (11.41) and (11.42) into (11.29), the real laser beam can be shown to propagate with a beam waist of

$$W^2(z) = W_0^2 + M^4 \frac{\lambda^2}{\pi^2 W_0^2}(z - z_0)^2, \qquad\qquad (11.44)$$

which is exactly how a Gaussian beam propagates except for the M^4 factor associated with the beam spread in the far field. Thus, if W_0, M, and z_0 (the location of the minimum beam waist) are known then you can predict the characteristics of the beam at any value of z, just as for a Gaussian beam. These expressions are valid for any multi-mode non-Gaussian (real) laser beam, which will be M^2 multiples of the diffraction limit. For a perfect Gaussian beam, $M^2 = 1$. Values of M^2 greater than unity represent laser beams with higher-order modes than the TEM_{00} Gaussian mode.

Therefore, in determining the way a beam propagates, one would first measure the M^2 number and W_0 of the real beam. Then convert that to the w_0 of a Gaussian beam, since $W_0 = Mw_0$. Then determine the propagation of the Gaussian beam using the propagation equations determined earlier in this chapter. The real beam will focus at the same location as the Gaussian beam passing through the same lens. It will also have the same Rayleigh range defined as

$$z_R = \frac{\pi w_0^2}{\lambda} = \frac{\pi W_0^2}{M^2 \lambda}. \tag{11.45}$$

The concept of M^2 is useful in measuring the mode quality of a laser beam. With an M^2 instrument it is possible to align laser mirrors or adjust the aperture of the beam to produce the desired mode quality. It is also possible to determine the aberrations induced in a beam after it has passed through an optical element, such as a lens, and to measure the astigmatism of a beam, since the M^2 value would be different in two orthogonal transverse beam directions.

11.4 PROPAGATION OF GAUSSIAN BEAMS USING *ABCD* MATRICES – COMPLEX BEAM PARAMETER

We have seen how the properties of a Gaussian beam can be completely predicted at any point in space if its beam waist and curvature are known at only one specific point, using (11.28) and (11.32). It would be convenient if there were a simple technique to calculate the propagation of such a beam through lenses, mirrors, refractive materials, and so on. Such a technique does exist, and it involves the use of *ABCD* matrices. The technique is based upon the use of the complex beam parameter q for Gaussian beams. This parameter q is defined as

$$\frac{1}{q} = \frac{1}{R(z)} - j\frac{\lambda_0}{\eta \pi w^2(z)}, \tag{11.46}$$

where q contains information about the wavelength λ as well as the beam curvature $R(z)$ and the beam waist $w(z)$ at the location z.

We can use the *ABCD* matrices to calculate the beam parameter at any point (2) if it is known at point (1) by using the relationship

$$q_2 = \frac{Aq_1 + B}{Cq_1 + D}. \tag{11.47}$$

In order to use q, which is defined in terms of $1/q$ in (11.46), we express (11.47) as

$$\frac{1}{q_2} = \frac{C + D(1/q_1)}{A + B(1/q_1)}. \tag{11.48}$$

Letting $1/q_2 = 1/q$, from (11.46) and (11.48) we have the following useful relationship:

$$\frac{C + D(1/q_1)}{A + B(1/q_1)} = \frac{1}{R(z)} - j\frac{\lambda_0}{\eta \pi w^2(z)}. \tag{11.49}$$

Thus, knowledge of q_1 leads to an exact description of q_2. The proof of this relation is beyond the scope of this text. We will only suggest that (11.49) is correct by offering two examples.

EXAMPLE

Consider the propagation of a Gaussian beam in free space. Assume that the beam has a planar wavefront at $z_1 = 0$. Determine the formula for such a propagation to a location z that originated at $z = z_1 = 0$ as described previously by (11.28).

For a planar wavefront at the minimum beam waist, $R(z_1) = \infty$ and $w(z_1) = w_0$. Using (11.46) and (11.48) and the $ABCD$ values ($A = 1$, $B = z$, $C = 0$, $D = 1$) for propagation of a beam in free space over a distance z from z_1, we have

$$\frac{1}{q_2} = \frac{0 + 1\left(\dfrac{1}{R(z_1)} - j \dfrac{\lambda_0}{\eta \pi w^2(z_1)} \right)}{1 + z\left(\dfrac{1}{R(z_1)} - j \dfrac{\lambda_0}{\eta \pi w^2(z_1)} \right)}, \tag{11.50}$$

which, using $R(z_1) = \infty$ and $w(z_1) = w_0$ and (11.49), reduces to

$$\frac{1}{q_2} = \frac{-j \dfrac{\lambda_0}{\eta \pi w_0^2}}{1 - zj \dfrac{\lambda_0}{\eta \pi w_0^2}} = \frac{1}{R(z)} - j \frac{\lambda_0}{\eta \pi w^2(z)}. \tag{11.51}$$

Separating the middle expression into real and imaginary parts leads to

$$\frac{1}{R(z)} - j \frac{\lambda_0}{\eta \pi w^2(z)}$$

$$= \frac{1}{z + \dfrac{1}{z}\left(\dfrac{\eta \pi w_0^2}{\lambda_0} \right)^2} - j \frac{1}{\left(\dfrac{\eta \pi w_0^2}{\lambda_0} + z^2 \dfrac{\lambda_0}{\eta \pi w_0^2} \right)}. \tag{11.52}$$

Equating real and imaginary parts, and defining $z_0 = \eta \pi w_0^2 / \lambda_0$, yields

$$R(z) = z\left[1 + \left(\frac{z_0}{z} \right)^2 \right] \tag{11.53}$$

for the propagation of the beam curvature and

$$w(z) = w_0\left[1 + \left(\frac{z}{z_0} \right)^2 \right]^{1/2} \tag{11.54}$$

for the propagation of the beam waist. Equations (11.53) and (11.54) can be compared to (11.32) and (11.28) to see that the $ABCD$ matrix representation for propagation of a Gaussian beam in terms of $1/q$ provides the correct equations.

EXAMPLE

Consider the situation of a large-diameter Gaussian beam with a planar wavefront arriving at a thin lens. Where does the minimum beam waist occur after it passes through the lens and is focused? Also, what is the minimum spot size (diameter) at the focus?

The combined $ABCD$ matrix for a thin lens and a subsequent translation over a distance z from that lens can be expressed as

$$\begin{bmatrix} 1 & z \\ 0 & 1 \end{bmatrix} \begin{bmatrix} 1 & 0 \\ -1/f & 1 \end{bmatrix} = \begin{bmatrix} 1-z/f & z \\ -1/f & 1 \end{bmatrix} = \begin{bmatrix} A & B \\ C & D \end{bmatrix}. \tag{11.55}$$

Inserting these values of A, B, C, and D into (11.48) gives the relationship

$$\frac{-1/f + 1(1/q_1)}{1 - z/f + z(1/q_1)} = \frac{1}{R(z)} - j\frac{\lambda_0}{\eta\pi w^2(z)}, \tag{11.56}$$

where

$$\frac{1}{q_1} = \frac{1}{R(z_1)} - j\frac{\lambda_0}{\eta\pi w^2(z_1)} = -j\frac{\lambda_0}{\eta\pi w^2(z_1)}, \tag{11.57}$$

since z_1 is the location where the beam enters the lens and has the minimum beam waist such that $R(z_1) = \infty$. Therefore, using (11.57) in (11.56), we have:

$$\frac{1}{R(z)} = \frac{-1/f + z(1/f^2 + 1/z_1^2)}{(1-z/f)^2 + (z/z_1)^2}; \tag{11.58}$$

$$\frac{\lambda_0}{\eta\pi w^2(z)} = \frac{1/z_1}{(1-z/f)^2 + (z/z_1)^2}. \tag{11.59}$$

Equation (11.58) can be used to predict $z = z_m$ where the beam goes through a minimum and begins to expand again. At that minimum point the beam wavefront would be planar such that $R(z) = \infty$ and $1/R(z) = 0$. Therefore, from (11.58), $z = z_m$ is given by

$$z_m = \frac{f}{1 + (f/z_1)^2}. \tag{11.60}$$

We see that the beam waist does not reach a minimum exactly at the focal point. However, since $z_1 = \pi w_0^2/\lambda_0$ (assume $\eta \cong 1$) and $z_1 \gg f$ for most reasonable values of z_1, it follows that $z_m \cong f$.

We can now use (11.59) to predict the focal spot size. We find that

$$\frac{\lambda_0}{\pi w^2(z_m)} = \frac{1/z_1}{(1 + z_m/f)^2 + (z_m/z_1)^2}, \tag{11.61}$$

from which we can describe $w(z_m)$ as

$$w(z_m) \cong \frac{\lambda_0 f}{\pi w_0} \quad \text{for } z_1 = \frac{\pi w_0^2}{\lambda_0} \gg f. \tag{11.62}$$

Also, since the lens diameter $d = 2w_0$ and the effective $(f\#) = f/d$, (11.62) can be expressed as

$$w(z_m) \cong \frac{2\lambda_0}{\pi}(f\#) = \frac{\lambda_0}{\pi(\text{NA})}, \tag{11.63}$$

in which the numerical aperture NA of the focusing element approximately equals $1/2(f\#)$. Hence the focal spot diameter $2w_0 \approx \lambda_0$ for a lens with NA $= 0.64$.

11.5 OPTIMIZATION OF OUTPUT COUPLING FOR A LASER CAVITY

A laser will operate satisfactorily with many possible combinations of mirror reflectivities, provided that the gain in a single pass through the amplifier is sufficiently large to equal or exceed the mirror transmission losses (and other losses). If that gain per pass were (for example) 20%, then any combination of mirror transmissions and other losses up to 20% – corresponding to mirror reflectivities ranging from 80% to 99+% – would produce laser output. Since lasers are devices that are constructed to provide useful beam output, it is desirable to choose a mirror reflectivity that will enable the maximum output. We can imagine that in this case a reflectivity of 80% would be too low, in that the laser would just barely be at threshold. A laser with reflectivity $\geq 99\%$ would have a very high intensity within the cavity, but very little would escape through the mirrors due to the low transmission factor. Thus there is an optimum mirror transmission between these two values that would provide the maximum power output for the laser beam.

We will obtain an expression for this optimum output "coupling" by considering a homogeneously broadened laser gain medium of length L surrounded by identical mirrors of transmission t. For a small signal gain coefficient of g_0 and a saturation intensity of I_{sat}, the gain coefficient (eqn. 7.71) can be expressed as

$$g = \frac{g_0}{1 + I/I_{\text{sat}}} = \frac{g_0}{1 + \beta}, \tag{11.64}$$

where I, the intensity at a given location within the laser cavity, represents an average intensity over the transverse dimension of the beam and

$$I/I_{\text{sat}} = \beta. \tag{11.65}$$

We also assume that the absorption and scattering losses for each pass through the amplifier can be represented by a coefficient a. The homogeneously broadened medium was chosen in order to simplify the analysis, since the waves traveling in opposite directions in such an amplifier will interact with the same atoms. We also assume that the beams traveling in opposite directions are of approximately equal intensity.

Small signal gain = g_0

Loss/pass = a

Figure 11-10. Output flux
from a recirculating beam
within a two-mirror cavity

Mirror transmission = t

$a + t \ll 1$

As the beam within the amplifier traveling in a specific direction is re-flected from one of the mirrors and re-enters the amplifier with an inten-sity I_1, we assume it exits the other end of the amplifier with an intensity I_2, as shown in Figure 11-10, such that

$$\beta_1 = I_1/I_{\text{sat}} \quad \text{and} \quad \beta_2 = I_2/I_{\text{sat}}. \tag{11.66}$$

For each pass through the amplifier, the gain is thus represented by β_2/β_1 and the loss is given by $(1-a-t)$. Hence the product of the gain per pass and the loss per pass can be equated to unity as was done in (7.58) and (7.60):

$$\frac{\beta_2}{\beta_1}(1-a-t) = 1. \tag{11.67}$$

For a small gain per pass, $a+t \ll 1$ and therefore

$$\text{gain} = \frac{\beta_2}{\beta_1} = \frac{1}{1-(a+t)} \cong 1+(a+t) = e^{gL} \cong 1+gL. \tag{11.68}$$

From this it follows that

$$a+t = gL. \tag{11.69}$$

Therefore, for steady-state conditions, the gain gL can be expressed as

$$gL = \frac{g_0 L}{1+\beta} \cong a+t \tag{11.70}$$

or

$$\beta = \frac{g_0 L}{a+t} - 1. \tag{11.71}$$

For mirrors of equal transmittance t, the output power emerging from each end of the laser is

$$I_t = \frac{1}{2}tI \quad \text{or} \quad \frac{2I_t}{I} = t. \tag{11.72}$$

The factor of $\frac{1}{2}$ is used since I represents the combined laser power trav-eling in both directions within the laser cavity, and only the portion of the beam approaching the mirror will contribute to the beam transmitted through the mirror.

Because $\beta = I/I_{\text{sat}}$ (from eqn. 11.65), we have

$$\frac{2I_t}{I_{\text{sat}}} = \beta t. \tag{11.73}$$

Thus, using (11.71), we have

$$\frac{2I_t}{I_{\text{sat}}} = \beta t = t\left(\frac{g_0 L}{a+t} - 1\right), \tag{11.74}$$

which can be solved for I_t to obtain

$$I_t = \frac{I_{\text{sat}}}{2} t\left(\frac{g_0 L}{a+t} - 1\right). \tag{11.75}$$

This is an expression for the laser output I_t through a mirror of transmission t in terms of the small signal gain $g_0 L$ of the amplifier, the saturation intensity I_{sat}, and the scattering and absorption losses a averaged over a single pass of the beam through the amplifier.

We would like to determine the optimum transmission coupling through the mirror of such a laser by optimizing the value of the mirror's transmission coefficient. We can do this by taking the derivative of I_t in (11.75) with respect to the transmission t and equating the result to zero:

$$\frac{dI_t}{dt} = \frac{I_{\text{sat}}}{2}\left[\left(\frac{g_0 L}{a+t} - 1\right) - \frac{tg_0 L}{(a+t)^2}\right] = 0. \tag{11.76}$$

The derivative will be zero for all values of I_{sat} only if the expression in brackets is equal to zero:

$$\left(\frac{g_0 L}{a+t} - 1\right) - \frac{tg_0 L}{(a+t)^2} = 0. \tag{11.77}$$

With some re-arranging this can be expressed as

$$g_0 L a - (a+t)^2 = 0. \tag{11.78}$$

Solving this equation for t will give us a specific, optimal value of t as

$$\boxed{t_{\text{opt}} = (g_0 L a)^{1/2} - a, \tag{11.79}}$$

or, in another form, as

$$\boxed{t_{\text{opt}} = (g_0 L a)^{1/2}\left[1 - \left(\frac{a}{g_0 L}\right)^{1/2}\right]. \tag{11.80}}$$

We can thus see that t_{opt} depends upon both the small signal gain $g_0 L$ and the absorption and scattering losses per pass a. Equations (11.79) and (11.80) give t_{opt} for the case of a symmetrical cavity in which t is the same for both mirrors; this case was chosen in order to simplify the derivation.

However, in most cases it is desirable to have the laser output occur at only one end of the cavity. Thus we would use one high-reflectivity mirror and design the other mirror to have an output transmission of $2t_{max}$.

We can now obtain an expression for the maximum power output $I_{t_{max}}$ from the laser by replacing t by t_{opt} in (11.75) such that

$$I_{t_{max}} = \frac{I_{sat}}{2} t_{opt} \left(\frac{g_0 L}{a + t_{opt}} - 1 \right). \tag{11.81}$$

Inserting the value of t_{opt} from (11.79) or (11.80), for $I_{t_{max}}$ we obtain

$$\frac{2I_{t_{max}}}{I_{sat}} = g_0 L \left[1 - \left(\frac{a}{g_0 L} \right)^{1/2} \right]^2 = \frac{t_{opt}^2}{a} \tag{11.82}$$

or

$$I_{t_{max}} = \left(\frac{t_{opt}^2}{2a} \right) I_{sat}, \tag{11.83}$$

respectively. Thus it is desirable to have the losses a be as low as possible and the transmission as high as possible. For example, with a mirror transmission of 5% and an absorption of 1%, the intensity $I_{t_{max}}$ would be 12.5% of I_{sat}.

EXAMPLE

A helium–neon laser has a small signal gain coefficient g_0 of 0.1/m at 632.8 nm and absorption and scattering losses of 0.5% per pass. For a discharge amplifier length of 0.2 m, what would be the optimum transmission for the output mirror, assuming that the other mirror is a high-reflecting mirror with a reflectivity of 99.9%?

From (11.79) we see that we need only the small signal gain and the absorption and scattering loss factor a in order to determine the optimum transmission. The small signal gain is the product of the small signal gain coefficient g_0 and the amplifier length L:

$$g_0 L = 0.1/\text{m} \times 0.2 \text{ m} = 0.02.$$

Inserting this value for $g_0 L$ and the given value of $a = 0.005$ into (11.79), we have

$$t_{opt} = (g_0 L a)^{1/2} - a = (0.02 \times 0.005)^{1/2} - 0.005 = 0.005.$$

Thus a transmission of 0.005 or 0.5% would be the optimum transmission from each of two mirrors. However, if we want the power to be transmitted from only one end of the laser, with a high-reflectivity mirror at the other end as described previously, we must double the output transmission (output coupling) of the other mirror to obtain $2 \times 0.5\% = 1\%$. Thus, the optimum output transmission for this laser is 1%. The difficult part of solving for this output transmission is in knowing what the absorption and scattering losses are in the laser cavity. It is often very difficult to measure small losses, which can occur as scattering from irregularities on the mirror surfaces, from the reflective coatings on the mirrors, or from the Brewster angle windows (if such windows are used in the optical cavity).

REFERENCES

H. Haus (1984), "Hermite–Gaussian beams and their transformations," in *Waves and Fields in Optoelectronics*. Englewood Cliffs, NJ: Prentice-Hall, Chapter 5.

W. W. Rigrod (1963), "Gain saturation and output power of optical masers," *Journal of Applied Physics* 34: 2602–9.

W. W. Rigrod (1965), "Saturation effects in high-gain lasers," *Journal of Applied Physics* 36: 2487–90.

A. Yariv and P. Yeh (1984), *Optical Waves in Crystals*. New York: Wiley, Chapter 2.

PROBLEMS

1. Determine the *ABCD* matrix for a beam translated a distance d_1, focused through a thin lens of focal length f_1, and then translated a distance d_2.

2. A collimated beam (plane waves) arrives at a concave mirror with a focal length of 0.15 m and is then focused by the mirror. Write out the *ABCD* matrix for this process.

3. Determine whether or not the following mirror arrangements lead to stability:

(a) two mirrors with radii of curvature of 1.8 m, separated by a distance of 2 m;

(b) one mirror with radius of curvature of 2 m and the other with radius 3 m, separated by a distance of 2.3 m;

(c) one mirror with radius of curvature 5 m and the other with radius 3 m, separated by a distance of 4 m;

(d) two mirrors with radius of curvature of 0.5 m, separated by a distance of 0.5 m.

4. A He–Ne laser beam of wavelength 632.8 nm operating with the TEM_{00} mode and a minimum beam waist of 0.2 mm propagates a distance of 2 m from the location of the minimum beam waist (within the laser cavity) and is incident upon a lens with focal length 50 mm. Determine the location of the focus of the beam using the complex beam parameter.

5. A Gaussian beam of wavelength 800 nm is measured to have a wavefront curvature of 3 m and is known to have a minimum beam waist somewhere of 0.75 mm. At that location, how far is the beam from its minimum?

6. How many modes are lasing in a He–Cd 325-nm laser with a confocal cavity, if the beam diameter is measured to be 0.1 m at a distance of 3 m from the center of the cavity and the minimum beam waist is 50 μm? (*Hint:* Use M^2.)

7. A laser cavity is constructed with a pair of mirrors having a 2-m radius of curvature, separated by a distance of 1.8 m. Determine if the cavity is stable. If a 0.2-m focal length thin lens is mounted directly in front of the laser output mirror, where will it focus and what will be the focal spot size? The laser is an Ar^+ laser operating at 514.5 nm.

8. Determine the beam waist of a Gaussian beam of wavelength 632.8 nm after being translated a distance of 5 m from the location of its minimum beam waist of 100 μm, transmitted into a planar piece of glass having index of refraction of 1.5 at normal incidence, and translated another 0.1 m within the glass.

9. A helium–cadmium laser containing a single isotope of Cd^{114} and operating at 441.6 nm has a gain coefficient of 20%/m, a confocal cavity length of 0.4 m, and an amplifier length of 0.3 m. Assume that there is a scattering loss at each of the Brewster windows of 0.2% and a scattering loss at each of the mirrors of 0.1%. What is the optimum mirror transmission if the output power is the same from each end of the laser? What would be the output intensity $I_{I_{max}}$ emerging from each mirror for this laser arrangement? Refer to (7.42) for the definition of I_{sat}. Also remember that we use the value of $\sigma_{ul}^H(\nu = \nu_0)$ from (7.16) in determining $I_{I_{max}}$, since we are assuming that the laser is operating at the center of the emission line ($\nu = \nu_0$) where the gain is the highest. To assist in determining $I_{I_{max}}$, use the energy-level diagram of the He–Cd laser shown in Figure 8-6. Note that the transitions shown in that figure are the only radiative decay channels for those upper and lower laser levels.

10. A CO_2 laser is measured to have an intensity of 10^5 W/m^2 emerging from one end of the laser, which has two identical mirrors each with a transmission of 10%. The gain of the laser ($g_0 L$) is also measured to be 0.05. What output mirror transmission would be required to double the intensity from the same laser?

11. Design a cavity for a CO_2 laser operating at 10.6 μm that would produce a beam 0.02 m in diameter ($2w$) at a distance of 2 m from the laser output mirror. State the cavity parameters that you would use for your design. Assume the laser has a gain length of 1 m.

SPECIAL LASER CAVITIES AND CAVITY EFFECTS 12

SUMMARY In this chapter we examine a number of special laser cavities that have been developed over the years to provide unique properties to a laser beam. Unstable resonators provide a high-power Gaussian-shaped output beam from a high-gain laser that has a wide gain medium. Q-switching allows the production of short laser pulses with power and energy per pulse that are many orders of magnitude higher than could be achieved on a steady-state basis in the same laser gain medium. Mode-locking is a technique that produces high-intensity ultrashort pulses from a laser that otherwise operates on a cw or quasi-cw basis. Pulse compression techniques for shorten-

ing mode-locked pulses are also described. Ring lasers are used to produce traveling- rather than standing-wave cavities, to avoid spatial hole burning, and to more efficiently extract energy. Several types of devices are also described for producing spectral narrowing of the laser output. These include diffraction gratings, Fabry–Perot etalons, and distributed feedback or distributed Bragg reflection with semiconductor lasers. The chapter then discusses astigmatically compensated cavities for operating lasers within a very small-diameter gain medium, and concludes with a description of waveguide cavities for gas lasers.

12.1 UNSTABLE RESONATORS

A type of laser resonator that can provide useful laser output with a reasonable beam quality, without meeting the criteria for stability as outlined in the previous chapter, is referred to as an unstable resonator. Such a class of resonators has been developed to obtain high-power output in a well-defined Gaussian-shaped beam from a laser medium that has a relatively high gain and a large amplifier width-to-length ratio (to take advantage of high energy storage and extraction due to the large volume).

A typical TEM_{00} mode from a laser cavity has a very narrow beam width w within the laser amplifier, as shown in the upper diagram of Figure 12-1. This can be understood by examining (11.37) for a confocal resonator in which $w = \sqrt{\lambda d/\pi}$ at the mirrors, where d is the separation between mirrors and λ is the laser wavelength. Thus, no matter how wide the laser amplifier is, in a *stable* resonator the energy will be extracted from the narrow region of diameter $2w$ only if a TEM_{00} mode Gaussian beam is desired, since $w \ll d$ for wavelengths in the optical spectral region. The remaining amplifier volume is essentially wasted, since either it doesn't contribute to laser output at all or it leads to higher-order modes that do not propagate in a well-defined manner and are therefore not as useful. In contrast, if we

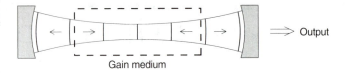

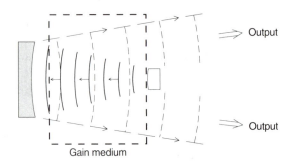

Stable resonator

Figure 12-1. Comparison of laser flux associated with a stable resonator and an unstable resonator

Gain medium

Unstable resonator

consider a cavity that is unstable, a portion of the beam will "spill" around the mirror as it diverges from the axis, as indicated in Figure 11-4(a) and in the lower portion of Figure 12-1. The expanding and diverging beam can be seen to pass through a large gain volume in contrast with the small gain region accessed by the stable mode. With a suitable unstable resonator cavity design, the "leakage" around the mirror can become a useful, intense laser beam that has a well-defined Gaussian-shaped profile.

The conditions and requirements associated with the desire to operate in the unstable region of the stability diagram are listed as follows.

(1) The volume of the amplifier associated with the region in which stable modes would occur is too restrictive compared with the amplifier dimensions (primarily the width).

(2) A high-power output from a large gain volume (with minimum amplifier length) is desired.

(3) The gain of the medium is high enough such that stable low-order modes do not have sufficient time to evolve before the gain is depleted. Such a high gain also makes it possible for the laser to operate in spite of the high diffraction losses of the unstable cavity.

If we consider the resonator stability diagram of Figure 11-5, the beam begins to diverge and its size becomes larger than the mirror size as parameters are moved from the stable regions to the unstable regions. When this happens, the finite size of the resonators must be considered. Unstable resonators take advantage of this situation by incorporating a very large-diameter mirror to intercept the diverging beam and thus make use of the deliberately diverging laser beam wavefronts associated with unstable resonator cavities.

The most common unstable resonator cavity has two mirrors of different diameters and radii of curvature. The large-diameter rear mirror has a radius of curvature of R_r and the smaller mirror at the output end of the cavity has a radius of curvature of R_o, as shown in the figure. We can also define the magnification ratio M as

$$M = \frac{R_r}{R_o}. \tag{12.1}$$

Such unstable resonator cavities have transverse mode envelopes that can readily fill large laser volumes yet also suppress higher-order transverse modes. The losses in such resonators are dominated by diffraction, which means that beam power will leak around the edges of the mirrors. This leakage can be thought of as the "output coupling," which can be advantageous if the unstable cavity is designed properly so that this beam output (leaking) is collimated when it emerges from the gain medium. Thus, the output coupling of such resonators is much larger than that of stable resonators because the diffraction around the edges of the output mirror provides the laser output. Also, because of this large diffraction loss, these resonators are not effective for low-gain systems.

The stability diagram of Figure 11-5 suggests that the constraints associated with unstable resonators are that $g_1 g_2 \geq 1$ or $g_1 g_2 \leq 0$. Such conditions are characterized by large diffraction losses within the laser cavity. For the present situation, $g_1 = 1 - d/R_1 = 1 - d/R_o$ and $g_2 = 1 - d/R_2 = 1 - d/R_r$. Unstable resonators can be classified as being either positive or negative branch according to whether

$$g_1 g_2 \geq 1 \quad \text{(positive branch)} \tag{12.2}$$

or

$$g_1 g_2 \leq 0 \quad \text{(negative branch)}. \tag{12.3}$$

Even though high diffraction losses exist within unstable resonators, such resonators can have well-defined output modes when the diffraction losses are matched to the single-pass gain of the amplifier medium.

For certain values of Fresnel number $a^2/\lambda d$ (where a is the mirror radius) and geometric magnification M (as defined in eqn. 12.1), it has been found that one mode will have substantially lower loss than all other transverse modes. Operation under these conditions yields good transverse mode discrimination. The value of M is used to determine the fractional output coupling of the beam. This value, along with the choice of Fresnel number, determines the mode diameter of the resonator. In order to obtain useful high-power output within the far-field central lobe, and thus better beam quality, it is desirable to use unstable resonators with high values of M.

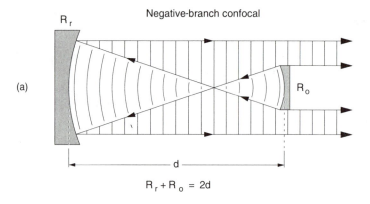

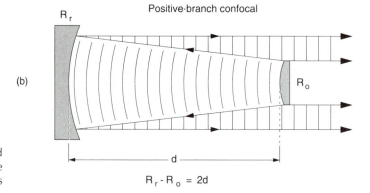

Figure 12-2. Negative- and positive-branch unstable resonators

Figure 12-2 shows two types of unstable resonators that produce a collimated output beam. These are referred to as *confocal* unstable resonators, and their cavities are two of the most common designs. The positive-branch unstable resonator has a constraint that

$$R_r - R_o = 2d, \tag{12.4}$$

whereas the negative-branch resonator requires that

$$R_r + R_o = 2d. \tag{12.5}$$

The intercavity focal point of the negative-branch unstable resonator can cause damage (in the case of a solid-state gain medium) or ionization and breakdown (in the case of a gaseous laser gain medium), both of which are very deleterious effects. Therefore, in most cases this negative-branch unstable resonator is of much less practical use than the positive-branch arrangement.

The conditions for a positive-branch confocal cavity (eqn. 12.2) can be rewritten as

$$R_o = 2d/(M-1) \tag{12.6}$$

and

$$R_r = 2Md/(M-1). \tag{12.7}$$

The geometrical output coupling can be given as

$$\delta = 1 - 1/M^2. \tag{12.8}$$

In the analysis of diffraction effects that identified stable transverse modes (see Section 10.2), the loss per pass through the resonator was graphed as a function of the Fresnel number $N = a^2/\lambda d$ in Figure 10-11 for a the radius of the limiting aperture, which would typically be either the mirror itself, the gain medium, or an aperture placed within the cavity. A similar analysis has been carried out for an unstable resonator. The losses were graphed versus an equivalent Fresnel number N_{eq}, which, for a symmetric double-ended unstable resonator, is

$$N_{eq} = \left[\frac{(M-1)}{2M^2}\right]\left(\frac{a^2}{\lambda d}\right). \tag{12.9}$$

The minimum diffraction losses were shown to occur for values of $N_{eq} = 0.5, 1.5, 2.5, \ldots$ For this case the output coupling (including diffraction considerations) has a value only slightly less than the geometric value of (12.8).

The condition for the confocal unstable resonator cavity arrangement can also be expressed as

$$R_o/2 + R_r/2 = d, \tag{12.10}$$

which can be translated into the stability condition of

$$g_1 + g_2 = 2g_1g_2. \tag{12.11}$$

This condition describes a contour that is unstable everywhere in the (g_1, g_2) plane except at the symmetric confocal point $g_1 = g_2 = 0$ and at the planar symmetric point where $g_1 = g_2 = 1$. Therefore, all asymmetric confocal resonators are unstable. Such a resonator design is of particular interest because one of the circulating beams in the resonator is always a collimated beam. This beam can therefore be used as the output beam of the resonator.

EXAMPLE

Design an unstable resonator cavity to operate in conjunction with a standard flashlamp-pumped Q-switched Nd:YAG laser rod.

The design of the unstable resonator cavity is based upon the cavity length, the output coupling, the rod diameter, and the mirror radii of curvature. It is convenient to choose a standard 6.3-mm–diameter, 50-mm–long Nd:YAG laser rod with the laser operating at $\lambda = 1.06 \, \mu$m. A positive-branch confocal unstable resonator cavity (Figure 12-2) will be used. The output coupling is thus determined to be only slightly less than that from (12.8): $\delta \approx 1 - 1/M^2$.

For the present cavity design we will choose to operate at $N_{eq} = 1.5$ and $M = 1 + 2d/R_o$ from (12.6). Substituting these parameters into (12.9), we have

EXAMPLE (cont.)

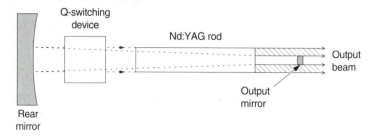

Figure 12-3. A Nd:YAG unstable resonator cavity

$$d = -\frac{R_o}{2} + \frac{a}{2}\left(\frac{R_o}{\lambda N_{eq}}\right)^{1/2};$$

R_r is then given by the expression in (12.7) for a specific value of M.

As an example, let $R_o = 0.5$ m, $N_{eq} = 1.5$, $a = 6.3/2 = 3.15$ mm, and $\lambda = 1.06$ μm; from the preceding formula it can be determined that $d = 0.633$ m. Using (12.1) we find that $R_r = 0.5M$. From (12.7) we can then find that $M = 3.53$ and $R_r = 1.77$ m. The geometrical output coupling can be obtained from (12.8) as $\delta = 1 - 1/M^2 = 1 - 1/12.46 = 0.92$. The output coupling as well as diffraction effects reduce this value to 0.84. Such a cavity, producing output of 750 mJ in a pulse duration of 12 ns at a repetition rate of 10 Hz, has provided the basis for a commercial laser system; see Figure 12-3. The reader may consult the reference on Nd:YAG unstable resonators for more details of this example.

12.2 Q-SWITCHING

GENERAL DESCRIPTION

Pumping begins when a laser is turned on, and the population inversion eventually reaches a steady-state value that is determined by the pumping rates to, and the decay rates from, the upper and lower laser levels (as described in Chapter 8). During this period of time, the laser beam begins to grow until it eventually reaches the saturation intensity and begins to extract energy from the medium – if the gain is high enough to overcome the cavity losses and the gain duration is long enough. As the beam grows, the population density N_u is reduced by stimulated emission according to (7.69), and consequently the inversion density (7.71) reaches a new, lower steady-state value such that the reduced gain equals the losses in the cavity. The time required to reach this new equilibrium value is the time t_s required for the developing beam to make m passes through the amplifier, as detailed in (7.64) and (7.65). This can be rewritten as

$$t_s = \frac{m}{c}[\eta_L L + \eta_C(d-L)], \tag{12.12}$$

where d is the distance between mirrors, L and η_L are the gain-medium length and index of refraction, and η_C is the index of refraction in the space within the cavity that does not include the gain medium. Values of t_s typically range from 1 to 1,000 ns.

In most types of laser gain media the upper laser level lifetime is shorter than t_s. In contrast, most solid-state lasers have an upper-level lifetime τ_u that is significantly longer than t_s. Thus, when such solid-state gain media are installed within a laser cavity, the laser output reaches I_{sat} long before the upper laser level population density N_u has reached its maximum potential, as determined from (9.1). In such cases the laser essentially operates with a gain that is much lower than could be achieved if the gain medium were pumped without the presence of a cavity. If it were possible to pump this solid-state gain media for the duration of τ_u without the cavity in place and then suddenly switch the cavity back into place, it would be possible to operate the laser with the highest possible gain and thus obtain a higher energy output. Of course, this would only be temporary: the laser output would be in the form of a giant pulse, since the gain is far above the steady-state conditions for gain in the cavity and would therefore be rapidly reduced by stimulated emission. In fact, such a process can be implemented. The technique is known as *Q-switching* since the cavity is rapidly changed from a low-Q to a high-Q state, where Q denotes the ratio of the energy stored to the energy dissipated within the cavity. This technique is used in many types of commercial solid-state lasers and CO_2 lasers to obtain a high-power pulsed output.

To produce the necessary high inversion density required for Q-switching, four requirements must be satisfied.

(1) The lifetime τ_u of the upper laser level u must be longer than the cavity buildup time t_s, so that the upper level can store the extra energy pumped into it over the extended pumping time:

$$\tau_u > t_s. \tag{12.13}$$

(2) The pumping flux duration T_P must be longer than the cavity build-up time, and preferably at least as long as the upper level lifetime τ_u:

$$T_P \geq \tau_u. \tag{12.14}$$

(3) The initial cavity losses must be high enough during the pumping duration τ_u to prevent beam growth and oscillation from occurring during that time:

[initial cavity losses] > [amplifier gain]. (12.15)

(4) The cavity losses must be reduced almost instantaneously, by suddenly introducing a high-Q cavity, so that the beam could then evolve and extract the extra energy that had previously built up in the upper laser level u of the gain medium.

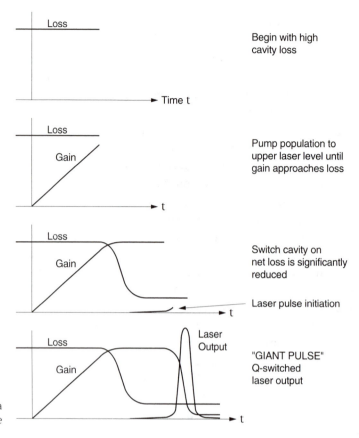

Loss

Begin with high
cavity loss

Time t

Loss

Gain

Pump population to
upper laser level until
gain approaches loss

t

Loss

Gain

Switch cavity on
net loss is significantly
reduced

Laser pulse initiation

t

Loss

Gain

Laser
Output

"GIANT PULSE"
Q-switched
laser output

t

Figure 12-4. Evolution of a
Q-switched laser pulse

Satisfying these requirements would produce a "giant pulse" laser output
or Q-switched laser output. The evolution of events dictated by these requirements is summarized in Figure 12-4, which shows a high initial cavity loss during the intense pumping and consequent gain build-up over the
time τ_u. The cavity loss is then rapidly reduced, producing a pulse output
within a time of the order of t_s after the cavity Q is changed.

THEORY

We now develop a theoretical basis for describing the time dependence of
the population inversion, and also of the output pulse duration, for various inversion densities above threshold before the Q-switch is activated.
Our goal is to obtain an expression for the total number of photons ϕ
within the laser cavity at the laser frequency ν, and also for the total population difference M within the laser gain medium at any instant during the
Q-switching procedure. (This M has no relation to the M factor described
in Chapter 11 for real laser modes or to the magnification factor M used in
Section 12.1 for the magnification of an unstable resonator.) We will relate
the number of photons to the laser intensity I. The population difference
is defined as $M = \Delta N_{ul} \cdot V$, where ΔN_{ul} is the population difference defined
by (7.12) and V is the mode volume within the laser gain medium.

We will first obtain an expression for the decay time of a laser beam bouncing back and forth within an optical cavity when there is no gain present. If a beam of energy E were instantaneously placed within such a cavity, the beam would subsequently reflect back and forth and allow part of the energy to escape via transmission through the mirrors. The energy would decay according to the following relationship:

$$\frac{dE}{dt} = -\frac{E}{t_C},$$ (12.16)

where t_C is the effective decay time of the optical cavity typically resulting from losses due to transmission of a portion of the light through the mirrors during each pass through the cavity.

If the fractional loss of E per pass in the cavity is given by L_F and $c/\eta d$ is the transit time per pass, then the decay time t_C can be expressed as

$$t_C = \frac{\eta d}{c L_F}$$ (12.17)

and the fractional loss of E per pass from (12.16) is

$$\frac{dE}{dt} = -\frac{c}{\eta d} L_F E.$$ (12.18)

Consider the round-trip steady-state equation for a laser (eqn. 7.60), which can be rewritten as

$$R_1 R_2 e^{(g_0-\alpha)2L} = 1.$$ (12.19)

Here we have ignored the fractional losses per pass of a_1 and a_2 that were incorporated into (7.60). This expression can be re-arranged to group all of the losses (mirror + absorption) together as follows:

$$R_1 R_2 e^{-2\alpha L} e^{2g_0 L} = 1.$$ (12.20)

The total loss can thus be expressed as an exponential factor L_F as follows:

$$R_1 R_2 e^{-2\alpha L} = e^{-2L_F},$$ (12.21)

where the factor of 2 is included because (12.19) involves a round trip within the cavity whereas L_F is the fractional loss for a single pass. From (12.21), the fractional loss per pass L_F can thus be written as

$$L_F = \alpha L - \ln(R_1 R_2)^{1/2}.$$ (12.22)

Using this relationship for L_F, we can now write a more explicit expression for t_C as

$$t_C = \frac{\eta d}{c[\alpha L - \ln(R_1 R_2)^{1/2}]}.$$ (12.23)

Our goal now is to develop an expression for the number of photons ϕ within the cavity at any specific time during the Q-switching process. In order to do this, we begin by considering the change in intensity per unit time, dI/dt, within the amplifier:

$$\frac{dI}{dt} = \frac{dI}{dz}\frac{dz}{dt}.$$

(12.24)

However, if we assume the distributed loss α in the expression for gain of (7.60) is small enough to neglect, we can write

$$\frac{dI}{dz} = g_0 I$$

(12.25)

and also

$$\frac{dz}{dt} = \frac{c}{\eta};$$

(12.26)

hence we can rewrite expression (12.24) as

$$\frac{dI}{dt} = \frac{g_0 c I}{\eta}.$$

(12.27)

We thus have an expression for the intensity change due to the amplifier.

We are now in a position to obtain the number of photons ϕ in the cavity at any specific time; this number is proportional to the instantaneous intensity I. We will also use the fact that only a fraction L/d of the photons are within the gain region and are therefore undergoing amplification at any given time. This suggests that the average intensity change in the cavity can be modified from (12.27) by the fraction L/d as follows:

$$\frac{dI}{dt} = \frac{g_0 c I}{\eta}\left(\frac{L}{d}\right).$$

(12.28)

Since $\phi \propto I$ and therefore $d\phi/dt \propto dI/dt$, the rate of change of the number of photons within the cavity due to the amplifier can now be expressed as

$$\frac{d\phi}{dt} = \frac{\phi g_0 c}{\eta}\left(\frac{L}{d}\right).$$

(12.29)

The decrease in the number of photons per unit time within the cavity due to the loss in photons that are transmitted through the mirrors is just ϕ/t_C, the product of the resonator decay rate $1/t_C$ (defined in eqn. 12.17), and the photon density ϕ. We can thus combine this decrease due to the resonator decay rate with the increase due to the amplifier of (12.29) to obtain an expression for the total change in photons within the cavity with time:

$$\frac{d\phi}{dt} = \frac{g_0 c \phi}{\eta}\left(\frac{L}{d}\right) - \frac{\phi}{t_C}.$$

(12.30)

Let us now define a new time variable τ in units of the cavity decay time t_C:

$$\tau = t/t_C.$$

(12.31)

The photon differential equation (eqn. 12.30) can then be written as

$$\frac{d\phi}{d\tau} = \phi\left[\left(\frac{g_0}{\eta d/cLt_C}\right) - 1\right] = \phi\left(\frac{g_0}{g_t} - 1\right).$$

(12.32)

Here g_t is defined as

$$g_t = \frac{\eta d}{cLt_C},$$ (12.33)

where g_t is the threshold gain, which is equal to the cavity losses.

Examining (12.32) we see that, for $g_0 = g_t$, $d\phi/d\tau = 0$. Hence g_t is equal to the steady-state gain of a continuous-wave (cw) laser, since there is no change in photon flux in the cavity with time for a laser operating under steady-state conditions.

We will now introduce the variable M in place of the gain value g_0 in (12.32). Since $g_0 = \sigma_{ul}\Delta N_{ul}$, g_0/g_t can be written as

$$\frac{g_0}{g_t} = \frac{M}{M_t},$$ (12.34)

where ΔN_{ul}^t and $M_t = \Delta N_{ul}^t V$ are the inversion per unit volume (inversion density) and the total inversion, respectively, *at threshold,* when the gain is equal to the cavity losses. We can thus rewrite (12.32) as

$$\frac{d\phi}{d\tau} = \phi\left(\frac{M}{M_t} - 1\right).$$ (12.35)

This equation is one of the two principal equations in the evolution of the Q-switched pulse. The positive term on the right-hand side, $\phi M/M_t$, represents the number of photons generated within the cavity by stimulated emission per unit of normalized time.

To obtain the second principal equation, we note that every time a photon is emitted in the stimulated emission process, M changes by a factor of 2. This occurs because for every atom in the upper laser level u that is sent to the lower laser level l by stimulated emission, the net difference in population is 2, since the upper level decreases by 1 and the lower level increases by 1. Thus we can write an expression for $dM/d\tau$ as

$$\frac{dM}{d\tau} = -2\left[\phi\frac{M}{M_t}\right].$$ (12.36)

This equation summarizes our statement that the decrease in the population inversion is twice that of the increase in the number of photons due to stimulated emission. Equation (12.36) is the second of the two equations we need to describe Q-switching. A plot of M and ϕ as a function of time is shown in Figure 12-5.

Dividing (12.35) by (12.36), we obtain

$$\frac{d\phi}{dM} = \left(\frac{M_t}{2M} - \frac{1}{2}\right) = \frac{1}{2}\left(\frac{M_t}{M} - 1\right).$$ (12.37)

Integrating this equation leads to

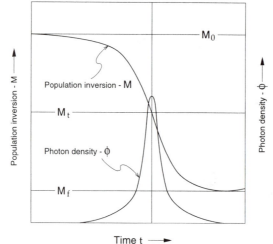

Population inversion - M

Figure 12-5. Temporal variation of the population inversion M and the photon density ϕ associated with a Q-switched pulse

$$\phi - \phi_0 = \frac{1}{2}\left[M_t \ln \frac{M}{M_0} - (M - M_0)\right],\tag{12.38}$$

where M_0 is the initial population difference and ϕ_0 is the initial photon density in the cavity. If we assume that there are no photons initially in the cavity (we neglect the few that are there due to spontaneous emission), then $\phi_0 \cong 0$ and we have

$$\phi = \frac{1}{2}\left[M_t \ln \frac{M}{M_0} - (M - M_0)\right].\tag{12.39}$$

This equation describes the relationship between the number of photons in the cavity and the inverted population M at any particular time. For a time $t_f \gg t_C$ there will also be no photons in the cavity, and we can thereby deduce the final inversion M_f by setting $\phi = 0$ in (12.39). We thus obtain the following relationship:

$$\frac{M_f}{M_0} = e^{(M_f - M_0)/M_t}.\tag{12.40}$$

The fraction of the original energy that is converted to laser energy is given as $(M_0 - M_f)/M_0$. From (12.40), this can be expressed as

$$\frac{M_0 - M_f}{M_0} = 1 - e^{(M_f - M_0)/M_t}.\tag{12.41}$$

The solution to this equation, graphed in Figure 12-6, can be seen to approach unity as M_0/M_t increases. This suggests that the higher above threshold M is before the cavity Q is switched, the more energy will be extracted from the cavity in the form of a giant pulse.

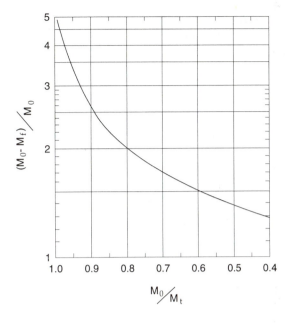

Figure 12-6. Fraction of original energy converted to laser energy in a Q-switched pulse

The instantaneous power output P can be approximated by multiplying the photon number by the energy $h\nu$ per photon and dividing by the cavity decay time t_C as follows:

$$P = \phi \frac{h\nu}{t_C}; \tag{12.42}$$

or, substituting in the expression for ϕ, we have

$$P = \frac{h\nu}{2t_C}\left[M_t \ln \frac{M}{M_0} - (M - M_0)\right]. \tag{12.43}$$

We can obtain P_{max}, the maximum value of P, by differentiating P with respect to M and equating the result to zero; P_{max} is then determined by setting $M = M_t$. Substituting this result into (12.43) yields

$$P_{max} = \frac{h\nu}{2t_C}\left[M_t \ln \frac{M_t}{M_0} - (M_t - M_0)\right]. \tag{12.44}$$

For values of $M_0 \gg M_t$, P_{max} can be expressed as

$$P_{max} \cong \frac{M_0 h\nu}{2t_C}, \tag{12.45}$$

which indicates that half of the inversion is dumped within the cavity decay time t_C.

Since $P = \phi h\nu/t_C$, the maximum number of photons stored inside the cavity is $M_0/2$. This is because, for $M_0 \gg M_t$, the pulse builds up in a time much less than t_C so that, when $M = M_t$, most of the photons are still in the cavity. Thus $(M_0 - M_t)/2$ is approximately equal to $M_0/2$.

EXAMPLE

What is the maximum Q-switched power output from a ruby laser with a chromium concentration of 1.6×10^{25} chromium ions per cubic meter? Assume a ruby laser rod 0.1 m long with a cavity length of 0.4 m and mirrors coated with a reflectivity of 90%. The index of refraction for ruby is 1.75. Assume also that the diameter of the multimode laser mode volume within the rod is approximately 2 mm. Determine the Q-switched power output from the laser if it is pumped to a factor of five times the threshold inversion density.

We will use (12.44) to compute the maximum power output P_{max} from the laser:

$$P_{max} = \frac{h\nu}{2t_C}\left[M_t \ln \frac{M_t}{M_0} - (M_t - M_0)\right].$$

We know that $M_0/M_t = 5$ and that, for the ruby laser at 694.3 nm, $h\nu = 2.86 \times 10^{-19}$ J. The cavity decay time t_C is slightly more complex than that given in (12.17) because the transit time is more complicated than just $c/\eta d$, since there are two different portions of the optical path within the cavity that have different indices of refraction. For a crystal of length L and a cavity with a mirror separation of d, the transit time per pass is given by

$$\left[\frac{\eta_L L + \eta_C(d - L)}{c}\right],$$

where η_L is the index of refraction within the rod and η_C is the index of refraction of air (i.e., of the medium between the mirrors but outside the amplifier). Thus the decay time is expressed as

$$t_C = \frac{[\eta_L L + \eta_C(d - L)]}{cL_F},$$

where L_F in this case is 0.1, since the mirrors have reflectivities of 90% and we assume there is an antireflection coating on the rod and thus no reflective loss at the interface between the rod and the air. We can compute t_C to be

$$t_C = \frac{1.75 \cdot 0.1 \text{ m} + 1.0(0.4 - 0.1) \text{ m}}{(3 \times 10^8 \text{ m/s}) \cdot 0.1} = 1.58 \times 10^{-8} \text{ s} = 15.8 \text{ ns}.$$

For P_{max} we thus have

$$P_{max} = \frac{2.86 \times 10^{-19} \text{ J}}{2 \cdot (1.58 \times 10^{-8} \text{ s})}\left[M_t \ln\left(\frac{1}{5}\right) - (M_t - 5M_t)\right]$$

$$= (9.05 \times 10^{-12} \text{ J/s})[-1.61M_t + 4M_t] = 2.16 \times 10^{-11}M_t \text{ J/s}.$$

Thus we need only calculate M_t for this laser. If we assume the three-level system (described in Section 8.2) for ruby and that essentially no population exists in level i due to the rapid decay out of that level into level u, then approximately half of the chromium ion population must be pumped into level u to achieve the inversion threshold. Thus, for a chromium ion concentration of 1.58×10^{25} m^{-3}, $M_t = [(1.58 \times 10^{25}$ m$^{-3})/2]V$, where V is the total volume occupied by the laser mode within the rod. For a rod 0.1 m long with mode diameter of 1 mm, the mode volume would be $V = \pi(0.001)^2$ m$^2 \cdot 0.1$ m $= 3.14 \times 10^{-7}$ m^3. Hence $M_t = (1.58 \times 10^{25})(3.14 \times 10^{-7})/2 = 4.96 \times 10^{18}$, so P_{max} may be estimated as

$$P_{max} = (2.16 \times 10^{-11})(4.96 \times 10^{18}) \text{ J/s} = 1.07 \times 10^8 \text{ W} \cong 100 \text{ MW}.$$

This is a substantial power output of a laser pulse!

METHODS OF PRODUCING Q-SWITCHING WITHIN A LASER CAVITY

ROTATING MIRRORS This was the first method used for Q-switching a laser. It consists of mounting a hexagonal-shaped mirror assembly on a rotating shaft and aligning the facets of the mirror with the laser cavity such that, for every sixth of a rotation of the shaft, a mirror would be aligned with the laser cavity as shown in Figure 12-7. This rotating mirror would serve as the rear mirror of the laser cavity, and an output mirror would be located at the other end of the cavity. For a ruby laser with an upper level lifetime of 3 ms, the shaft would have to rotate one sixth of a turn in 0.003 s – or a full turn in 0.018 s – requiring a shaft rotation speed of 3,333 revolutions per minute. This is a fairly high rotation speed, so the hexagonal mirror would have to be reasonably small to turn at that speed without developing vibration problems. Another problem with this technique is in obtaining the ideal mirror alignment for each facet.

ELECTRO-OPTIC SHUTTER When placed in the laser cavity, an electro-optic shutter can serve as a very fast optical device for switching from a high-loss to a low-loss cavity. Such a shutter is typically an electro-optic

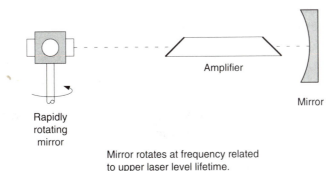

Rapidly
rotating
mirror

Amplifier

Mirror

Mirror rotates at frequency related
to upper laser level lifetime.

Figure 12-7. Q-switching with a rotating mirror as a shutter

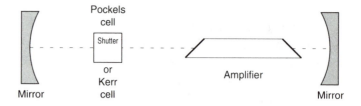

Voltage on -   Polarization rotated 45° for each pass through
             shutter. Beam returning to amplifier is rotated
             90° and thereby reflected from the cavity by
             the Brewster window, thereby producing high
             cavity loss.

Figure 12-8. *Q*-switching
with an electro-optic   Voltage off -   Beam passes through shutter undisturbed
shutter   providing a low loss cavity.

crystal that becomes birefringent when an electrical field is applied across
the crystal. Since these shutters operate by rotating the polarization of the
beam, the laser must have a polarizing element within the cavity, such as
Brewster angle windows on the amplifier, or a separate polarizing element
to force the laser to operate on only one polarization. When used as a *Q*-
switching cell, such a shutter might have the cavity arrangement shown in
Figure 12-8. In this case, the laser amplifier has Brewster angle windows
that allow a low-loss beam to transit the cavity polarized in the plane of
the paper on which Figure 12-8 is printed. When the voltage is on, the
shutter rotates the plane of polarization of the beam by 45° as it passes
through the cell. The beam reflects from the mirror and returns through
the cavity, traveling in the opposite direction, whereafter its plane is rotated by an additional 45° by the shutter. It thus arrives back at the laser
amplifier with a polarization of 90° with respect to that when it left the
amplifier. This 90° polarized beam is rejected by the Brewster window of
the amplifier and directed out of the cavity – in effect, introducing a high
loss in the cavity as needed according to Figure 12-4. When the voltage is
turned off, the polarization is no longer rotated and the beam returns to
the amplifier with no reflective loss at the Brewster window, to be further
amplified in the gain medium. Two types of electro-optic shutters that can
be used for *Q*-switching are the Kerr cell and the Pockels cell. The Pockels
cell is generally preferred over the Kerr cell because of the lower voltage
needed to produce the desired effect.

• POCKELS CELL   The Pockels effect is produced when an electric field is
applied to certain kinds of birefringent crystals to alter the indices of refraction of the crystals. This effectively provides a rotation of the polarization of the beam that is proportional to the thickness of the cell and the
magnitude of the field. The field is applied in the direction of the optic axis
of the crystal and in the direction of the optical beam, and the effect is
directly proportional to the magnitude of the applied electric field. The

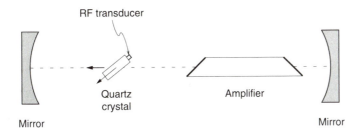

RF on - Beam deflected out of the cavity yielding high loss.

RF off - Beam transits the cavity with low loss.

Figure 12-9. *Q*-switching with an acousto-optic shutter

crystals most often used for this purpose are KD*P (potassium dihydrogen phosphate) and CD*A (cesium dihydrogen arsenate), since they require the least voltage to produce the desired effect.

• KERR CELL   A Kerr cell uses the Kerr electro-optic effect to rotate the laser beam. This effect is produced when an electric field is applied across a normally isotropic liquid to make it doubly refracting by aligning the molecules of the liquid. The field is applied in a direction transverse to that of the optical beam. Liquid nitrobenzene is a particularly good medium for use in a Kerr cell, since the effect can be produced at lower voltages than in other Kerr cell materials. The Kerr effect induces a degree of birefringence that is proportional to the square of the applied voltage.

ACOUSTO-OPTIC SHUTTER   An acousto-optic shutter typically uses a quartz crystal installed within the cavity, either at Brewster's angle or with low-loss antireflection coatings on the optical surfaces of the crystal, as shown in Figure 12-9. The quartz crystal has a piezoelectric transducer attached to the crystal that propagates strong acoustic waves within the crystal when the RF (radio-frequency) signal is applied to the transducer. When this field is applied during the time the amplifier is being pumped, the laser beam is deflected out of the laser cavity by an effective diffraction grating that is established by the acoustic waves propagating because of the RF signal. When the signal is turned off, the beam passes through the cavity undeflected and the *Q*-switched pulse develops in the cavity. The typical RF range is from 25 to 50 MHz.

SATURABLE ABSORBER   A saturable absorber can be used as a passive *Q*-switching element by placing it within the laser cavity as shown in Figure 12-10. It is also used for mode-locking as described in Section 12.3. The absorber would normally appear as a high-loss element (highly absorbing) at the laser frequency. As the laser crystal is pumped, the spontaneous emission at the laser frequency becomes more intense as the upper laser level population continues to increase during the pumping cycle. As the beam

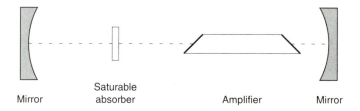

Figure 12-10. Q-switching with a saturable absorber

When the amplifier is pumped with sufficient energy, the emission from the amplifier saturates (bleaches) the absorber causing it to change from high loss to low loss thereby allowing a pulse of light to pass through the absorber.

develops and the spontaneous emission increases due to an increase in N_u, there is a moment at which the spontaneous emission passing down the length of the amplifier causes the absorber to "bleach through," thereby creating low loss in the absorber. At that time the beam will begin to traverse back and forth and begin to build up within the cavity. A good saturable absorber can be determined by ensuring that [the laser intensity at the time it arrives at the absorber] × [the absorption cross section of the absorber] *exceeds* [the intensity of the laser as it arrives at the gain medium] × [the stimulated emission cross section of the gain medium]. A liquid dye solution such as DODCI is typically used for such a Q-switching element. A thin film coating on one of the mirrors, in the form of a quantum well, can serve as a solid-state saturable absorber. The choice of quantum well is determined by the wavelength of the laser. Operating near the band edge of the quantum well increases the absorption cross section and thereby more effectively satisfies our criterion for an effective saturable absorber. The exact choice of saturable absorbers is thus determined by the laser wavelength and also the gain of the laser medium.

12.3 MODE-LOCKING

GENERAL DESCRIPTION

There are many uses of very short–duration laser pulses in the fields of digital communications, diagnostics of ultrafast processes, and ablation of materials without causing significant heating of the material. Consequently, much effort has been devoted to developing techniques for generating short pulses. In the previous section, we described how the process of Q-switching generated very intense short pulses. However, such Q-switched pulses are limited to minimum pulse durations of a few ns. Another technique that has allowed the generation of optical pulses as short as 6 fs (6×10^{-15} s) is known as *mode-locking*. For visible pulses of such a short duration, the electric field oscillates for only a few cycles and the actual pulse, if frozen in space, would be less than 2 μm in length or about 1/30th the thickness of a human hair.

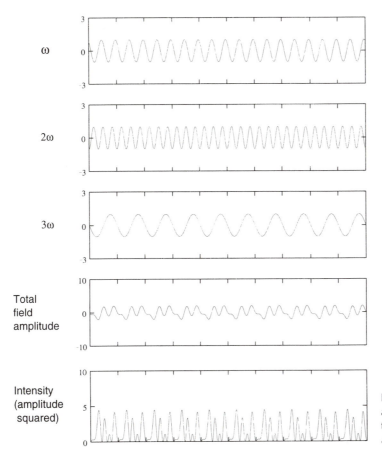

Figure 12-11(a). Amplitude and intensity of the sum of three out-of-phase waves added together

Mode-locking is achieved by combining *in phase* a number of distinct longitudinal modes of a laser, all having slightly different frequencies. When modes of electromagnetic waves of different frequencies but with random phases are added, they produce a randomly distributed, average output of both the electric field and the intensity in the time domain. An example of this is shown in Figure 12-11(a), where three out-of-phase sinusoidal waves of frequency ω, 2ω, and 3ω are added together to produce a randomly fluctuating total field amplitude. Squaring that amplitude provides the intensity variation with time, which is also shown. When the same three frequencies (modes) are added in phase – in other words, when all of their phases are zero at the same spatial location as shown in Figure 12-11(b) – they combine to produce a total field amplitude and intensity output that has a characteristic repetitive pulsed nature. Hence, achieving such phasing or mode-locking has become a powerful method for generating ultrashort pulses.

THEORY

The longitudinal mode separation of a laser cavity was shown in (10.31) to be $\Delta\nu = c/2\eta d$, where c/η is the velocity of the laser beam in a medium

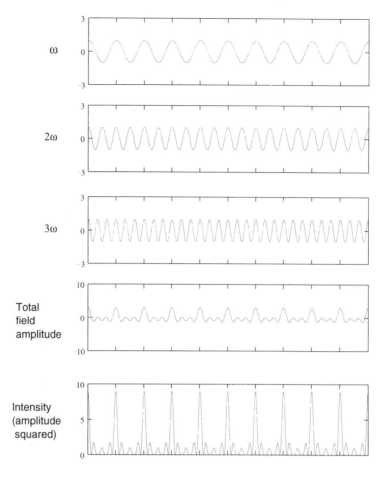

Figure 12-11(b). Amplitude and intensity of the sum of three in-phase waves added together

having index of refraction η and d is the separation between the cavity mirrors. If the gain bandwidth, which is associated with the emission linewidth of the laser transition, is greater than this mode separation, then it is possible to have more than one longitudinal mode oscillating simultaneously, as indicated in Figure 10-6. Normally, each of these modes oscillates independently of the other modes. For such a system, let the amplitude of the nth mode be expressed as

$$E(t) = E_n e^{i(\omega_n t + \phi_n)},\tag{12.46}$$

where ω_n is the frequency and ϕ_n is the phase of that mode. Assume that there are N modes of equal amplitude oscillating simultaneously in the cavity. The combined total amplitude of all of the modes can then be expressed as

$$E(t) = E_0 \sum_{n=0}^{N-1} e^{i(\omega_n t + \phi_n)}.\tag{12.47}$$

In this case, the difference in angular frequency between the modes is given as

$$\omega_{n+1} - \omega_n = \Delta\omega = 2\pi\Delta\nu = \frac{2\pi c}{2\eta d} = \frac{\pi c}{\eta d}.\tag{12.48}$$

Because these modes are oscillating in a random fashion, the total intensity is given by the absolute square of the total amplitude:

$$I(t) = |E(t)|^2 = E_0^2 \sum_{n=0}^{N-1} e^{i(\omega_n t + \phi_n)} e^{-i(\omega_n t + \phi_n)} = NE_0^2. \tag{12.49}$$

Thus, the intensity is equal to N times the intensity of the individual modes, which is not an unexpected result. This value can occasionally vary if a few modes randomly phase together, but for large N it does not vary significantly from the average value.

Let us now examine what would happen if we could make all the modes oscillate in phase; in other words, we lock all of the modes together. Not surprisingly, this technique is known as mode-locking. This effect can be expressed as

$$\phi_n = \phi_0 \quad \text{for all } n. \tag{12.50}$$

The combined field amplitude is then given as

$$E(t) = E_0 \sum_{n=0}^{N-1} e^{i(\omega_n t + \phi_0)} = E_0 e^{i\phi_0} \sum_{n=0}^{N-1} e^{i\omega_n t}. \tag{12.51}$$

Define ω_n as

$$\omega_n = \omega_{N-1} - n\Delta\omega. \tag{12.52}$$

We can then rewrite (12.51) as

$$E(t) = E_0 e^{i\phi_0} \sum_{n=0}^{N-1} e^{i(\omega_{N-1} - n\Delta\omega)t}, \tag{12.53}$$

which upon expansion becomes

$$E(t) = E_0 e^{i(\phi_0 + \omega_{N-1}t)} [1 + e^{-i\Delta\omega t} + e^{-2i\Delta\omega t} + \cdots + e^{-i(N-1)\Delta\omega t}]. \tag{12.54}$$

The expression in brackets is a finite series that has a value of

$$\frac{1 - e^{-iN\Delta\omega t}}{1 - e^{-i\Delta\omega t}}. \tag{12.55}$$

Therefore, the expression for $E(t)$ becomes

$$E(t) = E_0 e^{i(\phi_0 + \omega_{N-1}t)} \left[\frac{1 - e^{-iN\Delta\omega t}}{1 - e^{-i\Delta\omega t}} \right]. \tag{12.56}$$

We can now obtain the total intensity as

$$I(t) = E_0^2 \left| \left[\frac{1 - e^{-iN\Delta\omega t}}{1 - e^{-i\Delta\omega t}} \right] \right|^2 = E_0^2 \frac{\sin^2(N\Delta\omega t/2)}{\sin^2(\Delta\omega t/2)}. \tag{12.57}$$

This expression varies with t, but the maximum value occurs at

$$\frac{\Delta\omega t}{2} = 0, \pi, 2\pi, \ldots, n\pi. \tag{12.58}$$

Solving for t and taking the difference between two successive maxima at t_{n+1} and t_n, we have pulses at intervals Δt of

$$\Delta t_{sep} = t_{n+1} - t_n = \frac{2(n+1)\pi}{\Delta\omega} - \frac{2n\pi}{\Delta\omega} = \frac{2\pi}{\Delta\omega}$$

$$= \frac{2\pi}{2\pi\Delta\nu} = \frac{1}{\Delta\nu} = \frac{2\eta d}{c}. \tag{12.59}$$

The maximum value for $I(t)$ given by (12.57) can be determined by evaluating the function at any of the specific values given by (12.58), since the maximum is the same for all of those values. We will therefore choose to evaluate it at $\Delta\omega t/2 = 0$ (the simplest case) by taking the limit for $I(t)$ as $\Delta\omega t/2$ approaches zero:

$$I(t)_{lim} = \lim_{\Delta\omega t/2 \to 0} E_0^2 \frac{\sin^2 N\Delta\omega t/2}{\sin^2 \Delta\omega t/2}$$

$$= \lim_{\Delta\omega t/2 \to 0} E_0^2 \frac{N^2(\Delta\omega t/2)^2}{(\Delta\omega t/2)^2} = E_0^2 N^2. \tag{12.60}$$

The maximum value $I(t)_{max}$ is therefore given by

$$I(t)_{max} = E_0^2 N^2. \tag{12.61}$$

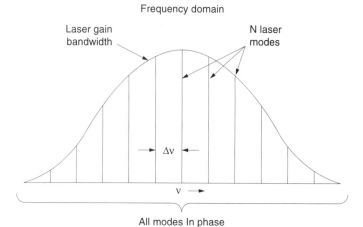

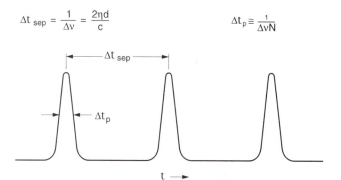

Figure 12-12. Parameters associated with mode-locked pulses

As shown in Figure 12-12, the pulses described by (12.58) are separated in time by $\Delta t_{sep} = 2\eta d/c$ according to (12.59) and they have a pulse width (FWHM) of $\Delta t_P = 2\eta d/Nc$. We can rewrite (12.48) as $\eta d/c = \pi/\Delta\omega$ and hence the expression for the pulse width as

$$\Delta t_P = \frac{2\eta d}{Nc} = \frac{2\pi}{\Delta\omega N} = \frac{1}{\Delta\nu N} = \frac{1}{\text{gain bandwidth}}. \qquad (12.62)$$

We can see that the predicted pulse width of each mode-locked pulse can be as short as the reciprocal of the bandwidth of the emission line, which is just $N\Delta\nu$. When such a pulsewidth is achieved experimentally it is referred to as a *bandwidth-limited pulse*.

EXAMPLE

Compute the mode-locked pulse width Δt_P and the separation between pulses Δt_{sep} for the following mode-locked lasers.

(a) A helium–neon laser operating at 632.8 nm with a mirror cavity spacing of $d = 0.5$ m.

(b) A Rh6G dye laser operating over its entire gain bandwidth (570–640 nm) with the cavity mirrors separated by 2 m. The index of refraction of a laser dye in a typical solvent is approximately 1.4.

(a) From (12.59) we have $\Delta t_{sep} = 2\eta d/c$. For a helium–neon gas laser, $\eta \cong 1$. Therefore, since $d = 0.5$ m, we have

$$\Delta t_{sep} = \frac{2 \cdot 1 \cdot 0.5 \text{ m}}{3 \times 10^8 \text{ m/s}} = 3.33 \times 10^{-9} \text{ s}.$$

From (12.62) we have $\Delta t_P = 1/(\text{gain bandwidth})$ for the mode-locked pulse width. For the He–Ne laser we assume that modes will lase over the FWHM emission linewidth of the 632.8-nm transition of 1.5×10^9 Hz from Table 4-1. Hence, the mode-locked pulse width is

$$\Delta t_P = \frac{1}{\Delta\nu_{FWHM}} = \frac{1}{1.5 \times 10^9 \text{ Hz}} = 6.67 \times 10^{-10} \text{ s}.$$

This He–Ne laser will thus produce pulses of 667-ps duration at intervals of just over 3 ns.

(b) Again from (12.59) we have $\Delta t_{sep} = 2\eta d/c$. Thus, for the conditions given,

$$\Delta t_{sep} = \frac{2\eta d}{c} = \frac{2 \cdot 1.4 \cdot 2}{3 \times 10^8} = 1.87 \times 10^{-8} \text{ s} = 18.7 \text{ ns}.$$

The bandwidth of the dye laser would be $(640 - 570) = 70$ nm. This can be converted to frequency width by the relationship

---

EXAMPLE (cont.)

$$\Delta\nu = \frac{c}{\lambda^2}\Delta\lambda = \frac{3\times10^8 \text{ cm/s}}{(6.05\times10^{-7}\text{ m})^2}(70\times10^{-9}\text{ m}) = 5.7\times10^{13}\text{ s}^{-1},$$

assuming that the average wavelength of the emission occurred at 605 nm. The bandwidth-limited pulse width from (12.62) would then be

$$\Delta t_P = \frac{1}{\text{gain bandwidth}} = \frac{1}{5.7\times10^{13}\text{ s}^{-1}} = 1.75\times10^{-14}\text{ s} = 17.5\text{ fs}.$$

TECHNIQUES FOR PRODUCING MODE-LOCKING

GENERAL CAVITY CONSIDERATIONS   The simplest form of mode-locking is achieved by operating a laser continuously in a stable two-mirror cavity, as shown in Figure 12-13. The gain bandwidth of the laser transition in the laser amplifier determines the ultimate minimum pulse duration as obtained from (12.62). Phasing the modes is achieved by placing an optical switch or shutter at one end of the cavity next to one of the mirrors. When the shutter is temporarily opened, the electric fields of all of the modes are simultaneously maximized and thus in phase. Then as each mode develops it generates sidebands that drive adjacent modes in phase if the cavity length is appropriately adjusted such that the round-trip time is equal to Δt_{sep}. The switch must be extremely fast to allow only a very short pulse to pass through the shutter, arrive at the mirror, and reflect back into the cavity. In actual practice two pulses exist simultaneously within the cavity, traveling in opposite directions and arriving at the shutter at the same time, so it is best to place the shutter at one end of the cavity. The shutter can be

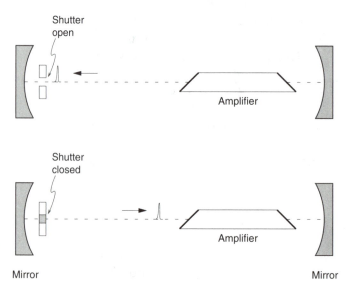

Figure 12-13. Arrangement for obtaining mode-locked pulses

either an active shutter, such as an acousto-optic device driven with RF power, or a passive nonlinear optical shutter that opens when an intense laser pulse arrives. An active shutter must be designed to open exactly when the laser pulse arrives, at a time interval of $\Delta t_{sep} = 2\eta d/c$ according to (12.59). For a shutter with fixed opening frequency, this interval Δt_P can be adjusted by altering the laser cavity length as indicated by the equation. A passive shutter opens automatically at the correct time, since it is opened by the pulse when it arrives at the shutter. A passive shutter may require a slight disturbance of the cavity, such as a vibration to initiate the pulse, in order to overcome the initial loss of the absorber when the beam is initially at low intensity.

ACTIVE SHUTTERS

• ACOUSTO-OPTIC SWITCHES   Acousto-optic modulators, as described in the previous section on Q-switching, are very useful as mode-locking switches for low-gain cw lasers. The switch consists of a quartz crystal with a piezoelectric transducer attached to one side. The crystal is located at one end of the laser cavity near the laser mirror, typically with the crystal surfaces positioned at Brewster's angle to minimize reflective losses from the quartz surfaces. An RF signal (typically 25–50 MHz) is applied to the transducer; this introduces an acoustic wave within the quartz crystal. This acoustic wave produces a periodic loss within the crystal, due to Bragg reflection, at twice the applied frequency of the RF signal. When the loss is not present (when the RF signal reduces to zero electric field in between the high losses), the shutter is effectively open, the mode phases are locked, and the short pulse passes through the shutter. The loss is not very large, so this technique is applicable only to low-gain lasers such as a cw Nd:YAG laser.

• SYNCHRONOUSLY PUMPED MODE-LOCKING   Another active-shutter technique used to produce mode-locking is that of switching on the gain for only a very short period of time while the short pulse is passing through the gain medium. This is done by pumping the gain medium with a long-pulse (1-ns–duration) mode-locked laser such as an argon ion laser. The cavity round-trip time of the mode-locked pump laser must be coincidental with that of the laser being mode-locked in order to maintain the necessary switch timing, as can be seen in Figure 12-14. Since the presence of gain serves as the switch to amplify the pulse, the lifetime of the laser transition of the gain medium must be significantly shorter than the round-trip time of the pulse within the cavity. This can also be seen in Figure 12-14 for the case of a dye laser. Even though the dye upper laser level is of the order of nanosecond duration, the mode-locked pulse can shorten to the pico-second time frame because the pulse extracts most of the pumped laser energy in the leading edge of the mode-locked pulse, as shown in the figure, thereby leaving no net gain for the trailing edge to be amplified.

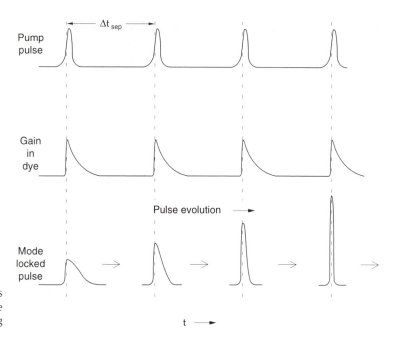

Figure 12-14. Synchronous pumping used to achieve mode-locking

PASSIVE SHUTTERS Passive mode-locking is a very desirable technique to produce ultrashort pulses, since it involves shutters that require no external controls and generally works without having to be serviced or replaced. The passive techniques that are most often used are colliding pulse mode-locking (CPM), additive pulse mode-locking (APM), and Kerr lens mode-locking (KLM).

• COLLIDING PULSE MODE-LOCKING (CPM) Colliding pulse mode-locking is produced by the interaction of two counterpropagating pulses in a thin saturable absorber installed within a ring laser cavity, as shown in Figure 12-15. The saturable absorber is located in an astigmatically compensated portion of the cavity so that the beam can be focused to a small spot within the saturable absorber. When the two counterpropagating pulses meet within the saturable absorber, they produce an increased intensity that begins to bleach out the medium. They also produce a transient regularly spaced variation in the population of the absorber molecules over the thickness of the absorber due to interference effects of the two beams. This interference produces a standing wave in the absorber region that effectively increases the intensity of the laser at that location, which thereby increases the bleaching effect even further. This in turn synchronizes, stabilizes, and shortens the laser pulses. This type of mode-locking, involving two pulses within the saturable absorber, has two advantages over techniques using only a single pulse. The first is that it avoids the fabrication problems associated with having the saturable absorber in optical contact with one of the mirrors; the second is that it provides a synchronizing function that yields shorter and more stable pulses.

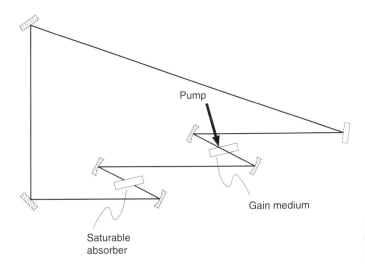

Pump

Gain medium

Saturable
absorber

Figure 12-15. Ring laser
arrangement for colliding
pulse mode-locking (CPM)

The requirements for producing CPM include: (1) a laser structure in which the difference in arrival time within the saturable absorber of the two interfering pulses is small compared to the pulse duration; and (2) a length of the optical path within the saturable absorber that is (less than or) of the order of the length of the pulse ($c\Delta t_P/\eta$). Thus, for a 100-fs pulse (10^{-13} s), the saturable absorber must be less than 20 μm in thickness. Synchronization of the two counterpropagating pulses within the saturable absorber occurs because in this case the minimum energy is lost from the beam to the absorber. CPM can also occur in a standing-wave cavity, but only if the saturable absorber is placed at precisely a submultiple m of the cavity length from one of the end mirrors. Stable pulses of duration less than 100 fs have been produced with this technique.

• ADDITIVE PULSE MODE-LOCKING (APM)   A mode-locked laser can be designed such that part of its output is sent through an optical fiber, as shown in Figure 12-16. If a mirror is located at the end of the fiber to return the beam to the cavity, the portion of the laser pulse that travels through the fiber is thus re-injected into the cavity. The laser essentially has two connected cavities, an arrangement that is referred to as a *coupled cavity*. The portion of the pulse within the fiber experiences self-phase modulation, which is described in more detail in the next section. This process red-shifts the leading edge of the pulse and blue-shifts the trailing edge of the pulse. If the length of the fiber is adjusted such that the pulse within the fiber returns to the main cavity at the appropriate time, then the leading edge of that returning pulse adds in phase with the main cavity pulse, thereby increasing its intensity, and the trailing edge adds out of phase with the cavity pulse, thereby decreasing its trailing edge. This effectively shortens the pulse, and continues to shorten it after every round trip until the pulse approaches the bandwidth limit of the gain medium. The nonlinearity required for phase modulation can be produced by amplitude

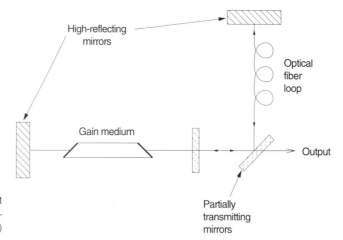

High-reflecting
mirrors

Optical
fiber
loop

Gain medium

Output

Partially
transmitting
mirrors

Figure 12-16. Arrangement for additive pulse mode-locking (APM)

modulation or by providing an intensity-dependent reflectivity of the end mirror of the fiber. Careful adjustment of the cavity length is required to properly time the arrival of the phase-modulated pulse with the laser pulse within the cavity. This generally requires making the optical path length of both cavities the same.

● KERR LENS MODE-LOCKING (KLM) Kerr lens mode-locking is a process that takes advantage of the nonlinear effect of self-focusing. This effect, described in more detail in Section 15.4, produces an intensity-dependent change in the refractive index of a material. Thus for a Gaussian-shaped beam passing through a material, with the beam more intense at the center than the edges, the index of refraction of the material will become higher at the center than at the edges of the beam, thereby effectively creating a lens that in turn slightly focuses the beam within the material. The path length that the beam traverses and the beam intensity will determine how much focusing occurs before the beam emerges from the material.

In KLM, the self-focusing effect is used to preferentially select the pulsed mode-locked set of modes that provide short pulses, rather than a single steady-state cw mode. This is accomplished by one of two possible techniques. Figure 12-17(a) shows a laser gain material such as $Ti:Al_2O_3$ and a cavity mirror with a normal cw mode shown as a dashed line within the laser cavity. An aperture, placed between the laser gain medium and the mirror, is small enough in diameter to provide a relatively high loss to the cw mode. However, if a pulse of light with higher intensity than the cw beam is generated within the gain medium, it will produce self-focusing as it propagates through the medium, which will reduce its diameter as it emerges from the medium. It will thus pass through the aperture with lower loss than the cw mode, thereby providing a more favorable environment for a pulsed laser than for a cw laser. This same aperturing effect can be achieved by making a smaller-diameter pump beam than the cw mode size, as shown in Figure 12-17(b). This in effect becomes a small aperture within

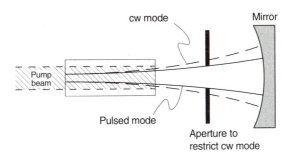

(a) Physical aperture

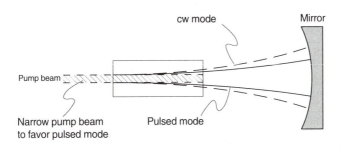

(b) Gain aperture

Figure 12-17. Two techniques for achieving Kerr lens mode-locking (KLM)

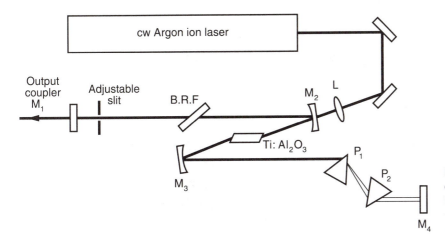

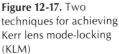

Figure 12-18. Laser cavity arrangement for implementing KLM (courtesy of Coherent, Inc.)

the gain medium, instead of an aperture external to the gain medium as in Figure 12-17(a).

The focusing effect within the gain medium also enhances self-phase modulation, which broadens the frequency spectrum of the pulse. This allows the pulse to be compressed to a much shorter duration than could normally be achieved with a mode-locked laser. Moreover, KLM automatically provides group velocity dispersion within the crystal at the same time that self-focusing is occurring, thereby providing the bandwidth for further pulse shortening (as described in what follows). A diagram of a laser cavity arrangement designed to use KLM is shown in Figure 12-18. It consists of a

Ti:sapphire laser in a standing-wave cavity, with a birefringent filter (BRF) and a prism pair to provide dispersion of the pulse. The Kerr lensing effect is achieved with the adjustable slit aperture located near the output coupler.

PULSE SHORTENING TECHNIQUES

Equation (12.62) suggested that the minimum pulse width of a mode-locked laser is determined by the gain bandwidth of the pulse. Thus, if the bandwidth could be increased beyond its original width associated with the laser transition, it might be possible to make the pulses even shorter in duration. This is accomplished by combining the following two processes: *Self-phase modulation* adds another bandwidth to the pulse, and *pulse compression* "squeezes" that additional bandwidth down to a narrower pulse width.

SELF-PHASE MODULATION   Self-phase modulation occurs via the interaction of a rapidly varying time-dependent laser pulse with the nonlinear intensity-dependent change in refractive index of an optical material. With this technique, another bandwidth is added to the pulse as it propagates within a nonlinear medium. The nonlinear change in refractive index, also associated with the effect of self-focusing as described in Section 15.4, occurs when a beam has sufficient intensity as it passes through an optical medium to vary the index of the medium according to the following equation:

$$\eta(\nu) = \eta_0(\nu) + \eta_2(\nu)I(\nu), \tag{12.63}$$

where η_0 is the normal index of refraction that varies with frequency ν, and η_2 is the intensity-dependent index. Note that η_2 is not dimensionless, since it is multiplied by the intensity to produce a dimensionless quantity. The pulse propagating through the medium will undergo self-phase modulation, which broadens the frequency spectrum of the pulse. This occurs because the leading edge of the pulse produces a transient increase in the refractive index (for the typical case in which η_2 is positive) which results in a red shift of the instantaneous frequencies of the rising portion of the pulse. Similarly, a blue shift occurs on the trailing edge of the pulse as the pulse intensity rapidly decreases. A physical interpretation of this is that a pulse arrives in an optical medium having a specific pulse width and center frequency as shown in Figure 12-19(a). When it undergoes self-phase modulation, on the rising portion of the pulse the index of refraction is increasing, thereby decreasing the velocity of the beam. This decrease in velocity means that the arrival of this rising portion of the electromagnetic wave on the exit side of the crystal is delayed, thereby effectively reducing the frequency as shown in Figure 12-19(b). On the trailing portion of the pulse the index of refraction is decreasing, thereby decreasing the optical path of the wave and speeding up its transit time through the crystal. In effect, this increases the frequency of the trailing edge, as shown in Figure 12-19(b). In other words, more waves arriving within the same time effectively increases

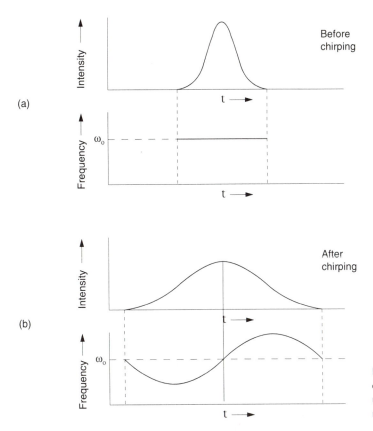

(a)

(b)

Figure 12-19. Pulse chirping as a result of undergoing self-phase modulation

the frequency of those waves. This process is also known as *pulse chirping* or *frequency chirping*.

PULSE COMPRESSION   Compression of a pulse of a given bandwidth begins by taking a pulse that has undergone self-phase modulation. This pulse is then either passed through a fiber that has anomalous dispersion or forced to interact with other dispersive elements such as prisms or diffraction gratings. Consider a pulse that has a time-dependent frequency distribution such as that shown in Figure 12-19(b). If such a pulse is subsequently passed through a fiber that has anomalous dispersion (see Section 2.3 and Figure 2-6) in the wavelength region of the pulse then the pulse will be shortened, since the red-shifted portion of the pulse will be slowed down (higher refractive index) and the blue-shifted portion of the pulse will be speeded up (lower refractive index). A pulse can also be compressed by reflecting it from a pair of gratings as shown in Figure 12-20(a) or by passing it through two pairs of prisms as shown in Figure 12-20(b).

For the grating compression, Figure 12-20(a) shows the red-shifted wavelengths of the pulse that arrive at the first grating are diffracted more than the blue-shifted wavelengths, and arrive at a different portion of the second grating than the blue wavelengths. Both the red and blue components of the beam then diffract again from the second grating and are both re-directed

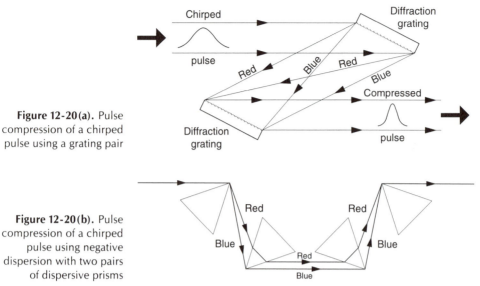

Figure 12-20(a). Pulse compression of a chirped pulse using a grating pair

Figure 12-20(b). Pulse compression of a chirped pulse using negative dispersion with two pairs of dispersive prisms

in the same direction, as shown in Figure 12-20(a). During this process the red portion of the pulse travels a longer path length than the blue portion of the beam. Hence, after diffracting from the second grating and recombining with the blue wavelengths, the total pulse has been compressed in time since the blue components have "caught up" with the red components.

The case for two prism pairs is shown in Figure 12-20(b). For appropriately located prisms as shown, the red and blue portions of the pulse are separated into two different optical paths, with the red-shifted path being longer than the blue-shifted path. The prism angle is selected to allow both the entering and exiting beams to occur at Brewster's angle as well as at the angle of minimum deviation of the prism. A similar compression to that of the grating pair is then obtained. The gratings are used for high-power pulses because they can be manufactured with a large surface area, which reduces the power loading per unit area on the grating. The prisms provide lower surface losses than the gratings, and also can be fabricated from different materials to provide varying degrees of dispersion.

12.4 RING LASERS

Lasers operating in a two-mirror optical cavity produce standing-wave patterns associated with the various longitudinal modes operating within the cavity, as indicated in Figure 10-5. If a laser is homogeneously broadened (as are most solid-state lasers), one might expect only a single longitudinal mode to dominate, since the mode with the highest gain will begin to oscillate first and reduce the gain over the entire gain bandwidth as the mode approaches the saturation intensity, thereby preventing other modes from reaching threshold as indicated in Figure 7-13. However, for a single longitudinal mode, the gain varies spatially in a sinusoidal pattern according to

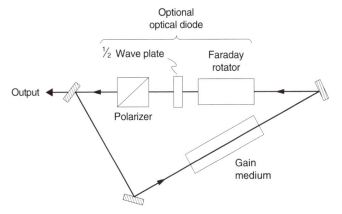

Figure 12-21. Example of a ring laser cavity, including an optional optical diode

the standing-wave variation within the cavity. This is referred to as spatial hole burning, as described in Section 10.3. At the null points of the standing wave of the dominant mode, stimulated emission into that mode does not occur and the gain is therefore not reduced in such spatial regions. It is therefore possible for other modes to rise above threshold and oscillate within the cavity by using the gain found at those null points of the fundamental mode. Thus, if it is desirable to have a single longitudinal and/or transverse mode operating within the cavity, it is often necessary to use a ring laser. In such a laser the gain medium is uniformly saturated by the beam. This makes single-mode operation more efficient, since one mode extracts all of the power available in the gain medium.

In contrast with the standing-wave cavity just described, ring lasers have the beam circulating in a loop rather than passing back and forth. For the beam to circulate in a loop, an optical cavity typically requires more than two mirrors. An example of such a ring laser cavity is shown in Figure 12-21. The laser beam within such a cavity actually consists of two beams traveling in opposite directions, with separate and independent resonances and hence no standing-wave modes.

• OPTICAL DIODE   In some instances a device known as an *optical diode* is placed within the ring laser cavity, as shown in Figure 12-21. This provides a unidirectional loss, thereby eliminating the portion of the beam traveling in one of the directions and allowing the other portion of the beam to propagate in the opposite direction. This is accomplished by a two-element optical diode consisting of (1) a faraday rotator that rotates the plane of polarization of each beam in the same direction, and (2) a half-wave plate that rotates the polarization of one of the beams back to its original position. This plate also rotates the polarization of the other beam (traveling in the opposite direction) further out of the original plane of polarization, thereby increasing the loss it sees as it propagates around the cavity. This increased loss effectively quenches the beam. Implicit in this design is that the original beam be polarized, either by using Brewster angle windows or by inserting a separate polarizer in the cavity. The laser output then consists of a single traveling wave, which means that no longitudinal standing-wave or traveling-wave modes exist within such a cavity. Such an arrangement eliminates the variation of the gain due to spatial hole burning, so the gain region tends to be more spatially homogeneous than that produced in a normal standing-wave cavity.

Ring lasers are useful for producing ultrashort mode-locked pulses. They are also used in laser gyroscopes as stable reference sources.

MONOLITHIC UNIDIRECTIONAL SINGLE-MODE Nd:YAG RING LASER

Most ring lasers use a resonator in which the beam lies entirely within a single plane. Operating in a nonplanar arrangement can yield an effect that is equivalent to rotation by a half-wave plate within the cavity, as just described for the optical diode. The monolithic unidirectional Nd:YAG ring laser achieves high single-mode output power with such a special unidirectional nonplanar resonator. Frequency stability is very good in this device because the entire laser and resonator is made from a single Nd:YAG laser crystal. Since there is only one unidirectional traveling wave, spatial hole burning is eliminated and high single-mode output power is achieved. Figure 12-22 shows such a monolithic laser, which is referred to as a MISER (monolithic isolated single-mode end-pumped ring) design. The polarization selection takes place at the partially transmitting face A, where the pump beam enters the crystal and where the output is also obtained. Total internal reflection occurs at points B, C, and D. The point C is on the top surface of the crystal, out of the plane of segments AB and DA. Unidirectional oscillation is then provided by applying a magnetic field to the crystal in the direction shown, so that magnetic rotation takes place along segments AB and DA. Other laser crystals can also be used for such devices.

TWO-MIRROR RING LASER

We have stated that a ring laser generally needs at least three mirrors to produce a circulating laser beam rather than a standing-wave mode pattern within the laser cavity. Figure 14-7 shows a diagram for a compact diode-

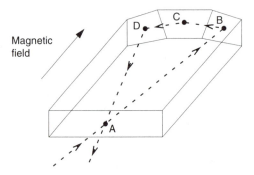

Figure 12-22. Monolithic isolated single-mode end pumped ring laser

pumped ring laser that requires only two mirrors. It takes advantage of the refractive effect of the Brewster angle cut of a Nd:YAG laser crystal to redirect the beam to one of the laser mirrors and so form the closed loop of the ring. Use of this design minimizes the number of optical components in the cavity and allows for compact packaging of the device.

12.5 CAVITIES FOR PRODUCING SPECTRAL NARROWING OF LASER OUTPUT

CAVITY WITH ADDITIONAL FABRY–PEROT ETALON FOR NARROW-FREQUENCY SELECTION

If the laser gain bandwidth is broader than the longitudinal mode spacing $\Delta\nu = c/2\eta d$, where d is the distance between the end mirrors of the cavity and η is the index of refraction of the gain medium as described in (10.31), then it is possible to have more than one longitudinal mode lasing within the cavity, as indicated in Figure 10-6. Sometimes it is desirable to have only a single longitudinal mode lasing within the cavity, and it may not be possible to shorten the cavity spacing d in order to produce a single mode according to the formula. However, it is still possible to insert within the laser cavity an additional Fabry–Perot cavity that serves as an extra frequency-selective loss element. This additional element is typically an optically transmissive material consisting of two parallel surfaces, both coated to achieve a specific reflectivity. The values of the mirror reflectivities are designed to give the desired additional loss at frequencies other than the desired laser frequency, as described by the Airy function of (10.18) and as indicated in Figure 10-4, thus quenching the gain at all of the modes except the desired one. Such a fixed Fabry–Perot device is shown in Figure 12-23 and is known as an *etalon*. It is typically a thick piece of quartz with optical surfaces that are especially flat, parallel, and of very high quality. Since the frequencies at which the Fabry–Perot resonances occur are proportional to the spacing between the mirrors, the etalon can be rotated as shown in the figure to tune the transmission to the desired frequency.

TUNABLE CAVITY

In lasers having a broad gain bandwidth – such as a dye laser, a $Ti:Al_2O_3$ laser, or lasers with multiple single-frequency transitions such as an argon ion laser – it is often desirable to tune or select any specific laser wavelength over that gain bandwidth without changing the cavity mirrors. A simple means of providing such a wide range of tunability is to install either a dispersive element (such as a prism) within the cavity or a variable-frequency high reflector (such as a diffraction grating) as one of the mirrors of the cavity. Tuning is accomplished by rotating the prism or the grating. Simple examples of these two cavity arrangements are shown in Figure 12-24. The use of two prisms, as shown in the figure, provides more dispersion than a

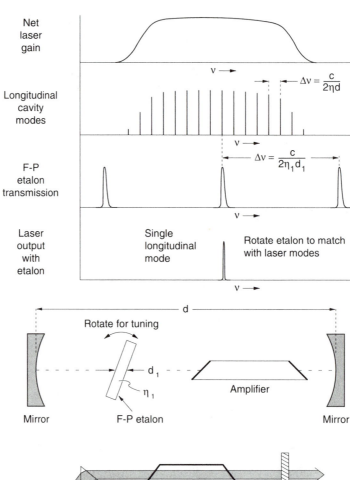

Net
laser
gain

Longitudinal
cavity
modes

$$\Delta v = \frac{c}{2\eta d}$$

F-P
etalon
transmission

$$\Delta v = \frac{c}{2\eta_1 d_1}$$

Laser
output
with
etalon

Single
longitudinal
mode

Rotate etalon to match
with laser modes

Rotate for tuning

Mirror

F-P etalon

Amplifier

Mirror

Figure 12-23.
Fixed-spacing Fabry–Perot
etalon for narrow-
frequency selection

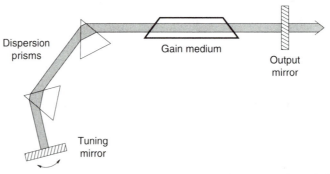

Dispersion
prisms

Gain medium

Output
mirror

Tuning
mirror

(a) Prism tuning

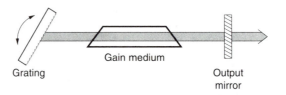

Grating

Gain medium

Output
mirror

Figure 12-24. Frequency
tuning using (a) prisms and
(b) a diffraction grating

(b) Grating tuning

single prism. In the case of diffraction grating, an etalon can also be used to provide additional dispersion. The prisms can be antireflection coated to minimize the reflection losses at the prism surfaces. The diffraction grating has higher dispersion than a prism and therefore offers more precise wavelength selection. However, such a grating also has a higher effective reflection loss and would therefore not be used for a lower-gain laser such as the argon ion laser. More complex tuning arrangements of this type are described in the reference at the end of the chapter.

BROADBAND TUNABLE cw RING LASERS

Figure 12-25 shows the schematic layouts of two versions of a commercial cw ring laser. Figure 12-25(a) shows the version that uses a tunable dye as the gain medium, and Figure 12-25(b) shows a similar layout that uses a titanium:sapphire crystal for the gain medium. The dye laser offers tunability from 370 nm to over 1 μm by changing dyes, whereas the titanium: sapphire laser operates from 680 nm to 1.1 μm. The ring in these lasers operates in a vertical plane with the pump laser beam (either an argon or krypton ion laser) incident from the left. Both lasers include an optical diode to produce a unidirectional ring as described previously. They also include a birefringent filter assembly that provides wavelength tunability by rotating the filters. This filter assembly is essentially a series of differently spaced Fabry–Perot etalons that select a narrow-frequency output

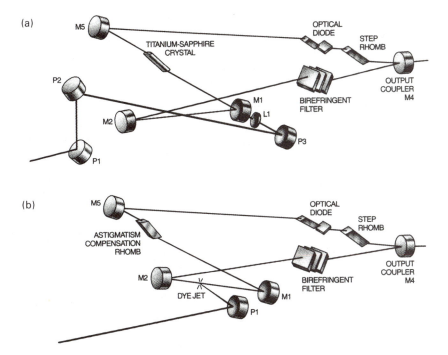

Figure 12-25. Schematic diagram of two versions of a commercial ring laser (courtesy of Coherent, Inc.)

from a broadband gain system. The step rhomb prism is a device used to displace the beam without introducing any other changes.

TUNABLE CAVITY FOR ULTRANARROW-FREQUENCY OUTPUT

By combining a diffraction grating in the Littrow mounting configuration with an intracavity beam-expanding telescope and a tilted Fabry–Perot etalon, it is possible to make a broadband tunable laser with a spectral bandwidth $\Delta\lambda$ of less than 0.0004 nm in the visible portion of the spectrum. This constitutes a fractional width of $\Delta\lambda/\lambda$ of less than one part in 10^6, which is an extremely narrow and useful tunable laser output. The cavity combines the frequency-selective features of a large-area diffraction grating with a Fabry–Perot intracavity etalon. A diagram of the arrangement is shown in Figure 12-26. A broadband laser medium such as a dye laser is side-pumped by a pulsed pump laser to produce a narrow-stripe gain medium in the gain cell. The pulsed output from the high-gain laser medium is expanded with the beam-expanding telescope to produce a wide collimated beam arriving at the high-dispersion echelle diffraction grating that serves as a reflecting mirror for the laser cavity. The tiltable Fabry–Perot etalon is also inserted in the cavity for additional wavelength narrowing and tunability. The fine tuning is achieved by rotating the diffraction grating.

For a dye laser amplifier, the output mirror has a broadband reflectivity of approximately 50% for good beam output. The gain cell must absorb the pump light in a very short depth to provide an effective point source for the telescope beam expander. Alternatively, a pinhole could be placed between the gain medium and the telescope. A Fabry–Perot etalon with an open space between the mirror surfaces can be used for ultraprecise tuning by evacuating the open space region and then gradually bleeding a gas into the cavity. This will adjust the index of refraction between the mirrors and thereby fine-tune the frequency, since the index of refraction of any gas is slightly higher than the index of a vacuum. Ultranarrow-frequency tunable

Figure 12-26. Cavity diagram of an ultranarrow-frequency tunable laser

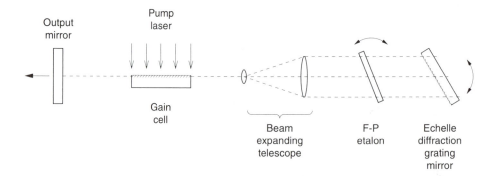

output powers of several mJ/pulse for 10-ns pulses have been achieved with this arrangement using dye gain media that operate in the visible and near-infrared portions of the wavelength spectrum.

DISTRIBUTED FEEDBACK (DFB) LASERS

The usual method of providing feedback in a laser oscillator is to have a mirror at each end of the cavity. With such a laser, standing waves are established as discrete longitudinal modes, since the boundary conditions at the reflective surfaces require the field to be zero at those locations. It is possible to provide the feedback and also the wavelength selection by effectively distributing this boundary condition throughout the gain medium in the form of a periodic variation in either the gain or the index of refraction of the medium, or by having the periodic index distribution located beyond the ends of the laser medium as described in the next section on distributed Bragg reflection. Such feedback structures are particularly effective in semiconductor lasers, where the gain is high and the fabrication of such periodic structures within the gain medium is not too difficult.

This distributed feedback effect can be understood in a simplified form by considering a periodic structure (within the gain medium) for which the variation in period is of length Λ. Thus, for a wave E^+ propagating in the forward (positive) direction, assume that a small portion Δ of the wave is scattered in the backward (negative) direction from each of the regular perturbations of the medium. As the wave E^+ moves in the forward direction and scatters from two successive perturbations, the scattering from the second perturbation will arrive back at the first perturbation with a phase delay of $2k\Lambda$, where k is the usual wave vector such that $k = 2\pi/\lambda$. Each additional scattering will also arrive out of phase by a factor of $n2k\Lambda$, where $n = 1, 2, 3, \ldots$. Thus, the total field in the backward direction could be significant if there were enough periodic ripples N, since that field would be represented by a factor of $E^- \cong N\Delta E^+$.

We can describe the periodic spatial distribution of the index of refraction by

$$\eta(z) = \eta + \eta_1 \cos kz, \tag{12.64}$$

and the variation of the gain coefficient by

$$g(z) = g + g_1 \cos kz, \tag{12.65}$$

where z is the dimension along the optic axis and η_1 and g_1 are the amplitudes of the spatial variation (or modulation) portion of the index of refraction and the gain coefficient, respectively. Oscillation will occur if the Bragg condition is satisfied such that

$$\frac{\lambda_0}{2\eta} = \Lambda. \tag{12.66}$$

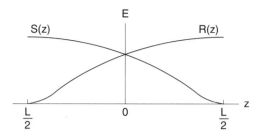

Figure 12-27. Forward and backward waves and index intensity profile of a distributed feedback laser

The threshold conditions and the spectral bandwidth can be derived from a coupled-wave analysis that is beyond the scope of this text. This analysis assumes that the field in the device is of the form

$$E = R(z)e^{-ikz/2} + S(z)e^{ikz/2}, \tag{12.67}$$

which involves two counterpropagating waves with complex amplitudes R and S, as shown in Figure 12-27. These waves grow because of the gain in the medium (eqn. 12.65) and feed energy into each other as a result of the spatial variation of η and g. The boundary conditions for the wave amplitudes are given by

$$R(-L/2) = S(L/2) = 0, \tag{12.68}$$

where L is the length of the gain medium.

Nonlinear calculations are involved in determining the final amplitudes of the waves, but the threshold conditions can be obtained from a linear analysis for conditions of large gains, $G = e^{2gL}$. The threshold for oscillation is

$$4g^2G = \left(\frac{\pi\eta_1}{\lambda}\right)^2 + \frac{g_1^2}{4}. \tag{12.69}$$

For the case in which only $\eta(z)$ is modulated, the threshold condition is

$$\eta_1 = \frac{\lambda_0}{L}\frac{\ln G}{\pi(G)^{1/2}}. \tag{12.70}$$

For the case of solely gain modulation, the threshold condition is

$$g_1/g = 4G^{-1/2}. \tag{12.71}$$

The frequency width over which oscillation occurs can be given by

$$\frac{\Delta\lambda}{\lambda_0} = \frac{\lambda_0}{4\pi\eta L}\ln G, \tag{12.72}$$

where $\Delta\lambda$ describes the frequency width for which oscillation is at threshold when the gain G at the center frequency exceeds threshold by a factor of 2.

EXAMPLE

Consider a Rh6G dye laser with a length $L = 10$ mm and a gain G of 100 operating at 630 nm. Compute the index of refraction variation that will produce oscillation and also the linewidth of the oscillation. The refractive index for the Rh6G dye dissolved in a typical solvent is approximately 1.4.

Equation (12.70) indicates that oscillation will occur if $\eta_1 = (\lambda_0/L) \times [\ln G/\pi(G)^{1/2}]$. Substituting our given values for λ_0, L, and G yields

$$\eta_1 = \frac{633\times 10^{-9}\text{ m}}{10^{-2}\text{ m}}\frac{\ln 100}{\pi(100)^{1/2}} = 9.3\times 10^{-6} \cong 10^{-5}.$$

The linewidth of the output for this device, according to (12.72), would be

$$\Delta\lambda = \frac{\lambda_0^2}{4\pi\eta L}\ln G.$$

Substituting the appropriate values for λ_0, η, L, and G leads to

$$\Delta\lambda = \frac{(633\times 10^{-9}\text{ m})^2}{4\pi\cdot 1.4\cdot 0.01\text{ m}}\ln 100 = 2.27\times 10^{-12}\text{ m}$$

$$= 2.3\times 10^{-12}\text{ m} = 2.3\times 10^{-3}\text{ nm}.$$

DISTRIBUTED BRAGG REFLECTION LASERS

Distributed feedback lasers have their frequency-selective feedback mechanism incorporated directly within the laser gain medium in the form of a corrugated index-varying grating structure. The distributed Bragg reflector cavity uses the same corrugated index structure in *two* regions, located at each end of the laser gain medium but beyond the gain medium. This is the semiconductor-laser equivalent of a more conventional laser, such as a He–Ne laser, with the mirrors located beyond the gas discharge tube. The advantage of a Bragg cavity is its ease of fabrication compared to a standard DFB laser, since it does not require epitaxial growth of laser material on top of the grating structure (see Section 14.9). Its disadvantage is that the coupling efficiency (the effective reflectivity) is lower than that for a DFB structure, owing to the discontinuity between the distributed Bragg regions and the gain region, as shown in Figure 14-25.

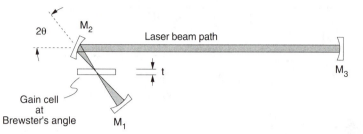

Figure 12-28.
Astigmatically
compensated three-mirror
laser cavity

12.6 LASER CAVITIES REQUIRING SMALL-DIAMETER GAIN REGIONS – ASTIGMATICALLY COMPENSATED CAVITIES

There are several types of lasers, including some dye lasers and solid-state lasers, for which it is necessary to have a small-diameter gain region, of the order of 10 μm, in order to produce sufficiently high pumping intensities for the gain medium. Such pumping is usually achieved by tightly focusing a pump laser into the gain medium, as suggested in Section 9.3 and indicated in Figure 9-16(g). With such lasers it is also generally desirable to have a long cavity length for tuning or for mode-locking. For small gain regions, an *astigmatically compensated* three-mirror laser cavity is very useful; see Figure 12-28.

In such a cavity, two of the curved laser mirrors (R_1 and R_2) are used to concentrate the laser beam into a small region containing the gain medium. A separate pump laser, also focused into this region, is not shown in the figure but is arranged in a manner similar to that shown in Figure 9-16(g). A third mirror R_3 can then be adjusted in length to obtain the desired round-trip time of the light within the laser cavity. For such a cavity arrangement, the center folding mirror R_2 introduces a reasonable amount of astigmatic distortion to the beam, since it is focusing off-axis with respect to the normal of the mirror. This can severely affect the beam quality of the laser. However, an appropriately designed gain medium can be used to compensate for the astigmatism of the mirror by incorporating a cell that has two parallel surfaces in the focal region of the mirror oriented at the Brewster angle as shown. This cell can simultaneously be used as an acousto-optic plate, a saturable absorber, et cetera; it is also possible to have more than one of these astigmatic mirror arrangements as part of the laser cavity.

A beam will be well focused if

$$2Nt = R_2 \sin\theta \tan\theta, \tag{12.73}$$

where R_2 is the radius of curvature of mirror M_2, θ is defined in Figure 12-28, t is the thickness of the Brewster cell (also shown in the figure), and N is given by

$$N = (\eta^2 - 1)\frac{(\eta^2 + 1)^{1/2}}{\nu^4}. \tag{12.74}$$

For lasers in which typical values of the index of refraction η of the gain medium range from 1.3 to 1.7, N varies from 0.396 to 0.446 (respectively). The value of N can of course be obtained from (12.74) for any specific value of the index of refraction. The reader is referred to the reference section for further information on this cavity arrangement.

12.7 WAVEGUIDE CAVITIES FOR GAS LASERS

By using a hollow dielectric waveguide structure, it is possible to construct a gas laser gain medium within a discharge bore that is smaller in cross section than necessary to support a Gaussian mode. This type of structure was developed to take advantage of the fact that the gain in a gas laser typically increases with pressure, provided that the product of pressure and tube diameter is kept constant. A discharge laser medium operated within such a structure can have very high gain and large power output in a very compact design. The laser beam propagates back and forth within the cavity via bouncing modes that reflect at small angles from the dielectric cavity walls. It has been shown that such a cavity has well-defined low-loss modes of propagation. These modes are not Gaussian shaped but instead are shaped more like inverted parabolas. Also, the bouncing modes of a waveguide laser also access the gain medium more effectively than would a normal Gaussian mode because, unlike a low-order Gaussian mode, the regions near the walls of the discharge for such a resonator contribute significantly to the laser gain and output. These waveguide lasers differ from the waveguide cavities used in semiconductor lasers and described in Chapter 14. In semiconductor lasers, waveguiding is effected by total internal reflection from a gain medium that has a higher index of refraction than the waveguide walls. In the waveguide gas laser the gain medium has an index of refraction near unity, and so it is the walls that have the higher value (approximately 1.5).

These features have been particularly useful in the development of CO_2 and excimer waveguide lasers, since both the laser gain and the stored energy increase with smaller bore diameter of the discharge. In such lasers the discharge is operated in the transverse direction, perpendicular to the laser beam axis. A simplified diagram of such a laser is shown in Figure 12-29.

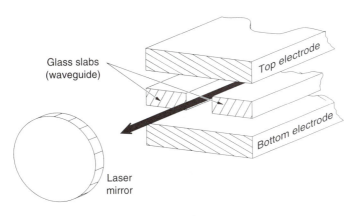

Figure 12-29. Waveguide gas laser electrode arrangement

REFERENCES

UNSTABLE RESONATORS

R. L. Herbst, H. Komine, and R. L. Byer (1977), "A 200 mJ unstable resonator Nd:YAG oscillator," *Optics Communications* 12: 5–7.

A. E. Siegman (1974), "Unstable optical resonators," *Applied Optics* 13: 353–67.

A. E. Siegman (1986), *Lasers.* Mill Valley, CA: University Science, Chapter 22.

Q-SWITCHING

A. E. Siegman (1986), *Lasers.* Mill Valley, CA: University Science, Chapter 26.

W. B. Wagner and B. A. Lengyel (1963), "Evolution of a giant pulse in a laser," *Journal of Applied Physics* 34: 2040-6.

MODE-LOCKING

R. L. Fork, B. I. Greene, and C. V. Shank (1981), "Generation of optical pulses shorter than 0.1 psec by colliding pulse mode locking," *Applied Physics Letters* 38: 671–2.

R. L. Fork, O. E. Martinez, and J. P. Gordon (1984), "Negative dispersion using pairs of prisms," *Optics Letters* 9: 150–2.

H. Hakatsuka, D. Grishkowsky, and A. Balant (1981), "Nonlinear picosecond-pulse propagation through optical fibers with positive group velocity dispersion," *Physical Review Letters* 47: 910–13.

H. Haus (1975), "Theory of mode locking with a slow saturable absorber," *IEEE Journal of Quantum Electronics* 11: 736–46.

J. Mark, L. Y. Liu, K. L. Hall, H. A. Haus, and E. P. Ippen (1989), "Femtosecond pulse generation in a laser with a nonliner external resonator," *Optics Letters* 14: 48–50.

D. K. Negus, L. Spinelli, N. Goldblatt, and G. Feugnet (1991), "Sub-100 femtosecond pulse generation by Kerr lens mode-locking in Ti:Al$_2$O$_3$," *OSA Proceedings on Advanced Solid-State Lasers* 10: 120–4.

A. E. Siegman (1986), *Lasers,* Mill Valley, CA: University Science, Chapters 27 and 28.

E. B. Treacy (1969), "Optical pulse compression with diffraction gratings," *IEEE Journal of Quantum Electronics* 5: 454–8.

RING LASERS

R. J. Kane and R. L. Byer (1985), "Monolithic, unidirectional single-mode Nd:YAG ring laser," *Optics Letters* 10: 65–7.

TUNABLE LASER CAVITIES

F. J. Duarte (1990), "Narrow-linewidth pulsed dye laser oscillators," in *Dye Laser Principles* (F. J. Duarte and L. W. Hillman, eds.). New York: Academic Press, Chapter 4.

T. W. Hansch (1972), "Repetitively pulsed tunable dye laser for high resolution spectroscopy," *Applied Optics* 11: 895–8.

DISTRIBUTED FEEDBACK

H. W. Kogelnik and C. V. Shank (1972), "Coupled wave theory of distributed feedback lasers," *Journal of Applied Optics* 43: 2327–35.

ASTIGMATICALLY COMPENSATED CAVITIES

H. W. Kogelnik, E. P. Ippen, A. Dienes, and C. V. Shank (1972), "Astigmatically compensated cavities for cw dye lasers," *IEEE Journal of Quantum Electronics* 8: 373-9.

WAVEGUIDE LASERS

P. W. Smith, O. R. Wood II, P. J. Maloney, and C. R. Adams (1981), "Transversely excited
waveguide gas lasers," *IEEE Journal of Quantum Electronics* 17: 1166-81.

PROBLEMS

1. In designing an unstable resonator for a 248-nm KrF excimer laser, for practical considerations it is desirable to have a relatively long gain medium such that
the separation between mirrors is 0.5 m. For a magnification factor of 5, what
would be the radii of curvature of both the output mirror and the rear mirror? If
the laser cavity were to be designed for minimum diffraction losses, what would be
the diameter of the output mirror?

2. Beginning with (12.4), the constraint for a positive-branch confocal resonator that $R_r - R_o = 2d$, obtain (12.10) and (12.11).

3. A 0.1-m–long Nd:YAG laser rod is installed in a 0.4-m–long laser cavity of
a Q-switched laser. Determine the approximate value for t_s and show that it is
significantly shorter than τ_u, the upper laser level lifetime.

4. A 0.1-m–long ruby laser rod has the mirrors coated on the ends of the rod.
The output mirror has a reflectivity of 90% and the rear mirror a reflectivity of
99.9%. If there is no excited state absorption (i.e. $\alpha = 0$) in the laser, what would
be the fractional loss per pass in the laser cavity, and what would be the decay
time of the cavity?

5. What would be the maximum Q-switched power output from the ruby
laser given in Problem 4 if the mode occupies a diameter of 2 mm within the rod?
Assume that the laser gain is 100 times the threshold value. *Hint:* Determine ΔN_{ul}
from the equations of Section 12.7.

6. Obtain (12.38), the expression for photon density, from (12.37). Obtain the
value of M where ϕ is a maximum.

7. What would be the minimum pulse duration of a mode-locked alexandrite
laser? If the separation between mirrors is 2 m and the alexandrite laser rod is
0.1 m long, what would be the separation between mode-locked pulses?

8. For a gallium arsenide distributed feedback laser in which only the index
of refraction is modulated, what would be the necessary index modulation in
order for the laser to operate with this type of resonator? Assume that the laser
cavity is 0.5 mm in length. What would be the bandwidth of the laser output?

SPECIFIC LASER SYSTEMS

LASER SYSTEMS INVOLVING LOW-DENSITY GAIN MEDIA

13

SUMMARY Gaseous (low-density) gain media are used in approximately half of the existing commercial lasers. This fraction will probably decrease in the future owing to the compactness and potential increased reliability of solid-state lasers, but many applications of gaseous lasers will no doubt remain for years to come. Certainly for very short–wavelength lasers, gaseous or plasma media will always be the dominant media. This chapter summarizes the important features of the various lasers using gaseous media. Each laser is described in terms of general features, laser structure, excitation mechanism, and applications. Also included is a table for each laser, listing its general properties and characteristics.

13.1 HELIUM–NEON LASER

GENERAL DESCRIPTION

The helium–neon laser was one of the first lasers ever developed, and is still one of the most widely used lasers. The lasers are trouble-free and have extremely long operating lifetimes. They operate in a low-pressure mixture of helium and neon gases, and the laser transitions occur within the neutral atomic species. The most common wavelength is the 632.8-nm transition in the red portion of the spectrum. Additional wavelengths have become available in recent years: in the green at 543.5 nm, the yellow at 594 nm, the orange at 612 nm, and the infrared at 1.523 μm. There are numerous other infrared wavelengths that lase but have generally not found significant commercial applications. One particularly strong transition at 3.39 μm competes with the 632.8-nm transition, having a common upper laser level, and therefore must be quenched by creating high loss in the laser cavity at that wavelength to allow the 632.8-nm transition to lase. These lasers generally produce powers in the range of 0.5 to 50 mW in the red, with much lower powers on the other transitions. The lasers operate continuous wave (cw) and have a very stable low-noise output. The gain bandwidth follows the Doppler-broadened emission linewidth with a FWHM (full width half maximum) of 1.5 GHz. A laser cavity length of 0.3 m would thus allow several longitudinal modes to lase simultaneously within the laser cavity. The complete laser assembly is typically 0.15–0.5 m in length, with lateral dimensions of the order of several tens of millimeters. It is not uncommon for such lasers to operate for lifetimes of up to 50,000 hours.

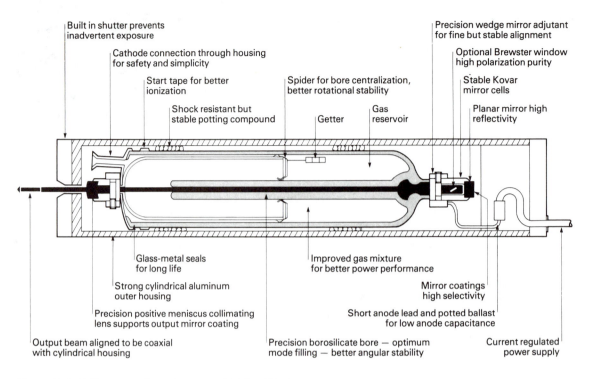

Figure 13-1. Helium–neon laser (courtesy of Melles-Griot)

LASER STRUCTURE

The mixture of helium and neon gases (approximately 15% neon) are introduced into a glass laser enclosure consisting of a narrow-bore tube that comprises the gain region: an anode region where a high-voltage electrode exists and a cathode region where a large aluminum cathode is located, as shown in Figure 13-1. The optimum gas pressure is determined by the product of the pressure and the tube-bore diameter, so that the optimum product is 3.6–4.9 torr-mm with a He : Ne pressure ratio of 5 : 1. Thus, for a discharge-bore diameter of 1.5 mm, the optimum total gas pressure would be 2.5 torr (2.5/760 = 0.0033 of an atmosphere). Also located in the anode and cathode regions are the internally mounted laser mirrors that comprise the optical cavity. If a polarized output is desired from the laser, a Brewster angle window is also inserted into the laser cavity within either the anode or cathode region of the tube. A separate power supply is used to supply the discharge current necessary to provide excitation of the upper laser level. The power supply generally requires a high voltage of up to 1,000 V and a discharge current of a few to tens of milliamps. A higher starting voltage is used to produce the initial ionization within the discharge; this starting voltage is typically applied to the discharge tube for only a few microseconds (or less) after the starting switch is turned on.

The laser cavity might include a semiconfocal cavity as indicated in Figure 11-1, where the power is emitted from a 1% transmitting flat mirror (approximately 99% reflecting) at the cathode end of the laser and the anode has a 99.9% reflecting spherical mirror whose curvature is determined by the desired cavity length.

EXCITATION MECHANISM

A general energy-level diagram for the helium–neon laser was shown in Figure 9-4, along with more details in Figure 9-8. It was indicated in Section 9.2 that the helium–neon laser operates via transfer of energy from the helium metastable energy levels to specific excited neutral neon energy levels that are energetically close to the helium metastable levels. When the discharge is initiated within the laser bore region, the discharge electrons, which comprise the discharge current, collide with both helium and neon atoms and provide excitation to the excited levels of both helium and neon. The helium metastable levels accumulate population because they have no radiative decay to the helium ground state. As a result, the helium metastable levels in a helium–neon laser discharge acquire population densities of the order of 10^{17} per cubic meter, which is approximately three orders of magnitude higher than that of other excited energy levels of the atoms within the discharge. This population is then transferred to neon energy levels that are energetically within a few kT of the helium metastable energy levels via collisions of the He singlet metastable levels with ground-state neon atoms, leaving ground-state helium atoms and excited Ne atoms in the upper laser level as shown in Figure 9-4. This process can be described by the following diagram:

$$He(1s2s\,^1S)^* + Ne(2p^6) \rightarrow He(1s^2) + Ne(2p^53s)^*,$$

where the asterisk (*) indicates an excited state.

The excitation rate Γ_{0u}, which in this case occurs from the intermediate state q (the He singlet metastable level) and is thus represented as Γ_{qu}, can be described according to (9.7) as

$$\Gamma_{qu} = N_{Ne}\bar{v}_{(He-Ne)}\sigma_{(He-Ne)},$$

where N_{Ne} is the Ne ground-state density, $\bar{v}_{(He-Ne)}$ is the average relative velocity of the He metastable atoms and the Ne ground-state atoms, and $\sigma_{(He-Ne)}$ is the excitation cross section that describes the probability of energy transfer from the helium singlet metastable atoms to the excited neon atoms. Using typical numbers of $N_{Ne} = 1.4 \times 10^{22}$ m^{-3} (based upon 1/6 of 2.5 torr total pressure), $\bar{v}_{(He-Ne)} = 2.6 \times 10^3$ m/s, and $\sigma_{(He-Ne)} \approx 10^{-20}$ m^2, we can estimate an excitation rate of

$$\Gamma_{qu} = (1.4 \times 10^{22}\ m^{-3})(2.6 \times 10^3\ m/s)(10^{-20}\ m^2)$$
$$= 3.6 \times 10^5\ s^{-1}.$$

Using the data indicated in Figure 4-2(b) and ignoring the 60-nm transition (since it would be radiation trapped), the decay rate A_u of the upper

TABLE 13-1

TYPICAL HELIUM–NEON LASER PARAMETERS

Laser Wavelengths (λ_{ul})	632.8 nm	543.5 nm
Laser Transition Probability (A_{ul})	3.4×10^6/s	2.83×10^5/s
Upper Laser Level Lifetime (τ_u)	1.7×10^{-7} s	
Stimulated Emission Cross Section (σ_{ul})	3×10^{-17} m^2	2×10^{-18} m^2
Spontaneous Emission Linewidth and Gain Bandwidth, FWHM ($\Delta \nu_{ul}$)	1.5×10^9 Hz (Doppler)	1.5×10^9 Hz
Inversion Density (ΔN_{ul})	5×10^{15}/m^3	
Small Signal Gain Coefficient (g_0)	0.15/m	
Laser Gain-Medium Length (L)	0.1–1.0 m	
Single-Pass Gain ($e^{\sigma_{ul}\Delta N_{ul}L}$)	0.03	
Gas Pressure	2.5 torr	
Gas Mixture	He : Ne at 5 : 1	
Index of Refraction of Gain Medium	≈ 1.0	
Pumping Method	electrical discharge	
Electron Temperature	10–11 eV	
Gas Temperature	500 K	
Mode of Operation	cw	
Output Power	0.5–100 mW	
Mode	TEM$_{00}$	

laser level can be estimated to be 7.3×10^6 s^{-1}. For a typical He singlet metastable density of $N_{\text{He}}^M = 1 \times 10^{17}$ m^{-3} for the conditions of a He–Ne laser, we can thus compute an approximate value of the upper laser level density of

$$N(3s)^* = \frac{N_{\text{He}}^M \Gamma_{qu}}{A_u}$$

$$= \frac{(1 \times 10^{17} \text{ m}^{-3})(3.6 \times 10^5 \text{ s}^{-1})}{7.3 \times 10^6 \text{ s}^{-1}} \approx 5 \times 10^{15} \text{ m}^{-3}.$$

The stimulated emission cross section for the 632.8-nm transition can be calculated from (7.28) to obtain a value of 3×10^{-17} m^2, assuming that the gas temperature is of the order of 400 K and a single isotope of neon (Ne^{20}) is used to keep the gain bandwidth to a minimum and thereby increase the gain.

Using the value determined here for the stimulated emission cross section and assuming that $N_u \gg N_l$, thereby replacing ΔN_{ul} by N_u, the small signal gain at the center of the Doppler-broadened emission line for the 632.8-nm transition can be estimated as

$$g_{ul} = \sigma_{ul}\Delta N_{ul} \cong \sigma_{ul}N_u$$
$$\cong (3 \times 10^{-17} \text{ m}^2)(5 \times 10^{15} \text{ m}^{-3}) = 0.15 \text{ m}^{-1},$$

which is very close to the measured value. Using a natural mixture of neon will reduce the maximum gain by approximately 10%, as can be seen in

Figure 4-13. Additional modes will then only develop from the Ne^{22} isotope if the much smaller gain in the frequency range of that isotope exceeds the losses within the laser cavity.

APPLICATIONS

Applications for the helium–neon laser include interferometry, laser printing, bar-code reading, and as pointing and directional reference beams. In interferometry, the He–Ne laser provides the very stable, single-transverse-mode reference beam necessary to identify optical properties of materials such as surface figure and smoothness. In laser printing, the well-characterized beam is used as a writing source on photosensitive material to provide detailed print patterns. Most supermarkets and other stores now use He–Ne lasers as check-out or inventory scanners to read the digitally encoded bar codes located on products. Pointing applications include reference beams for alignment by pipe layers, three-dimensional right-angle reference beams in the construction industry, reference beams for surveying, and target-aiming devices for guns. The 1.523-μm laser is used for measurements of optical fiber transmission lines, which have a minimum loss in that wavelength region.

13.2 ARGON ION LASER

GENERAL DESCRIPTION

The argon ion laser is one of a class of noble-gas ion lasers that operate in the visible and ultraviolet spectral regions. Historically these lasers have been known as *ion lasers*. Other lasers, such as the He–Cd laser, also operate in an ionized species of atoms but are most often referred to as metal vapor lasers. Thus, ion lasers consist primarily of lasers operating in ionized species of the noble gases of argon, krypton, and xenon. The argon and krypton ion lasers are primarily cw lasers although a few of these lasers are also commercially available as pulsed lasers. The lasers incorporate electron excitation resulting when the plasma electrons within the gas discharge collide with the laser species.

The argon ion laser can provide approximately 25 visible wavelengths ranging from 408.9 to 686.1 nm, and more than 10 ultraviolet wavelengths ranging from 275 to 363.8 nm. Wavelengths as short as 229 nm are also produced by intracavity frequency doubling of a visible argon ion laser. In the visible spectral region, cw powers of up to 100 W are available, with the output concentrated on a few strong lines (including the 488.0-nm and the 514.5-nm transition). The emission and gain bandwidth of these lasers is determined primarily by Doppler broadening, with a width on each laser transition of the order of 2.5 GHz. The lasers operate at an argon gas pressure of approximately 0.1 torr. The discharge tubes have an operating life that ranges from 2000 to 5000 hours. High discharge currents and low gas pressure lead to very high plasma electron temperature, producing a significant amount of heat. Hence these lasers have a need for significant cooling, which in most cases takes the form of water cooling. The lasers also are provided with an axial magnetic field to prevent the electrons from prematurely escaping from the gain region and colliding with the discharge tube wall (an effect that would produce additional unwanted heat).

LASER STRUCTURE

Argon ion lasers operate in high-temperature plasma tubes with a bore diameter of 1–2 mm and lengths ranging from 0.1 m to approximately 1.8 m. A diagram of a commercial argon ion laser is shown in Figure 13-2. It includes a series of tungsten disk bore segments held in place by a copper support ring, as shown in the figure. The bore region is segmented, since the tungsten bore disks are conductive and hence a single tungsten tubing for the bore would cause the electrical current to be conducted through the tungsten rather than through the discharge. Heat is conducted away from the bore by the copper support ring. Provision for equalizing the gas pressure in every region of the tube is provided by including gas return holes in the outer region of the copper rings. The series of tungsten bore disks are

Figure 13-2. Cutaway diagram of an argon ion laser (courtesy of Coherent, Inc.)

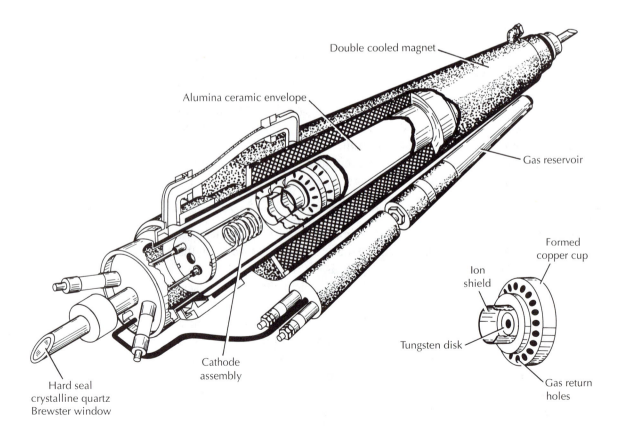

installed within an alumina ceramic envelope that is surrounded by a water jacket and combined with a magnetic coil. The magnet provides a magnetic field in the direction of the discharge current that slows the electron movement toward the tube walls, since collision with the walls leads to undesirable heating and bore erosion. A heated cathode is provided for efficient electron emission into the discharge. Plasma tubes of this type are designed to operate with bore temperatures of the order of 1,000°C, which are produced by the very high power inputs required for effective laser operation.

Commercial argon lasers are generally produced in three sizes: the high-power, large-frame and the medium-power, small-frame water-cooled lasers; and the low-power air-cooled lasers. Large-frame lasers offer cw powers of up to 100 W at several wavelengths and as much as 10 W on the strong transition at 514.5 nm. These lasers require input powers of 60 kW or more and water cooling at flow rates of 5 gal/min at a pressure of 60 lb/in^2 gauge. The lasers are approximately 2 m long and have a separate power supply. The small-frame lasers provide output powers of up to 5 W multi-line and 2 W single-line. They require up to 10 kW of input power and cooling rates of 2 gal/min at a pressure of 25 lb/in^2 gauge. Both of these lasers require an axial magnetic field, which confines the electrons for a longer time within the discharge plasma before they strike the walls of the plasma tube. The air-cooled argon ion laser produces approximately 10 mW TEM$_{00}$ mode at 488.0 nm. The beam amplitude noise is less than 2% peak to peak and less than 0.1% rms at lower frequencies. The lasers require input powers of the order of 1 kW and have a laser head with dimensions of the order of 35 cm in length by 15 cm square.

EXCITATION MECHANISM

An energy-level diagram for the argon ion laser transitions is shown in Figure 9-5. Excitation occurs in a different way for the pulsed argon ion laser and the cw argon laser. For the visible laser transitions, the pulsed laser excitation occurs via a single-step process, where the electrons collide with argon atoms in the 3p^6 ground state and excite them directly to the set of 3p^{4}4p upper laser levels in the Ar$^+$ ion. In the cw argon laser the excitation is a two-step process, as shown in Figure 9-5, where the electrons first collide with ground-state neutral argon atoms to produce ground-state argon single ions. The second step involves electron collisional excitation from the argon ion ground state to the upper laser levels. The population inversion is established between the upper laser levels involving the 3p^{4}4p configuration and the lower laser levels of the 3p^{4}4s electronic configuration. A population inversion is produced on transitions between all of these upper levels and lower laser levels, owing to the rapid decay of the lower laser levels as indicated in Figure 9-5.

The two-step excitation process leads to a laser power output that is proportional to the square of the discharge current. Thus, it is very desirable to operate these lasers at very high discharge currents. There is, however, a maximum discharge current (electron density) at which the electrons begin to depopulate the upper laser level detrimentally (see Section 8.4 for a description of the collisional depopulation process).

KRYPTON ION LASER

The krypton ion laser resembles the argon ion laser in that the energy level arrangements are similar and consequently the laser operation is similar. In fact, krypton lasers are generally made by replacing the argon gas with krypton gas in the same discharge tube. The optimum gas pressure is slightly different and, of course, the laser mirrors must be changed to take into account the wavelengths and gain of the krypton laser transitions. The reason for using a krypton laser is that it provides different laser wavelengths than the argon laser. Those wavelengths range from 406.7 to 676.4 nm, with the dominant laser outputs occurring at 406.7, 413.1, 530.9, 568.2, 647.1, and 676.4 nm; the strongest transition occurs in the red at 676.4 nm. These lasers thus offer a broader spectrum of wavelengths of moderate to high power over the visible spectrum than do the argon lasers. Some manufacturers provide a laser with a mixture of argon and krypton gases, which provides the strong blue and green transitions of argon as well as the strong red transition of krypton. Such lasers are often used for multicolor displays.

APPLICATIONS

Argon and krypton ion lasers are used primarily for phototherapy of the eye, pumping dye lasers, laser printers, cell cytometry, and stereolithography. Phototherapy of the eye involves the dissolution of small streamers of blood that develop within the eyes of people with diabetes. If these blood streamers are not removed, the patient can become blind. The blood streamers occur within the vitreous humor near the light-sensitive surface, the *retina*. To remove these streamers, a physician directs an argon laser at them through the lens of the eye. The blue and green wavelengths are highly absorbed by the blood streamers and thus dissolve them, but they are not absorbed by the transparent regions of the eye through which the beam passes before reaching the streamers.

Argon lasers are also very effective in pumping cw dye and titanium sapphire lasers. The laser wavelengths are in the pump absorption region of both of these high-density laser materials, and the laser has the power to provide sufficient gain for pumping such lasers using the techniques indicated in Figure 9-16(g). The high intensity of the argon laser and the blue and green wavelengths make it suitable for use in printers. The laser can be focused to a small spot, and the high intensity makes possible very fast scanning rates for printing purposes. Cell cytometry involves the use of laser light to count various types of living cells. The laser beam is directed into a flowing sample of the material containing the cells to be counted, and the scattered light that is detected when a cell passes through the region of illumination is then registered on a detector and counted.

TABLE 13-2

TYPICAL ARGON ION LASER PARAMETERS

Laser Wavelengths (λ_{ul}) Most Often Used	488.0 nm	514.5 nm
Laser Transition Probability (A_{ul})	7.8×10^7/s	
Upper Laser Level Lifetime (τ_u)	1.00×10^{-8} s	
Stimulated Emission Cross Section (σ_{ul})	2.5×10^{-16} m^2	
Spontaneous Emission Linewidth and Gain Bandwidth, FWHM ($\Delta\nu_{ul}$)	5×10^9 Hz	
Inversion Density (ΔN_{ul})	2×10^{15}/m^3	
Small Signal Gain Coefficient (g_0)	0.5/m	
Laser Gain-Medium Length (L)	0.1–1.0 m	
Single-Pass Gain ($e^{\sigma_{ul}\Delta N_{ul}L}$)	0.1–0.5	
Gas Pressure	0.1 torr or less in bore region	
Index of Refraction of Gain Medium	≈ 1.0	
Pumping Method	electrical discharge	
Electron Temperature	20–50 eV	
Gas Temperature	1,200°C	
Mode of Operation	cw	
Output Power	100 mW to 50 W	
Mode	TEM$_{00}$ or multi-mode	

Stereolithography is a technique that generates three-dimensional samples of any object that is capable of being described by computerized graphics. With this technique, a layer of UV-sensitive liquid plastic is poured into a chamber. The UV laser light is then used to irradiate a thin layer of the plastic at the surface of the liquid. The plastic hardens where the liquid is irradiated, and then a second thin layer of the plastic is poured upon the previous layer. This layer is then irradiated to produce additional hardened features of the model. After many layers, the hardened plastic model is fully developed and can be removed from the unhardened liquid. Such models are used in the development of prototypes ranging from automobile parts to perfume bottles. The advantages of this technique include the opportunity to check immediately the tolerance of parts that must be fit together and to observe the appearance of a particular design of an object prior to mass production.

13.3 HELIUM–CADMIUM LASER

GENERAL DESCRIPTION

The cw He–Cd laser is perhaps the best-known and most widely used metal vapor laser. It produces visible laser output in the blue at 441.6 nm and ultraviolet output at 354.0 nm and 352.0 nm. The 352-nm transition produces significantly more power output than the 354-nm transition. A number of other lasers in the green and red portions of the spectrum have been observed in various types of He–Cd discharges, but are not presently available commercially. The He–Cd laser that operates in a positive-column type of DC discharge, similar to that of the He–Ne laser, is not capable of producing laser output at these other visible transitions and so those transitions will not be dealt with further in this discussion.

Doppler broadening dominates over natural broadening for the He–Cd laser transitions. However, for this laser, isotope broadening is the dominant broadening process as discussed in detail in Chapter 4. The laser is typically operated with a natural mixture of isotopes that includes Cd 106, 108, 110, 111, 112, 113, 114, and 116. Each of the even isotopes has a 1.1–1.5-GHz emission linewidth on the laser transitions, and the series of even isotopes are separated by approximately 1.5 GHz to produce an emission spectrum as indicated in Figure 4-14(a). Also included in that spectrum are the distributed spectrums of the odd isotopes Cd 111 and Cd 113. The odd isotopes have a nuclear spin and thus each divides into three spectral components with a distribution sufficiently spread out that those isotopes do not have a major impact upon the emission spectrum at any specific wavelength, as indicated in Figure 4-14(a). The laser can also be operated with a single even isotope of Cd to produce approximately twice the power output and a much narrower emission bandwidth corresponding to that of the single even isotope. The higher power is obtained because the gain is higher and all of the energy is concentrated in only a few longitudinal modes.

These lasers operate within a cylindrically shaped glass discharge tube with a bore of 1–2 mm. The tube contains He gas at a pressure of several torr and a partial pressure of Cd that is approximately 1% of that of He. A DC discharge is operated within the tube at currents in the range of 60–100 mA. The Cd is distributed within the bore by a process known as *cataphoresis:* the Cd is heated and vaporized at the anode end (positive potential) of the discharge and is transported toward the cathode end (negative potential) of the discharge by the electric field acting upon the Cd ions; it then condenses in the cathode region. This process is effective because of the relatively high fractional ionization of the Cd within the discharge (1%), which allows the Cd ions to be pulled toward the cathode. The lasers typically operate for a lifetime of 4,000 to 5,000 hours with an initial loading of several grams of Cd metal in the anode reservoir.

Helium is gradually lost within the laser system, either by leaking through the glass or by being buried by the Cd deposition at the cathode end of the laser tube. Thus a He reservoir is included with a sensor and a feedback control system that maintains a constant He pressure within the laser tube over the lifetime of the laser.

LASER STRUCTURE

There are two versions of this laser: one with a long discharge tube having a bore length (gain region) of 60–75 cm, and a shorter version with a bore length of 25–35 cm. The longer version consists of either one long bore

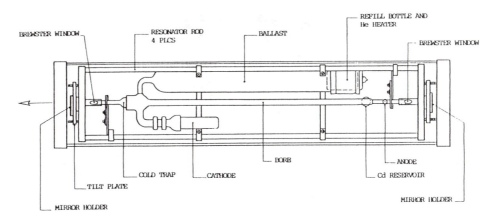

Figure 13-3. Helium–cadmium laser (courtesy of LiCONix, Inc.)

or two shorter bores within the same laser cavity. The lasers are installed within air-cooled housings (some use cooling fans) with a separate power supply. The housing has lateral dimensions of the order of 15 mm, and is approximately 40 cm longer than the bore length. The larger lasers produce up to 200 mW multi-mode at 441.6 nm and 100 mW multi-mode at 325 nm. A 10 mW single TEM_{00} mode blue laser is also produced from a 25–30 cm bore length. A diagram of the small version of a commercial He–Cd laser is shown in Figure 13-3. The heated Cd reservoir is located at the anode end of the laser discharge bore region. A metal vapor trap is located beyond the cathode to prevent metal vapor from condensing on the laser window or mirror. The *ballast* is a region within the tube that provides a large reservoir of He gas to help maintain the proper He pressure. Beyond the reservoir is a bottle containing He at high pressure. The He in that bottle is diffused into the reservoir region as needed to maintain proper He pressure. The rigid cavity structure is used to maintain the mirror alignment with respect to the fixed location of the laser bore.

EXCITATION MECHANISM

The Cd neutral ground state has an outer electronic configuration of $4d^{10}5s^2$ (see Section 3.5). The removal of a 5s electron leaves the atom in the ground state of the ion or the state of $4d^{10}5s\,^2S_{1/2}$ as indicated in Figure 8-6. The electronic structure of the upper laser levels is relatively unique in that it involves the removal of an inner shell d electron from the Cd atom, leaving the atom ionized in the $4d^95s^2\,^2D_{5/2}$ state, as seen in the figure. The laser transition involves decay from the upper laser level to the $4d^{10}5p\,^2P_{3/2}^o$ excited-ion state, which then rapidly decays to the ion ground state as shown in Figure 8-6. Three mechanisms have been proposed for the excitation of the Cd laser. The first involves the transfer of energy via Penning ionization from He metastable levels to the Cd upper laser levels. This is similar to the process described for excitation of the He–Ne laser except that – because the Cd upper laser level is an ion level, the exact energy coincidence

TABLE 13-3

TYPICAL HELIUM–CADMIUM LASER PARAMETERS

Laser Wavelengths (λ_{ul})	441.6 nm	353.6 nm	325.0 nm
Laser Transition Probability (A_{ul})	1.4×10^6/s	1.6×10^5/s	7.8×10^5/s
Upper Laser Level Lifetime (τ_u)	7.1×10^{-7} s ($^2D_{5/2}$), 1.1×10^{-6} s ($^2D_{3/2}$)		
Stimulated Emission Cross Section (σ_{ul})	9×10^{-18} m^2		
Spontaneous Emission Linewidth and Gain Bandwidth, FWHM ($\Delta\nu_{ul}$)	1.1×10^9/s	1.4×10^9/s	1.5×10^9/s
Inversion Density (ΔN_{ul})	4×10^{16}/m^3		
Small Signal Gain Coefficient (g_0)	0.3/m		
Laser Gain-Medium Length (L)	0.25–1.5 m		
Single-Pass Gain ($e^{\sigma_{ul}\Delta N_{ul}L}$)	0.07–0.45		
Gas Pressure	5–10 torr He		
Gas Mixture	He : Cd at 100 : 1		
Index of Refraction of Gain Medium	≈ 1.0		
Operating Temperature	tube bore 350°C, Cd 260°C		
Pumping Method	electrical discharge		
Electron Temperature	≈ 5–7 eV		
Gas Temperature	300°C		
Mode of Operation	cw		
Output Power	10–200 mW		
Mode	TEM$_{00}$ or multi-mode		

is not required since the electron emitted in the ionization process takes up the energy difference between the He metastable energy and the Cd upper laser level energy (approximately 12 eV). The second process is electron excitation from the Cd ion ground state to the upper laser level. The third process, which can be shown to represent only a small percentage of the total excitation, involves photoionization of the Cd atoms to populate the upper laser level via emission from the 50–60-nm transitions of the He atoms in the discharge.

APPLICATIONS

Applications of these lasers include stereolithography, printing, microchip inspection, flow cytometry, lithography, and fluorescence analysis. Stereo-lithography was described previously for the argon ion laser. The He–Cd laser has lower power and thus produces samples at a slower speed, but it does not require water cooling and typically has a longer operating life than the high-power UV argon ion lasers. The blue wavelength is used for printing on photosensitive materials because the short-wavelength blue photons react more readily with many photosensitive materials. In flow cytometry, the He–Cd laser is sometimes preferred where the shorter wavelength is desired from the standpoint of either scattering or absorption of light by the material. For fluorescence analysis also, the shorter wavelengths of the helium–cadmium laser are often more useful than those of the argon ion laser. He–Cd lasers are also used extensively for inspection of electronic circuit boards. They are used in generating the regular patterns associated with DFB lasers and distributed Bragg reflection lasers (see Section 12.5) and in direct-write lithographic applications. They are also used in compact-disk, laser-disk, and CD-ROM mastering.

13.4 COPPER VAPOR LASER

GENERAL DESCRIPTION

The copper vapor laser (CVL) is the most useful in the class of pulsed metal vapor lasers, which are inherently high-gain, self-terminating lasers. In addition to the copper vapor laser, two other lasers of importance in this category are the gold vapor laser and the lead vapor laser. All of these lasers require rapid electrical excitation of the heated metal vapors contained within a cylindrically shaped discharge tube filled with a buffer gas of either helium or neon. The excitation process produces a pulsed laser output ranging from 10 to 50 ns at repetition rates of up to 100 kHz. A simple energy-level diagram of these three lasers is shown in Figure 8-11, indicating the wavelengths of the laser transitions associated with each of these atomic species. The primary wavelengths for the copper vapor laser are at 510.5 and 578.2 nm. The gold laser operates primarily with an orange output of 627.8 nm, and the lead vapor laser operates in the near infrared at 722.9 nm. These lasers produce pulsed energies of the order of 1 mJ or higher, leading to average powers of 10–100 W for pulse repetition rates of 10–100 kHz. Since the lasers operate on atomic transitions, the gain bandwidth is of the order of the atomic emission linewidth, which is typically 2–3 GHz.

The optimum pressure for these lasers is of the order of 1 torr of metal vapor and 40–50 torr of buffer gas. This high metal vapor pressure requires laser bore temperatures of the order of 1,100°C for the lead laser, 1,500°C for the copper laser, and 1,650°C for the gold laser. The vapor is provided by inserting high-purity bars of the appropriate metal into the laser tube, closing the tube, and pumping to evacuate the air. The buffer gas is then inserted and the excitation is initiated. Most of the lasers are self-heated in that the discharge current produces the necessary heat to vaporize the metal. This process is controlled by the repetition rate and the amount of insulation surrounding the discharge tube. The metal charge is gradually used up since it migrates slowly from the central bore region to the cool end regions of the laser. A typical charge of copper metal will last for 500–1,000 hours before a new charge is necessary.

LASER STRUCTURE

These lasers are generally constructed with refractory ceramic discharge tubes having bore diameters of up to 6 cm and lengths of 1–3 m. A simplified

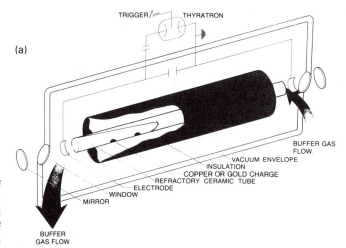

TRIGGER / THYRATRON

(a)

BUFFER GAS FLOW.
VACUUM ENVELOPE
INSULATION
COPPER OR GOLD CHARGE
REFRACTORY CERAMIC TUBE
ELECTRODE
WINDOW
MIRROR

BUFFER GAS FLOW

Figure 13-4. Diagram of a copper vapor laser: (a) schematic; (b) cutaway (courtesy of Oxford Lasers, Inc.)

(b)

INDUSTRIAL STANDARD GAS HANDLING SYSTEM
Ensures precise sequencing and control of gas flow. All metal construction means high quality, no contamination and ultra low gas consumption.

FAIL SAFE WATER COOLING ENVELOPE
Safely absorbs all residual heat in case of cooling water failure.

COOLING SYSTEM
Cooling water is always isolated from the laser tube. Corrosion resistant wetted parts allow low cost tap water to be used.

HIGH STABILITY CAVITY MIRRORS
Fast and simple change of optics is possible at all times.

ERGONOMICALLY STYLED ENCLOSURE
This combines ease of access with compliance to all relevant safety standards.

HIGH VOLTAGE MODULATOR UNIT
Contains all HV switching components in a single, RFI screened, energy-efficient sealed unit.

DOUBLE RFI SCREENINIG
Eliminates interference with sensitive instrumentation and ensures compliance with RFI/EMI standards.

layout of a CVL is shown in Figure 13-4(a). It shows the laser tube, the copper metal charge, the electrodes, the vacuum envelope, and the windows and mirrors; the thyratron triggering device is also shown. Figure 13-4(b) shows the laser integrated into a commercial product. The laser structure includes a water coolant system, a high-voltage modulator, and a gas-handling system. The gain in this type of metal vapor laser can be as high as 6 dB/cm (lead laser) and thus the lasers operate with a relatively high transmission output coupler and reach the saturation intensity after only a few passes through the amplifier. With such short pulses and high gain, the lasers operate in a highly multi-mode configuration.

EXCITATION MECHANISM

The relevant energy levels for these lasers include the neutral ground state; the *resonance* level, or first excited state above the ground state, which is optically connected to the ground state; and one or more intermediate levels, between the resonance level and the ground state, that have the same parity as the ground state and are thereby metastable to radiative decay. With such an arrangement, the strongest radiative emission normally takes place from the resonance level to the ground state (high transition probability) with somewhat weaker emission (intermediate transition probability) to the intermediate levels. Thus, when a pulsed discharge current is applied across the discharge tube, electron collisional excitation occurs strongly from the ground state to the resonance level and only very weakly to the intermediate levels, since excitation cross sections are known to be proportional to radiative transition probabilities (see Section 8.4). The preferred excitation to the upper laser level produces a large temporary population inversion between the resonance and intermediate levels. Decay via either spontaneous or stimulated emission from the upper laser level to the lower metastable laser level as well as continued excitation of the lower laser level rapidly quenches the gain and terminates the laser's pulse. In order for this process to work, the density of the ground state must be high enough for radiation trapping (see Section 8.4) to occur between the resonance level and the ground state; such trapping inhibits the rapid radiative decay that would prevent the inversion from occurring. This excitation process follows the model described in Section 8.3 for transient three-level lasers, which assumes radiation trapping with the ground state by excluding radiative decay on the resonance transition.

Theoretical modeling of the discharge process has suggested that, if the electron temperature is lower than approximately 2 eV, then the lower laser level receives a larger percentage of the excitation than the upper laser level in spite of the much lower excitation cross section. Thus, it is desirable to have the electron temperature in the 2–10 eV range in order to produce a population inversion. When the discharge pulse is initiated, the electron temperature begins at a low value and rises above the 2-eV threshold for gain in about 100 ns, thus causing a delay in the occurrence of the laser pulse to a point several hundred nanoseconds later in the current pulse.

After the laser pulse and current pulse cease, the metastable lower laser level must be depleted of its population before the next current pulse can be applied. Otherwise, the lower laser level would remain filled and no inversion could be obtained. Depopulation of the lower laser level occurs either by diffusion to the walls, which is a relatively slow process (10–100 μs), or by the more rapid electron collisional decay. In either case, the population is transferred to the ground state. The electron collisional decay process can exceed the diffusion process by as much as a factor of 100 if the electron density and temperature are high. Discharge currents of up to 1,000 A produced by voltages up to 20 kV are used to provide the excitation pulse.

COPPER VAPOR LASER The copper vapor laser is useful because of its visible wavelengths and high efficiency (as high as 1–2%), yielding average power output of 100 W or more. The relevant copper energy levels are the $3d^{10}4s\,^2S_{1/2}$ neutral ground state, the $3d^{10}4p\,^2P^o_{3/2,1/2}$ resonance levels, and the $3d^94s^2\,^2D_{5/3,3/2}$ metastable lower laser levels. Both resonance levels have a rapid radiative decay – to the ground state of $1.4\times10^8\,s^{-1}$ on transitions of 324.8 and 327.4 nm – which must be quenched by radiative trapping. Using the trapping formula of (8.43), it can be shown that trapping will begin at a density of approximately 3×10^{17} per cubic meter for a bore radius of 2 cm, which corresponds to a vapor pressure of 10^{-5} torr. This pressure is well below the optimum vapor pressure of 1 torr of copper atoms for a CVL, which means that radiative decay to the ground state will be strongly radiatively trapped under the normal operating conditions and thus that decay process will be effectively quenched. The radiative rates of the laser transitions are $2\times10^6\,s^{-1}$ for the 510.5-nm transition and $1.6\times10^6\,s^{-1}$ for the 578.2-nm transition. These rates are much lower than the resonance transition rates (when trapping is not present), and thereby provide a slow decay rate. This allows accumulation of population in the upper laser level before the stimulated emission begins to dominate and thereby deplete the population of those levels later in the current pulse.

GOLD VAPOR LASER The gold vapor laser operates in a similar way to that of the copper vapor laser but at a different wavelength, 627.8 nm (in the orange portion of the spectrum). The gold vapor laser operates at a somewhat higher laser tube temperature of approximately 1,650°C to provide the necessary gold vapor pressure for efficient laser operation. This, of course, requires slightly more power input to keep the gold laser operating, but the discharge tube design is similar to that of the copper vapor laser. The power output is typically one fourth that of the copper vapor laser for an equivalent input power.

LEAD VAPOR LASER The lead vapor laser, operating primarily at 722.9 nm, was the first laser of this type to be developed and has the highest gain – of the order of 1.5 cm^{-1}. It has found limited use primarily because its wavelength is beyond the range of the normal sensitivity of the human eye and photosensitive detectors.

Energy levels for these lasers are shown in Figure 8-11.

APPLICATIONS

Applications for the copper vapor laser include high–repetition rate pumping of tunable dye lasers (in particular for uranium isotope enrichment),

TABLE 13-4

TYPICAL COPPER VAPOR LASER PARAMETERS

Laser Wavelengths (λ_{ul})	510.5 nm	578.2 nm
Laser Transition Probability (A_{ul})	2×10^6/s	1.65×10^6/s
Upper Laser Level Lifetime (τ_u)	5×10^{-7} s	6.1×10^{-7} s
Stimulated Emission Cross Section (σ_{ul})	8.6×10^{-18} m^2	1.25×10^{-17} m^2
Spontaneous Emission Linewidth and Gain Bandwidth, FWHM ($\Delta\nu_{ul}$)	2.3×10^9 Hz (Doppler broadening)	
Inversion Density (ΔN_{ul})	8×10^{17}/m^3	
Small Signal Gain Coefficient (g_0)	5/m	
Laser Gain-Medium Length (L)	1.0–2.0 m	
Single-Pass Gain ($e^{\sigma_{ul}\Delta N_{ul}L}$)	1,000–20,000	
Gas Pressure	40 torr Ne, 0.1–1.0 torr Cu vapor	
Gas Mixture	Ne:Cu at 400:1 to 40:1	
Index of Refraction of Gain Medium	≈ 1.0	
Pumping Method	electrical discharge	
Electron Temperature	5 eV	
Gas Temperature	1,500°C	
Mode of Operation	pulsed	
Output Power	1 MW/pulse, 20 kHz rep. rate	
Mode	high-order multi-mode	

high-speed flash photography, large-image projection television, and material processing. The high average power output and high peak power of the copper vapor laser allow it to pump tunable dye lasers, in turn which provide a high laser power at a precise wavelength that is absorbed by specific uranium isotopes, thereby ionizing them. The ions of those isotopes are then collected by a charge collector to produce an enriched isotopic species. In high-speed flash photography, the laser pulsing at repetition rates of up to 20 kHz with submicrosecond-duration pulses can be used for stroboscopic illumination of various rapidly moving objects (such as bullets or explosive fragments) that can be captured on film. For projection television, the high power of the copper vapor laser combined with suitable beam-scanning hardware and software is sufficient to produce a very large video image. The laser is used to ablate materials in which very precise edges are desired (e.g., drilling precision holes). The gold vapor laser is used for photodynamic therapy, which is associated with the treatment of some types of cancer. The treatment involves ingestion of a photo-absorptive material that accumulates primarily at a cancerous tumor site within the body. The gold laser is then used to irradiate that area of the body, and the laser energy is absorbed by and thus destroys the cancerous tissue.

13.5 CARBON DIOXIDE LASER

GENERAL DESCRIPTION

The carbon dioxide laser is one of the most powerful and efficient lasers available. It operates in the middle infrared on rotational–vibrational transitions in the 10.6-μm and 9.4-μm wavelength regions. Both pulsed and cw laser output occurs in several different types of gas discharge configurations in a mixture of carbon dioxide, nitrogen, and helium gases, typically with a $CO_2 : N_2$ ratio of about $0.8 : 1$ and somewhat more helium than N_2. These lasers have produced cw powers of greater than 100 kW and pulsed energies of as much as 10 kJ. The gain occurs on a range of rotational-vibrational transitions that are dominated by either Doppler broadening or pressure broadening, depending upon the gas pressure. The lasers range from small cw waveguide-type systems of the order of 0.35 m long to much larger pulsed laser amplifiers – that a person could walk through – designed for laser fusion. One of the most useful CO_2 lasers for materials application is a cw version with a cavity length of 1–2 m producing one or more kW of power. Another laser of this class is the CO laser, which emits at approximately half the wavelength of the CO_2 laser in the 5–6 μm wavelength region.

LASER STRUCTURE

There are a number of different types of laser structures used for the CO_2 laser. These structures are depicted in Figure 13-5 and can be summarized as follows.

LONGITUDINALLY EXCITED LASERS   These lasers are operated as conventional gas discharge lasers in the form of long, narrow, cylindrically shaped glass enclosures with electrodes at opposite ends from which the discharge excitation current is introduced, as seen in Figure 13-5(a). These lasers can be either pulsed or cw and can have lengths of up to several meters. In some versions the discharge enclosure is sealed off, requiring periodic replacement of the tube because the gases eventually degrade owing to break-

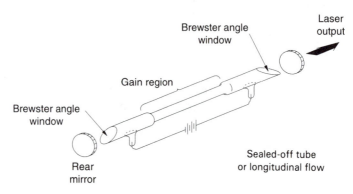

Figure 13-5(a).
Longitudinal discharge
CO_2 laser

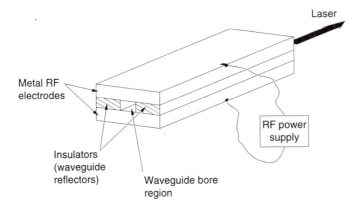

Figure 13-5(b). Waveguide CO_2 laser

NOTE: Mirrors are not shown.

down of the CO_2 which produces oxygen that can corrode the electrodes. In other versions, the gas is flowed through the tube longitudinally and can be recirculated to conserve the gases. Discharge currents of hundreds of milliamps and tube-bore diameters of several centimeters are common for this type of laser. A water coolant jacket usually surrounds the discharge region.

WAVEGUIDE LASERS This type of structure laser is probably the most efficient way to produce a compact cw CO_2 laser. It consists of two transverse radio-frequency (RF) electrodes separated by insulating sections that form the bore region, as shown in Figure 13-5(b). An RF power supply is connected to the electrodes to provide a high-frequency alternating field across the electrodes within the bore region. The lateral dimensions of the bore are up to a few millimeters (generally square), which is sufficient to propagate the beam in waveguide modes reflecting from the insulating materials but too "lossy" for normal Gaussian modes. The waveguide modes access the entire gain volume, since the modes reflect off of the discharge walls in a zigzag fashion. The small bore allows high-pressure operation and provides rapid heat removal, both of which lead to high gain and high power output. This type of laser produces cw powers of up to 100 W from a tube not much larger than a helium–neon laser. (See Section 12.7.)

TRANSVERSELY EXCITED LASERS These lasers operate at high total gas pressures of 1 atmosphere or more in order to benefit from obtaining a much higher energy output per unit volume of gas, since there are more laser species in the gas at higher pressures. Operating at high pressure in a longitudinal discharge is extremely difficult because extremely high voltages are required initially to ionize the gas and thereby initiate the discharge process. Also, arcing tends to form within the discharge, an effect that causes the discharge current to follow an irregular path such as that produced by a bolt of lightning. In a transverse discharge, the two electrodes are placed parallel to each other over the length of the discharge,

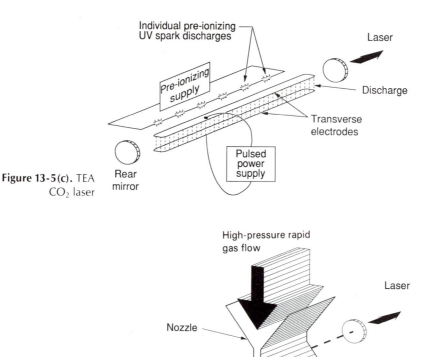

Figure 13-5(c). TEA CO_2 laser

Figure 13-5(d). CO_2 gas dynamic laser

separated by a few centimeters or more, and a high voltage is applied across the electrodes as shown in Figure 13-5(c). Prior to the application of the high voltage, a form of pre-ionization is used to ionize the space between the electrodes uniformly and thereby fill it with electrons. With this pre-ionization, the discharge can then proceed in a uniform fashion over the entire electrode assembly rather than forming a narrow high-current arc at just one location. The pre-ionization is produced by flashes of ultraviolet light from a row of pre-ionizing UV spark discharges, which uniformly ionize a portion of the gas between the electrodes. Such lasers can produce many joules of energy for each liter of discharge volume. This configuration has resulted in some of the highest-energy pulsed lasers yet produced. When this transverse laser excitation technique was first developed it was known as a TEA (transverse excitation atmospheric) laser. Such a configuration has also been used very successfully for excimer lasers, which will be described in Section 13.6.

GAS DYNAMIC LASERS   In gas dynamic lasers the gas is flowed in the transverse direction to the laser axis, as in the transverse laser mentioned above.

TABLE 13-5

TYPICAL CARBON DIOXIDE LASER PARAMETERS

Laser Wavelengths (λ_{ul})	10.6 μm	9.4 μm
Laser Transition Probability (A_{ul})	0.25/s	
Upper Laser Level Lifetime (τ_u)	4 s	
Stimulated Emission Cross Section (σ_{ul})	3×10^{-22} m^2	
Spontaneous Emission Linewidth and Gain Bandwidth, FWHM ($\Delta\nu_{ul}$)	6×10^7 Hz	
Inversion Density (ΔN_{ul})	3×10^{21}/m^3	
Small Signal Gain Coefficient (g_0)	0.9/m	
Laser Gain-Medium Length (L)	0.3–2.0 m	
Single-Pass Gain ($e^{\sigma_{ul}\Delta N_{ul}L}$)	0.3–5.0	
Gas Pressure	50–760 torr	
Gas Mixture	CO_2:N_2 at 0.8:1 (plus He)	
Index of Refraction of Gain Medium	$\approx$1.0	
Pumping Method	electrical excitation, collisional transfer	
Electron Temperature	3–4 eV	
Gas Temperature	200–400°C	
Mode of Operation	cw or pulsed	
Output Power	1–10,000 W	
Mode	TEM$_{00}$ or multi-mode	

Excitation occurs as a result of heat input into the gas (electrically or thermally) to populate the upper laser level. A small amount of water vapor is also added to the gas mixture. The rapidly flowing gas is then allowed to expand supersonically through an expansion nozzle into a low-pressure region using high-speed pumps, as shown in Figure 13-5(d). This expansion causes the gas to supercool and thereby provide rapid relaxation of the lower laser level from the highest rotational states to the lowest rotational states, leaving a population inversion of those empty higher-lying rotational states with respect to the upper laser level. Lasers of this design have produced cw output powers greater than 100 kW. This type of excitation was developed primarily for military applications, but lower-power versions have found applications in material processing.

EXCITATION MECHANISM

Detailed energy-level diagrams of the CO_2 laser were shown in Figures 5-6(a)–(c). Those figures showed the close energy coincidence between the N_2 $v = 1$ vibrational level and the $(0, 0, 1)$ vibrational level (upper laser level) of CO_2. Although laser action can be produced in pure CO_2 gas, such action is very inefficient with low gain. When N_2 is added to the discharge the laser becomes one of the most efficient lasers, with "wall plug" efficiencies as high as 30%. The nitrogen $v = 1$ vibrational level is efficiently excited but is metastable to radiative decay. It thus accumulates population, which is then collisionally transferred to the CO_2 $(0, 0, 1)$ vibrational level. As pointed out in Section 5.1, the upper laser level is the Σ_u level involving the asymmetric stretch mode of vibration, whereas the lower laser level is the $(1, 0, 0)$ symmetric stretch Σ_g level. The radiative decay rate from the lower laser level is approximately 20 times faster than the decay from the upper laser level, which establishes the population inversion as suggested by (8.20).

APPLICATIONS

Probably the most significant area in which the CO_2 laser is used is in the general field of material processing. This includes cutting, drilling, material removal, etching, melting, welding, cladding, alloying, submelting, annealing, hardening, and so forth. The other principal area is in medical applications, where it is used for cutting and cauterizing. The advantage of using a laser for these applications is that a very intense heating source can be applied to a very small area. The material ablation applications can be described in terms of cw or pulsed operations. For cw applications, the process is a thermal one in which the laser source serves as a cutting or heating tool. For pulsed operations, such as the ablation of thin films of material or drilling of small holes, the laser must reach an intensity on the target of 10^{12} to 10^{13} W/m^2.

13.6 EXCIMER LASERS

GENERAL DESCRIPTION

Excimer lasers are a group of pulsed lasers that incorporate electronic transitions within short-lived molecules. Such lasers are most often composed of the combination of a rare-gas atom (such as Ar, Kr, or Xe) and a halogen atom (Fl, Cl, Br, or I). The word "excimer" is a contraction of the phrase "excited dimer." The excimer molecule exists only as an excited molecule because the ground state is extremely short-lived owing to the repulsive force between the two atoms of the molecule in its ground state, as discussed in Section 5.1. Excimer lasers typically emit in the ultraviolet spectral region, although some operate in the visible spectrum. The principal excimer laser transitions occur in XeF at 353 nm, XeCl at 308 nm, KrF at 248 nm, ArF at 193 nm, and F_2 at 153 nm. A typical excimer laser, the KrF laser, operates with a total gas pressure of 2 atmospheres and with partial pressures of 10 torr F_2, 30 torr Ar, and approximately 1,400 torr He. The pulsed output of these lasers has a duration lasting anywhere from 10 to 50 ns with a pulse energy of 0.1–1 J, and can be operated at a repetition rate of up to several hundred hertz. The gain bandwidth can be greater than 1 nm in width but the output tends to occur near the peak of the emission spectrum over a narrower width of 0.3 nm, as can be seen in Figure 13-6. It is also possible to inject a narrowed beam into such an amplifier to

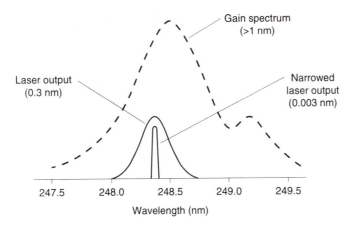

Figure 13-6. Gain spectrum and laser output of a krypton–fluoride laser, including a line-narrowed spectrum

obtain a very narrow laser output of the order of 0.003 nm, as shown in the figure, for use in such applications as microlithography. Excimer lasers typically have a gain medium of 0.5–1.0 m in length and have a transverse discharge electrode configuration as described for the CO_2 laser. Most of these lasers require a recirculating gas system to provide purified gas constituents to the gain region on a regular basis, since the reactive nature of the halogen species tends over time to induce breakdown and the formation of other undesired species during operation. Excimer lasers have very high gain and thus normally produce a high-order multi-mode output. Their high pulse energy and ultraviolet output make them attractive for materials processing applications.

LASER STRUCTURE

A typical excimer laser structure is shown in Figure 13-7. Due to the corrosive nature of the halogen species, the entire structure is made of stainless steel with polyvinyl and teflon components. The discharge is transverse (Figure 13-5) and the electrodes are long, flat metal pieces that have a rounded shape so that when voltage is applied, the electric field is uniform everywhere between the electrodes; this leads to uniform excitation rather than undesired arc formation. A pre-ionization pulse is also needed to provide the initial electron "seeding" in the region between the electrodes, enabling a uniform excitation. The pre-ionization pulse is typically produced by a row of miniature UV spark discharges referred to as a *flash-board,* as shown for the transversely excited CO_2 laser of Figure 13-5(c). These sparks emit enough UV radiation to produce ionization within the gain region, thereby increasing the electrical conductivity of the gaseous medium. The discharge current is provided by discharging a high-voltage capacitor using a thyratron switching device. The output generally is obtained by using a high-reflecting mirror at the rear of the laser and a quartz flat as the output coupling mirror. Since the gain is typically of the order of 0.8 m^{-1}, the low reflectivity of the quartz flat is sufficient to provide the necessary optical feedback within the medium for efficient energy extraction.

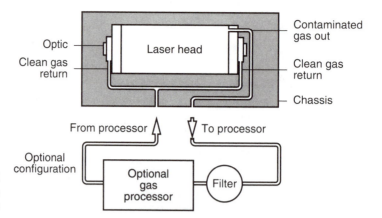

Figure 13-7. Typical
excimer laser structure
(courtesy of Questek)

Waveguide excimer lasers, similar to the waveguide CO_2 lasers described previously, use dielectric discharge tubes with bore regions of lateral dimensions of less than 1 mm. The metal transverse electrodes are external to the bore and provide both a high-voltage pre-ionizing pulse and an RF main pulse to the tube. As in the CO_2 waveguide lasers, the RF excitation in the excimer is very efficient. The corrosive halogen gases used in this laser come into contact only with the glass discharge tube and thus very little corrosion or gas degradation occurs. Hence these lasers, unlike other excimer lasers, can be operated in a sealed-off configuration for extended periods without requiring gas flow or gas replacement. The electrodeless microwave type of discharge produces homogeneous excitation of the small volume of the gain region, with little tendency toward the formation of arcs and other inhomogeneities typically associated with pulsed discharge devices. A XeCl waveguide laser has produced over 1 W of average power at a repetition rate of 1,000 Hz.

EXCITATION MECHANISM

The excimer laser is produced by forming excited-state species of the excimer molecule in the upper laser level. Since the ground state of these molecules is not stable, it is not possible to produce the upper laser level by direct pumping from such a ground state. Therefore, indirect pumping is necessary. The laser medium includes noble-gas atoms such as argon, and fluorine molecules in the form of F_2. When an electrical discharge or an electron beam is initiated within the gain medium, the noble-gas atoms are excited and ionized, leaving for example Ar* or Ar^+, and the fluorine molecules are dissociated to produce F atoms. Many of the F atoms rapidly collect a free electron (produced by ionizing the Ar) to form F^- negative ions. The excited excimer molecules of ArF* in the upper laser level are then produced by collisions between these two species via the following reactions:

$$Ar^+ + F^- \rightarrow ArF^*,$$
$$Ar^* + F^* \rightarrow ArF^*.$$

TABLE 13-6

TYPICAL EXCIMER LASER PARAMETERS

Laser Wavelengths (λ_{ul})	XeF, 351 nm; XeCl, 308 nm; KrF, 248 nm; ArF, 193 nm
Laser Transition Probability (A_{ul})	0.8 to 2.5×10^8/s
Upper Laser Level Lifetime (τ_u)	4–12 ns
Stimulated Emission Cross Section (σ_{ul})	2.4 to 4.5×10^{-20} m^2
Spontaneous Emission Linewidth and Gain Bandwidth, FWHM ($\Delta \nu_{ul}$)	10^{13} Hz
Inversion Density (ΔN_{ul})	10^{20}/m^3
Small Signal Gain Coefficient (g_0)	2.6/m
Laser Gain-Medium Length (L)	0.5–1.0 m
Single-Pass Gain ($e^{\sigma_{ul}\Delta N_{ul}L}$)	4–14
Gas Pressure	1,500 torr
Gas Mixture (example KrF)	10 torr F$_2$, 30 torr Kr, 1,460 torr He
Index of Refraction of Gain Medium	≈ 1.0
Pumping Method	electrical discharge, recombination
Electron Temperature	$\approx$2–5 eV
Gas Temperature	200°C
Mode of Operation	pulsed
Laser Pulse Energy	0.5–1.0 J
Laser Pulse Duration	30–50 ns
Output Power	1 J/pulse at 100 Hz = 100 W
Mode	high-order multi-mode

The ArF* molecules subsequently radiate, leaving them in the ground state of ArF, which immediately dissociates to produce Ar and F atoms. The process then begins again when the next pulse of electrons is produced within the gain medium.

The stimulated emission cross section for these laser molecules is typically of the order of 2–5×10^{-20} m^2. The decay rate of the upper laser level is typically of the order of 10^8 s^{-1}, which corresponds to an upper laser level lifetime of the order of $\tau_u \approx 10^{-8}$ s. Using (9.6) as a guide, the population of the upper laser level can be expressed as

$$N_u = N_N N_H V_R \sigma_{NH} \tau_u,$$

where N_N is the population density of the excited noble-gas atom Ar* or ion Ar$^+$ and N_H is the population density of F$^-$ or F* when these species are formed in the discharge. The V_R term denotes the relative velocity between the noble-gas species and the halogen species as they collide. This velocity is approximately the average velocity of those species as determined by the temperature from (4.59). The relative collision cross section σ_{NH} for the reactions expressed is typically 10^{-20} m^2. For the gas pressures used in excimer lasers, if we assume that a significant portion of the noble-gas atoms and fluorine molecules are converted to excited atoms and ions that can contribute to the upper laser level density N_u, then we can infer that N_N is of the order of 10^{23} m^{-3} and N_H is approximately one third of that. A typical value for V_R is 500 m/s. Using our equation for the pumping of the upper laser level leads to a value for N_u of 1–2×10^{20} m^{-3}. This value of N_u is also the value of ΔN_{ul}, since N_l is negligible owing to the unstable ground state of these molecules. Thus, the gain coefficient can be expressed as

$$g_0 = \sigma_{ul} \Delta N_{ul} = (3 \times 10^{-20} \text{ m}^2)(2 \times 10^{20} \text{ m}^{-3}) = 6 \text{ m}^{-1}.$$

APPLICATIONS

Excimer lasers are used primarily for materials processing, medical applications, photolithography, and pumping of dye lasers. In the materials processing area, excimer lasers are advantageous because of their wavelength and energy per pulse. Materials processing requires pulse intensities of the order of 10^{13} W/m^2, which are easily obtainable with an excimer laser. Also, the absorption coefficient for most materials is much greater for ultraviolet wavelengths than for visible and near-infrared wavelengths. Hence excimer laser beams are absorbed over a much shorter depth when materials are irradiated with such lasers, thereby producing sharper edges in cutting processes. This is also advantageous for such medical applications as laser surgery and corneal sculpting, which provides optical correction without the need for eyeglasses. In the lithography field, the excimer laser provides a good ultraviolet illumination source at 248 nm and eventually at 193 nm for producing microchip features potentially as small as 0.18–0.25 μm. Excimer lasers are also used for the pumping of dye lasers, since all dyes have an extended absorption into the ultraviolet even though their peak absorption might occur at longer wavelengths.

13.7   NITROGEN LASER

GENERAL DESCRIPTION

The pulsed molecular nitrogen laser is a gas discharge laser that produces ultraviolet laser output at 337.1 nm. This laser operates in N_2 on the $C^3\Pi_u \rightarrow B^3\Pi_g$ vibronic system, which involves a change in both electronic and vibrational energy levels. The laser produces peak power outputs of up to 100 kW over a duration of 10 ns, corresponding to a pulse energy of 100 mJ per pulse. It operates at N_2 pressures ranging from 20 torr to atmospheric pressure in a sealed chamber at repetition rates of up to 200 Hz. The laser has high gain and thereby produces a highly multi-mode output that is useful for pumping dye lasers, probably its most frequent application. With the more recent development of excimer lasers that produce several hundred millijoules per pulse at a much higher efficiency, the N_2 laser has seen a significant decrease in usage in recent years. One version of this laser,

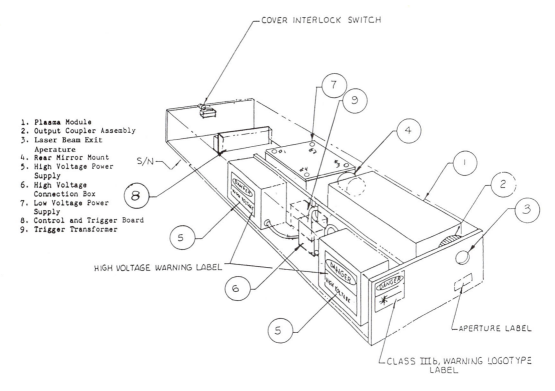

Figure 13-8. Compact molecular nitrogen laser system (courtesy of Laser Science, Inc.)

which is a very small structure of 0.2–0.25 m in length (including power supply, as shown in Figure 13-8), has found use in pumping very compact tunable dye lasers.

LASER STRUCTURE AND EXCITATION MECHANISM

The laser is typically excited with a transverse pumping arrangement in a high-pressure chamber similar to the CO_2 TEA lasers. The short upper laser level lifetime of 40 ns requires an extremely rapid pumping flux, which occurs largely via direct pumping from the N_2 ground state. The lower laser level has a much longer lifetime of 10 μs and so the laser is of the self-terminating type similar to that of the copper vapor laser. The laser output duration ranges from 2 to 20 ns. A high-energy laser pulse can be produced using very low–reflectivity cavity mirrors, although a high-reflecting mirror is typically used as the rear mirror along with a 4% reflecting front mirror. Flat mirrors are adequate in this case because only a few passes of the beam occur within the gain medium; therefore, alignment and diffraction losses are not very important.

APPLICATIONS

Nitrogen lasers are used primarily for pumping compact tunable dye lasers. They can be manufactured in a relatively small size and have a sealed-off

TABLE 13-7

TYPICAL NITROGEN LASER PARAMETERS

Laser Wavelengths (λ_{ul})	337.1 nm
Laser Transition Probability (A_{ul})	2.5×10^7/s
Upper Laser Level Lifetime (τ_u)	40 ns
Stimulated Emission Cross Section (σ_{ul})	4×10^{-17} m^2
Spontaneous Emission Linewidth and Gain Bandwidth, FWHM ($\Delta\nu_{ul}$)	2.6×10^{11}/s
Inversion Density (ΔN_{ul})	2.5×10^{17}/m^3
Small Signal Gain Coefficient (g_0)	10/m
Laser Gain-Medium Length (L)	0.1–1.0 m
Single-Pass Gain ($e^{\sigma_{ul}\Delta N_{ul}L}$)	up to 10^5
Gas Pressure	20–760 torr (sometimes with 10% He)
Index of Refraction of Gain Medium	≈ 1.0
Pumping Method	fast electrical discharge
Electron Temperature	>8 eV
Gas Temperature	300 K
Mode of Operation	pulsed
Pulse Width	1–10 ns
Pulse Energy	250 μJ to 10 mJ
Peak Output Power	250 kW to 1 MW
Mode	high-order multi-mode

laser tube that operates for a long lifetime. They can be combined with a compact dye system that includes a diffraction grating providing pulsed tunable laser output over the entire visible spectrum.

13.8 FAR-INFRARED GAS LASERS

GENERAL DESCRIPTION

With the exception of free-electron lasers and some recent developments in far-infrared semiconductor lasers (see Chapter 14), most of the far-infrared lasers involve molecular transitions in gases, sometimes between rotational-vibrational levels but more usually between rotational levels. Far-infrared gas lasers include transitions in 18 molecules covering a wide wavelength region from just under 30 μm to about 1.8 mm. Pumping schemes involve either electrical excitation or optical pumping by other infrared lasers. The output powers range from a few milliwatts for cw lasers to up to 30 kW for pulsed lasers. The gain bandwidth can be quite large (greater than 10^{12} Hz) owing to collisional broadening. Two of the first far-infrared lasers were the water vapor laser and the HCN laser. Both were operated in a low-pressure gas discharge with discharge currents of the order of a few milliamps in large-bore discharge tubes. Most of the far-infrared lasers are produced

TABLE 13-8

TYPICAL FAR-INFRARED LASER PARAMETERS

Laser Wavelengths (λ_{ul})	28 transitions, ranging from 99 to 373 μm
Laser Transition Probability (A_{ul})	typically 1/s or less
Upper Laser Level Lifetime (τ_u)	1–10 μs (collisional deactivation)
Stimulated Emission Cross Section (σ_{ul})	2×10^{-15} m^2
Spontaneous Emission Linewidth and Gain Bandwidth, FWHM ($\Delta\nu_{ul}$)	10^5 Hz
Inversion Density (ΔN_{ul})	2×10^{14}/m^3
Small Signal Gain Coefficient (g_0)	0.5/m
Laser Gain-Medium Length (L)	1–4 m
Single-Pass Gain ($e^{\sigma_{ul}\Delta N_{ul}L}$)	1.6–7.4
Gas Pressure	0.7 torr
Gas Mixture (example)	$N_2 : CH_4$ at 1:1
Index of Refraction of Gain Medium	≈ 1.0
Pumping Method	electrical discharge or optical pumping
Gas Temperature	300°C
Mode of Operation	some pulsed, some cw
Output Power	up to 1 kW (pulsed), up to 100 mw (cw)
Mode	TEM$_{00}$

in low-pressure molecular gases in the pressure range of 0.01 to 1 torr, and are optically pumped by pulsed middle-infrared lasers such as the CO_2 laser. Some of these molecular species include CH_3OH, $C_2H_2F_2$, CH_3OH, and CH_3F.

LASER STRUCTURE

Far-infrared lasers consist of long (1–3 m), large-bore (5–10 mm) glass tubes. In the case of discharge excited lasers, electrodes are installed at both ends of the tube for the introduction of the discharge current. By operating the laser gain medium at high pressure to obtain pressure-broadened emission, continuous tuning of the laser over the entire molecular emission bandwidth can be achieved.

EXCITATION MECHANISM

Gas discharge–pumped far-infrared lasers involve direct electron excitation of the relevant rotational–vibrational levels. Optical pumping is achieved by using a middle-infrared laser such as a CO_2 laser to pump the upper laser level directly (with no direct excitation of the lower laser level). Inversion is established by favorable decay processes associated with the upper and lower laser levels according to (8.20).

APPLICATIONS

Applications include plasma diagnostics, spectroscopy of semiconductors, atmospheric spectroscopy, heterodyne sources for far-infrared astronomy, imaging, and radar. The use of far-infrared lasers in plasma diagnostics allows the probing and diagnosis of a specific range of plasma electron densities also associated with the laser wavelength. The excited bands of semiconductors are also probed with far-infrared lasers. In atmospheric spectroscopy, these lasers are used in wavelength regions where far-infrared radiation can be transmitted to probe remote regions of the atmosphere owing to the reduced absorption by the atmosphere at such wavelengths.

13.9   CHEMICAL LASERS

GENERAL DESCRIPTION

Chemical lasers are lasers in which the pumping energy is obtained from a chemical reaction. These lasers typically operate on molecular transitions, although there is one atomic chemical laser operating in atomic iodine. Most chemical lasers operate in the near- to middle-infrared portion of the spectrum. The most well known chemical lasers are those operating on vibrational transitions of hydrogen fluoride and deuterium fluoride. Chemical lasers have been developed primarily for military and space applications, where pumping power – in the form of electrical energy – might not be available. Chemical lasers have been developed that emit powers up to several megawatts for antimissile defense. With the ending of the cold war, efforts to develop such lasers have been significantly reduced.

The hydrogen fluoride laser emits in the wavelength range of 2.6–3.3 μm, a region of high absorption within the atmosphere. By using deuterium instead of hydrogen, the laser wavelength range is shifted to 3.5–4.2 μm, a region where the atmospheric absorption is low. Other chemical lasers include the HBr laser operating at 4.0–4.7 μm, the CO laser at 4.9–5.8 μm, and the CO_2 laser at 10.0–11.0 μm.

LASER STRUCTURE

Chemical lasers are typically devices in which the chemical vapor is mixed and then flowed through the gain region in a direction transverse to the laser beam axis. Toxic and combustible chemicals are generated in such lasers, and the residual gases must be either neutralized or collected for later disposal. Special nozzles are used to generate the appropriate flow conditions within the gain region in order to maximize the efficiency of the lasers.

EXCITATION MECHANISM

Chemical lasers are generally initiated by a chemical reaction within which the excited laser species is produced. The HF and DF lasers are produced by mixing hydrogen with fluorine gas. There are a few commercial HF/DF lasers that provide the fluoride atoms by dissociating SF_6 within a gas discharge and then flowing those species into a reaction chamber with hydrogen. Oxygen can be added to attract the free sulfur left over within the discharge.

The atomic iodine laser, which operates at a wavelength of 1.3 μm, is produced via the generation of excited molecular oxygen by reacting molecular chlorine with hydrogen peroxide. This long-lived excited molecular oxygen in turn transfers its energy to atomic iodine, producing excitation to the upper laser level.

APPLICATIONS

As mentioned previously, the primary applications for chemical lasers are for high-power weapons at remote locations such as on a battlefield or in space. The HF laser wavelengths could be suitable for propagation in space whereas the DF lasers could serve as ground-based devices that would propagate through the atmosphere.

13.10 X-RAY LASERS

GENERAL DESCRIPTION

X-ray lasers are a category of laser media in which gain has been demonstrated at various discrete wavelengths ranging from 3.56 nm to 46.9 nm. Until very recently the population inversions have generally been produced in extremely high–temperature laser-produced plasmas using very large, high-power pumping lasers. The X-ray lasers generated within such a plasma are typically obtained by focusing the pumping laser, using a cylindrical lens to produce an elongated plasma, onto a solid target composed of material comprising the laser gain medium in a way similar to that shown in Figure 9-16(j). Recently a laser of this type has also been produced within a cylindrical discharge plasma formed by a low-inductance, high-voltage current pulse.

The maximum output achieved from such lasers is a few millijoules but more typically only a few microjoules or less. The first X-ray laser occurred in highly ionized selenium with 24 electrons removed (Se^{24+}) at 20.6 and 20.9 nm. The electronic configuration of that ionization stage of Se is similar to that of neutral neon (Ne-like), and the transition is similar to a known laser transition of neutral neon. Such a comparison is referred to as isoelectronic scaling of energy levels and radiative transitions, as discussed

TABLE 13-9

X-RAY LASER WAVELENGTHS AND SPECIES

Wavelength (nm)	Ion Species	Transition
H-like		
3.879	Al^{12+}	$3 \rightarrow 2$
4.553	Mg^{11+}	$3 \rightarrow 2$
5.419	Na^{10+}	$3 \rightarrow 2$
8.091	F^{8+}	$3 \rightarrow 2$
10.243	O^{7+}	$3 \rightarrow 2$
18.210	C^{5+}	$3 \rightarrow 2$
Li-like		
8.73	Si^{11+}	$5d \rightarrow 3p$
8.89	Si^{11+}	$5f \rightarrow 3d$
10.57	Al^{10+}	$5f \rightarrow 3d$
12.989	Si^{11+}	$4f \rightarrow 3d$
15.466	Al^{10+}	$4f \rightarrow 3d$
Be-like		
12.35	Al^{9+}	$5d \rightarrow 3p$
17.78	Al^{9+}	$4f \rightarrow 3d$
Ne-like		
8.156	Ag^{37+}	$3p \rightarrow 3s$
9.936	Ag^{37+}	$3p \rightarrow 3s$
10.038	Ag^{37+}	$3p \rightarrow 3s$
10.508	Ag^{37+}	$3p \rightarrow 3s$
10.64	Mo^{32+}	$3p \rightarrow 3s$
12.300	Ag^{37+}	$3p \rightarrow 3s$
13.1	Mo^{32+}	$3p \rightarrow 3s$
13.27	Mo^{32+}	$3p \rightarrow 3s$
13.94	Mo^{32+}	$3p \rightarrow 3s$
14.16	Mo^{32+}	$3p \rightarrow 3s$
15.50	Y^{29+}	$3p \rightarrow 3s$
15.71	Y^{29+}	$3p \rightarrow 3s$
15.98	Sr^{28+}	$3p \rightarrow 3s$
16.41	Sr^{28+}	$3p \rightarrow 3s$
16.490	Y^{29+}	$3p \rightarrow 3s$
16.65	Sr^{28+}	$3p \rightarrow 3s$
16.867	Se^{24+}	$3p \rightarrow 3s$
17.455	Sr^{28+}	$3p \rightarrow 3s$
18.243	Se^{24+}	$3p \rightarrow 3s$
19.606	Ge^{22+}	$3p \rightarrow 3s$
20.638	Se^{24+}	$3p \rightarrow 3s$
20.978	Se^{24+}	$3p \rightarrow 3s$
21.217	Zn^{20+}	$3p \rightarrow 3s$
21.884	As^{23+}	$3p \rightarrow 3s$
22.028	Se^{24+}	$3p \rightarrow 3s$
22.111	Cu^{19+}	$3p \rightarrow 3s$

Wavelength (nm)	Ion Species	Transition
22.256	As^{23+}	$3p \to 3s$
22.49	Sr^{28+}	$3p \to 3s$
23.224	Ge^{22+}	$3p \to 3s$
23.626	Ge^{22+}	$3p \to 3s$
24.670	Ga^{21+}	$3p \to 3s$
24.732	Ge^{22+}	$3p \to 3s$
25.111	Ga^{21+}	$3p \to 3s$
26.232	Zn^{20+}	$3p \to 3s$
26.294	Se^{24+}	$3p \to 3s$
26.723	Zn^{20+}	$3p \to 3s$
27.931	Cu^{19+}	$3p \to 3s$
28.467	Cu^{19+}	$3p \to 3s$
28.646	Ge^{22+}	$3p \to 3s$
32.65	Ti^{12+}	$3p \to 3s$
46.9	Ar^{8+}	$3p \to 3s$
Co-like		
4.607	Ta^{46+}	$3d^84d \to 3d^84p$
5.176	Yb^{43+}	$3d^84d \to 3d^84p$
Ni-like		
3.560	Au^{51+}	$4d \to 4p$
4.318	W^{46+}	$4d \to 4p$
4.483	Ta^{45+}	$4d \to 4p$
5.023	Yb^{42+}	$4d \to 4p$
5.097	Ta^{45+}	$4d \to 4p$
5.611	Yb^{42+}	$4d \to 4p$
6.583	Eu^{35+}	$4d \to 4p$
7.100	Eu^{35+}	$4d \to 4p$
7.24	W^{46+}	$4d \to 4p$
7.3	Sm^{34+}	$4d \to 4p$
7.442	Ta^{45+}	$4d \to 4p$
7.535	W^{46+}	$4d \to 4p$
7.747	Ta^{45+}	$4d \to 4p$
8.107	Yb^{42+}	$4d \to 4p$
8.440	Eu^{35+}	$4d \to 4p$
8.441	Yb^{42+}	$4d \to 4p$
10.039	Eu^{35+}	$4d \to 4p$
10.456	Eu^{35+}	$4d \to 4p$

in detail in Section 3.1. The first gas discharge–pumped device produced laser output in Ne-like Ar (Ar^{8+}) at 46.9 nm.

Other isoelectronic scaling sequences for producing X-ray lasers include hydrogen-like, lithium-like, beryllium-like, cobalt-like, and nickel-like scaling sequences. Other X-ray laser transitions are listed in Table 13-9 along with the scaling sequence they are associated with. The Ni-like isoelectronic scaling sequence has produced the shortest-wavelength lasers to date. Most of the transitions in Table 13-9 have provided gains of only e^2 to e^4 over the plasma lengths studied, although the Se and Ge lasers have reached I_{sat}. Higher input laser energies would be necessary to provide longer gain lengths and thus higher laser output. The emission linewidths are typically Doppler-broadened but are quite large, of the order of 5×10^{12} Hz (0.01 nm), owing to the wavelength and temperature scaling of Doppler broadening as indicated in (4.59).

Some new techniques such as optical field ionization (OFI) are currently being used in an attempt to produce lasers with much lower threshold energies than those mentioned here. Such techniques use ultrashort pulsed lasers with durations of less than a picosecond to produce highly "nonequilibrium" plasmas, which are more easily inverted than are plasmas formed under equilibrium conditions.

LASER STRUCTURE

The general configuration for producing X-ray lasers is that of generating a line focus of a powerful Nd:glass laser onto a solid target material over a length of 1–5 cm and a width of a few hundred microns, as shown in Figure 13-9. This focusing arrangement and laser target are located inside an evacuated chamber surrounded by sophisticated diagnostics, as shown in

Figure 13-9. Experimental arrangement of a soft–X-ray laser pumped by the Novette laser (courtesy of Lawrence Livermore National Laboratories)

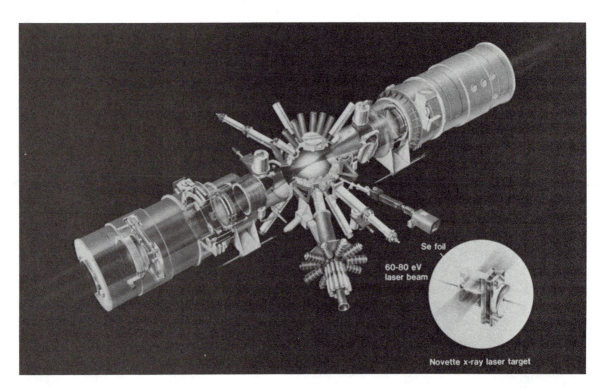

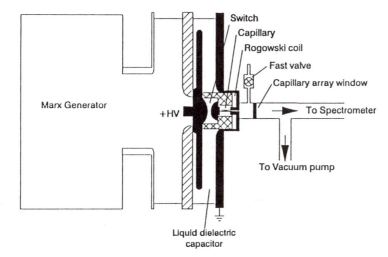

Figure 13-10. Discharge-pumped EUV laser (courtesy of Colorado State University)

the figure. The circular mirror located on the front of the target, as depicted in the closeup view of Figure 13-9, is placed there for beam alignment. An intense plasma of the ions of the laser species is produced within the focal region of the initiating laser, and the electrons are stripped to obtain the desired ionization stage. When the desired ionization stage is reached, the plasma electrons produce excitation to the upper laser level, much as in the case of the argon ion laser but at a much higher ionization stage.

Optical cavities are not particularly useful with these lasers for several reasons. First, the gain duration is so short that only a few passes will occur through the amplifying medium, depending upon how close the mirrors are placed to the medium. Second, if the mirrors are placed too close while attempting to obtain more passes through the amplifier, the intense X-ray radiation from the plasma will damage the mirrors even before they have a chance to operate. Third, mirrors are not yet available that can withstand the saturation intensities of these lasers.

In the plasma discharge device shown in Figure 13-10, the plasma is produced within a 4-mm bore quartz tube with lengths of up to 0.12 m by using the very rapid discharge pulse made possible by the low-inductance system. The discharge device includes a low-inductance parallel-plate liquid dielectric capacitor, a high-voltage spark-gap switch, and the capillary discharge tube. The capacitor is pulse charged by a Marx generator, which generates very high voltages, and then rapidly discharged through the capillary by triggering the spark gap pressurized with SF_6 gas. The Rogowski coil is used to monitor the discharge current.

EXCITATION MECHANISM

When a high-power laser – such as a high-power frequency-doubled Nd: glass laser system – is cylindrically focused onto a solid target, the target material is vaporized to form ions of the appropriate ionization stage containing the laser transition. A similar ionizing process can take place in a discharge device where the rapidly rising electrical current produces the required ionization. The appropriate ionization stage is reached by sequentially removing electrons from the atom. For example, in generating Ne-like Se (Se^{24+}), 24 electrons must be removed from the Se atom. This requires a plasma temperature approaching 1 keV or 10^7 K. For the Ne-like Ar (Ar^{8+}), a plasma temperature of 100 eV is sufficient to produce the required ionization, since 16 fewer electrons need to be removed from the atom. Excitation of the upper laser level then occurs, most often by collisions of the plasma electrons that are already present with ground-state ions of that ion stage, in a manner similar to the excitation of an Ar^+ laser but occurring at much higher energies (shorter wavelengths). The requirements for producing such an X-ray laser are an input laser energy of tens to thousands of joules within a duration of the order of a few nanoseconds or less and a plasma electron density of the order of 10^{25} per cubic meter or more.

MIMINUM PUMP ENERGY   We will now obtain an expression for the minimum pump energy required to produce such a laser versus the laser wavelength. We will assume that the pump energy per atom, E_P, required to place the atom in the upper laser level u must be at least as great as the laser wavelength:

$$E_P \cong h\nu_{ul} = h\frac{c}{\lambda_{ul}}.$$

The atoms pumped to level u will then remain in that level only for the lifetime τ_u of level u. Hence, if we assume that $\tau_u \cong 1/A_{ul}$, we can write an expression for the pump power per atom required to have that atom in the upper laser level u as

$$P_P = \frac{E_P}{\tau_u} = E_P \cdot A_{ul}.$$

We can then obtain the pump power per unit volume as

$$\frac{\text{pump power}}{m^3} = \frac{\text{pump power}}{\text{atom}} \times \frac{\text{number of atoms}}{m^3}$$

$$= P_P \cdot N_u = \frac{hc}{\lambda} A_{ul} \cdot N_u.$$

Assuming that there are no mirrors available for short-wavelength lasers, we will use the single-pass requirement for the gain from (7.56) of $\sigma_{ul}\Delta N_{ul} \approx 10$, where we have chosen the value of 10 (instead of 12 or 15) to simplify the expression since we seek only an estimate of the pump power. We also assume that $\Delta N_{ul} \approx N_u$ by ignoring the lower laser level population. We then have the constraint that $\sigma_{ul} N_u L = 10$, from which we obtain an expression for N_u as

$$N_u \approx \frac{10}{\sigma_{ul}L}.$$

TABLE 13-10

TYPICAL X-RAY LASER PARAMETERS

Laser Wavelengths (λ_{ul})	3.56–46.9 nm
Laser Transition Probability (A_{ul})	typically 10^{11}/s
Upper Laser Level Lifetime (τ_u)	typically 10^{-11} s
Stimulated Emission Cross Section (σ_{ul})	10^{-19}–10^{-20} m^2
Spontaneous Emission Linewidth and Gain Bandwidth, FWHM ($\Delta\nu_{ul}$)	1 to 5×10^{12}/s
Inversion Density (ΔN_{ul})	10^{20}–10^{21}/m^3
Small Signal Gain Coefficient (g_0)	400/m
Laser Gain-Medium Length (L)	0.01–0.05 m (laser plasma excitation), up to 0.12 m (discharge excitation)
Single-Pass Gain ($e^{\sigma_{ul}\Delta N_{ul}L}$)	10–10^6
Gas Density	10^{25}–10^{26} ions/m^3
Index of Refraction of Gain Medium	≈ 1.0
Pumping Method	laser-produced plasma or fast-discharge plasma
Electron Temperature	100–1,000 eV
Plasma Temperature	100–500 eV
Mode of Operation	pulsed
Output Pulse Duration	500 ps to 10 ns
Output Energy/Pulse	10 nJ to 1 mJ
Maximum Peak Output Power	1–2 MW
Mode	high-order mode (ASE)

Substituting this value for N_u into the preceding formula, we have

$$\frac{\text{pump power}}{\text{m}^3} = \frac{hc}{\lambda_{ul}} A_{ul} \cdot \frac{10}{\sigma_{ul}L}.$$

If we assume that the dominant broadening mechanism is Doppler broadening, then we can use the expression for σ_{ul} of (7.28) and obtain

$$\frac{\text{pump power}}{\text{m}^3} = \frac{hc \cdot A_{ul} \cdot 10 \cdot (16\pi^3)^{1/2}\Delta\nu_D}{\lambda_{ul}(\ln 2)^{1/2}\lambda_{ul}^2 A_{ul}L}.$$

We can now incorporate the following expression for Doppler broadening from (4.59):

$$\Delta\nu_D = 2\nu_{ul}\left(\frac{2(\ln 2)kT}{Mc^2}\right)^{1/2} = \frac{2c}{\lambda_{ul}}\left(\frac{2(\ln 2)kT}{Mc^2}\right)^{1/2}.$$

Using this expression in our previous formula yields

$$\frac{\text{pump power}}{\text{m}^3} = \frac{80hc(2\pi^3kT/M)^{1/2}}{\lambda_{ul}^4 L}$$

$$\approx \frac{10^{-20}}{\lambda_{ul}^4 L}\sqrt{\frac{T}{M_N}},$$

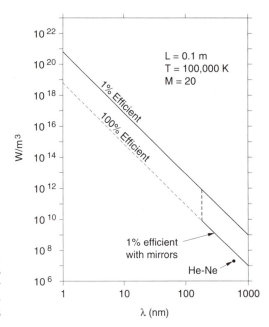

Figure 13-11. Minimum pumping requirements versus wavelength for producing short-wavelength lasers

where λ_{ul} is the laser wavelength in meters, L is the length of the gain medium in meters, T is the plasma temperature in degrees Kelvin, and M_N is the mass number (the total number of neutrons and protons per ion of the laser species; see the Appendix).

We can see that the pump power per cubic meter depends upon the fourth power of the wavelength, as well as on the temperature and mass of the atoms and the effective gain length. The pumping flux is seen to depend very strongly upon the laser wavelength (λ^4) but only weakly on the temperature and mass, since they involve only a square-root factor. Our calculation assumes that 100% of the pump energy is converted to laser output, which is obviously unrealistic. We have graphed the pump power versus wavelength for an amplifier length of 0.1 m, an average temperature of 100,000 K, and an average mass of $M_N = 20$ in Figure 13-11 for an efficiency of 100% (the dashed line) and 1% (the solid line). An efficiency of 0.1% would be a factor of 10 higher than the 1% line, and so forth. We have also shown a reduced pumping flux of a factor of 100 above a wavelength of 200 nm (ultraviolet) which is obtained by assuming an effective gain length 100 times the 0.1-m length of the graph. This is a region for which good mirrors are available, so the mirror factor m of (7.62) is assumed to be 100. An experimental point is also shown for the power input (20 W) of a small helium–neon laser operating at 632.8 nm, which agrees quite well with the graph that is associated with mirrors since the He–Ne efficiency is just over 0.1%. The temperature of a He–Ne laser is substantially less than 10^5 K, but the temperature dependence in our equation varies with the square root of the temperature and is thus not very significant on this graph.

It can be seen in Figure 13-11 that at a wavelength of 10 nm – a wavelength region where significant gain has been demonstrated in soft–X-ray lasers, the pump power per cubic meter is nearly 10^{17} W/m^3 for a 1% efficient laser, which is a power that could be achieved only in a very small volume and for very short time durations with any pumping technique short of a nuclear explosion. Most X-ray lasers have efficiencies that are closer to 10^{-7}, so that the pumping powers are even much higher than shown in the graph. Thus, for a laser operating at 10 nm with a plasma volume of 10^{-9} m^3, a duration of 10^{-9} s, and an efficiency of 10^{-6}, the pump energy would be 800 J (the power multiplied by the pump duration), which is approximately the laser energy used to produce the first X-ray laser in Se^{24+} ions.

APPLICATIONS

X-ray lasers are one of the most recently developed types of lasers, and are not as yet commercially available. The X-ray lasers produced to date operate at a pulse repetition rate of only a few pulses per day and are expensive to operate. Thus, applications are currently limited to those that can justify the expense. One such application is X-ray holography of living biological materials. With a laser pulse of a few hundred picoseconds' duration, this process can generate a "stop action" three-dimensional image of living species with the high resolution possible with short illumination wavelengths. Other applications include X-ray microscopy, crystallography, medical radiology, microprobing, atomic physics studies, plasma diagnostics, radiation chemistry, photolithography, and metallurgical studies. As more compact lasers are developed, more applications will no doubt emerge.

13.11 FREE-ELECTRON LASERS

GENERAL DESCRIPTION AND EXCITATION MECHANISM

Conventional lasers are based upon the concept of creating population inversions between energy levels of discrete bound states of materials. Free-electron lasers differ in that they involve electrons oscillating in a vacuum, void of any material gain medium other than the electrons themselves. When oscillating, the electrons radiate in the typical dipole radiation pattern that maximizes in a direction perpendicular to the direction of oscillation. For this situation, the electrons can be thought of as making transitions between continuous states rather than between bound states. These transitions are produced by sending electrons at very high velocities through an alternating magnetic field structure to produce the necessary oscillations. The relativistic speed of the electrons causes the oscillating frequency to shift from the low frequency produced in the electron rest frame to a very high frequency observed in the laboratory frame. The frequency of the radiation is determined by the speed (kinetic energy) of the electrons and by the period of the alternating magnetic field structure. Laser mirrors are placed at opposite ends of the magnetic structure and normal to the direction of the electrons in order to send a portion of the radiated energy back through the magnetic structure so that a standing-wave pattern of radiation is established between the mirrors, much as occurs in a traditional laser. This standing-wave pattern of the electric field further contributes to the oscillation of the electrons traversing this path and thus stimulates additional radiation in the desired direction, thereby producing a strong optical beam within and beyond the laser cavity. The electrons are prevented from impacting (and thereby damaging) the mirrors by turning them with the use of bending magnets.

Free-electron lasers have produced stimulated emission over a wide range of frequencies ranging from 248 nm to 8 mm and output powers of up to 1 GW per pulse for a 60-ns pulse at a wavelength of 8 mm. The wavelength region of possible laser emission for these devices is determined largely by the energy range of the electrons they produce. The quality of the laser output relies heavily upon the quality of the electron beam. The spread in energy of the electron beam creates broadening, which in turn broadens the laser radiation. The angular divergence of the electron beam also leads to reduced gain. Non-uniformities in the magnet can also produce broadening of the emission. Pulsed electron beams lead to a pulsed laser output and continuous beams lead to continuous laser output. As the pulses are made longer, the emission lineshape becomes much narrower.

LASER STRUCTURE

A typical free-electron laser structure is shown in Figure 13-12. A high-energy electron beam is injected into the region between the laser mirrors using bending magnets. The permanent magnet array between the mirrors

Figure 13-12. Free-electron laser

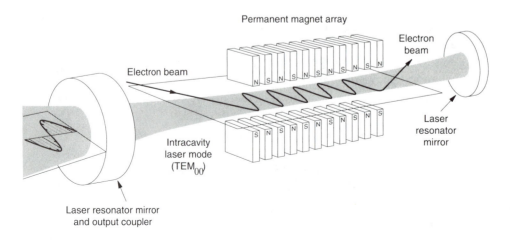

Permanent magnet array

Electron beam

Electron beam

Laser resonator mirror

Intracavity laser mode (TEM$_{00}$)

Laser resonator mirror and output coupler

TABLE 13-11

TYPICAL FREE-ELECTRON LASER PARAMETERS

Laser Wavelengths (λ_{ul})	248 nm to 8 mm
Fractional Laser Bandwidth	10^{-3} to 10^{-7}
Gain Per Pass	1–300%
Laser Gain-Medium Length (L)	1–25 m
Pumping Method	high-energy electron beam
Electron Beam Peak Current	0.1–800 A
Electron Beam Energy	200 kV to 1 GeV
Electron Beam Pulse Length	2 ps to cw
Undulator Magnet Period	5 mm to 0.2 m
Magnetic Field Strength	0.02–1.0 T
Output Power	up to 1 GW (pulsed), up to 10 W (cw)
Mode	TEM_{00}

forces the electrons to oscillate and radiate. The optical field is reflected back into the region where the electrons are present to further stimulate them in phase. The source of electrons can be a storage ring, a radio-frequency linear accelerator (linac), a microtron, an induction linac, an electrostatic accelerator, or a pulse line accelerator and modulator. Recirculating the electron beam within a storage ring can significantly improve the efficiency of these lasers as well as narrow the linewidth by producing a more continuous beam output. However, such lasers generally produce relatively low-power output. The high electron-beam currents of induction linacs and pulse line accelerators yield much higher laser power than that from synchrotron storage rings. An RF linac-based free-electron laser has operated at 10.6 μm with a gain of 7% and a power output of 7 kW.

APPLICATIONS

Numerous applications of free electron lasers have been suggested, but such lasers have not yet been developed to the point of having a significant impact in the commercial arena. Possible applications include biomedical and photochemical interactions, laser isotope separation, materials processing, and physics research. High-power free-electron lasers also include a number of military applications, owing to the very high potential efficiency of such lasers.

REFERENCES

GAS LASERS

C. S. Willett (1974), *Introduction to Gas Lasers: Population Inversion Mechanisms.* Oxford: Pergamon.

MOLECULAR LASERS

A. J. DeMaria (1973), "Review of cw high-power CO_2 lasers," *Proceedings of the IEEE* 61: 731–48.

M. Obara and F. Kannari (1991), "Rare gas halide lasers," in *Encyclopedia of Lasers and Optical Technology* (R. A. Meyers, ed.). New York: Academic Press.

M. Rokni, J. A. Mangano, J. H. Jacob, and J. C. Hsia (1978), "Rare gas fluoride lasers," *IEEE Journal of Quantum Electronics* 14: 464–81.

R. Sauerbrey (1994), "Ultraviolet, vacuum-ultraviolet, and X-ray lasers," in *Electro-Optics Handbook* (R. Waynant and M. Ediger, eds.). New York: McGraw-Hill, Chapter 3.

O. R. Wood II (1974), "High-power pulsed molecular lasers," *Proceedings of the IEEE* 62: 355–97.

FAR-INFRARED GAS LASERS

T. Y. Chang (1974), "Optically pumped submillimeter-wave sources," *IEEE Transactions on Microwave Theory and Techniques* 22: 983–8.

X-RAY LASERS

R. C. Elton (1990), *X-Ray Lasers*. New York: Academic Press.

D. Matthews (1991), "X-ray lasers," in *Handbook of Laser Science and Technology, Supplement I: Lasers*. Boca Raton, FL: CRC Press.

J. J. Rocca, V. Shlyaptsev, F. G. Tomasel, O. D. Cortazar, D. Hargshorn, and J. L. A. Chilla (1994), "Demonstration of a discharge pumped table-top soft–X-ray laser," *Physical Review Letters* 73: 2192–5.

S. Suckewer and C. H. Skinner (1990), "Soft–X-ray lasers and their applications," *Science* 247: 1553–7.

FREE-ELECTRON LASERS

H. P. Freund and R. K. Parker (1991), "Free-electron lasers," in *Encyclopedia of Lasers and Optical Technology* (R. A. Meyers, ed.). New York: Academic Press.

J. A. Pasour (1994), "Free electron lasers," in *Electro-Optics Handbook* (R. Wayant and M. Ediger, eds.). New York: McGraw-Hill, Chapter 8.

LASER SYSTEMS INVOLVING HIGH-DENSITY GAIN MEDIA   **14**

SUMMARY This chapter provides a summary of the various types of lasers that use high-density gain media, including dye lasers, solid-state lasers, and semiconductor lasers. Most of the important commercial lasers are described in this section. The sequence of presenting them, beginning with dye lasers and ending with semiconductor lasers, in no way represents the order of their importance. Rather, it follows the sequence used in Chapter 5, where the energy levels for these systems were described. At the time of this writing, both solid-state and semiconductor lasers are emerging as the important lasers of the future because of their compactness, efficiency, reliability, and – in many cases – low cost. Not all solid-state and semiconductor lasers are described in this chapter, but only those having a significant commercial market. Furthermore, by the time this book is published, other new solid-state laser materials may have emerged. As in Chapter 13 on low-density gain media, a table listing the properties of each laser gain medium is included in the appropriate section.

14.1 DYE LASERS

GENERAL DESCRIPTION

Dye lasers, often referred to as organic dye lasers, are produced in liquid gain media. The dyes have very broad emission and gain spectrums that lead to tunable laser output and short-pulse (mode-locked) laser output. The laser gain medium consists of strongly absorbing and emitting organic dyes dissolved in a solvent, as described in Section 5.2. The typical dye concentration is a 10^{-4} to 10^{-3} molar solution, that is, from 10^{24} to 10^{25} dye molecules per cubic meter. A typical dye laser can operate over a wavelength range of 30–40 nm. The gain region is slightly narrower than the emission bandwidth owing to ground-state absorption, as described in Section 8.4 and indicated in Figure 8-16. There are over 200 laser dyes that when used sequentially can produce tunable laser output over wavelengths ranging from 320 to 1,200 nm (0.32–1.2 μm). A listing of several laser dyes, covering a broad range of wavelengths, is given in Table 14-1. There are three main types of dye lasers: pulsed dye lasers, cw (continuous wave) dye lasers, and mode-locked dye lasers. Pulsed dye lasers pumped by other lasers (such as excimer, nitrogen, or frequency-multiplied Nd:YAG lasers) can produce output pulses of up to tens of millijoules in a 10-ns pulse in a beam of a few millimeters in diameter at repetition rates of up to 1 kHz. Flashlamp-pumped pulsed dye lasers have produced up to 400 J of output

TABLE 14-1

A SELECTION OF LASER DYES COVERING THE VISIBLE SPECTRUM

Dye Name	Molecular Weight	Wavelength Range (nm)	Wavelength of Maximum Gain (nm)
Polyphenyl 2	542	363–410	383
Stilbene 1	569	391–435	415
Stilbene 3	435	409–465	435
Coumarin 102	255	460–515	477
Coumarin 30	347	485–535	518
Coumarin 6	350	506–558	535
Rhodamine 110	367	528–580	540
Rhodamine 6G	479	570–640	593
Dicyanomethylene	303	610–705	661

in a 10-μs pulse; cw lasers produce powers of up to 2 W and can be made to have very narrow emission linewidths of less than 1 kHz. Mode-locked dye lasers have produced pulses of 200 fs without dispersive prisms or gratings in the cavity and as short as 6 fs, the shortest optical pulses ever produced, using gratings as dispersive elements. Dye lasers have been produced in sizes ranging from small structures of the order of 0.1 m in length to elaborate mode-locked systems that occupy several optical tables.

LASER STRUCTURE

We now describe the structure of four categories of dye lasers as follows.

LASER-PUMPED PULSED TUNABLE DYE LASERS This class of dye lasers provides a tunable ultranarrow-frequency pulsed laser output over wavelengths ranging from 190 nm to 4.5 μm. Figure 14-1 shows a diagram of the basic dye laser that covers the wavelength range from 380 to 900 nm when pumped by either a frequency-doubled or -tripled Nd:YAG laser or an excimer laser. The extended frequencies at both the shorter and longer wavelengths are produced by sum and difference frequency mixing of the basic range of wavelengths, using additional optical components not shown. The output pulse energies range from tens to hundreds of millijoules (depending upon the wavelength) over pulse durations of the order of 10 ns. The pump beam (either doubled or tripled YAG or excimer) is split off into several separate beams, as shown in the figure. One beam is used to pump a narrow-frequency–output tunable oscillator containing a beam expander, a grating, and an etalon, similar to that described in Section 12.5. The other beams are used to either side-pump or end-pump a series of dye amplifiers. The dye laser output can be as narrow as 10 GHz; with an air-spaced etalon, it can be reduced to 1.5 GHz.

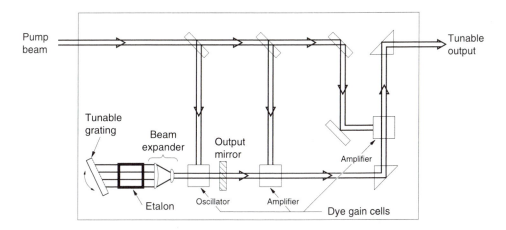

Figure 14-1. Laser-pumped tunable dye laser

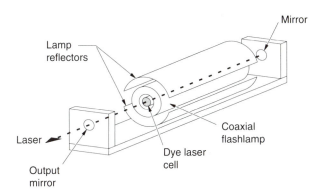

Figure 14-2. Flashlamp-pumped pulsed dye laser

FLASHLAMP-PUMPED PULSED DYE LASER A diagram patterned after a commercial flashlamp-pumped dye laser is shown in Figure 14-2. The dye laser cell is shown surrounded by a coaxial cylindrically shaped flashlamp. Provision for water cooling is made between the flashlamp and the dye cell. Reflectors surrounding the flashlamp redirect its output toward the dye cell. Such a laser produces 5 J per pulse in a 1.5-μs pulse duration or a peak power of over 3 MW per pulse. The laser has an input power of 1000 J and is capable of operating at a 0.5-Hz repetition rate. Such a laser uses the Rh6G dye and is tuned to operate at the 585-nm wavelength for medical applications.

cw TUNABLE DYE LASERS Tunable cw dye lasers are generally pumped by other cw lasers such as the argon ion laser. A basic standing-wave cavity arrangement for such a laser is shown in Figure 14-3. It consists of a three-mirror cavity in which two of the mirrors focus the beam into the thin flowing dye region (referred to as a *jet stream*) that is oriented at Brewster's angle as shown. The dye would flow in a direction perpendicular to the printed page. The dye is end-pumped, either by pumping at a slight angle into the jet stream with an additional mirror (as shown in the figure) or by

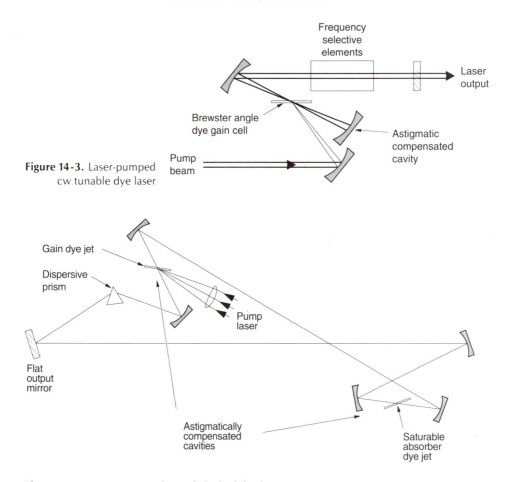

Figure 14-3. Laser-pumped cw tunable dye laser

Figure 14-4. Arrangement of a mode-locked dye laser

pumping directly through the third mirror. This type of cavity was described in detail in Section 12.6. Such a cavity can be modified to provide tuning by inserting a prism and rotating the laser mirror located beyond the prism, as shown in Figure 12-24(a). The pump beam can also enter the cavity through the prism, since it is at a shorter wavelength than the dye output and can therefore enter the prism at a different angle than the laser beam. A more sophisticated cw ring dye laser is shown in Figure 12-25(b) and described in the related text.

MODE-LOCKED DYE LASER Figure 14-4 is a diagram of a passively mode-locked ring dye laser cavity, a common configuration for generating femtosecond dye laser pulses. Such a cavity must be designed with an odd number of beam waists within the cavity, in this case created by the five curved mirrors, to ensure a uniform spectral distribution of the laser bandwidth within the cavity. Also, the optical path length between the two dye jets is one fourth of the optical path around the entire ring. Since there are two counterpropagating pulses within the cavity, this ensures that both pulses will have equal properties by allowing the gain recovery time to be the same for each. The exact dimensions of such a cavity are critical in producing the shortest pulses. The reader is referred to the references for further information.

EXCITATION MECHANISM

The efficient radiating states of dye molecules when dissolved in liquid solvents have energies that can be described by the harmonic oscillator model. Each electronic level contains a series of vibrational levels with an energy separation determined by the normal-mode oscillation frequency. They also contain rotational levels within each vibrational level, similar to that described for much simpler molecules in Section 5.1. Since these molecules are in a liquid solution, the rotational and vibrational levels smear together owing to collisional interactions with the solvent (cf. T_2 broadening in Section 4.3) to form a continuous spectrum of energies for each electronic state. The populations within each electronic level are distributed according to the Boltzmann distribution. Thus, at room temperature most of the population within the ground electronic state is at the bottom of that state.

The energy-level diagram for a typical dye molecule was shown in Figure 5-12, which indicates the singlet series of levels S_0 and S_1 and the triplet series of T_1 and T_2. Since S_1 is a very broad energy state, the range of absorbing wavelengths (the absorption spectrum) that can pump population into that state S_1 (the upper laser level) is also very broad, as shown in Figure 5-10. After pumping is initiated using the range of wavelengths described by the absorption spectrum, the population within the excited singlet state S_1 rapidly relaxes, within 10^{-12} to 10^{-13} seconds, to the lowest levels of S_1 according to a Boltzmann distribution, since S_1 has a relatively long relaxation time ranging from 2 to 5×10^{-9} s. A population inversion is thus established between the population in the lowest-lying levels of S_1 and the higher-lying levels of S_0 that are not occupied. The spontaneous emission rate between various sublevels of S_1 and those of S_0 is typically 2 to 5×10^8 s^{-1}. As the population decays (by either spontaneous or stimulated emission) from S_1 to a specific sublevel of S_0 it decays rapidly by collisions back to the lowest-lying sublevels of S_0, which in effect provides the rapid decay from the lower laser level similar to that for the four-level laser model described in Section 8.2.

If the singlet energy levels were the only levels involved in the operation of dye lasers, such lasers would be much more practical devices than they presently are. The triplet levels cause most of the problems associated with dye laser gain media. Not all of the decay from the S_1 state goes back to the S_0 state. A small portion (approximately 1 in 1,000) of the excited molecules during each excitation cycle decays over to the triplet T_1 state in a process referred to as *intersystem crossing*. This is a spin-forbidden transition that provides a trap for excited molecules and consequently leads to two problems. First, absorption from the triplet T_1 state to the triplet T_2 state provides an absorptive loss in the laser gain medium, as described by

TABLE 14-2

TYPICAL DYE LASER PARAMETERS

Laser Wavelengths (λ_{ul})	320–1,200 nm
Laser Transition Probability (A_{ul})	2 to 5×10^{8}/s
Upper Laser Level Lifetime (τ_u)	2 to 5×10^{-9} s
Stimulated Emission Cross Section (σ_{ul})	1 to 4×10^{-20} m^2
Spontaneous Emission Linewidth	30–40 nm
Gain Bandwidth, FWHM ($\Delta\nu_{ul}$)	25–30 nm
Inversion Density (ΔN_{ul})	2×10^{22}/m^3
Small Signal Gain Coefficient (g_0)	500/m
Laser Gain-Medium Length (L)	0.001–0.01 m
Single-Pass Gain ($e^{\sigma_{ul}\Delta N_{ul}L}$)	2–1,000
Dye Concentration	10^{-3} to 10^{-4} molar (10^{24} to 10^{25}/m^3)
Solvent	typically methanol, ethanol, dimethylformamide, or water
Index of Refraction of Gain Medium	≈ 1.4
Operating Temperature	room temperature
Pumping Method	optical (laser or flashlamp)
Pumping Wavelength and Bandwidth	maximum at a wavelength 30 nm shorter than gain maximum, with a 30-nm bandwidth
Output Power	up to 10^9 W/pulse or 400 J/pulse, up to 2 W (cw)
Mode	single-mode or multi-mode

the coefficient α in (7.60) and also indicated in Figure 8-16. Second, molecules in T$_1$ tend to react and convert to other molecular species, causing a gradual degradation of the dye. Thus, with cw dye lasers it is necessary to flow the dye to prevent triplet absorption and also to avoid significant degradation. Thermal effects due to non-uniform pumping can also degrade the optical properties of the dye by producing index distortions caused by thermal gradients.

APPLICATIONS

Most applications of dye lasers are for research-oriented experiments in which either narrow-band tunable laser emission or ultrafast optical pulses are desired. The dye laser was the first laser to be continuously tunable over a broad wavelength range, and was thus a very important new tool for spectroscopists. Applications have ranged from studies of atomic physics to photochemistry to emission and absorption spectroscopy in solids. Perhaps the most well-known application in spectroscopy is in isotope separation: a tunable dye laser, pumped by a copper vapor laser, is used to selectively excite specific uranium isotopes leading to an enriched species of uranium (as described in Section 13.4). Dye lasers have also been used to slow down atoms to very slow speeds never before achieved, by tuning the lasers to specific absorption frequencies in atoms. Ultrafast pulses have been used to study the dynamics of excited states of semiconductors and other solids. In such experiments, a femtosecond-duration pulse is separated into two pulses, one that can be delayed with respect to the other. The first pulse is used to produce an excited state in a material and the second (delayed) pulse is used to probe the decay of the excited state, with a time resolution of less than a picosecond. In the field of medicine, dye lasers are used to remove birthmarks, treat malignancies by selective absorption (as described for the case of the gold vapor laser in Section 13.4), and shatter kidney stones or gall stones by directing the beam through an optical fiber catheter into either the kidneys or the gall bladder. The development of tunable solid-state lasers such as the titanium sapphire laser has reduced the need for dye lasers, but they still play an important role in the tunable laser market.

14.2 RUBY LASER

GENERAL DESCRIPTION

The ruby laser was the first laser. It operates in a pulsed mode at 694.3 nm with an emission linewidth of 0.53 nm. The ruby laser rod consists of a sapphire crystal (Al_2O_3) with chromium ions (Cr^{3+}) doped in at a typical concentration of 0.05% by weight. At this concentration there are approximately 10^{25} Cr ions per cubic meter. The ruby laser is typically flashlamp-pumped with pulse durations ranging from a fraction of a millisecond to a few milliseconds. The laser can also be Q-switched. The pumping absorption bands occur at 400 nm and 550 nm with an approximate bandwidth of 50 nm at each wavelength, and thus match very well with the pumping spectrum of xenon flashlamps. The laser is a three-level system (as described in Section 8.2) and therefore has a much higher pumping threshold than such four-level laser systems as the Nd:YAG laser. However, the ruby laser does have an exceptionally long upper laser level lifetime of 3 ms, which gives it an unusually high energy-storage capability. Pulse energies of up to 100 J are possible, although the pulse repetition rate is low (1–2 pulses per second) thereby limiting the average power. This limitation is due to the excess heat produced by the high pumping flux required to produce laser output. Relaxation oscillations, as described in Section 7.7, often occur in this laser. The ruby crystal is hard and durable, has good thermal conductivity, and is chemically stable. Ruby laser crystals, from which laser rods are cut, can be grown with very high optical quality.

LASER STRUCTURE

Ruby laser rods can be grown in diameters of up to 25 mm and rod lengths of up to 0.2 m. The rods are generally placed in a double elliptical cavity

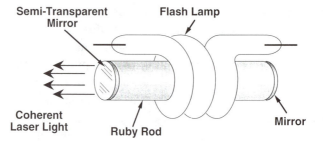

Semi-Transparent Mirror

Flash Lamp

Coherent Laser Light

Ruby Rod

Mirror

Figure 14-5. The first ruby laser (courtesy of Hughes Research Laboratories)

with two linear flashlamps for the pumping source, as indicated in Figure 9-16(c). The mirrors are typically external to the rod, and usually flat or slightly concave. Thermal lensing of the rod itself during the pumping cycle is compensated by the slightly curved mirrors. The rear mirror typically has high reflectivity and the output mirror is partially transmitting. The laser can also be operated in an oscillator–amplifier configuration. In this case the beam that is input to the amplifier should be at or above the saturation intensity in order to extract energy efficiently from the amplifier. Because of the high pumping flux required for this laser, water is generally flowed in the region of the pumping cavity to remove excess heat from the laser rod. Since the water is transparent in the wavelength region of the pumping bands, it has little effect upon reducing the pumping flux from the flashlamp before it is absorbed by the rod. A diagram of the first ruby laser is shown in Figure 14-5.

EXCITATION MECHANISM

As described in Section 5.3 and indicated in Figure 5-13, the energy-level diagram for ruby involves a $^4A_2$ ground state with a statistical weight of $g = 4$, two excitation bands of $^4F_2$ accessed by pumping with green light and $^4F_1$ accessed with blue light. These two energy bands rapidly decay in a time of the order of 0.1 μs to two ^2E levels. The upper of these two levels is the $2\bar{A}$ level with a statistical weight of $g = 2$ and the lower level, the $\bar{E}$ level with $g = 2$, serves as the upper laser level. The transition from the $\bar{E}$ level to the $^4A_2$ level is the laser transition at 694.3 nm.

We have stated that the lifetimes of the ^4F energy bands are extremely short. Thus, when excitation is applied to a ruby laser rod, most of the population resides either in one of the ^2E levels or in the $^4A_2$ ground state. The optical transitions $\bar{E} \rightarrow {}^4A_2$ and $2\bar{A} \rightarrow {}^4A_2$ are referred to as the R_1 and R_2 lines, occurring at wavelengths of 694.3 nm and 692.9 nm (respectively). In thermal equilibrium, the populations of these two levels are determined by a Boltzmann distribution according to (6.11). Hence, if the crystal is maintained at room temperature during the pumping process, there would generally be approximately 15% more population in the $\bar{E}$ level than in the $2\bar{A}$ level at any given time.

From (8.13) we can determine the threshold pumping rate to be

$$\Gamma_{0i} \equiv \Gamma_{li} > \gamma_{ul} = A_{ul} = 1/\tau_u = 3.3 \times 10^2 \text{ s}^{-1}.$$

TABLE 14-3

TYPICAL RUBY LASER PARAMETERS

Laser Wavelength (λ_{ul})	694.3 nm
Laser Transition Probability (A_{ul})	333/s
Upper Laser Level Lifetime (τ_u)	3.0 ms
Stimulated Emission Cross Section (σ_{ul})	2.5×10^{-24} m^2
Spontaneous Emission Linewidth and Gain Bandwidth, FWHM ($\Delta\nu_{ul}$)	3.3×10^{11}/s ($\Delta\lambda_{ul} = 0.53$ nm)
Inversion Density (ΔN_{ul})	8×10^{24}/m^3
Small Signal Gain Coefficient (g_0)	20/m
Laser Gain-Medium Length (L)	0.1 m
Single-Pass Gain ($e^{\sigma_{ul}\Delta N_{ul}L}$)	7.5
Doping Density	0.05% by weight (10^{25}/m^3)
Index of Refraction of Gain Medium	≈ 1.76
Operating Temperature	room temperature
Thermal Conductivity of Laser Rod	42 W/m-K at 300 K
Thermal Expansion Coefficient of Laser Rod	5.8×10^{-6}/°C
Pumping Method	flashlamp
Pumping Bands	404 nm and 554 nm, with bandwidths of 50 nm each
Output Power	up to 100 J/pulse
Mode	single-mode or multi-mode

APPLICATIONS

One of the most important applications of the ruby laser has been holography. Because of its very high pulse energy and reasonable coherence length, a high-power version of the ruby laser can be used to record large-volume holograms in a single laser pulse. Recording such holograms is much more difficult with other, lower-power lasers because multiple pulses must be used. Ruby lasers are used in measuring such plasma properties as electron density and temperature. They are also used to remove tattoos and also skin lesions resulting from excess melanin.

14.3 NEODYMIUM YAG AND GLASS LASERS

GENERAL DESCRIPTION

The Nd ion when doped into a solid-state host crystal produces the strongest emission at a wavelength just beyond 1 μm. The two host materials most commonly used for this laser ion are yttrium aluminum garnate (YAG) and glass. When doped in YAG, the Nd:YAG crystal produces laser output primarily at 1.064 μm; when doped in glass, the Nd:glass medium lases at wavelengths ranging from 1.054 to 1.062 μm, depending upon the type of glass used. Nd also lases at 0.94 μm and at 1.32 μm from the same upper laser level as the 1.064 μm transition, although these transitions have lower gain.

The Nd laser incorporates a four-level system and consequently has a much lower pumping threshold than that of the ruby laser. The upper laser level lifetime is relatively long (230 μs for Nd:YAG and 320 μs for Nd: glass), so population can be accumulated over a relatively long duration during the pumping cycle when the laser is used either in the Q-switching mode or as an amplifier. The emission and gain linewidths are .45 nm for YAG and 28 nm for glass. These lasers can be pumped either by flashlamps or by other lasers. Diode pumping is a relatively recent technique that has led to the development of much more compact Nd lasers, both at the fundamental wavelength of 1.06 μm and at the frequency-doubled wavelength of 0.53 μm.

The Nd:YAG crystal has good optical quality and high thermal conductivity, making it possible to provide pulsed laser output at repetition rates of up to 100 Hz. The crystal size is limited to lengths of approximately 0.1 m and diameters of 12 mm, thereby limiting the power and energy output capabilities of this laser. Doping concentrations for Nd:YAG crystals are typically of the order of 0.725% by weight, which corresponds to approximately 1.4×10^{26} atoms per cubic meter.

For Nd:glass laser gain media, very large–size laser materials have been produced. Rods of up to 2 m long and 0.075 m in diameter and disks of up to 0.9 m in diameter and 0.05 m thick have been successfully demonstrated. The large-diameter disks have been used as amplifiers to obtain laser pulse energies of many kilojoules. The drawback of Nd:glass laser materials is their relatively poor thermal conductivity, which restricts these lasers to relatively low pulse repetition rates. For example, a large Nd:glass laser system constructed at Lawrence Livermore National Laboratories, the NOVA laser, produces output pulses of up to several kilojoules per pulse but can be pulsed at a repetition rate of only several pulses per day. If the large glass amplifiers are not allowed to cool down completely between pulses, thermal and optical distortion severely reduce the energy of the subsequent pulse, with the risk (due to self-focusing effects) of damaging the amplifiers.

LASER STRUCTURE

Neodymium-doped lasers range from small diode-pumped versions with outputs of a few milliwatts, doubled into the green, up to high–average-power lasers with average powers of up to several kilowatts. The high-average-power lasers use either oscillator–amplifier combinations or slab lasers as described in Figure 9-16(d). The laser structure for Nd lasers can be categorized as follows: flashlamp-pumped Nd:YAG lasers, both cw and pulsed; diode-pumped Nd:YAG lasers; and large Nd:glass amplifiers.

FLASHLAMP-PUMPED Q-SWITCHED Nd:YAG LASERS One of the most useful types of Nd:YAG lasers is the flashlamp-pumped Q-switched oscillator-

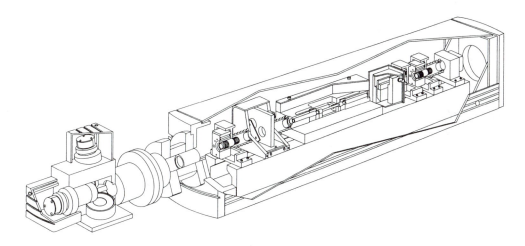

Figure 14-6. Flashlamp-pumped, *Q*-switched Nd:YAG laser (courtesy of EXCEL/Control Laser, Inc.)

amplifier system. A diagram of such a system is shown in Figure 14-6. The system includes a double elliptical pumping arrangement, as shown in Figure 9-16(c) consisting of two linear flashlamps and a 6-mm–diameter, 0.1-m–long Nd:YAG laser rod with antireflection coatings on both ends of the rod. This laser road assembly is installed inside an unstable resonator laser cavity along with a Pockels cell *Q*-switching device. The unstable resonator cavity for this configuration was discussed in the boxed example of Section 12.1 and depicted in Figure 12-3; the Pockels cell *Q*-switching arrangement was described in Section 12.2. The amplifier typically increases the oscillator output energy by up to a factor of 10.

FLASHLAMP-PUMPED cw Nd:YAG LASERS This heading is really a misnomer, for two reasons. First, when a lamp is operated cw rather than pulsed, it is generally referred to as an "arc" lamp rather than a "flash" lamp. Second, although the laser gain medium is pumped by a cw arc lamp, the laser is actually not a true cw laser but instead is *Q*-switched at a high repetition rate, of the order of 20 kHz, to produce 200–300-ns pulses with an average power of as much as 15 W. This high–average-power long-pulse laser is distinguished from the laser described in the previous section, which is flashlamp pumped to produce high–peak-power, *Q*-switched pulses of only 10-ns duration. The longer pulse duration obtained from the arclamp-pumped laser is used extensively for material processing applications. This type of laser is also very effective for frequency doubling the 1.06 μm light to the green at 0.53 μm. (See Chapter 15 for a discussion of frequency multiplication.)

DIODE-PUMPED Nd:YAG LASERS A generic version of a diode-pumped solid-state laser was shown in Figures 9-16(h) and 9-20. A number of laser manufacturers are currently developing compact diode-pumped Nd:YAG lasers. The GaAs laser diode has the ideal pumping wavelength for the

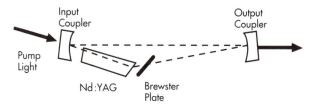

Input
Coupler

Output
Coupler

Pump
Light

Nd:YAG

Brewster
Plate

Figure 14-7. Diode-pumped Nd:YAG ring laser (courtesy of Coherent, Inc.)

Nd^{3+} ion in the region of 0.8 μm. Many of the applications for a diode-pumped Nd:YAG laser are in the green portion of the spectrum, making it necessary to frequency-double the 1.06 μm laser output to 0.53 μm, a wavelength near the peak of the response of the human eye. The cavity arrangement of a diode-pumped, frequency-doubled Nd:YAG laser is shown in Figure 14-7. This unique arrangement allows the operation of a two-mirror ring laser (see Section 12.4) by taking advantage of the refraction of the laser beam at the output of the Nd:YAG laser rod at the Brewster angle to direct the beam toward the output mirror. The doubling crystal (see Section 15.5) is mounted intracavity for high conversion efficiency. This laser produces single-frequency output of greater than 10 mW and a linewidth of less than 2 MHz with a consequently large longitudinal coherence length (see Section 2.4) of greater than 150 m.

Nd:GLASS AMPLIFIERS   Nd:glass laser materials are primarily used as amplifiers for very large pulsed lasers. They have played a major role in the development of the laser fusion program. Lawrence Livermore National Laboratories developed the NOVA laser, consisting of eight separate glass amplifier beamlines arranged in parallel and all fed by a single oscillator. Each beamline consists of approximately 20 amplifier stages, each using 3–5 flashlamp-pumped glass amplifier slabs arranged similarly to that shown in Figure 9-16(e). Each beamline has an input energy per pulse of approximately 100 nJ from the common oscillator and a final output energy of approximately 10 kJ over a pulse duration of 1–2 ns. Thus, a total gain of 10^{11} is achieved in each amplifier beamline, producing a combined energy from the eight beamlines of 80 kJ and a peak power of up to 8×10^{13} W! An even larger laser, referred to as the National Ignition Facility (NIF), is presently planned for completion in the year 2002. It would consist of 192 beamlines or *arms,* each consisting of a series of approximately 20 Nd:glass amplifier stages. The glass in these amplifiers would be Nd-doped phosphate glass. Each slab in the amplifiers would have a gain coefficient similar to that of the slabs in the NOVA laser, 0.55% per millimeter, yielding a total gain similar to that of the NOVA laser system. The final output stage of each beamline would consist of large glass slabs with an effective aperture nearly 0.4 m square, pumped by large-bore xenon flashlamps as shown in Figure 14-8. The parts would be modular to simplify servicing of the laser. Each beamline would produce an output energy of 15 kJ in a pulse duration of 3.5 ns for a total energy of nearly 3 MJ!

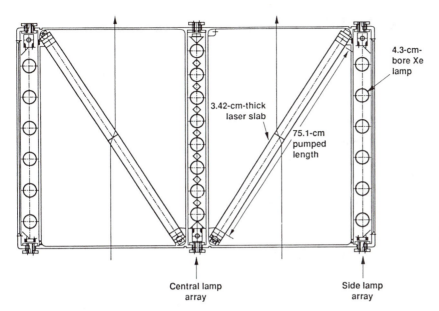

Figure 14-8. Laser amplifier slabs and flashlamps for the proposed National Ignition Facility (courtesy of Lawrence Livermore National Laboratory)

EXCITATION MECHANISM

The Nd laser energy levels doped in both YAG and glass were shown in Figures 5-14(a) and (b). The $^4F_{3/2}$ upper laser level is split into two components, referred to as R_1 and R_2. The R_2 component is the upper laser level and is approximately 0.011 eV above the R_1 level. Because these levels are closely coupled, their populations generally exist in a Boltzmann distribution (eqn. 6.11) and hence at room temperature only about 40% of their combined population is in the upper laser level. Thus, as laser action occurs and stimulated emission depletes the upper laser level, it is rapidly replenished by the population in the R_1 level until both are depleted. As seen from the figures just referred to, there is a broad range of pumping wavelengths in the Nd:YAG system over the region from 0.3 to 0.9 μm. This range includes two strong bands, at 0.75 μm and 0.81 μm, that make these lasers attractive for pumping with efficient diode lasers. The lower laser level is 0.26 eV above the ground state. As a result, there is essentially no significant thermal population in the lower laser level at room temperature and hence the lower laser level decays extremely rapidly by thermalizing collisions. This makes the pumping threshold very low, as suggested in (8.30). Thus, excitation can be achieved with broadband xenon flashlamps as well as with other lasers, in particular the GaAs diode laser that can be matched with the 0.81-μm pump band.

APPLICATIONS

The Nd laser probably has more different kinds of applications than any other type of laser. Perhaps the major application is in various forms of

TABLE 14-4

TYPICAL NEODYMIUM:YAG AND NEODYMIUM:GLASS LASER PARAMETERS

	Nd:YAG	Nd:Glass
Laser Wavelengths (λ_{ul})	1.064 μm	1.054–1.062 μm
Laser Transition Probability (A_{ul})	4.3×10^3/s	2.9 to 3.4×10^3/s
Upper Laser Level Lifetime (τ_u)	230 μs	290–340 μs
Stimulated Emission Cross Section (σ_{ul})	6.5×10^{-23} m^2	2.9 to 4.3×10^{-24} m^2
Spontaneous Emission Linewidth and Gain Bandwidth, FWHM ($\Delta\nu_{ul}$)	1.2×10^{11}/s ($\Delta\lambda_{ul} = 0.45$ nm)	7.5×10^{12}/s ($\Delta\lambda_{ul} = 28$ nm)
Inversion Density (ΔN_{ul})	1.6×10^{23}/m^3	8×10^{23}/m^3
Small Signal Gain Coefficient (g_0)	10/m	3/m
Laser Gain-Medium Length (L)	0.1–0.15 m	0.1 m
Single-Pass Gain ($e^{\sigma_{ul}\Delta N_{ul}L}$)	2–20	1.3
Doping Density	1.4×10^{26}/m^3	4.6×10^{26}/m^3
Index of Refraction of Gain Medium	1.82	1.50–1.57
Operating Temperature	300 K	300 K
Thermal Conductivity of Laser Rod	13 W/m-K	$\approx$1 W/m-K
Thermal Expansion Coefficient of Laser Rod	6.9×10^{-6}/K	8.5 to 14.0×10^{-6}/K
Pumping Method	optical (flashlamp or laser)	
Pumping Bands	300–900 nm, with strongest peaks at 810 nm and 750 nm (peaks wider in glass)	
Output Power	1 J/pulse	up to 10 kJ/pulse in large amplifiers
Mode	single-mode or multi-mode	

material processing: drilling, spot welding, and laser marking. Because they can be focused to a very small spot, the lasers are also used in resistor trimming and in circuit mask and memory repair, as well as in cutting out specialized circuits. Medical applications include many types of surgery, such as membrane cutting (associated with cataract surgery), gall-bladder surgery, and cauterizing gastrointestinal bleeding. Many medical applications take advantage of low-loss optical fiber delivery systems that can be inserted within the body to deliver laser energy to the appropriate location.

Neodymium lasers are also used in military applications such as rangefinding and target designation. High-power pulsed versions are also used for X-ray production by focusing the laser onto a solid target from which a high-temperature plasma is produced that radiates in the X-ray and soft–X-ray spectral regions. They are also used in large laser systems for studies of inertial confinement fusion. Another large market is in scientific and general laboratory use. Frequency multiplication into the green and ultraviolet makes these lasers good pumping sources for pumping tunable dye lasers and other types of laser probes and diagnostics.

14.4 ALEXANDRITE LASER

GENERAL DESCRIPTION

The alexandrite laser uses a chromium-doped chrysoberyl crystal, designated as $Cr^{3+}:BeAl_2O_4$, in which the active ion is Cr^{3+}. The doping concentration of Cr^{3+} is approximately 0.1 atomic percent, corresponding to a chromium ion density of 3.5×10^{25} per cubic meter. This laser was originally operated on the R line as in the case of ruby, but was later shown to have gain on a vibronic sideband over a wavelength range of 700 to 820 nm. It can be operated with high average power, and has a high slope efficiency (see Section 9.3) and a low gain threshold comparable to that of Nd:YAG. The laser can be operated cw or pulsed and can also be Q-switched and mode-locked. The upper level lifetime is 260 μs at a temperature slightly above room temperature, making the laser very amenable to flashlamp pumping. The pump absorption band occurs over most of the visible spectrum, and the laser has been pumped with xenon and mercury flashlamps and by argon and krypton ion lasers. The crystal has very good thermal conductivity, almost twice that of YAG, and has a very high optical damage threshold. It also has good chemical stability, mechanical strength, and high hardness. The rods can be grown of very high optical quality in sizes similar to that of Nd:YAG (6-mm diameter) and up to 0.12 m long.

The $BeAl_2O_4$ host crystal is an optically biaxial crystal (see Section 15.1). This leads to a strong polarization dependence of the absorption and emission spectra. Emission and laser output occurs mainly along the b axis of the crystal. This naturally occurring birefringence dominates over any possible thermal birefringence, which prevents depolarization loss during laser operation. Without this natural birefringence, both thermal- and stress-induced birefringence could mix the two optical polarizations as the beam propagates within the crystal, resulting in large losses if polarization discriminating optical elements (such as Pockels cells or Brewster angle windows) are located within the laser cavity.

LASER STRUCTURE

Flashlamp-pumped versions of this laser are similar to those of the Nd:YAG lasers (summarized in Section 14.3); they use linear flashlamps and a double elliptical pumping cavity. The gain is lower in alexandrite owing primarily to the broad emission linewidth. Thus the output-mirror reflectivity must be higher than for a Nd:YAG laser. Although the laser can be operated cw by pumping with another laser, most alexandrite lasers are flashlamp-pumped pulsed lasers (either Q-switched or long-pulse). These lasers can be operated with average powers of several tens of watts and can

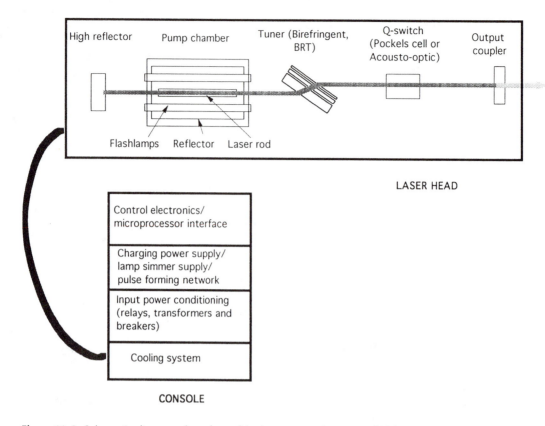

High reflector   Pump chamber   Tuner (Birefringent, BRT)   Q-switch (Pockels cell or Acousto-optic)   Output coupler

Flashlamps   Reflector   Laser rod

LASER HEAD

Control electronics/ microprocessor interface

Charging power supply/ lamp simmer supply/ pulse forming network

Input power conditioning (relays, transformers and breakers)

Cooling system

CONSOLE

Figure 14-9. Schematic diagram of an alexandrite laser system (courtesy of Light Age, Inc.)

be frequency-doubled into the near ultraviolet. They can also be pumped with short-wavelength diode lasers. However, diode lasers at wavelengths in the 630–650-nm region are still under development and consequently not as efficient as the GaAs 0.8-nm diodes used to pump Nd:YAG lasers. An alexandrite laser is shown in Figure 14-9.

EXCITATION MECHANISM

The energy-level diagram of alexandrite was shown in Figure 5-15. It has a $^4A_2$ ground state consisting of a series of vibrational levels that are populated according to a Boltzmann distribution (eqn. 6.11). Thus, as with a dye laser gain medium, there is very little population in the higher-lying vibrational levels at room temperature. Strong absorption of the pump radiation occurs from this $^4A_2$ state to the $^4T_2$ state (upper laser level), which is a band of vibronic levels, much as in the case of a dye laser. The pump band ranges from 380 to 630 nm, which matches well with flashlamps. Pumping within this wavelength region produces population in the $^4T_2$ band, which rapidly relaxes to the lowest levels of the band. Laser action subsequently occurs via transitions from the lowest levels of the $^4T_2$ state back to higher levels of the $^4A_2$ state (similar, again, to the case of a

TABLE 14-5

TYPICAL ALEXANDRITE LASER PARAMETERS

Laser Wavelengths (λ_{ul})	700–820 nm
Laser Transition Probability (A_{ul})	3.8×10^3/s
Upper Laser Level Lifetime (τ_u)	260 μs at 298 K
Stimulated Emission Cross Section (σ_{ul})	1×10^{-24} m^2
Spontaneous Emission Linewidth and Gain Bandwidth, FWHM ($\Delta\nu_{ul}$)	2.6×10^{13}/s ($\Delta\lambda_{ul} = 50$ nm)
Inversion Density (ΔN_{ul})	6×10^{24}/m^3
Small Signal Gain Coefficient (g_0)	4–20/m
Laser Gain-Medium Length (L)	0.12 m
Single-Pass Gain ($e^{\sigma_{ul}\Delta N_{ul}L}$)	1.6–11
Doping Density	1.75 to 10×10^{25}/m^3
Index of Refraction of Gain Medium	1.74
Operating Temperature	500 K
Thermal Conductivity of Laser Rod	23 W/m-K
Thermal Expansion Coefficient of Laser Rod	6×10^{-6}/K
Pumping Method	optical (flashlamp or laser)
Pumping Bands	380–630 nm, with peaks at 410 nm and 590 nm
Output Power	up to 1.2 J/pulse
Mode	single-mode or multi-mode

dye laser). The $^4T_2$ has a relatively short lifetime of 6.6 μs, but it is strongly coupled to the longer-lived ^2E state located 0.1 eV below the $^4T_2$ state. In fact, most of the decay from the $^4T_2$ state occurs to the ^2E state. Thus, when optical pumping of the $^4T_2$ state occurs, the population rapidly equilibrates with the ^2E state, which thereby serves as a storage state and effectively provides the upper laser level (the $^4T_2$ state) with a long lifetime. The population between the two states is determined by a Boltzmann distribution (eqn. 6.11); at room temperature, approximately 60% of the population is in the ^2E state and 40% is in the $^4T_2$ state. The upper laser level population could thus be increased at higher temperatures, thereby increasing the effective laser gain. However, the lower laser level populations will also increase with increasing temperatures. The net of these effects is that the gain can be increased by increasing the laser rod temperature, but only at laser wavelengths above 740 nm. The optimum rod temperature for these longer wavelengths is approximately 200°C. The lower laser levels decay back to the ground state via vibronic transitions, thereby maintaining a thermal distribution of the population in this state. Thus, any excess population that is dumped into a specific vibronic lower laser level is rapidly redistributed according to the thermal distribution, with most of it going to the lowest-lying levels of the $^4A_2$ state.

APPLICATIONS

Perhaps the largest application for the alexandrite laser is a general medical procedure referred to as *selective photothermalysis*. This process uses the wavelengths and short-pulse capabilities of the alexandrite laser to irradiate tissue and produce selective absorption in undesirable cells without harming the surrounding normal cells. This process can be used for applications ranging from cancer therapy to tattoo removal. In some cases the undesirable cells are "tagged" with a photosensitizer (absorber) at the alexandrite wavelength; thus the laser is not significantly absorbed by the normal cells. The laser beam size is not critical since the undesirable cells are self-selected by the beam. The use of a short laser pulse allows heat to destroy the bad cells without the heat spreading to the surrounding cells. In the case of tattoo removal the tattoo itself is absorbing and therefore requires no tagging. Other applications include pollution detection and removal of kidney stones.

14.5   TITANIUM SAPPHIRE LASER

GENERAL DESCRIPTION

The titanium sapphire laser (Ti:Al$_2$O$_3$) is the most widely used tunable solid-state laser. It can be operated over a wavelength range of 660–1,180 nm and thus has the broadest gain bandwidth of any laser. It also has a relatively large stimulated emission cross section for a tunable laser. Titanium ions are typically doped into a sapphire (aluminum oxide) crystal at a concentration of 0.1% by weight. The laser has achieved cw outputs of nearly 50 W and also terawatts of peak power from 100-fs–duration mode-locked pulses. The laser can be flashlamp pumped, but the technique is not efficient owing to the unusually short upper laser level lifetime of the laser crystal, 3.8 μs at room temperature, that does not match well with the longer pulse duration of typical flashlamps. Commercial titanium sapphire lasers are therefore typically pumped with either argon ion lasers (for cw operation) or frequency-doubled Nd:YAG or Nd:YLF lasers (for pulsed operation). The pump absorption band covers the range from less than 400 nm to just beyond 630 nm and peaks around 490 nm, as can be seen in Figure 14-10. The titanium sapphire crystal has high thermal conductivity, good chemical inertness, good mechanical rigidity, and high hardness.

LASER STRUCTURE

An example of a cw Ti:Al$_2$O$_3$ laser using an X-cavity design is shown in Figure 14-11(a). It uses an astigmatically compensated cavity for the Ti:Al$_2$O$_3$ laser crystal (see Section 12.6). In such a cavity design, the crystal typically ranges from 2 to 10 mm in length, depending upon the dopant level, and

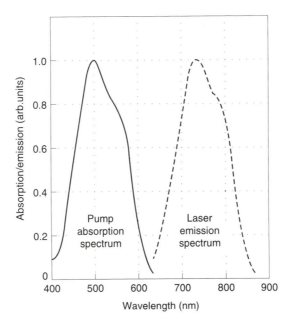

Figure 14-10. Absorption and emission spectra of titanium:sapphire laser amplifier rod

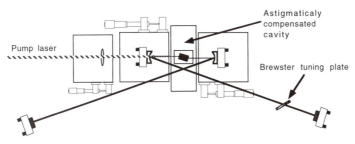

(a) cw Ti:sapphire - X fold configuration

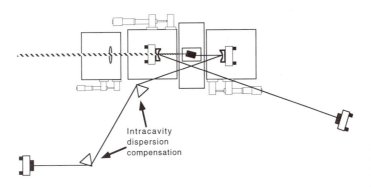

(b) Femtosecond mode-locked Ti:sapphire

Figure 14-11. Diagrams of (a) a cw titanium sapphire laser and (b) a femtosecond mode-locked titanium sapphire laser (courtesy of CREOL)

is arranged with the output faces of the crystal at Brewster's angle. The longer-length crystal lengths with lower doping concentrations are used with higher pumping flux intensities in order to obtain higher power output. Generally, either a cw argon ion laser or a doubled Nd:YAG laser is used as the pumping source. The pump beam enters the cavity from the left, as shown in the figure. A birefringent filter, installed within the cavity at Brewster's angle, can be rotated for wavelength tuning. A modified version of this cavity, shown in Figure 14-11(b), is used to produce mode-locked pulses. It includes two prisms for intracavity dispersion compensation and uses the Kerr lens mode-locking (KLM) technique described in Section 12.3. The necessary aperture within the $Ti:Al_2O_3$ crystal to produce KLM is provided by a separate aperture located next to the crystal (not shown), or simply by the aperturing effect associated with the small diameter of the pump beam. Extremely precise adjustments and alignment of the mirrors and cavity dimensions are essential to maintain a stable mode-locked output of this laser.

EXCITATION MECHANISM

The energy-level structure of this laser is similar to that of a dye laser. The ground state, a $^2T_2$ state, has a broad sequence of overlapping vibrational or vibronic levels extending upward from the lowest level, as shown in Figure 5-16. The first excited state is a ^2E state that also extends upward with a series of overlapping vibronic levels. This energy-level structure is unique for laser crystals in that there are no d-state energy levels above the upper laser level. Thus, the simple energy-level structure involving a 3d electron eliminates the possibility of excited-state absorption, an effect that reduces the tuning range of other tunable solid-state lasers (see Section 8.4).

Excitation therefore occurs from the lowest vibronic levels of the $^2T_2$ ground state (those that are sufficiently populated at room temperature) to the broad range of excited vibronic levels of the ^2E excited state. The population pumped to all of the vibrational levels of the broadband excited state rapidly relaxes to the lowest levels of that state. It then decays back to any one of the vibronic levels of the ground state in a manner that is similar to a dye laser, but with a much lower radiative rate. When the population reaches the excited vibronic levels of the ground state, it very rapidly relaxes to the lowest-lying levels, leaving a distribution dictated by the Boltzmann relationship of (6.11).

The laser energy-level arrangement is effectively a four-level system, as in a dye laser, in which all of the higher-lying vibronic levels of the ^2E state serve as level i of the four-level system described in Section 8.2. The lowest vibrational levels of the ^2E state serve as the upper laser level u. These levels decay to any of the excited vibrational levels of the ground state $^2T_2$, any of which can be considered as the lower laser level l. These levels then rapidly relax to the lowest levels of $^2T_2$ serving as the ground state 0.

APPLICATIONS

Titanium sapphire lasers are used in infrared spectroscopy of semiconductors and in laser radar, rangefinders, and remote sensing. They are used in

TABLE 14-6

TYPICAL TITANIUM SAPPHIRE LASER PARAMETERS

Laser Wavelengths (λ_{ul})	660–1,180 nm
Laser Transition Probability (A_{ul})	2.6×10^5/s
Upper Laser Level Lifetime (τ_u)	3.8 μs
Stimulated Emission Cross Section (σ_{ul})	3.4×10^{-23} m^2
Spontaneous Emission Linewidth and Gain Bandwidth, FWHM ($\Delta\nu_{ul}$)	1.0×10^{14}/s ($\Delta\lambda_{ul} = 180$ nm)
Inversion Density (ΔN_{ul})	6×10^{23}/m^3
Small Signal Gain Coefficient (g_0)	20/m
Laser Gain-Medium Length (L)	0.1 m
Single-Pass Gain ($e^{\sigma_{ul}\Delta N_{ul}L}$)	7–10
Doping Density	3.3×10^{25}/m^3
Index of Refraction of Gain Medium	1.76
Operating Temperature	300 K
Thermal Conductivity of Laser Rod	3.55 W/m-K
Thermal Expansion Coefficient of Laser Rod	5×10^{-6}/K
Pumping Method	optical (flashlamp or laser)
Pumping Bands	380–620 nm
Output Power	up to 50 W (cw), 10^{12} W for 100-fs pulse
Mode	single-mode or multi-mode

medical applications such as photodynamic therapy. They are also used to produce short pulses of X-rays by focusing the mode-locked pulses onto solid targets from which high-density and high-temperature radiating plasmas are produced, plasmas that in turn emit large fluxes of X-rays.

14.6 CHROMIUM LiSAF AND LiCaF LASERS

GENERAL DESCRIPTION

The chromium-doped lithium strontium aluminum fluoride (Cr:LiSAF) and lithium calcium aluminum fluoride (Cr:LiCaF) lasers are broadband tunable lasers in the same category as the alexandrite and titanium sapphire lasers. The Cr:LiSAF laser can be tuned over a wavelength ranging from 780 to 1,010 nm and the Cr:LiCaF laser can be tuned from 720 to 840 nm. Both cw and pulsed output have been obtained from both lasers: a cw output of up to 1.2 W, and a pulsed output of over 10 J with a slope efficiency of 5%. These lasers have relatively long upper-level lifetimes of 67 μs for Cr:LiSAF and 170 μs for Cr:LiCaF, so both can be effectively flashlamp pumped. They have also been laser pumped with AlGaAs diode lasers and argon ion lasers. Both of these fluoride laser materials can be doped to very high concentrations (up to 15% Cr) without affecting the upper laser level

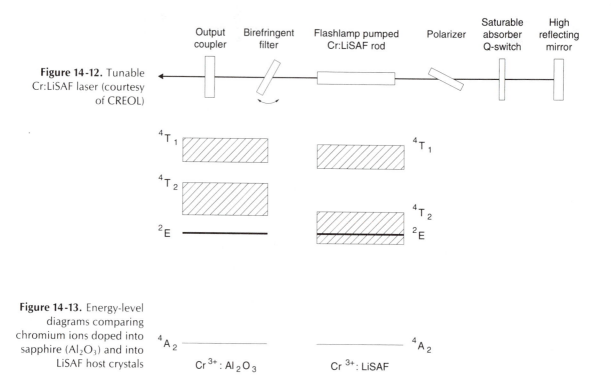

Figure 14-12. Tunable Cr:LiSAF laser (courtesy of CREOL)

Figure 14-13. Energy-level diagrams comparing chromium ions doped into sapphire (Al_2O_3) and into LiSAF host crystals

lifetime, which results in more uniform flashlamp pumping as indicated in Figure 9-18(c). These lasers have also been used as regenerative amplifiers leading to very short–pulse amplification. The laser crystals are chemically stable when treated properly. Their thermal properties are closer to those of Nd:glass than to Nd:YAG. They are durable but not as hard as YAG. Their average power handling capabilities are not as good as $Ti:Al_2O_3$.

LASER STRUCTURE

A diagram of a flashlamp-pumped Cr:LiSAF laser is shown in Figure 14-12. This laser uses a LiSAF rod of 6-mm diameter and 0.1-m length. A birefringent filter is used to tune the wavelength from 750 to 1,000 nm, and a saturable absorber is used as a Q-switching element. The laser is capable of producing 10 MW pulses of 10–20-ns duration at a repetition rate of 10 Hz. This flashlamp-pumped tunable laser is competitive with the alexandrite laser in that it operates in a partially overlapping (but mostly a different) spectral region.

EXCITATION MECHANISM

The Cr:LiSAF laser energy-level diagram is shown in Figure 14-13 compared with the energy-level diagram for the ruby laser crystal described in Section 14.2. The same energy levels are involved: those of the Cr^{3+} ion, including the $^4A_2$ ground state; the $^4T_1$ and $^4T_2$ pumping bands; and the ^2E laser level. In the case of ruby, the ^2E level lies below the pumping bands

TABLE 14-7

TYPICAL CHROMIUM-LiSAF AND CHROMIUM-LiCaF LASER PARAMETERS

	LiSAF	LiCaF
Laser Wavelengths (λ_{ul})	780–1,010 nm	720–840 nm
Laser Transition Probability (A_{ul})	1.5×10^4/s	5.9×10^3/s
Upper Laser Level Lifetime (τ_u)	67 μs	170 μs
Stimulated Emission Cross Section (σ_{ul})	4.8×10^{-24} m^2	1.3×10^{-24} m^2
Spontaneous Emission Linewidth and Gain Bandwidth, FWHM ($\Delta\nu_{ul}$)	8.3×10^{13}/s ($\Delta\lambda_{ul} = 200$ nm)	6.4×10^{13}/s ($\Delta\lambda_{ul} = 130$ nm)
Inversion Density (ΔN_{ul})	3×10^{24}/m^3	7×10^{24}/m^3
Small Signal Gain Coefficient (g_0)	16/m	9/m
Laser Gain-Medium Length (L)	up to 0.15 m	
Single-Pass Gain ($e^{\sigma_{ul}\Delta N_{ul}L}$)	up to 10	up to 4
Doping Density	up to 15% Cr ions	
Index of Refraction of Gain Medium	1.4	1.39
Operating Temperature	300 K	300 K
Thermal Conductivity of Laser Rod	2.91 W/m-K	4.58 W/m-K (χ_1) 5.14 W/m-K (χ_3)
Thermal Expansion Coefficient of Laser Rod	18.8×10^{-6}/K (χ_1) -10.0×10^{-6}/K (χ_3)	22.0×10^{-6}/K (χ_1) 3.6×10^{-6}/K (χ_3)
Pumping Method Pumping Bands	optical (flashlamp or laser) peak at 620 nm ($\Delta\lambda = 120$ nm) peak at 420 nm ($\Delta\lambda = 90$ nm) peak at 280 nm ($\Delta\lambda = 30$ nm)	
Output Power	up to 10-MW pulses of 10-ns duration	
Mode	single-mode or multi-mode	

and thus the populations, when pumped to those bands, rapidly relaxes by nonradiative collisions to the ^{2}E laser level. In the case of Cr^{3+} doped in LiSAF, the ^{2}E level lies within the ^{4}T$_2$ energy, owing to the different forces acting upon the Cr^{3+} ion in the LiSAF crystal host matrix. Hence the upper laser level effectively becomes the broadened ^{4}T$_2$ level, because the radiative rate – and hence the stimulated emission cross section – from the ^{4}T$_2$ level to the ^{4}A$_2$ ground state is much greater than from the embedded ^{2}E level to the ground state. Consequently, the lifetime of the upper laser level is much shorter for Cr:LiSAF (67 μs) than for ruby (3 ms). Taking advantage of ^{4}T$_2$ as the upper laser level gives LiSAF its broad tunability.

APPLICATIONS

CrLiSAF and Cr:LiCaF lasers have many of the same applications that Ti:Al$_2$O$_3$ has. These materials do not have the bandwidth of Ti:Al$_2$O$_3$, but they do have an advantage in that they can be pumped with flashlamps. They may be attractive amplifiers for ultrahigh-power short-pulse generation, since they can be made in large diameters and store large amounts of energy.

14.7 FIBER LASERS

GENERAL DESCRIPTION

The first fiber laser was operated in a Nd-doped glass fiber and was transversely pumped by coiling the fiber around a flashlamp. Later, end pumping was successfully demonstrated, which made pumping with semiconductor lasers a feasible technique. The fiber laser of current interest is the erbium-doped fiber amplifier. Its operating wavelength is 1.53 μm, a wavelength that is very close to the optimum wavelength for fiber-optic communication, 1.55 μm, where fibers have the lowest transmission losses.

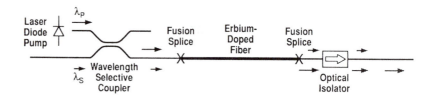

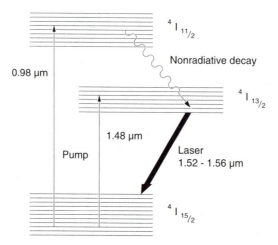

Figure 14-14. Erbium-doped fiber amplifier installed in a long-distance fiber-optic transmission line (courtesy of AT&T Bell Laboratories)

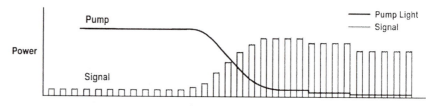

Figure 14-15. Energy-level diagram of an erbium-doped fiber laser

TABLE 14-8

TYPICAL ERBIUM FIBER LASER PARAMETERS

Laser Wavelengths (λ_{ul})	1.53–1.56 μm
Laser Transition Probability (A_{ul})	90/s
Upper Laser Level Lifetime (τ_u)	11 ms
Stimulated Emission Cross Section (σ_{ul})	7×10^{-25} m^2
Spontaneous Emission Linewidth and Gain Bandwidth, FWHM ($\Delta\nu_{ul}$)	3×10^{12} Hz ($\Delta\lambda_{ul} = 25$ nm)
Inversion Density (ΔN_{ul})	2×10^{24}/m^3
Small Signal Gain Coefficient (g_0)	1.35/m
Laser Gain-Medium Length (L)	12 m
Single-Pass Gain ($e^{\sigma_{ul}\Delta N_{ul}L}$)	10^7
Index of Refraction of Gain Medium	1.5
Operating Temperature	300 K
Pumping Method	optical (laser diodes)
Pumping Wavelengths	980 nm ($\Delta\lambda = 18$ nm) and 1,480 nm

LASER STRUCTURE

Figure 14-14 shows a diagram of an erbium-doped fiber amplifier incorporated into a section of a long-distance optical fiber transmission line. It consists of a laser diode pump laser, a wavelength-selective coupler that allows the pump wavelength to enter the fiber transmission system without disturbing the signal, an erbium-doped fiber amplifier spliced into the optical fiber transmission system, and an optical isolator. The diode laser continually pumps the fiber, and as pulses pass through the system they are amplified as shown in the figure while the pump light is depleted.

EXCITATION MECHANISM

An energy-level diagram for the erbium-doped is shown in Figure 14-15. Optical pumping can be seen to occur from the ground state $^4I_{15/2}$ to the $^4I_{11/2}$ state at a wavelength of 0.98 μm or directly to the $^4I_{13/2}$ upper laser state at 1.48 μm. When the pumping occurs to the $^4I_{11/2}$ level, rapid relaxation occurs to the upper laser levels. When pumping is direct to the upper laser state, rapid relaxation occurs to the lowest-lying levels of that state from which laser action is produced. The laser output then occurs in the region of 1.53–1.56 μm. The lower level is similar to that of a dye laser in that it is part of the ground state within which the population is distributed according to the thermal distribution described in (6.11). Consequently, the higher-lying levels of that state are not significantly populated and can therefore serve as the lower level of a population inversion. The population then rapidly decays nonradiatively to the lower levels of that state.

APPLICATIONS

The primary application for the erbium-doped fiber laser is in long-distance communications over fiber-optic networks. It is especially useful in undersea communication links through which the amplification of the optical signal can be accomplished directly by inserting pieces of erbium fiber at appropriate locations in the fiber network, as shown in Figure 14-14. The optically transmitted signal will thus obtain a direct optical "boost" or amplification as it travels from one continent to another. This supersedes the earlier technique of converting the optical signal to an electrical one, amplifying it, and then re-inserting it into the fiber for further optical transmission. Such amplifiers are also being used in ultrashort pulse production, where fiber solitons with pulsewidths as short as 100 fs and pulse repetition frequencies of up to 10 GHz are under development. Because of their simplicity, fiber lasers are currently under consideration as possible replacements for distributed-feedback semiconductor lasers.

14.8 COLOR CENTER LASERS

GENERAL DESCRIPTION

Color center lasers are broadly tunable solid-state lasers that operate in the near infrared at wavelengths ranging from 0.8 to 4 μm. Tuning throughout this range is achieved by using several different color center crystals in sequence as in the case of dye lasers. These lasers are optically pumped by other lasers. The color center laser materials are made by producing *point defects* in an alkali–halide crystal lattice. These defects involve halide ion vacancies in combination with trapped electrons, and can also include doped alkali ion impurities. The laser species is produced by irradiating an alkali-halide crystal with X-rays to produce the point defects, referred to as F centers, F_A centers, F_2 centers, and F_2^+ centers. In each of these cases the point defect produces laser energy levels at which the laser emission occurs in the near infrared. Table 14-9 lists the crystals and their centers (active sites) and other characteristics. The emission and gain broadening is dominated primarily by collisions due to the crystal lattice vibrations (phonon broadening). Since the vibrations are random, the broadening is homogeneous with a typical bandwidth of 100 nm. This large emission bandwidth allows for efficient mode-locking and has produced subpicosecond pulses.

LASER STRUCTURE

A typical color center laser consists of an alkali–halide crystal that has been irradiated with X-rays to produce the F centers (color centers) at random locations within the crystal. The number of color centers produced is

TABLE 14-9

CRYSTAL CHARACTERISTICS FOR COLOR CENTER LASERS

Host Material	Center	Pump Wavelength (μm)	Tuning Range	Maximum Power (W)	Operational Lifetime
LiF	F_2^+	0.647	0.82–1.05	1.8	Days
NaF	$(F_2^+)^*$	0.87	0.99–1.22	0.4	Weeks
KF	F_2^+	1.06	1.22–1.50	2.7	Days
NaCl	F_2^+	1.06	1.4–1.75	1	Days
NaCl:OH	$F_2^+:O^{2-}$	1.06	1.42–1.85	3	Years
KCl:Tl	$Tl^0(1)$	1.06	1.4–1.64	1	Years
KCl:Na	$(F_2^+)_A$	1.32	1.62–1.95	0.05	Months
KCl:K$_2$O	$F_2^+:O^{2-}$	1.32	1.7–1.85	0.06	Months
KCl:Li	$(F_2^+)_A$	1.32	2.0–2.5	0.4	Months
KI:Li	$(F_2^+)_A$	1.7	3.0–4.0	0.006 (pulse)	Unknown
KCl:Na	$F_B(II)$	0.514	2.25–2.65	0.05	Years
KCl:Li	$F_A(II)$	0.514	2.3–3.0	0.2	Years
RbCl:Li	$F_A(II)$	0.647	2.6–3.6	0.1	Years
KCl	N_2	1.06	1.27–1.35	0.04 (pulse)	Months

determined by how long the crystal is irradiated. These color centers remain within the crystal for durations ranging from days to years, depending upon the particular crystal (see Table 14-9), so provision must be made in the laser design to re-irradiate the crystals from time to time. A typical cavity arrangement, shown in Figure 14-16, consists of a color center crystal installed at Brewster's angle between two curved mirrors. The beam reflects from one of the mirrors, passes through a dichroic mirror from which the pump beam is inserted, and passes to a prism and third mirror as shown. The third mirror can then be rotated to provide wavelength tuning over the gain bandwidth of the color center emission profile. The curved mirrors on either side of the crystal focus the pump beam to a spot on the laser crystal 20–35 μm in diameter, in order to produce a gain region of that dimension within the crystal and to achieve the high-pump intensity required to produce significant gain. The laser mode is consequently similar in size to the pump beam in that region of the cavity. Typical laser crystals range from 1 to 3 mm in thickness. The laser crystal is usually installed in a vacuum to allow operation at cryogenic temperature and also to prevent condensation of moisture on the crystal.

EXCITATION MECHANISM

Color center lasers can be associated with a four-level pumping scheme. Figure 14-17 shows a typical energy-level diagram for an F-center laser.

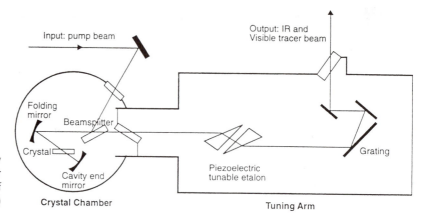

Figure 14-16. Laser cavity arrangement of a color center laser (courtesy of Burleigh Instruments, Inc.)

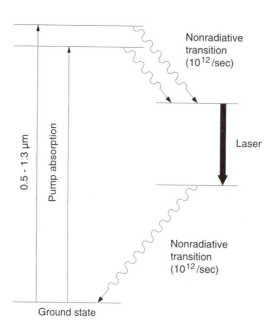

Figure 14-17. Typical energy-level diagram of an F-center color center laser

The pump absorption band typically occurs in the wavelength region of 0.5–1.3 μm. Relaxation to the upper laser level occurs in a time frame of the order of 10^{-12} s. The lifetime of the upper laser level is of the order of a few hundred nanoseconds. The lower laser level relaxes back to the ground state in a time of the order of 10^{-12} s. Consequently, the majority of the population is in either the ground state or the upper laser level, as is typical for many four-level lasers. These lasers require pump intensities of the order of 10^5 W/cm^2.

APPLICATIONS

Tunable middle-infrared lasers are very useful for applications that involve sensing and probing the vibrational transitions of molecules. Color center

TABLE 14-10

TYPICAL COLOR CENTER LASER PARAMETERS

Laser Wavelengths (λ_{ul})	0.8–4 μm
Laser Transition Probability (A_{ul})	5×10^6/s
Upper Laser Level Lifetime (τ_u)	200 ns
Stimulated Emission Cross Section (σ_{ul})	$\approx 10^{-20}$ m^2
Spontaneous Emission Linewidth and Gain Bandwidth, FWHM ($\Delta\nu_{ul}$)	20×10^{12}/s ($\Delta\lambda_{ul} = 200$ nm)
Inversion Density (ΔN_{ul})	3×10^{22}/m^3
Small Signal Gain Coefficient (g_0)	300/m
Laser Gain-Medium Length (L)	1.5–2.0 mm
Single-Pass Gain ($e^{\sigma_{ul}\Delta N_{ul}L}$)	1.6
Doping Density	2×10^{23}/m^3
Index of Refraction of Gain Medium	1.49
Operating Temperature	77 K
Pumping Method	Nd:YAG or argon ion laser
Output Power	up to 500 mW (cw)
Mode	TEM$_{00}$

lasers are attractive from the standpoint of their high power output (up to 150 mW) combined with tunability. Their limited use is largely due to their size, the requirement for an additional pump laser, and cryogenic cooling.

14.9 SEMICONDUCTOR DIODE LASERS

GENERAL DESCRIPTION

Semiconductor lasers are small, efficient laser devices with typical dimensions of less than a millimeter. They operate on wavelengths ranging from 0.6 to 1.55 μm, depending upon the materials of the laser medium. Lasers operating in the 0.5-μm-wavelength region are also under development; however, at the time of this writing these lasers are not yet commercially available. Semiconductor lasers generally operate on a cw basis. Pulsed operation is possible but does not produce significantly higher power than the cw lasers. The typical gain bandwidth, corresponding to the recombination emission linewidth, is of the order of 20 nm, but the laser bandwidth can be significantly reduced by cavity or resonator effects as we shall describe (see also Chapters 10 and 12). Although most semiconductor lasers can be optically pumped, electrical pumping is much more practical. Hence, all commercial semiconductor lasers are operated by passing an electrical current through the laser medium.

The small size of semiconductor lasers is made possible by the extremely large population inversion densities ΔN_{ul} that can be produced on a steady-state basis by locating two specially doped semiconducting materials directly adjacent to each other to form a junction and placing a forward-bias voltage between them, as indicated in Figures 5-23 and 5-24. These two doped materials include one with an excess of electrons (n-doping) and one with an excess of holes (p-doping), as discussed in detail in Section 5.4. In all commercial semiconductor lasers, thin layers of material (referred to as *cladding layers*) are also added to the junction, as shown in Figure 5-25. These cladding layers, typically about 1 μm thick, are composed of materials having both a wider energy bandgap and a lower index of refraction than the materials of the junction region. Such lasers are referred to as *double heterostructure* (DH) devices. The purpose of the additional layers is to: (1) more effectively guide the laser mode through the material by forming a waveguide; (2) reduce the lateral dimensions of the region where the current flows, thereby reducing heat dissipation in the material; and (3) adapt the crystalline lattice spacing of the doped laser material to the slightly different lattice spacing of the substrate on which the junction is grown. (See Section 5.4 for more details.) When the electrical current flows through the layers, most of the energy deposition occurs in the active region of the junction where the electrons are injected and the gain is produced, since that is the area where the highest electrical resistance occurs. Thus the added layers provide additional desired features without producing any significant extra heat loss.

The first useful semiconductor devices were composed of GaAs materials that emit in the near infrared at 0.8 μm. Since then, laser emission wavelengths have been extended further into the infrared to 1.55 μm, based upon InP/InGaAsP–layered semiconductor materials (discussed in Section 5.4) for use in optical communications. Laser wavelengths have also been extended to 0.63 μm or (630 nm) based upon GaAs/AlGaAs–layered materials for applications such as laser pointers and bar-code readers. Also, a new class of semiconductor lasers has recently been demonstrated in the green and blue in the region around 0.5 μm (500 nm) using Zn, Cd, Se, and S materials, although these lasers are not yet commercially available at the time of this writing.

The gain coefficient (g_0) in a semiconducting laser is generally between 5,000 and 10,000 m^{-1}. Thus, even for a typical gain length of 1 mm or less, the gain per pass is large enough to overcome the large inherent distributed loss (of the order of 2,000 m^{-1}) within the gain medium. The gain bandwidth of a semiconductor laser is typically of the order of 10^{13} Hz or 20 nm, corresponding to the emission linewidth of the recombination radiation; the exception is quantum-well lasers, which have a somewhat narrower bandwidth of the order of 5 nm. The recombination emission broadening within the laser gain region (the junction) is homogeneous. Therefore, the laser wavelength will generally occur at the peak of the recombination emission profile unless special frequency-selective techniques (described in what follows) are used to select the desired laser frequency or wavelength. Figure 14-18 shows a typical output of a semiconductor laser versus current

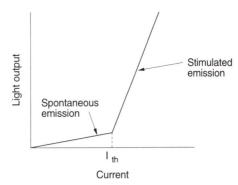

Figure 14-18. Output flux from a semiconductor laser versus laser current

flowing through the laser. At low currents where the absorption exceeds the gain, the recombination emission from the junction region increases linearly with current. At the threshold current I_{th} (typically quoted in mA) where the gain exceeds the losses mentioned previously, laser action begins and the output increases dramatically, as shown in the figure.

LASER STRUCTURE

The length in the axial or laser direction of the semiconductor laser is typically about 0.2–1 mm, and the material providing the gain is uniform over that length. In one of the two transverse directions of a DH laser, the additional semiconductor materials are grown in layers of various thicknesses ranging from 5 to 1,000 nm to form the heterojunction. The additional layers are used to confine the electron–hole recombination region (so as to minimize the generation of excess heat) and also to confine the optical beam. In the other lateral direction, material boundaries are fabricated with lithographic techniques to widths of the order of a few hundred microns to provide confinement and guiding of the electrical current and/or the optical beam, as shown in Figure 14-19. The heterojunction is typically grown upon a substrate material such as either GaAs or InP (see Chapter 5), with a metalization layer under the substrate to provide for electrical contact. Above the heterojunction is another metalization layer for electrical contact. The current density flowing perpendicular to the plane of the junction is of the order of 800 A/cm^2, whereas the maximum current that can be used before extensive thermal damage occurs is of the order of 50 mA. Thus the current must be confined in the lateral direction in the plane of the junction, by devising a narrow electrical contact that minimizes the region where current flows. This electrical contact takes the form of either a narrow strip of metal above the heterojunction or a broad uniform layer under which an insulator is applied, leaving a narrow stripe region without insulation through which current can flow as indicated in Figure 14-19.

The laser cavity can be produced in one of two ways. The first is the typical Fabry–Perot cavity, but in this case the mirrors are produced at the ends of the laser gain medium by cleaving the ends of the semiconductor

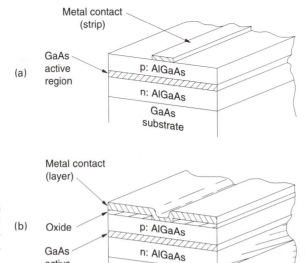

Figure 14-19. Two examples of lateral confinement of the laser current in a semiconductor laser: (a) metal stripe electrode; (b) oxide insulating layer

crystal perpendicular to the optical axis of the crystal (a natural cleavage plane). Because of the high index of refraction of the material (typically $\eta \approx 3.5$), the reflectivity at those cleaved interfaces is of the order of 30%. The reflectivities can be altered by adding dielectric coatings to the surfaces. In commercial semiconductor lasers, the laser cavity length L typically ranges from 0.2 to 1.0 mm. This cleaved type of cavity produces a relatively broad spectrum laser output that can consist of many longitudinal modes, although typically a single mode at the wavelength of the highest gain will dominate because of homogeneous broadening. The second type of cavity is one that produces a very narrow–spectrum, single-frequency output. This can be achieved by one of several techniques that are described in a later paragraph entitled "Frequency Control of Laser Output."

There are several general types of semiconductor laser structures. These include both gain- and index-guided structures, using either the traditional heterojunctions or quantum-well active regions; semiconductor arrays; and surface-emitting lasers. We next consider these various structures in more detail.

STRUCTURE DESIGN IN THE DIRECTION PERPENDICULAR TO THE PLANE OF THE JUNCTION   In the direction perpendicular to the plane of the junction, the laser medium consists of several very thin layers of various types of semiconducting material of varying thicknesses. These thin layers, referred to as *epitaxial layers,* are grown on top of the substrate by one of three techniques: (1) liquid phase epitaxy (LPE); (2) molecular beam epitaxy (MBE); or (3) metallorganic chemical vapor deposition (MOCVD). LPE is used for growing typical heterojunction structures, but the thinner quantum-well layers must be grown by the more precise methods of either

MBE or MOCVD. (See the references listed at the end of this chapter for more detailed discussion of semiconductor growth techniques.)

• HOMOJUNCTION LASERS   Homojunction lasers consist of a single junction of the n- and p-doped materials. Because of the large amount of heat dissipation and the gradual tapering off of the gain in directions away from the junction, this type of device can be effectively operated only at very low temperatures (well below room temperature). The first semiconductor lasers were operated with this junction arrangement, but they are not practical devices and so will not be discussed further.

• HETEROJUNCTION LASERS   Heterojunction lasers consist of several layers of various materials: semiconductor materials, both doped and undoped; insulating layers, usually in the form of oxides; and metallic layers for conduction of current. A single layer in the center of these layers – the active layer where gain is produced – is a direct bandgap material that is an efficient radiator, while the adjacent layers (cladding layers) can be indirect bandgap material (see Section 5.4). Heterojunction or heterostructure devices are fabricated from a range of lattice-matched semiconductor materials, which are also described in Chapter 5. The materials most often used are either $GaAs/Al_xGa_{1-x}As$ or $In_{1-x}Ga_xAs_yP_{1-y}/InP$, which are all III–V semiconductor alloys (obtained from the third and fifth columns of the periodic table of the elements). The x and y indicate the fraction or concentration of impurity material in the layer. The $GaAs/Al_xGa_{1-x}As$–based systems are closely lattice matched to and therefore grown upon a GaAs substrate. They typically have either a pure Ga/As active region or an $Al_xGa_{1-x}As$ active region with $x < 0.15$. These junction materials provide laser wavelengths ranging from 0.78 to 0.87 μm.

The $In_{1-x}Ga_xAs_yP_{1-y}/InP$–based systems can be lattice matched to InP substrates. In these lasers, the active region is $In_{1-x}Ga_xAs_yP_{1-y}$, with InP forming the cladding layers and substrate. The choice of the fractions x and y must be selected to achieve the appropriate lattice matching and the laser wavelength. The typical wavelength range for these lasers is from 1.1 to 1.65 μm. This range includes the optimum wavelengths for transmission of laser pulses through optical fibers of 1.3 μm, where the minimum material dispersion occurs in a quartz-based fiber, and 1.55 μm, where the lowest loss occurs. Extending the wavelengths into the visible spectrum has been achieved by using $(Al_xGa_{1-x})_{1/2}In_{1/2}P$ lasers that are lattice matched to GaAs. The higher concentrations of Al are used for the cladding layers, and either lower concentrations of Al or concentrations of $Ga_{1/2}In_{1/2}P$ are used as the active region. Such devices have achieved laser operation at wavelengths as short as 0.63 μm.

Lasers in II–VI compounds such as ZnSe have recently been produced at wavelengths ranging from 0.460 to 0.530 μm at cryogenic temperatures. A continuous room-temperature device has also recently been produced in

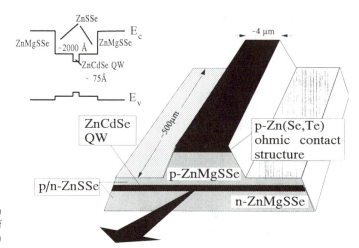

Figure 14-20. Blue–green laser diode (courtesy of Brown University)

the wavelength range from 0.500 to 0.510 μm with an output power of up to 10 mW. This achievement relied upon the manipulation of the electronic and optical properties of these wide-bandgap semiconductors (which are more similar to insulators – see Figure 5-18) by use of quantum-well structures and superlattices, as shown in Figure 14-20. Extensive efforts are under way in a number of research groups to address the reliability issues associated with these lasers.

Another class of wide-bandgap semiconductor materials based upon the GaN compound have led to the development of efficient broadband blue light-emitting diodes (LEDs). These devices will no doubt be used for diode pumping of many types of solid-state lasers, and their spectral characteristics will most likely be extended into the ultraviolet. They also may eventually lead to blue and ultraviolet semiconductor lasers.

Longer-wavelength semiconductor lasers at 2.2 μm have been operated continuously at 30°C with GaInAsSb/AlGaAsSb active regions lattice matched to a GaSb substrate. A new type of laser, referred to as a *quantum cascade* laser, has been made to operate in the 4–10-μm wavelength region. These lasers emit on transitions between quantum-well levels, such as between levels 2 and 1 or 3 and 2 of Figure 5-26. Such lasers have narrow emission linewidths due to the narrow nature of the quantum energy levels. At longer wavelengths, another new type of laser is the *intervalence band* laser operating in p-type semiconductors. It involves transitions within the valence band, typically in p-doped germanium, from one energy location to another. The laser transition is homogeneously broadened, and since there is a continuous spectrum of energies, the output is continuously tunable from 50 to 250 μm.

• QUANTUM-WELL LASERS   In Section 5.4 we pointed out that reducing the active-layer thickness (in the direction perpendicular to the plane of the junction) to dimensions of the order of 10 nm or less changes the conduction and valence bands from being normal, parabolic, continuous bands (as shown in Figure 5-17) to step-function bands with discrete energies (Figure 5-26). The net effects of this in a semiconductor laser device are (1) to decrease the threshold current by increasing the density of states associated with the upper laser level population, and (2) to minimize the volume within which heat is generated, since most of the heating occurs within the active region. The laser wavelength can also be changed by varying the quantum-well thickness. Varying the composition of the quantum well can also produce different laser wavelengths by changing the energy bandgap as described in Section 5.4. This is accomplished by using materials of different lattice dimension than that of the substrate, thereby disrupting the regular alignment of the atoms. The layers are effectively "strained," so this type of layer is referred to as a *strained quantum well*. Although making lasers with ultrathin quantum-well layers requires more precise control of the fabrication process, the advantages of these lasers are so great that most new laser designs incorporate such quantum-well gain regions.

STRUCTURE DESIGN IN THE LATERAL DIRECTION PARALLEL TO THE PLANE OF THE JUNCTION There are two types of structures used to determine the transverse or lateral dimension of the useful gain region of a semiconductor laser. In one case, the region over which gain is produced is limited and thus serves to define the laser width; this is referred to as *gain guiding*. Alternatively, an index-of-refraction change is fabricated into the laser so that the beam is confined by reflection at that interface; this is referred to as *index guiding*.

• GAIN-GUIDED STRUCTURES In heterostructure lasers, confinement of the current to a narrow strip (Figure 14-19) will limit the amount of current flowing in the laser and thus prevent thermal damage to the semiconductor. Such a localized current region also limits the laser gain region and thus the laser mode dimension in the lateral direction. Thus, when the current flows through the semiconductor layers and into the active region, it produces gain only in a narrow Gaussian-shaped stripe, as shown in Figure 14-21. When excitation is applied, the laser mode develops where the gain exists. This minimizes the amount of current required to provide a specific gain by restricting the lateral extent of the current flow. A disadvantage of this type of device is that the effective width of the gain region broadens as the current applied to the structure is increased. In many applications it is desirable to control the width independently of the current, as described in what follows.

• INDEX-GUIDED STRUCTURES Confinement of the laser mode in the lateral direction can also be achieved by fabricating stripes of material of lower index of refraction adjacent to the gain region, as shown in the ridge waveguide structure of Figure 14-22(a). A ridge of p-type material such as p:AlGaAs is formed above the active region (GaAs, for example), with the oxide coating adjacent to the ridge as shown. This provides a low-index-

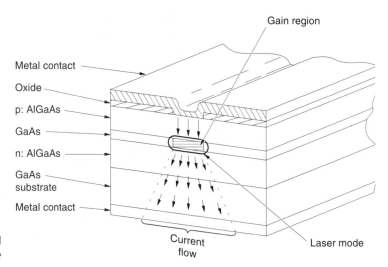

Figure 14-21. Gain-guided laser structure

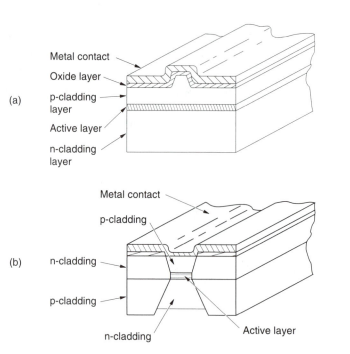

Figure 14-22. Two examples of index-guided laser structures

of-refraction material adjacent to the higher-index p-type material; the laser mode extending above the active region is confined by this index barrier. Thus the laser operates in a waveguide mode that is determined by the width of the index barrier. This type of laser has many patented configurations relating to the details of the index barrier. The buried heterostructure laser is shown in Figure 14-22(b). Other types not shown include the channeled-substrate planar laser, the buried-crescent laser, and the dual-channel planar-buried heterostructure laser.

HIGH-POWER SEMICONDUCTOR DIODE LASERS High-power output from semiconductor lasers can be classified into two categories: *single-mode* operation, with powers exceeding 100 mW, and *multi-mode* output of many watts, achieved in semiconductor arrays.

• SINGLE-MODE LASERS   It might be expected that high-power single-mode operation could easily be achieved by increasing the gain volume and by increasing the current to produce more gain per unit volume. However, such high power operation in both single longitudinal and transverse modes is limited by three effects: (1) spatial hole burning (see Section 10.3), which leads to multimode operation; (2) catastrophic damage of the cleaved facets (usually the laser mirrors) due to absorption of the laser beam; and (3) temperature increase in the active region as the current is increased, which limits the maximum output power.

The design of the structure must therefore take into account these three limitations, which can be overcome by the following techniques. The active-layer width in the direction perpendicular to the plane of the junction can be reduced from a conventional value of 0.2 μm to 0.03 μm. This makes the gain and thus the beam waist smaller in the center of the laser, thereby reducing the chance for multimode operation, but also makes the beam much larger at the mirror facets, reducing the power density at the mirrors. Absorption at the facets can be eliminated by using nonabsorbing mirrors (NAM). These can be effected by depositing a different material from that of the gain medium in the mirror region and so prevent the absorption due to having the gain medium adjacent to the mirrors (see Section 8.4). Also, reflectivity of the facet mirror can be reduced to lower the beam intensity within the cavity by using dielectric coatings, consisting of alternate layers of high- and low-index material. Finally, temperature increases in the active region may be controlled by proper heat sinking.

Geometries of several high-power commercial lasers include the quantum-well ridge (QWR) waveguide laser, the twin-ridge structure (TRS), the buried twin-ridge structure (BTRS), the current-confined constricted double-heterostructure large–optical cavity (CC-CDH-LOC) laser, and the buried V-groove–substrate inner strip (BVSIS) laser. These lasers all have a large spot size at the mirrors, low threshold current, high quantum efficiency, and a contribution of low- and high-reflectivity mirror coatings. They also have a thin active layer that in many cases is of quantum-well dimensions (i.e., less than 10 nm). The reader is referred to the first reference on semiconductor lasers for further information on these devices.

• MULTI-MODE ARRAYS   This type of semiconductor laser consists of a row of heterojunction lasers fabricated adjacent to each other. When the distance between them is from 5 to 10 μm, the lasers can all be made to operate in phase owing to the overlap of modes from adjacent lasers, as shown in Figure 14-23. All of the lasers operating simultaneously produces a stripe of coherent laser output. When the lasers are separated by greater distances, the output from each individual laser is phased randomly with respect to the other lasers. The wider separation allows for better heat removal

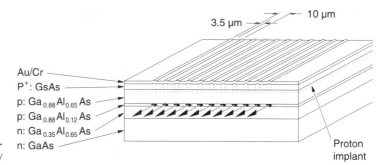

10 µm

3.5 µm

Au/Cr

P⁺: GsAs

p: $Ga_{0.88}Al_{0.65}As$

p: $Ga_{0.88}Al_{0.12}As$

n: $Ga_{0.35}Al_{0.65}As$

n: GaAs

Proton implant

Figure 14-23.
Semiconductor laser array

and is thus more favorable for high-power output. Continuous-wave power of up to 10 W has been achieved from a laser with dimensions of the order of 1 mm³. The principal problem in the operation of such a laser is removing heat generated by the ohmic heating loss in the junction regions of all of the individual lasers. This is accomplished by designing special heat sinks that are attached to the laser.

SURFACE-EMITTING LASERS Semiconducting lasers emitting in a direction normal to the axis of the laser gain medium are referred to as surface emitting lasers (SELs). They can be fabricated in large two-dimensional arrays, which proves useful from two standpoints. First, if such lasers can be individually turned on and off and so act as individual amplifiers, they can be used in applications associated with optical memory, optical computing, and optical data storage. Second, when a large number of diode lasers within an array are lasing simultaneously, the extremely high-power output makes such lasers suitable for optically pumping other solid-state lasers, such as the Nd:YAG laser.

Surface-emitting lasers are fabricated in two different types of structures. One is a semiconductor laser with a distributed grating coupler, where the grating redirects a portion of the laser beam out of the cavity in a direction normal to the gain axis of the laser. The other is a vertical cavity laser that is fabricated with the laser gain axis mounted perpendicular to the semiconductor substrate, with the mirrors attached at the ends of the gain medium as part of the fabrication process.

An example of the structure of a distributed grating surface-emitting laser is shown in Figure 14-24(a). If a second-order grating is fabricated within the laser then the first-order diffraction occurs in a direction perpendicular to the grating surface, thereby providing the laser output as indicated in the figure. Such gratings are presently fabricated by creating holographic interference patterns on the laser surface using Ar ion and He–Cd lasers. Electron-beam writing can also be used to generate such patterns. The vertical cavity laser is shown in Figure 14-24(b). In this laser the gain length is limited to very short dimensions. High-reflectivity mirrors are therefore necessary to allow the gain to exceed the losses within the cavity, as suggested by (7.60). Both thin metal films and alternate high- and low-index dielectric layers have been used to create such mirrors. One technique

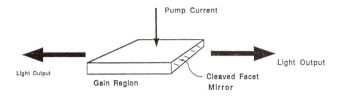

Conventional Edge-Emitting Diode Laser

(a)

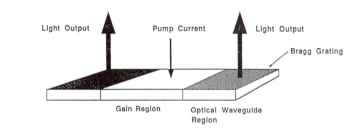

Grating Surface–Emitting Diode Laser

(b)

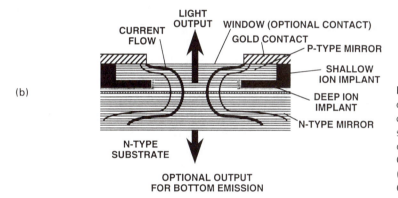

Figure 14-24. Two types of surface-emitting semiconductor lasers: (a) grating surface-emitting (courtesy of David Sarnoff Research Center); (b) vertical cavity (courtesy of Vixel Corporation)

for making such lasers is to deposit the various layers of material uniformly over the surface of the entire structure and then etch away the material from around the individual lasers to produce structures similar to that shown in Figure 14-24(b).

FREQUENCY CONTROL OF LASER OUTPUT Narrow-frequency laser output in semiconductor lasers is achieved by forcing operation in a single longitudinal mode. Because the cavity length in such lasers is typically somewhat less than 1 mm, the longitudinal mode separation is 50–100 GHz according to (10.31). This is significantly greater than the spacings for most other types of lasers, in which the cavity length ranges from 0.1 to 1 m. Therefore, frequency-selective elements have fewer modes among which to discriminate. Thus, for standing-wave cavities, the frequency is typically selected by use of a coupled cavity as shown in Figure 14-25(a). Such an arrangement has two separate Fabry–Perot cavities, which is equivalent to installing a Fabry–Perot etalon within a longer cavity as described in Section 12.5. The second means of frequency discrimination involves using either distributed feedback or distributed Bragg reflection as indicated

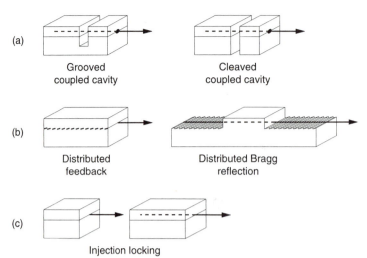

(a)

Grooved
coupled cavity

Cleaved
coupled cavity

(b)

Distributed
feedback

Distributed Bragg
reflection

Figure 14-25. Three types
of frequency selection for
narrow-frequency
semiconductor lasers

(c)

Injection locking

in Figure 14-25(b), both of which were also described in Section 12.5. The third technique is injection locking, which consists of inserting a narrow-frequency laser into the existing laser as indicated in Figure 14-25(c). The intensity of the injection laser must be higher than the spontaneous emission of the existing laser so that the frequency of the injecting laser will dominate. Control of the transverse modes can be accomplished by altering the geometry of the laser gain medium – in particular, by control of the lateral dimensions of the active region, or the index-guiding regions, if applicable – or by use of an external cavity.

EXCITATION MECHANISM

Excitation is provided when a forward-bias voltage of a few volts is applied across the semiconductor layers in a direction normal to the layers. The resulting electric field attracts additional electrons from the n-type material and holes from the p-type material into the junction region, where they produce recombination radiation. This radiation is equivalent to the spontaneous emission of other laser gain media. The upper laser level density is the density of electron–hole pairs entering the junction region before the pairs have recombined. The density N_l of the lower laser level is the density of existing electron–hole pairs in the junction region before current is applied to the junction. These pairs contribute to a large internal absorption α_i within the laser gain medium that must be overcome before net gain is produced. From (7.60) we can therefore obtain an expression for the threshold gain coefficient g_{th} of the semiconductor laser as

$$g_{\text{th}} = \alpha_i + \frac{1}{2L} \ln\left(\frac{1}{R_R R_F}\right),$$

where R_R and R_F are the reflectivities of the rear and front mirrors of the laser (typically the cleaved end surfaces of the semiconducting material) and L is the length of the gain medium. The scattering losses (a_1 and a_2 of eqns. 7.60 and 7.61) that occur in a semiconductor laser are distributed within the gain medium and can therefore be combined with the free-carrier absorption to contribute to α_i, which is of the order of 0.2 m^{-1}. Note that the threshold gain is related to the absorption losses as well as to the reflectivities of the mirrors and the cavity length. The inverse relationship of the gain to the cavity length occurs because for longer cavity lengths the beam will build up to a higher intensity (owing to the increased gain length) as it passes between the mirrors before it is reduced by the mirror transmission losses.

When an electrical current is applied to the semiconductor, the recombination emission (spontaneous emission) increases linearly with current until a threshold current I_{th} is reached, at which point the gain equals the threshold gain g_{th}. As the current is increased above this point, stimulated emission is produced and emitted from the laser such that the output increases linearly with current, but at a much higher rate of increase than the spontaneous emission, as shown in Figure 14-18.

The objective of semiconductor laser design is to minimize the electrical current required to produce sufficient gain to overcome the large inherent losses of the material. Minimizing this current is achieved by reducing the lateral dimensions of the recombination region, both in the direction of the layers and orthogonal to the layers. Current can be minimized in the direction of the layers by reducing the active-layer region to dimensions of 100–200 nm in heterojunction devices. In recent years this dimension has been reduced even further, to 5–10 nm in quantum-well devices. Quantum-well devices also provide a narrower emission bandwidth, which in turn increases the effective laser gain for a given current. Quantum-well devices have been made to operate at threshold currents as low as 0.5 mA.

APPLICATIONS

Perhaps the application that was initially the driving force in the development of semiconductor lasers is in the field of long-distance communications. It was shown in the early 1970s that optical fibers, free of impurities, could transmit certain wavelengths of light at very low loss and were also relatively inexpensive to fabricate, since their raw materials consisted of SiO_2 (ordinary sand). Extensive development of such lasers in the wavelength region of 1.3 μm (where the fibers have minimum dispersion) and 1.55 μm (where they have the lowest losses) has led to undersea fiber-optic communication systems all over the world. The use of semiconductor lasers as sources in local area networks (LANs) is also a growing field. These will include high-speed computer networks, avionic systems, satellite networks, and high-definition television. It is also envisioned that someday all homes will be interconnected with fiber-optic cables for processing of video, audio, and computer signals. Because the interconnects, switches, couplers, et cetera associated with such systems will have losses, higher-

TABLE 14-11

TYPICAL SEMICONDUCTOR DIODE LASER PARAMETERS

Laser Wavelengths (λ_{ul})	0.5–1.55 μm
Laser Transition Probability (A_{ul})	$\approx 10^9$/s
Upper Laser Level Lifetime (τ_u)	10^{-9} s
Stimulated Emission Cross Section (σ_{ul})	10^{-19} m^2
Spontaneous Emission Linewidth and Gain Bandwidth, FWHM ($\Delta\nu_{ul}$)	2.5 to 10×10^{12}/s ($\Delta\lambda_{ul} = 5$ to 20 nm)
Inversion Density (ΔN_{ul})	10^{23} to 10^{24}/m^3
Small Signal Gain Coefficient (g_0)	10^4–10^5
Laser Gain-Medium Length (L)	200–500 μm
Single-Pass Gain ($e^{\sigma_{ul}\Delta N_{ul}L}$)	10^3–10^4
Index of Refraction of Gain Medium	3.4
Operating Temperature	300 K
Pumping Method	electrical
Output Power	milliwatts to watts
Mode	high-order (elliptical)

power semiconductor lasers will be needed as sources. All of these fiber-optic systems will provide a large market for semiconductor lasers.

Although the field of communications was envisioned as the largest application of semiconductor lasers, at the present time the use of such lasers in compact disk players constitutes their largest single market. There are millions of such players in homes around the world. The semiconductor laser is the "needle" or probe that reads the information from the compact disk, as a result of irradiating the small grooves in the disk. The digitally modulated radiation that is scattered from the disk is collected by an optical sensor and processed through the audio amplifier system. This disk technique is also used for information storage and retrieval in computers and other systems. Other uses of semiconductor lasers include high-speed printing, free-space communication, pump sources for other solid-state lasers (including fiber amplifiers), laser pointers, and various medical applications.

REFERENCES

DYE LASERS

F. J. Duarte and L. W. Hillman, eds. (1990), *Dye Laser Principles.* New York: Academic Press.

SOLID-STATE LASERS

W. Koechner (1992), *Solid State Laser Engineering,* 3rd ed. New York: Springer-Verlag.
P. F. Moulton (1992), "Tunable solid state lasers," *Proceedings of the IEEE* 80: 348–64.

SEMICONDUCTOR LASERS

S. W. Corzine, R. H. Yan, L. H. Coldren, and P. Zory, eds. (1993), *Quantum Well Lasers*. Orlando: Academic Press.

P. J. Delfyett and C. H. Lee (1991), "Semiconductor injection lasers," in *Encyclopedia of Lasers and Optical Technology*. New York: Academic Press.

P. Derry, L. Figueroa, and C. S. Hong (1995), "Semiconductor lasers," in *Handbook of Optics,* vol. I, 2nd ed. (M. Bass, ed.). New York: McGraw-Hill, Chapter 13.

"Far Infrared Semiconductor Lasers," a collection of papers in this special edition of the *Journal of Optical and Quantum Electronics* 23, 1991.

SECTION VI

FREQUENCY MULTIPLICATION OF LASER BEAMS

FREQUENCY MULTIPLICATION OF LASERS AND OTHER NONLINEAR OPTICAL EFFECTS

15

SUMMARY Lasers generate coherent radiation at many wavelengths ranging from the millimeter wavelength region to the soft–X-ray spectral region. During the time that lasers were under development, many wavelengths were not available from lasers. A major effort consequently developed to use the nonlinear optical effects in materials (mostly solid optical materials) to generate frequencies that were not available and to develop the capability of generating tunable coherent beams. Thus an entire field evolved associated with the generation of coherent wavelengths by using lasers as the initiating source to produce other useful wavelengths via nonlinear interactions of the laser beam with various materials. This field involves mostly second- and third-order nonlinear processes, which are described briefly in this chapter. Such processes have been used very successfully in many commercial products. Frequency doubling and tripling of near-infrared solid-state lasers is a particularly significant application of these processes. The reader may consult the references at the end of the chapter for more detailed discussions.

15.1 WAVE PROPAGATION IN AN ANISOTROPIC CRYSTAL

The optical properties of most materials are isotropic: a beam of light interacting with the material will be affected in the same way no matter which angle the beam is directed with respect to the material. This is true of gases, liquids, and most solids. However, there are a number of crystalline solids that are anisotropic. The two types of anisotropic crystals are uniaxial and biaxial crystals, where each category is a function of symmetry. Crystals of cubic symmetry are all isotropic. Crystals that have trigonal, tetragonal, and hexagonal symmetries are all uniaxial, whereas crystals with orthorhombic, monoclinic, and triclinic symmetries are always biaxial.

In an isotropic crystal the index of refraction of the material is the same in every direction. Thus, the velocity of a beam passing through the material also is the same in every direction. In an anisotropic crystal, there exists an orientation of the crystal for which two orthogonal polarizations of a beam would "see" two different indices of refraction in the material and thereby propagate at two different velocities. In a uniaxial crystal, there exists one orientation of the crystal for which the index of refraction and wave velocity are independent of the polarization of the beam; this defines the direction of the optic axis (*c*-axis) of the crystal. For biaxial crystals, there is no orientation for which the index is the same for two orthogonal polarizations.

We recall the following expression for a plane wave from Chapter 2:

$$\mathbf{E} = \mathbf{E}_0 e^{-i(\mathbf{k}\cdot\mathbf{r}-\omega t)}. \tag{2.48}$$

In this expression, $\mathbf{k}$ is the vector that defines the direction of propagation of the wave. This vector can be expressed in three orthogonal components as $\mathbf{k} = k_x\hat{i} + k_y\hat{j} + k_z\hat{k}$. Also, $\mathbf{r}$ is the position vector indicating a specific spatial location defined as $\mathbf{r} = x\hat{i} + y\hat{j} + z\hat{k}$. From Chapter 2 we remember that the phase velocity of the wave is defined as $v = \omega/k$. Thus, since the magnitude of $\mathbf{k}$ is associated with the velocity, we can conclude that for an isotropic crystal the values of k_x, k_y, and k_z would all be the same. We have stated that anisotropic crystals fall into two categories. For a *uniaxial* crystal there is a specific orientation of the crystal in which the k vectors (and thus the indices of refraction) associated with two of the orthogonal directions are identical, while the other k vector and its index of refraction, associated with the third orthogonal direction, has a different value from that of the other two directions. The direction of that third vector defines the optic axis of such a crystal. If the index of refraction associated with the two identical axes is smaller than that of the third axis, then the crystal is labeled positive uniaxial; if that index is greater than the index of the third direction, the crystal is referred to as negative uniaxial. A *biaxial* crystal, when oriented properly, has three different orthogonal k vectors and thus three different indices of refraction. Many optical components take advantage of the nonsymmetric properties of these types of crystals.

Just as we described the relationship of momentum and wavelength for an electron in (3.15), we also have a similar relationship for a photon:

$$\lambda = \frac{h}{p}, \tag{15.1}$$

where p is the momentum of the photon. Solving for p and using the relationship $k = 2\pi/\lambda$ we find that

$$p = \frac{h}{\lambda} = \frac{hp}{2\pi} = \hbar k. \tag{15.2}$$

Since momentum has directional properties and is thus a vector quantity, we can write

$$\mathbf{p} = \hbar\mathbf{k}. \tag{15.3}$$

For a wave incident upon a material boundary, such as a beam of light passing from air into a crystal, conservation of the momentum of the wave must be considered. If the beam is incident at an angle with respect to the normal to the surface of the material, it can be divided into vector components that are normal to the surface and parallel to the surface. Conservation of momentum dictates that both the normal component and the transverse (parallel) component of the momentum must be separately conserved at the boundary. The transverse component of the momentum $\mathbf{p}$ must therefore be continuous across the boundary. In other words, there cannot be an abrupt jump or change in the transverse momentum as the beam enters the medium nor likewise any abrupt change in the transverse component of $\mathbf{k}$. We will use this property when describing the process of phase matching later in this chapter.

15.2 POLARIZATION RESPONSE OF MATERIALS TO LIGHT

In Chapter 2 we defined the polarization vector $\mathbf{P}$ as the dipole moment per unit volume that is induced in a material in response to an electric field applied to the material. The relationship between $\mathbf{P}$ and the applied electric field $\mathbf{E}$ was expressed in (2.6) as

$$\mathbf{P} = \chi \epsilon_0 \mathbf{E}. \tag{2.6}$$

The factor χ was suggested to be à useful parameter when considering the material response to electric fields oscillating in the optical frequency range (10^{13}–10^{15} Hz). For symmetric materials such as glass, χ is a simple scalar quantity. For anisotropic materials, χ is a tensor in order to account for the various polarization responses of the material to different directions of the applied field with respect to the crystal axis.

When the electric field is increased significantly, nonlinear interactions begin to occur within the material to the extent that (2.6) is no longer sufficient to describe the observed effects. Equation (2.6) must therefore be generalized to express the polarization $\mathbf{P}$ as a power series in the field strength E as follows:

$$P = \epsilon_0(\chi_1 E + \chi_2 E^2 + \chi_3 E^3 + \cdots). \tag{15.4}$$

The polarization makes a significant contribution to the nonlinear optical process, since it results from charges oscillating within the medium that are produced by the incident oscillating electric field and re-radiate, thereby adding to that field. For simplicity, in (15.4) we have expressed P and E as scalar quantities. We refer to the χs as *susceptibilities*; χ_1 is known as the linear susceptibility, and χ_2 and χ_3 are referred to as the second- and third-order nonlinear optical susceptibilities. Equation (15.4) is a time-dependent equation in which we have assumed that the response of the material is instantaneous to the variation of the applied field. For this assumption to be true, the medium must have no loss or dispersion. From our discussion in Section 2.3 regarding the response of a dielectric material to an applied field, we know that this is usually not the case. Nevertheless, it is useful to simplify the discussion at this point to explain some of the general properties of the nonlinear material response to an applied field. We can also define the second- and third-order nonlinear polarizations as

$$P_2 = \epsilon_0 \chi_2 E^2 \tag{15.5}$$

and

$$P_3 = \epsilon_0 \chi_3 E^3, \tag{15.6}$$

respectively. We do this because we will find that the second- and third-order processes are distinctly different from each other. For example, materials that are centrosymmetric (i.e., have inversion symmetry) produce no second-order processes but still provide significant third-order processes. Such materials include liquids, gases, and amorphous solids such as glass. Third-order processes occur for both centrosymmetric and noncentrosymmetric materials.

The magnitudes of the nonlinear susceptibility coefficients are such that the second- and third-order polarizations become comparable to the linear polarization term P_1 when the applied electric field E is of the order of the electric field produced between the electron and proton of a hydrogen atom, $E \approx e/(4\pi\epsilon_0 a_H^2)$, as deduced from (3.3). Thus the second- and third-order susceptibilities are of the order of

$$\chi_2 \approx 2 \times 10^{-11} \text{ m/V} \tag{15.7}$$

and

$$\chi_3 \approx 4 \times 10^{-23} \text{ m/V}, \tag{15.8}$$

respectively. In the following sections we will briefly describe a number of nonlinear processes that use laser radiation to generate new frequencies of coherent light and produce other interesting nonlinear effects.

15.3 SECOND-ORDER NONLINEAR OPTICAL PROCESSES

SECOND HARMONIC GENERATION

Probably the simplest second-order process is that of second harmonic generation. In this process, an intense laser beam of angular frequency $\omega_1 = 2\pi\nu_1$ is passed through a crystal having a nonzero value of χ_2 such that the beam emerging from the crystal contains the angular frequencies ω_1 of the input beam and also $\omega_2 = 2\omega_1$, twice the frequency of the input beam. This can be shown to occur by considering the second-order nonlinear polarization term (15.5). We will assume an oscillating electromagnetic field of frequency ω incident upon the material. This field can be described by the equation

$$E = E_0 e^{-i\omega t} + E_0^* e^{+i\omega t}, \tag{15.9}$$

where the second term on the right is just the complex conjugate of the first term. The second-order nonlinear polarization can now be computed from (15.5) as follows:

$$P_2 = \epsilon_0 \chi_2 E^2 = \epsilon_0 \chi_2 [E^*E + EE^* + E^2 e^{-2i\omega t} + E^2 e^{+2i\omega t}]. \tag{15.10}$$

This can be rewritten as

$$P_2 = 2\epsilon_0 \chi_2 E^*E + 2\epsilon_0 \chi_2 E^2 e^{-i(2\omega)t} + 2\epsilon_0 \chi_2 (E^*)^2 e^{i(2\omega)t}. \tag{15.11}$$

The first term is a time-independent factor that produces no oscillating electromagnetic radiation. It is a DC electric field response referred to as

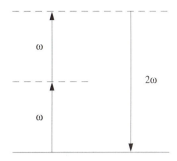

Figure 15-1. Second harmonic generation

optical rectification. The second and third terms can be seen to contain frequencies that are exactly twice the frequency of the fundamental wave. These terms can lead to a significant output from the material at an angular frequency of 2ω if certain other conditions are satisfied. Such output is referred to as *second harmonic generation*. The goal when attempting to generate useful second harmonic radiation is to maximize the value of the latter two terms in (15.11). Under certain conditions it is possible to convert nearly all of the original frequency of the beam to the second harmonic frequency.

We have described a simple picture of second harmonic generation from the wave standpoint in (15.11). However, it can also be considered from the photon standpoint. Consider the energy-level diagram of Figure 15-1. It shows two photons of the fundamental frequency ω, each with energy $\hbar\omega$, combining to produce an energy of $2\hbar\omega$. The energy levels $\hbar\omega$ and $2\hbar\omega$ are shown as dashed lines because they are not eigenstates of the material in which the second harmonic radiation is generated but are instead levels of the combined material–photon beam system. In that sense they are known as *virtual* levels, since they are not levels that accrue population. Instead, two photons of frequency ω are destroyed and one photon of frequency 2ω is simultaneously created, as indicated in the figure.

SUM AND DIFFERENCE FREQUENCY GENERATION

The previous discussion considered the combination (addition) of two photons of the same frequency to produce a single photon of twice the frequency. We can now generalize this process to allow for the case in which the two photons have different frequencies ω_1 and ω_2. In a similar fashion to our treatment of two photons with the same frequency, let us write the expression for the field as

$$E = E_1 e^{-i\omega_1 t} + E_1^* e^{i\omega_1 t} + E_2 e^{-i\omega_2 t} + E_2^* e^{i\omega_2 t}, \tag{15.12}$$

which includes the complex conjugates of both E_1 and E_2. We again compute the second-order nonlinear polarizability as

$$\begin{aligned}
P_2 &= \epsilon_0 \chi_2 E^2 \\
&= \epsilon_0 \chi_2 [E_1^2 e^{-i(2\omega)t} + (E_1^*)^2 e^{-i(2\omega)t} + E_2^2 e^{-i(2\omega)t} + (E_2^*)^2 e^{-i(2\omega)t} \\
&\quad + 2E_1 E_1^* + 2E_2 E_2^* + 2E_1 E_2 e^{-i(\omega_1 + \omega_2)t}
\end{aligned}$$

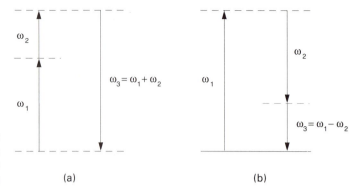

(a)   (b)

$$+2E_1^*E_2^*e^{-i(\omega_1+\omega_2)t}+2E_1E_2^*e^{-i(\omega_1-\omega_2)t}$$
$$+2E_1^*E_2e^{-i(\omega_1-\omega_2)t}].\qquad\qquad(15.13)$$

We have conveniently grouped common terms together. These include DC terms, second harmonic terms (involving $2\omega_1$ and $2\omega_2$), and two new terms involving $\omega_1+\omega_2$ and $\omega_1-\omega_2$. The new term involving $\omega_1+\omega_2$ generates a new frequency that is the sum of the two original frequencies and is thus known as *sum frequency generation*. The term involving the difference between the two frequencies, $\omega_1-\omega_2$, is referred to as *difference frequency generation*. From the standpoint of the photon picture (instead of the wave picture), these two additional processes are shown in Figure 15-2 with the virtual energy levels of the combined material–photon system. In the sum frequency generation, when a new photon $\omega_3=\omega_1+\omega_2$ is created, the frequencies ω_1 and ω_2 are destroyed. In the difference frequency generation, also shown in Figure 15-2, the photon of higher frequency ω_1 is destroyed while both ω_2 and ω_3 are created. Since ω_2 is already present as one of the input beams, this suggests that ω_2 is amplified in the process – that is, photons are added to the beam at the frequency ω_2.

The processes of sum and difference frequency generation can be very useful, not only in generating specific frequencies at new wavelengths but also in generating new tunable frequencies at new wavelengths. For example, an intense single-frequency laser can serve as the ω_1 in these processes and a lower-power tunable laser can provide the ω_2 and so obtain strong signals at either $\omega_1+\omega_2$ or $\omega_1-\omega_2$.

Based on the foregoing considerations, it might be expected that all of the frequencies $2\omega_1$, $2\omega_2$, $\omega_1+\omega_2$, and $\omega_1-\omega_2$ might simultaneously be generated when two input beams of frequency ω_1 and ω_2 are transmitted through the optical material. As it turns out, this is not the case. There is a constraint known as *phase matching* that must be implemented in order to efficiently generate any of those additional frequencies, yet phase matching can be applied to only one of those processes at a time. We will describe the phase matching process later (see Section 15.6), since it is applicable to all of the nonlinear processes.

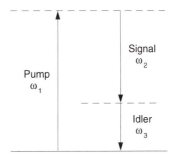

Figure 15-3. Optical parametric oscillation

OPTICAL PARAMETRIC OSCILLATION

In describing the process of difference frequency mixing, we realized that the frequency ω_2 is amplified while the frequency ω_3 is being generated. This amplification can be enhanced by placing the optical harmonic crystal within an optical cavity in which the mirrors are specifically made reflective at the frequency of ω_2. Thus the intensity at that frequency will build up within the cavity, as described in the discussion of the Fabry–Perot interferometer (Section 10.1), and this increased intensity of ω_1 will also generate more intensity at ω_3. Such an amplification process is known as an *optical parametric oscillator* (OPO). Of course, either ω_1 or ω_2 can be a tunable laser to generate amplified tunable output.

A diagram of an OPO is shown in Figure 15-3. The intense input beam at frequency ω_1 is known as the *pump* frequency, the desired amplified frequency at ω_2 is the *signal* frequency, and the unwanted frequency ω_3 is the *idler* frequency. This process is used most often in the infrared frequency range, where tunable lasers are not as readily available as in the visible portion of the frequency spectrum.

15.4 THIRD-ORDER NONLINEAR OPTICAL PROCESSES

We will now consider the third-order nonlinear polarization processes suggested in (15.6). As stated previously, these processes can be applied to all optical materials, centrosymmetric or noncentrosymmetric. We will consider the process in which an intense laser beam of angular frequency $\omega_1 = 2\pi\nu_1$ is transmitted through an optical material, which could consist of a liquid, a solid, or a gas. Emerging from that material would be a beam consisting of not only the fundamental frequency ω_1 but also a frequency three times that frequency, or $3\omega_1$. We will thus investigate the third-order nonlinear polarization term of (15.6), which involves the applied electric field taken to the third power. If we assume that the input beam is a sinusoidal electromagnetic wave of a single frequency ω_1, we can write this electric field as

$$E = E_1 \cos \omega_1 t. \tag{15.14}$$

Using this expression in (15.6) for the third-order nonlinear polarization, we obtain

$$P_3 = \chi_3 \epsilon_0^3 E_3^3$$
$$= \chi_3 \epsilon_0^3 E_1^3 \cos^3 \omega_1 t = \chi_3 \epsilon_0^3 E_1^3 [\tfrac{1}{4} \cos 3\omega_1 t + \tfrac{3}{4} \cos \omega_1 t]. \tag{15.15}$$

THIRD HARMONIC GENERATION

The first term of (15.15) contains an electromagnetic wave of frequency $3\omega_1$. This is the term associated with the process of third harmonic generation. The energy-level diagram of this process, from the standpoint of the photon picture rather than the wave picture, is shown in Figure 15-4. It indicates three virtual energy levels of the combined material–photon system, and shows three photons of the fundamental frequency ω_1 being destroyed while one photon of the third harmonic frequency $3\omega_1$ is being created. This process has been successfully demonstrated in gaseous, liquid, and solid materials.

When the input beam consists of photons of three different frequencies (ω_1, ω_2, and ω_3), the process becomes much more complex in that there are many more terms to consider in writing the expression for the third-order nonlinear polarization. Two of these processes are shown in the energy-level diagrams of Figure 15-5. The first is simply the equivalent of third harmonic generation as described previously, and the second involves the

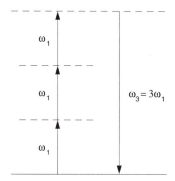

Figure 15-4. Third harmonic generation

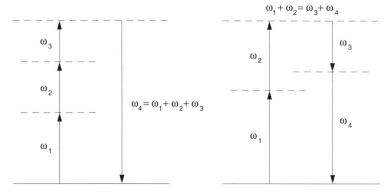

Figure 15-5. Four-wave mixing

creation of two photons of different frequencies at the expense of two other photons, also of different frequencies. This process, along with its many ramifications, is known as *four-wave mixing*.

INTENSITY-DEPENDENT REFRACTIVE INDEX – SELF-FOCUSING

When a beam of sufficient intensity passes through an optical material, the index of refraction of the material can be altered by the intensity of the beam. This occurs in a manner indicated by the following relationship:

$$\eta(\omega) = \eta_0(\omega) + \eta_{2I}(\omega) I(\omega). \tag{15.16}$$

In this expression $\eta_0(\omega)$ is the fundamental refractive index of the material at any specific frequency $\omega = 2\pi\nu$ that might be measured with an optical beam of low intensity, $\eta_{2I}(\omega) I(\omega)$ is the effective nonlinear refractive index at that frequency associated with the beam of intensity $I(\omega)$, and $\eta(\omega)$ is the total index of refraction at that frequency. Note that, unlike $\eta_0(\omega)$ and $\eta(\omega)$, in this definition $\eta_{2I}(\omega)$ is *not* a dimensionless quantity. The coefficient $\eta_{2I}(\omega)$ of the nonlinear index of refraction term $\eta_{2I}(\omega) I(\omega)$ of (15.16) can be expressed as

$$\eta_{2I}(\omega) \cong \frac{9\pi}{\eta_0^2(\omega)} \chi_3. \tag{15.17}$$

This change in the index of refraction of a material with an optical field of intensity $I(\omega)$ is known as the *optical Kerr effect*. It is present in almost all optical materials. The value of $\eta_{2I}(\omega)$ is typically of the order of 10^{-20} m^2/W. This effect can lead to self-focusing in a laser or amplifier rod because a typical laser beam profile would have a variation of intensity within the rod, and thus the index of refraction would vary over the cross section of the rod while the beam is present. For example, if the beam has a Gaussian cross section, the intensity would be the greatest on the axis of the rod and thus the index of refraction would be greater on the axis than off the axis. This is the equivalent of making a lens within the rod that can cause the beam to focus inside of the rod if either the rod is sufficiently long or the intensity sufficiently high (or some combination of both). If the intensity of this focused beam is sufficiently high, then the optical material will be damaged at the point of focus within the rod. The amount of self-focusing that will occur can be determined by the B integral, which is defined as

$$B = \frac{2\pi}{\lambda} \int_0^L \eta_{2I} I(z)\, dz, \tag{15.18}$$

where the variation in the intensity with distance z within the amplifier material is taken into account. The value of the B integral that leads to severe

TABLE 15-1

SECOND-ORDER NONLINEAR OPTICAL MATERIALS

Crystal	Transparency region (μm)
LiB_3O_5	0.16–2.6
β-BaB_2O_4	0.19–2.5
$KNbO_3$	0.40–5.5
$LiNbO_3$	0.40–5.0
$Ba_2NaNb_5O_{15}$	0.37–5.0
KDP	0.20–1.4
KTP	0.35–4.4

self-focusing and damage to the rod is approximately 3 to 5, and thus must be kept below that range (see page 287).

15.5 NONLINEAR OPTICAL MATERIALS

Because of the effectiveness in generating new frequencies from existing lasers via harmonic generation and sum and difference frequency generation, there has been an extensive effort in recent years to identify effective materials for such processes. In addition to having a large nonlinearity, these materials must be transparent not only at the laser frequency but also at the newly generated frequency. They must (1) be resistant to optical damage, (2) have high mechanical hardness, (3) exhibit good thermal and chemical stability, (4) be capable of being grown in useful sizes, and (5) have the appropriate phase-matching properties.

We recall that second harmonic crystals must have no inversion symmetry, which suggests that only crystalline materials can be used, whereas third harmonic crystals can have inversion symmetry (can be *centrosymmetric*). Bulk second-order nonlinear materials are generally inorganic crystals. A number of semiconductor compounds are useful for second harmonic generation when used in waveguides, and also some organic materials have been under investigation in recent years. Table 15-1 lists a number of second-order nonlinear optical materials that have high nonlinear conversion coefficients as well as the necessary transparency.

15.6 PHASE MATCHING

DESCRIPTION OF PHASE MATCHING

In Chapter 2 we considered the polarization response of material to an electromagnetic wave passing through that material. The polarization response consisted of a phased array of dipoles of the material that produced an electromagnetic wave in phase with and at the same frequency as the incident electromagnetic wave. In the preceding analysis in this chapter we saw that for higher beam intensities, the polarization response also included other frequencies. In order to find out which of these other possible frequencies could be significantly enhanced, we must consider the phases of the waves at these other frequencies. For the nonlinear case, the phases of the radiating dipoles of the material are determined not by a single wave at one frequency (as was the case for a low-intensity beam) but rather by the relative phases of two or more waves of different frequencies interacting with each other. Also, the additional wave that is generated, such as a second harmonic wave, has a velocity that is associated with the velocity of light (the index of refraction) of the material at that new frequency. Thus, the re-radiated dipole radiation will rapidly decay if the phase of the dipoles is not kept in step with the phase of the incident electromagnetic wave. In other words, the dipole response of the material must consist of a phased array of radiating antennas if it is to propagate a significantly intense beam at that newly generated frequency.

In order to obtain the wave equation expressed in (2.75), we used the fact that $\mathbf{V} \cdot \mathbf{E} = 0$ when considering a linear response of the polarization of the electric field. This assumption is valid provided that the properties of the medium are spatially uniform. For higher field intensities in which the response is no longer linear, $\mathbf{V} \cdot \mathbf{E}$ is no longer zero and the differential equation becomes more complex than that of (2.75). In performing such an analysis, let us consider the situation wherein two collinear beams of frequencies ω_1 and ω_2 are incident upon a nonlinear lossless anisotropic material in an attempt to generate significant conversion to a frequency $\omega_3 = \omega_1 + \omega_2$. A solution to a wave equation similar to that of (2.65) then leads to a differential equation of the form

$$\frac{d^2 E_3}{dz^2} + 2ik_3 \frac{dE_3}{dz} = -\frac{\mu_0^2 \chi_2 \omega_3^2}{2\pi^2} E_1 E_2 e^{i(k_1 + k_2 - k_3)z}. \tag{15.19}$$

In this expression, E_1, E_2, E_3 are the wave amplitudes and k_1, k_2, k_3 are the amplitudes of the propagation vectors of the three waves. Generally, the first term on the left can be neglected when compared to the second term (the slowly varying approximation). Let us also define

$$\Delta k = k_1 + k_2 - k_3. \tag{15.20}$$

Then (15.19) can be approximated as

$$\frac{dE_3}{dz} = \frac{i\mu_0^2 \chi_2 \omega_3^2}{4\pi^2 k_3} E_1 E_2 e^{i\Delta kz}. \tag{15.21}$$

This is a *coupled wave equation,* indicating how the amplitudes of waves of frequency ω_1 and ω_2 couple to produce a wave of frequency ω_3.

For the special case in which $\Delta k = 0$, the amplitude E_3 of (15.21) will increase linearly with distance z into the material. This condition is known as *ideal* phase matching; it is the case described previously in which the radiating dipoles of the material are phased with the field of the incident wave and thus continually add coherently to the beam as it propagates through the material. The total power emitted at ω_3 thus increases as the square of the number of atoms that participate in the interaction.

The intensity I_3 emitted at frequency ω_3 can be obtained by solving (15.21) for E_3 by integrating over a length from 0 to L and using it in the expression for the Poynting vector of (2.53). This leads to the following expression:

$$I_3 = \frac{c\mu_0^4\eta_3\chi_2^2\omega_3^4|E_1|^2|E_2|^2}{32\pi^5 k_3^2}\left|\frac{e^{i\Delta kL}-1}{\Delta k}\right|^2. \tag{15.22}$$

Rewriting this formula in terms of the product ΔkL, we have

$$I_3 = \frac{c\mu_0^4\eta_3\chi_2^2\omega_3^4|E_1|^2|E_2|^2 L^2}{32\pi^5 k_3^4}\left|\frac{e^{i\Delta kL}-1}{\Delta kL}\right|^2. \tag{15.23}$$

The expression within the absolute-value bars can be rewritten as

$$\left|\frac{e^{i\Delta kL}-1}{\Delta kL}\right|^2 = \frac{\sin^2(\Delta kL/2)}{(\Delta kL/2)^2}, \tag{15.24}$$

and is known as the *phase mismatch factor*. Converting the amplitude factors in (15.23) to intensities and converting the frequencies to wavelengths leads to the expression

$$I_3 = \frac{c^3\mu_0^4\chi_2^2 I_1 I_2 L^2}{2\pi\eta_1\eta_2\eta_3\lambda_3^2}\frac{\sin^2(\Delta kL/2)}{(\Delta kL/2)^2}. \tag{15.25}$$

Figure 15-6 shows a plot of the phase mismatch factor of (15.24) and (15.25) as a function of $\Delta kL/2$. From the figure it can be seen how rapidly the factor reduces from unity to zero for nonzero values of $\Delta kL/2$. Of course, the most important factor is in making $\Delta k = 0$, since for $L = 0$ the

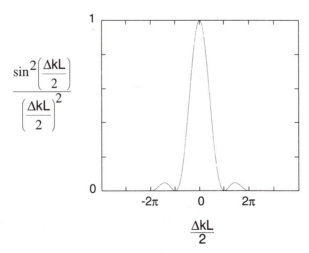

Figure 15-6. Phase mismatch factor versus $\Delta kL/2$

beam has not yet entered the material. Making $\Delta k = 0$ is thus the process of phase matching. We can therefore write

$$\Delta k = k_1 + k_2 - k_3 = 0. \tag{15.26}$$

For a wave of velocity v we have

$$k = \frac{\omega}{v} = \frac{\omega}{c/\eta} = \frac{\eta \omega}{c}, \tag{15.27}$$

so we can rewrite (15.26) as

$$\frac{\eta_1 \omega_1}{c} + \frac{\eta_2 \omega_2}{c} - \frac{\eta_3 \omega_3}{c} = 0 \tag{15.28}$$

with the additional constraint that $\omega_1 + \omega_2 = \omega_3$.

For second harmonic generation, $\omega_1 = \omega_2$. Therefore, from (15.28) we have

$$2\eta_1 \omega_1 = \eta_3 \omega_3. \tag{15.29}$$

But $\omega_3 = 2\omega_1$, so (15.29) suggests that, for ideal phase matching ($\Delta k = 0$) for second harmonic generation,

$$\eta_1 = \eta_3. \tag{15.30}$$

However, from (2.89) and (2.90) we learned that the index of refraction is not independent of frequency. In fact, it tends to increase with frequency over long frequency ranges between absorption resonances, as indicated in Figure 2.7. Hence, phase matching can be achieved only by using either (1) birefringent crystals in which the velocity for two different polarizations is different, or (2) the effect of anomalous dispersion in the region of absorption resonances. Method (2) can be used only over a very limited frequency range, whereas method (1) can be applied more generally.

ACHIEVING PHASE MATCHING

There are two principal methods for achieving phase matching in birefringent optical materials: angle tuning and temperature tuning of the crystals. For harmonic generation in shorter-wavelength regions where there are no optically transparent materials, gases are used for harmonic generation. In that region, phase matching is achieved by using two different gases with different indices of refraction.

ANGLE TUNING   In order to describe angle tuning, we will consider the case of a uniaxial birefringent crystal. For such a crystal the *ordinary* ray is one that is polarized perpendicular to the plane that includes the optic axis (c axis) and the propagation vector **k**, as shown in Figure 15-7. The *extraordinary* ray is one that is polarized in the plane of the optic axis and

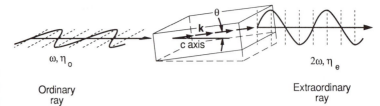

Figure 15-7. Phase matching using angle tuning

Ordinary ray

Extraordinary ray

the propagation vector. The ordinary ray has an index of refraction η_o; the extraordinary ray has an index of refraction η_e that varies with the angle θ between the propagation vector $\mathbf{k}$ and the optic axis. This variation is given by

$$\frac{1}{\eta_e^2(\theta)} = \frac{\sin^2\theta}{\bar{\eta}_e^2} + \frac{\cos^2\theta}{\eta_o^2}, \qquad (15.31)$$

where $\bar{\eta}_e$ is the principal value of the extraordinary refractive index. (Note that $\eta_e(\theta) = \bar{\eta}_e$ for $\theta = 90°$ and $\eta_e(\theta) = \eta_o$ for $\theta = 0°$.) Thus, by rotating the crystal in the plane of the optic axis and the incident beam (the direction of the k vector) it is possible to select a range of values of η_e from which the condition of (15.30) can be satisfied. For negative uniaxial crystals, in which $\eta_e < \eta_o$, we would select the fundamental frequency (frequencies) as the ordinary ray and the harmonic frequency as the extraordinary ray that is tuned to the phase-matching condition by rotating the crystal. For positive uniaxial crystals ($\eta_o < \eta_e$), the opposite is true: the fundamental frequency (or frequencies) is the extraordinary ray and the harmonic frequency is the ordinary ray.

TEMPERATURE TUNING Whenever the k vector and the optic axis are neither parallel nor perpendicular to each other, the ordinary and extraordinary waves tend to diverge from each other since the direction of energy flow (the Poynting vector) and the k vector are not parallel. This limits the crystal length over which harmonic conversion can occur, and hence also the amount of harmonic power that can be generated according to (15.25). The extraordinary ray essentially diverges from the beam path, so the effect is referred to as *walkoff*. A second means of producing phase matching for conditions in which walkoff is present is that of temperature tuning. In some crystals the index of refraction is dependent upon the crystal temperature. Thus, to achieve phase matching at a specific wavelength, the temperature of the crystal is adjusted to a specific value and held constant to obtain the optimum harmonic conversion.

TYPES OF PHASE MATCHING

Two types of phase matching can be accomplished in solid-state optical materials. Type-I phase matching is that in which both ω_1 and ω_2 have the same polarization. Of course, this would occur automatically for the process of second harmonic generation in which $\omega_1 = \omega_2$. For a positive uniaxial crystal ($\eta_e > \eta_o$), the conditions for type-I phase matching would be $\eta_3^o \omega_3 = \eta_1^e \omega_1 + \eta_2^e \omega_2$; for a negative uniaxial crystal ($\eta_e < \eta_o$), the conditions would be $\eta_3^e \omega_3 = \eta_1^o \omega_1 + \eta_2^o \omega_2$. Type-II phase matching occurs when the polarizations of ω_1 and ω_2 are orthogonal to each other, as might be the case for sum and difference frequency generation. Here the phase matching conditions would be $\eta_3^o \omega_3 = \eta_1^o \omega_1 + \eta_2^e \omega_2$ for positive uniaxial crystals and $\eta_3^e \omega_3 = \omega_1^e \omega_1 + \eta_2^o \omega_2$ for negative uniaxial crystals.

15.7 SATURABLE ABSORPTION

A nonlinear process that can be associated with real (rather than virtual) energy levels and population changes in those levels is that of saturable absorption. This is a process in which a material can be highly absorbing at a specific wavelength when a low-intensity beam is incident upon the material, yet an extremely intense beam (at that same wavelength) will pass through the medium with little change in intensity. In effect, we demonstrated this process in Section 8.1 when we showed that a population inversion could not be achieved in a two-level system. In Section 7.7 we obtained the following intensity-dependent expression for the gain of a system:

$$g(\nu) = \frac{g^0(\nu)}{1 + (I/I_{\text{sat}})} = \frac{\sigma_{ul}(\nu)\Delta N_{ul}^0}{1 + (I/I_{\text{sat}})}, \tag{7.71}$$

where $\sigma_{ul}(\nu)$ is the stimulated emission cross section, ΔN_{ul}^0 the population difference without the beam present, and I_{sat} the intensity at which significant population change occurs owing to the presence of the beam. The term ΔN_{ul}^0 was defined as $\Delta N_{ul}^0 = [N_u - (g_u/g_l)N_l]$, where N_u and N_l denote the populations of the upper and lower levels of the transition. If the population in the lower level l is significantly greater than the population of the upper level u, then we showed that it was convenient to express the product $\sigma_{ul}(\nu)\Delta N_{ul}$ in terms of an absorption coefficient $\alpha_{ul}(\nu)$ as described in (7.32). We can thus rewrite (7.71) as

$$\alpha(\nu) = \frac{\alpha_0(\nu)}{1 + I/I_{\text{sat}}}, \tag{15.32}$$

where $\alpha_0(\nu)$ is the absorption coefficient $\alpha_{ul}(\nu)$ described previously and associated with Beer's law in (7.31).

We must remember that the subscripts l and u in (7.32) were used in reference to the lower and upper laser levels, respectively. However, (7.32) and the associated value of $\sigma_{ul}(\nu)$ given in Chapter 7, as well as (15.32), are applicable to any transition and not just laser transitions. Thus, according to (15.32), if the intensity of a beam of light incident upon a material is sufficiently high that the ratio $I/I_{\text{sat}} \gg 1$, then the absorption coefficient will be rapidly driven to zero and the beam will be transmitted through the material, even though the material might be highly absorbing for a beam of low intensity. This effect is also referred to as *bleaching*. When bleaching occurs, a small portion of the leading edge of the beam will be absorbed owing to the energy required to produce the bleaching of the material in response to that wavelength.

Saturable absorbers are used in Q-switching and mode-locking, as described in Chapter 12. Saturable absorbers are generally designed to have a fast recovery time, so that emission does not get through the absorber in the time period after the desired pulse passes through. For example, in mode-locking it is desirable to have a high enough absorption to keep the low-level spontaneous emission of the amplifier from bleaching the absorber during the time between pulses arriving at the absorber. Useful saturable absorber materials were described in Section 12.2.

OPTICAL BISTABILITY   A saturable absorber can be used to make a bistable element within an optical system. A bistable element allows two levels of a beam of light to pass through the system and so provide switching from one level to the other by means of the optical beam itself. Such a bistable system is shown in the graph of Figure 15-8, where the output intensity of the beam is plotted against the input intensity. It can be seen that for low values of the input intensity, the output intensity has a value of approximately $I_1(\text{out})$. Then, as the input intensity of the beam is increased, eventually an intensity of $I_2(\text{in})$ is reached at which the output intensity jumps significantly to a value of $I_2(\text{out})$. Then, as the intensity is decreased, the output intensity remains high until a much lower value $I_1(\text{in})$ of the input intensity is reached. At that point the output beam returns to its much lower value $I_1(\text{out})$, as shown in the diagram.

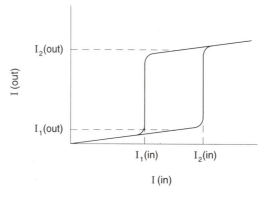

Figure 15-8. Output signal versus input signal, showing bistability

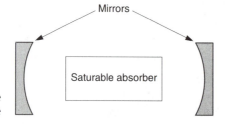

Figure 15-9. Bistable optical device

A bistable element can be made by placing a saturable absorber between two mirrors of a Fabry–Perot interferometer, as shown in Figure 15-9. When a low-intensity light beam is directed through the interferometer, the beam is absorbed and therefore essentially no light emerges from the interferometer. As the input beam intensity is increased, there will be an intensity, corresponding to $I_2(\text{in})$ of Figure 15-8, at which the light bleaches through the interferometer and most of the signal emerges from the device, corresponding to $I_2(\text{out})$ of the figure. Then, as the intensity is decreased, the transmitted power remains at a high level because the bleaching will remain in effect in the saturable absorber at a lower intensity than $I_2(\text{in})$. This occurs owing to the presence of the stored beam photons within the cavity, which keep the intensity high even though the input beam is lower. As $I(\text{in})$ is reduced even further, at some point the intensity reaches a point $I_1(\text{in})$ below the value required to maintain the bleaching, and the beam transmission drops precipitously to the lower value $I_1(\text{out})$. The characteristics of the saturable absorber (such as the energy-level lifetimes and the magnitude of the absorption coefficient) would have to be matched with those of the Fabry–Perot cavity (such as the mirror reflectivities and the cavity spacing) in order to make the system operate properly with the desired saturation level and time constant. Such a system can be used as an optical switch or as a binary memory cell.

15.8 TWO-PHOTON ABSORPTION

Another nonlinear process that involves the transfer of energy to excited energy levels of the material, as in the case of a saturable absorber, is the process of two-photon absorption. This effect can be shown by the photon picture in Figure 15-10. In this case the destruction (absorption) of two photons results in the population of an excited state of the absorbing medium. In a typical normal absorption process, the absorption coefficient (eq. 7.32) is a constant, independent of the input beam intensity. This coefficient depends upon factors such as the wavelength, the transition probability, the linewidth of the absorber, and the concentration of the absorbing species. In contrast, the two-photon absorption process, as with the

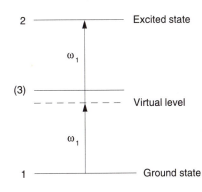

Figure 15-10.
Two-photon absorption

case of the saturable absorber, has an absorption cross section σ_{abs} that depends upon the intensity I:

$$\sigma_{abs} = \sigma_{TP}I. \tag{15.33}$$

It should be noted that σ_{TP} does not have units of length^2 as do most cross sections, since the product of σ_{TP} and I must have units of length^2. The transition rate, equivalent to A_{ul} for a normal electric dipole transition, is proportional to the square of the incident photon intensity. The two-photon transition rate can be significantly enhanced if an intermediate level is located near the energy of the first photon, as shown by the dashed line in Figure 15-10.

15.9 STIMULATED RAMAN SCATTERING

Stimulated Raman scattering is a process by which a photon is absorbed by an optical medium, with the emission of a lower-energy photon and the simultaneous absorption of the balance of the original photon energy by the medium, thereby placing the medium in an excited state. This is illustrated in the material–photon energy-level diagram of Figure 15-11, where a photon of frequency ω_1 incident upon the material is absorbed to a virtual level with the simultaneous emission of a photon of lower frequency ω_2 and absorption by the optical medium of a photon at a frequency of ω_V, leaving the molecule in an excited level (2). The absorbed frequency ω_V uses the subscript V because this process was first demonstrated for molecular systems in which the V refers to a molecular vibrational level. Significant coherent energy can thus be generated at ω_2 if the Raman gain coefficient is sufficiently high and the intensity at the input frequency of ω_1 is sufficiently intense.

15.10 HARMONIC GENERATION IN GASES

Second harmonic generation is not possible using gases as the nonlinear medium, since they are isotropic materials. It is possible to generate higher-order harmonics in gases. However, for wavelengths in the infrared, visible,

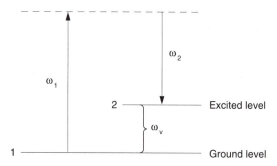

Figure 15-11. Stimulated Raman scattering

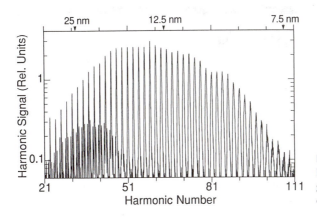

Figure 15-12. High-order harmonics generated in a gaseous medium (courtesy of Stanford University)

and ultraviolet, the use of crystals to generate a nonlinear response is far more effective than using gases for harmonic generation. Nonetheless, third and higher-order harmonics (up to more than the 100th harmonic) have been generated in gases and vapors. An example is shown in Figure 15-12 for the generation of up to the 109th harmonic in neutral neon atoms by focusing 35-mJ, 125-fs–duration Ti:Al$_2$O$_3$ laser pulses at 806.5 nm into a thin layer of neon gas. Other nonlinear processes have also been used to produce tunable coherent radiation of reasonable intensities at many wavelengths less than 200 nm when there are no other effective means for producing such radiation. In many cases the intensity of the generating beam is sufficiently high that the interaction length is short enough that phase matching is not required. In other cases, two gases of different refraction indices at the appropriate wavelengths are used to produce phase matching. The reader may consult the references for further information on this topic.

REFERENCES

N. Bloembergen (1965), *Nonlinear Optics.* New York: Benjamin.

R. W. Boyd (1992), *Nonlinear Optics.* New York: Academic Press.

J. J. Macklin, J. D. Kmetec, and C. L. Gordon III (1993), "High-order harmonic generation using intense femtosecond pulses," *Physical Review Letters* 70: 766–9.

J. Reintjes, R. C. Eckardt, C. Y. She, N. E. Karangelen, R. C. Elton, and R. A. Andrews (1976), "Generation of coherent radiation at 53.2 nm by fifth harmonic conversion," *Physical Review Letters* 37: 1540–3.

C. L. Tang (1995), "Nonlinear optics," in *Handbook of Optics,* vol. II (M. Bass, ed.). New York: McGraw-Hill, Chapter 38.

F. Zernike and J. E. Midwinter (1973), *Applied Nonlinear Optics.* New York: Wiley.

APPENDIX

ATOMIC CHARACTERISTICS

Atomic number (Z)	Element	Symbol	Ionization potential (eV)	Most common isotope mass number (M_N)
1	Hydrogen	H	13.595	1
2	Helium	He	24.580	4
3	Lithium	Li	5.390	7
4	Beryllium	Be	9.320	9
5	Boron	B	8.296	11
6	Carbon	C	11.264	12
7	Nitrogen	N	14.54	14
8	Oxygen	O	13.614	16
9	Fluorine	F	17.418	19
10	Neon	Ne	21.559	20
11	Sodium	Na	5.138	23
12	Magnesium	Mg	7.644	24
13	Aluminum	Al	5.984	27
14	Silicon	Si	8.149	28
15	Phosphorus	P	10.55	31
16	Sulfur	S	10.357	32
17	Chlorine	Cl	13.01	35
18	Argon	Ar	15.755	40
19	Potassium	K	4.339	39
20	Calcium	Ca	6.111	40
21	Scandium	Sc	6.56	45
22	Titanium	Ti	6.83	48
23	Vanadium	V	6.74	51
24	Chromium	Cr	6.764	52
25	Manganese	Mn	7.432	55
26	Iron	Fe	7.90	56
27	Cobalt	Co	7.86	59
28	Nickel	Ni	7.633	58
29	Copper	Cu	7.724	63
30	Zinc	Zn	9.391	64
31	Gallium	Ga	6.00	69
32	Germanium	Ge	7.88	74
33	Arsenic	As	9.81	75
34	Selenium	Se	9.75	80
35	Bromine	Br	11.84	79
36	Krypton	Kr	13.996	84
37	Rubidium	Rb	4.176	85
38	Strontium	Sr	5.692	88
39	Yttrium	Y	6.377	89

Atomic number (Z)	Element	Symbol	Ionization potential (eV)	Most common isotope mass number (M_N)
40	Zirconium	Zr	6.835	90
41	Niobium	Nb	6.881	93
42	Molybdenum	Mo	7.131	98
43	Technetium	Tc	7.23	
44	Ruthenium	Ru	7.365	102
45	Rhodium	Rh	7.461	103
46	Palladium	Pd	8.33	106
47	Silver	Ag	7.574	107
48	Cadmium	Cd	8.991	114
49	Indium	In	5.785	115
50	Tin	Sn	7.332	120
51	Antimony	Sb	8.639	121
52	Tellurium	Te	9.01	130
53	Iodine	I	10.44	127
54	Xenon	Xe	12.127	132
55	Cesium	Cs	3.893	133
56	Barium	Ba	5.210	138
57	Lanthanum	La	5.61	139
58	Cerium	Ce		140
59	Praseodymium	Pr		141
60	Neodymium	Nd	6.3	142
61	Promethium	Pm		
62	Samarium	Sm	5.6	152
63	Europium	Eu	5.67	153
64	Gadolinium	Gd	6.16	158
65	Terbium	Tb		159
66	Dysprosium	Dy		164
67	Holmium	Ho		165
68	Erbium	Er		166
69	Thulium	Tm		169
70	Ytterbium	Yb	6.22	174
71	Lutecium	Lu	6.15	175
72	Hafnium	Hf	5.5	180
73	Tantalum	Ta	7.7	181
74	Tungsten	W	7.98	184
75	Rhenium	Re	7.87	187
76	Osmium	Oa	8.7	192
77	Iridium	Ir	9.2	193
78	Platinum	Pt	9.0	195
79	Gold	Au	9.22	197
80	Mercury	Hg	10.434	204
81	Thallium	Tl	6.106	205
82	Lead	Pb	7.415	208
83	Bismuth	Bi	7.287	209
84	Polonium	Po	8.43	
85	Astatine	At		
86	Emanation	Em	10.745	
87	Francium	Fr		
88	Radium	Ra		
89	Actinium	Ac		

Atomic number (Z)	Element	Symbol	Ionization potential (eV)	Most common isotope mass number (M_N)
90	Thorium	Th		232
91	Protoactinium	Pa		
92	Uranium	U	4	238
93	Neptunium	Np		
94	Plutonium	Pu		
95	Americium	Am		
96	Curium	Cm		
97	Berkelium	Bk		
98	Californium	Cf		
99	Einsteinium	E		
100	Fermium	Fm		
101	Mendeleevium	Mv		

Notes: The atomic number Z denotes the number of protons in the nucleus. The mass number M_N is also referred to as the atomic weight of an atom.

ABBREVIATIONS

APM	additive pulse mode-locking
A	ampere
ASE	amplified spontaneous emission
Å	Angstrom
cm	centimeter
CPM	colliding pulse mode-locking
cw	continuous wave
CVL	copper vapor laser
C	coulomb
dB	decibel
°C	degree Celsius
DC	direct current
DFB	distributed feedback
DH	double heterostructure
eV	electron volt
ESA	excited-state absorption
EUV	extreme ultraviolet
F	farad
fs	femtosecond
FWHM	full width at half maximum
gal	gallon
GHz	gigahertz
GW	gigawatt
H	henry
Hz	hertz
in	inch
J	joule
K	(degree) Kelvin
KLM	Kerr lens mode-locking
kHz	kilohertz
km	kilometer
kV	kilovolt
kW	kilowatt
linac	linear accelerator
MHz	megahertz
MJ	megajoule
MW	megawatt
m	meter
μm	micron (micrometer)
μs	microsecond
mA	milliamp
mJ	millijoule
mm	millimeter
mrad	milliradian
ms	millisecond
mW	milliwatt
min	minute
nm	nanometer
ns	nanosecond
NA	numerical aperture
OFI	optical field ionization
OPO	optical parametric oscillator
ps	picosecond
rad	radian
RF	radio-frequency
rms	root mean square
s	second
T	tesla
TEM	transverse electromagnetic
TEA	transverse excitation atmospheric
UV	ultraviolet
VUV	vacuum ultraviolet
V	volt
W	watt

INDEX

THRESHOLD PUMPING CONDITIONS FOR INVERSION

Traditional three-level laser

$$\Gamma_{li} > \gamma_{ul}\left(1 + \frac{\gamma_{il}}{\gamma_{iu}}\right) \tag{8.12}$$

For radiative decay and when γ_{il}/γ_{iu} is small

$$\Gamma_{li} > A_{ul} \tag{8.13}$$

Three-level gas laser

$$\frac{N_u}{N_l} \cong \frac{1}{(1+\Gamma_{0l}/\Gamma_{0u})} \frac{A_{l0}}{A_{ul}} \overset{?}{>} 1 \tag{8.20}$$

Four-level laser

$$\Gamma_{0i} > \frac{\gamma_{0l}\gamma_{ul}}{\gamma_{l0}} = e^{-\Delta E_{l0}/kT}\gamma_{ul} \tag{8.30}$$

Transient inversion

$$\frac{N_u}{N_l} = \frac{1 - e^{-t/\tau_u}}{\left[\left(\frac{\Gamma_{0l}}{\Gamma_{0u}}\right)+1\right]\frac{t}{\tau_u} - (1 - e^{-t/\tau_u})} \overset{?}{>} 1 \tag{8.35}$$

RADIATION TRAPPING

$$\sigma_{0l}N_0 b = 1.46 \quad [b \text{ is bore radius}] \tag{8.43}$$

ELECTRON COLLISIONAL THERMALIZATION

$$n_e^{\max} = \frac{0.13\sqrt{T_e}}{\lambda_{ul}^3} \text{ m}^{-3} \quad [T_e \text{ in K, } \lambda \text{ in m}] \tag{8.52}$$

LASER MODES

Longitudinal modes

Separation between modes

$$\Delta\nu = \frac{c}{2\eta d} \tag{10.31}$$

Mode frequency

$$\nu = n\left(\frac{c}{2\eta d}\right) \quad [n \text{ a positive integer}] \tag{10.39}$$

$$\nu = n\left(\frac{c}{2}\left[\frac{1}{\eta_C(d-L)+\eta_L L}\right]\right) \tag{10.40}$$

Transverse modes

$$U_{pq}(x, y) = H_p\left(\frac{\sqrt{2}x}{w}\right)H_q\left(\frac{\sqrt{2}y}{w}\right)e^{-(x^2+y^2)/w^2}$$

$$[p, q \text{ are integers}] \tag{10.53}$$

where

$$H_0(u) = 1 \qquad H_1(u) = 2u$$

$$H_2(u) = 2(2u^2 - 1)$$

$$H_m(u) = (-1)^m e^{u^2}\frac{d^m(e^{-u^2})}{du^m} \tag{10.54}$$

MODE LOCKING

$$\Delta t_{\text{sep}} = \frac{1}{\Delta\nu} = \frac{2\eta d}{c} \tag{12.59}$$

$$\Delta t_P = \frac{2\eta d}{Nc} = \frac{1}{\Delta\nu N}$$

$$= \frac{1}{\text{gain bandwidth}} \tag{12.62}$$

GAUSSIAN BEAMS

Beam waist

$$w(z) = w_0\left[1 + \left(\frac{\lambda z}{\pi w_0^2}\right)^2\right]^{1/2} \tag{11.28}$$

$$w(z) = w_0\left[1 + \left(\frac{\lambda(z-z_0)}{\pi w_0^2}\right)^2\right]^{1/2} \tag{11.29}$$

$$w(z) = w_0\left(1 + \frac{z^2}{z_R^2}\right)^{1/2} \tag{11.30}$$

$$z_R = \frac{\pi w_0^2}{\lambda} \tag{11.31}$$

Wavefront curvature

$$R(z) = z\left[1 + \left(\frac{\pi w_0^2}{\lambda z}\right)^2\right] \tag{11.32}$$

Angular spread

$$\theta(z) = \frac{2\lambda}{\pi w_0} = 0.64\frac{\lambda}{w_0} \tag{11.33}$$

Complex beam parameter

$$\frac{1}{q} = \frac{1}{R(z)} - j\frac{\lambda_0}{\eta\pi w^2(z)} \tag{11.46}$$

$$\frac{1}{q_2} = \frac{C+D(1/q_1)}{A+B(1/q_1)} \tag{11.48}$$

$$\frac{C+D(1/q_1)}{A+B(1/q_1)} = \frac{1}{R(z)} - j\frac{\lambda_0}{\eta\pi w^2(z)} \tag{11.49}$$

OPTIMUM MIRROR TRANSMISSION AND POWER OUTPUT

$$t_{\text{opt}} = (g_0 La)^{1/2} - a \tag{11.79}$$

$$t_{\text{opt}} = (g_0 La)^{1/2}\left[1 - \left(\frac{a}{g_0 L}\right)^{1/2}\right] \tag{11.80}$$

$$I_{t_{\max}} = \left(\frac{t_{\text{opt}}^2}{2a}\right)I_{\text{sat}} \tag{11.83}$$

GENERAL PHYSICAL CONSTANTS

Constant	Symbol	Value	MKS units	CGS units
Speed of light in vacuum	c	2.99792457	10^8 m/s	10^{10} cm/s
Elementary charge	e	1.60210	10^{-19} C	10^{-20} cm$^{1/2}$-g$^{1/2}$
Avogodro constant	N_A	6.02252	10^{23} mole^{-1}	10^{23} mole^{-1}
Electron rest mass	m_e	9.1091	10^{-31} kg	10^{-28} g
Proton rest mass	M_p	1.67252	10^{-27} kg	10^{-24} g
Planck constant	h	6.6256	10^{-34} J-s	10^{-27} erg-s
		4.1354	10^{-15} eV-s	
Rydberg constant	R_H	1.0967758	10^7 m^{-1}	10^5 cm^{-1}
First Bohr radius	a_H	5.29172	10^{-11} m	10^{-9} cm
Boltzmann constant	k	1.38054	10^{-23} J/K	10^{-16} erg/K
Stefan–Boltzmann constant	σ	5.6697	10^{-8} W/m^2-K^4	10^{-5} erg/cm^2-s-K^4
Electron volt	eV	1.60209	10^{-19} J	10^{-12} erg
Permittivity of the vacuum	ϵ_0	8.854	10^{-12} F/m	
Permeability of the vacuum	μ_0	4π	10^{-7} H/m	

VECTOR IDENTITIES

TRIPLE PRODUCTS

(1) $\mathbf{A} \cdot (\mathbf{B} \times \mathbf{C}) = \mathbf{B} \cdot (\mathbf{C} \times \mathbf{A}) = \mathbf{C} \cdot (\mathbf{A} \times \mathbf{B})$

(2) $\mathbf{A} \times (\mathbf{B} \times \mathbf{C}) = \mathbf{B}(\mathbf{A} \cdot \mathbf{C}) - \mathbf{C}(\mathbf{A} \cdot \mathbf{B})$

PRODUCT RULES

(3) $\nabla(fg) = f(\nabla g) + g(\nabla f)$

(4) $\nabla(\mathbf{A} \cdot \mathbf{B}) = \mathbf{A} \times (\nabla \times \mathbf{B}) + \mathbf{B} \times (\nabla \times \mathbf{A}) + (\mathbf{A} \cdot \nabla)\mathbf{B} + (\mathbf{B} \cdot \nabla)\mathbf{A}$

(5) $\nabla \cdot (f\mathbf{A}) = f(\nabla \cdot \mathbf{A}) + \mathbf{A} \cdot (\nabla f)$

(6) $\nabla \cdot (\mathbf{A} \times \mathbf{B}) = \mathbf{B} \cdot (\nabla \times \mathbf{A}) - \mathbf{A} \cdot (\nabla \times \mathbf{B})$

(7) $\nabla \times (f\mathbf{A}) = f(\nabla \times \mathbf{A}) - \mathbf{A} \times (\nabla f)$

(8) $\nabla \times (\mathbf{A} \times \mathbf{B}) = (\mathbf{B} \cdot \nabla)\mathbf{A} - (\mathbf{A} \cdot \nabla)\mathbf{B} + \mathbf{A}(\nabla \cdot \mathbf{B}) - \mathbf{B}(\nabla \cdot \mathbf{A})$

SECOND DERIVATIVES

(9) $\nabla \cdot (\nabla \times \mathbf{A}) = 0$

(10) $\nabla \times (\nabla f) = 0$

(11) $\nabla \times (\nabla \times \mathbf{A}) = \nabla(\nabla \cdot \mathbf{A}) - \nabla^2 \mathbf{A}$

CONVERSION FROM SPHERICAL TO CARTESIAN COORDINATES

$x = r \sin \theta \cos \phi$

$y = r \sin \theta \sin \phi$

$z = r \cos \theta$

$r = \sqrt{x^2 + y^2 + z^2}$

$\theta = \tan^{-1}(\sqrt{x^2 + y^2}/z)$

$\phi = \tan^{-1}(y/x)$

$\hat{i} = \sin \theta \cos \phi \hat{r} + \cos \theta \cos \phi \hat{\theta} - \sin \phi \hat{\phi}$

$\hat{j} = \sin \theta \sin \phi \hat{r} + \cos \theta \sin \phi \hat{\theta} + \cos \phi \hat{\phi}$

$\hat{k} = \cos \theta \hat{r} - \sin \theta \hat{\theta}$

$\hat{r} = \sin \theta \cos \phi \hat{i} + \sin \theta \sin \phi \hat{j} + \cos \theta \hat{k}$

$\hat{\theta} = \cos \theta \cos \phi \hat{i} + \cos \theta \sin \phi \hat{j} - \sin \theta \hat{k}$

$\hat{\phi} = -\sin \phi \hat{i} + \cos \phi \hat{j}$

PARAMETERS DETERMINING THE STABILITY OF SOME TWO-MIRROR LASER CAVITIES

1. *Both mirrors are concave:* Either the center of curvature of each mirror lies beyond the other mirror, or the center of curvature of each mirror lies between the other mirror and the center of curvature of the other mirror.

2. *One mirror is convex and one mirror is concave:* The center of curvature of the concave mirror lies between the convex mirror and the center of curvature of the convex mirror.